Lecture Notes in

T0237837

Editorial Board

R. Beig, Wien, Austria
W. Beiglböck, Heidelberg, Germany
W. Domcke, Garching, Germany
B.-G. Englert, Singapore
U. Frisch, Nice, France
P. Hänggi, Augsburg, Germany
G. Hasinger, Garching, Germany
K. Hepp, Zürich, Switzerland
W. Hillebrandt, Garching, Germany
D. Imboden, Zürich, Switzerland
R. L. Jaffe, Cambridge, MA, USA
R. Lipowsky, Potsdam, Germany
H. v. Löhneysen, Karlsruhe, Germany
I. Ojima, Kyoto, Japan
D. Sornette, Nice, France, and Zürich, Switzerland
S. Theisen, Potsdam, Germany
W. Weise, Garching, Germany
J. Wess, München, Germany
J. Zittartz, Köln, Germany

The Lecture Notes in Physics

The series Lecture Notes in Physics (LNP), founded in 1969, reports new developments in physics research and teaching – quickly and informally, but with a high quality and the explicit aim to summarize and communicate current knowledge in an accessible way. Books published in this series are conceived as bridging material between advanced graduate textbooks and the forefront of research and to serve three purposes:

- to be a compact and modern up-to-date source of reference on a well-defined topic

- to serve as an accessible introduction to the field to postgraduate students and nonspecialist researchers from related areas

- to be a source of advanced teaching material for specialized seminars, courses and schools

Both monographs and multi-author volumes will be considered for publication. Edited volumes should, however, consist of a very limited number of contributions only. Proceedings will not be considered for LNP.

Volumes published in LNP are disseminated both in print and in electronic formats, the electronic archive being available at springerlink.com. The series content is indexed, abstracted and referenced by many abstracting and information services, bibliographic networks, subscription agencies, library networks, and consortia.

Proposals should be sent to a member of the Editorial Board, or directly to the managing editor at Springer:

Christian Caron
Springer Heidelberg
Physics Editorial Department I
Tiergartenstrasse 17
69121 Heidelberg / Germany
christian.caron@springer.com

M. Lemoine
J. Martin
P. Peter (Eds.)

Inflationary Cosmology

 Springer

Editors

Martin Lemoine
CNRS - Université Pierre et Marie Curie
Institut Astrophysique de Paris (IAP)
98 bd. Arago
75014 Paris, France
lemoine@iap.fr

Jerome Martin
CNRS - Université Pierre et Marie Curie
Institut d'Astrophysique de Paris (IAP)
98 Boulevard Arago
75014 Paris, France
jmartin@iap.fr

Patrick Peter
CNRS - Universit Pierre et Marie Curie
Institut d'Astrophysique de Paris (IAP)
98 Boulevard Arago
75014 Paris, France

M. Lemoine et al. (Eds.), *Inflationary Cosmology*, Lect. Notes Phys. 738 (Springer, Berlin Heidelberg 2008), DOI 10.1007/ 978-3-540-74353-8

ISSN 0075-8450
ISBN 978-3-642-09376-0   e-ISBN 978-3-540-74353-8

This work is subject to copyright. All rights are reserved, whether the whole or part of the material is concerned, specifically the rights of translation, reprinting, reuse of illustrations, recitation, broadcasting, reproduction on microfilm or in any other way, and storage in data banks. Duplication of this publication or parts thereof is permitted only under the provisions of the German Copyright Law of September 9, 1965, in its current version, and permission for use must always be obtained from Springer. Violations are liable for prosecution under the German Copyright Law.

Springer is a part of Springer Science+Business Media
springer.com
© Springer-Verlag Berlin Heidelberg 2007
Softcover reprint of the hardcover 1st edition 2007

The use of general descriptive names, registered names, trademarks, etc. in this publication does not imply, even in the absence of a specific statement, that such names are exempt from the relevant protective laws and regulations and therefore free for general use.

Cover design: eStudio Calamar S.L., F. Steinen-Broo, Pau/Girona, Spain

Preface

A seminal paper, dated 1981, marked the birth of what was to become the most successful paradigm in modern cosmology following that of the big bang itself: inflation. Its 25th birthday offered a welcome opportunity to celebrate the phenomenological success of inflation and to gather the world leading scientists engaged in forefront research in this field. Such was the objective of the XXII IAP colloquium, which took place at "Institut d'Astrophysique de Paris" (IAP) in June 2006. During this meeting and the immediately following two-week workshop, scientists from the world over and from both observational and theoretical communities gathered to discuss the present status, the achievements and the shortcomings as well as the future of the theory of inflation. The numerous discussions that took place offered solid ground for the publication of regular proceedings. However, inflationary cosmology encompasses different disciplines of physics, from high energy physics to observational astrophysics, and it has also become a field of research in its own right. Therefore it was felt that a more pedagogical text, containing exhaustive discussions of the ins and outs of inflation, would be more useful. This is precisely what this present volume of the Lecture Notes in Physics series is aiming at.

As is by now well known, cosmic inflation corresponds to an episode of accelerated expansion in the very early Universe which solves the handful of puzzles that plague the standard hot big bang cosmology, namely the flatness, horizon, monopole excess problems, and, in some models, the problem of the primordial singularity. These achievements even come with a bonus: the production of density perturbations to the level needed to explain the origin of large scale structure of the Universe. The first chapter of this volume, by A. Linde, introduces this framework, offers a historical overview of this subject and develops the present status of the theory. This is followed L. Kofmann's discussion on preheating which describes how matter and radiation can have been produced during this period which smoothly connects inflation with the standard big bang phase.

Any cosmological model needs to be implemented in a particle physics context. The contribution of D. Lyth shows how this can be done in the most reasonable extensions of the standard particle physics model, namely those based on supersymmetry. This chapter is followed by the discussion of R. Kallosh on the embedding of inflation in string theories.

As of today, there are various ways of implementing inflation. One such framework is "eternal inflation", in which different parts of the Universe undergo an episode of inflation at different times, the Universe being eternally inflating and self-reproducing. This particular scenario is discussed in length by S. Winitzki

As shown by J. Martin in a subsequent chapter, the production of density perturbations during inflation is akin to the production of charged particles out of the vacuum in a strong electric field. This analogy is developed in full detail in order to explain the inflationary origin of primordial density fluctuations. The numerical implementation of the calculation of these perturbations, which is required in order to compare these results to the high accuracy data of cosmic microwave background fluctuations, is then discussed by C. Ringeval.

The next chapter, by D. Wands, discusses the models containing more than one scalar field, in particular their dynamics and the observational predictions; the curvaton model is here reviewed as an alternative to the pure inflationary production of perturbations. Then, A. Riotto shows that the measurement of non-Gaussianities in the spectrum of inflationary perturbations could offer a way of discrimating the different models.

Finally the possibility of finding alternative scenarios to inflation is a major but unanswered issue. The old contender, in which topological defects seed the primordial density fluctuations has been shown to disagree with cosmic microwave background data. However, as M. Sakellariadou argues, such topological defects might still be present in our Universe as they should be produced in convincing models of inflation. Their contribution to the observed fluctuations might open a window on physics of an otherwise inaccessible energy scale. R. Brandenberger concludes this volume by presenting a radically different perspective in which string gas cosmology plays the main role and by pointing out some shortcomings of inflation which may argue for the need of a broader conceptual framework.

Paris,   *Martin Lemoine, Jérôme Martin & Patrick Peter.*
April 2007

Contents

3 Particle Physics Models of Inflation

4 Inflation in String Theory

5 Predictions in Eternal Inflation

1

Inflationary Cosmology

Andrei Linde

Department of Physics, Stanford University, Stanford, CA 94305, USA
alinde@stanford.edu

Abstract. This chapter presents a general review of the history of inflationary cosmology and of its present status.[1]

1.1 Brief History of Inflation

Since the inflationary theory is more than 25 years old, perhaps it is not inappropriate to start this chapter with a brief history of its development, and some personal recollections.

Several ingredients of inflationary cosmology were discovered in the beginning of the 1970s. The first realization was that the energy density of a scalar field plays the role of the vacuum energy/cosmological constant [1], which was changing during the cosmological phase transitions [2]. In certain cases these changes occur discontinuously, due to first-order phase transitions from a supercooled vacuum state (false vacuum) [3].

In 1978, we with Gennady Chibisov tried to use these facts to construct a cosmological model involving exponential expansion of the universe in the supercooled vacuum as a source of the entropy of the universe, but we immediately realized that the universe becomes very inhomogeneous after the bubble wall collisions. I mentioned our work in my review article [4], but did not pursue this idea any further.

The first semi-realistic model of inflationary type was proposed by Alexei Starobinsky in 1979–1980 [5]. It was based on the investigation of a conformal anomaly in quantum gravity. His model was rather complicated, and its goal was somewhat different from the goals of inflationary cosmology. Instead of attempting to solve the homogeneity and isotropy problems, Starobinsky considered the model of the universe which was homogeneous and isotropic from the very beginning, and emphasized that his scenario was "the extreme opposite of Misner's initial 'chaos'."

---

[1] Based on a talk given at the 22nd IAP Colloquium, "Inflation+25", Paris, June 2006.

A. Linde: *Inflationary Cosmology*, Lect. Notes Phys. **738**, 1–54 (2008)
DOI 10.1007/978-3-540-74353-8_1   © Springer-Verlag Berlin Heidelberg 2008

On the other hand, the Starobinsky model did not suffer from the graceful exit problem, and it was the first model to predict gravitational waves with a flat spectrum [5]. The first mechanism of production of adiabatic perturbations of the metric with a flat spectrum, which are responsible for galaxy production, and which were found by the observations of the CMB anisotropy, was proposed by Mukhanov and Chibisov [6] in the context of this model.

A much simpler inflationary model with a very clear physical motivation was proposed by Alan Guth in 1981 [7]. His model, which is now called "old inflation," was based on the theory of supercooling during the cosmological phase transitions [3]. Even though this scenario did not work, it played a profound role in the development of inflationary cosmology since it contained a very clear explanation of how inflation may solve the major cosmological problems.

According to this scenario, inflation is described by the exponential expansion of the universe in a supercooled false vacuum state. False vacuum is a metastable state without any fields or particles but with a large energy density. Imagine a universe filled with such "heavy nothing." When the universe expands, empty space remains empty, so its energy density does not change. The universe with a constant energy density expands exponentially, thus we have inflation in the false vacuum. This expansion makes the universe very big and very flat. Then the false vacuum decays, the bubbles of the new phase collide, and our universe becomes hot.

Unfortunately, this simple and intuitive picture of inflation in the false vacuum state is somewhat misleading. If the probability of the bubble formation is large, bubbles of the new phase are formed near each other, inflation is too short to solve any problems, and the bubble wall collisions make the universe extremely inhomogeneous. If they are formed far away from each other, which is the case if the probability of their formation is small and inflation is long, each of these bubbles represents a separate open universe with a vanishingly small Ω. Both options are unacceptable, which has lead to the conclusion that this scenario does not work and cannot be improved (graceful exit problem) [7, 8, 9].

The solution was found in 1981–1982 with the invention of the new inflationary theory [10], see also [11]. In this theory, inflation may begin either in the false vacuum, or in an unstable state at the top of the effective potential. Then the inflaton field ϕ slowly rolls down to the minimum of its effective potential. The motion of the field away from the false vacuum is of crucial importance: density perturbations produced during the slow-roll inflation are inversely proportional to $\dot{\phi}$ [6, 12, 13]. Thus the key difference between the new inflationary scenario and the old one is that the useful part of inflation in the new scenario, which is responsible for the homogeneity of our universe, does *not* occur in the false vacuum state, where $\dot{\phi} = 0$.

Soon after the invention of the new inflationary scenario it became so popular that even now most of the textbooks on astrophysics incorrectly describe inflation as an exponential expansion in a supercooled false vacuum state during the cosmological phase transitions in grand unified theories. Unfortunately, this scenario was plagued by its own problems. It works only if the effective potential of the field ϕ has a very flat plateau near $\phi = 0$, which is somewhat artificial. In most versions of this scenario the inflaton field has an extremely small coupling constant, so it could not be in thermal equilibrium with other matter fields. The theory of cosmological phase transitions, which was the basis for old and new inflation, did not work in such a situation. Moreover, thermal equilibrium requires many particles interacting with each other. This means that new inflation could explain why our universe was so large only if it was very large and contained many particles from the very beginning [14].

Old and new inflation represented a substantial but incomplete modification of the big bang theory. It was still assumed that the universe was in a state of thermal equilibrium from the very beginning, that it was relatively homogeneous and large enough to survive until the beginning of inflation, and that the stage of inflation was just an intermediate stage of the evolution of the universe. In the beginning of the 1980s these assumptions seemed most natural and practically unavoidable. On the basis of all available observations (CMB, abundance of light elements) everybody believed that the universe was created in a hot big bang. That is why it was so difficult to overcome a certain psychological barrier and abandon all of these assumptions. This was done in 1983 with the invention of the chaotic inflation scenario [15]. This scenario resolved all problems of old and new inflation. According to this scenario, inflation may begin even if there was no thermal equilibrium in the early universe, and it may occur even in the theories with simplest potentials such as $V(\phi) \sim \phi^2$. But it is not limited to the theories with polynomial potentials: chaotic inflation occurs in *any* theory where the potential has a sufficiently flat region, which allows the existence of the slow-roll regime [15].

1.2 Chaotic Inflation

1.2.1 Basic Model

Consider the simplest model of a scalar field ϕ with a mass m and with the potential energy density $V(\phi) = \frac{m^2}{2}\phi^2$. Since this function has a minimum at $\phi = 0$, one may expect that the scalar field ϕ should oscillate near this minimum. This is indeed the case if the universe does not expand, in which case the equation of motion for the scalar field coincides with the equation for the harmonic oscillator, $\ddot{\phi} = -m^2\phi$.

However, because of the expansion of the universe with Hubble constant $H = \dot{a}/a$, an additional term $3H\dot{\phi}$ appears in the harmonic oscillator equation:

$$\ddot{\phi} + 3H\dot{\phi} = -m^2\phi \ . \tag{1.1}$$

The term $3H\dot{\phi}$ can be interpreted as a friction term. The Einstein equation for a homogeneous universe containing a scalar field ϕ looks as follows:

$$H^2 + \frac{k}{a^2} = \frac{1}{6}(\dot{\phi}^2 + m^2\phi^2) \ . \tag{1.2}$$

Here $k = -1, 0, 1$ for an open, flat or closed universe respectively. We work in units $M_{\rm pl}^{-2} = 8\pi G = 1$.

If the scalar field ϕ initially was large, the Hubble parameter H was large too, according to the second equation. This means that the friction term $3H\dot{\phi}$ was very large, and therefore the scalar field was moving very slowly, as a ball in a viscous liquid. Therefore at this stage the energy density of the scalar field, unlike the density of ordinary matter, remained almost constant, and the expansion of the universe continued at a much greater speed than in the old cosmological theory. Due to the rapid growth of the scale of the universe and the slow motion of the field ϕ, soon after the beginning of this regime one has $\ddot{\phi} \ll 3H\dot{\phi}$, $H^2 \gg \frac{k}{a^2}$, $\dot{\phi}^2 \ll m^2\phi^2$, so the system of equations can be simplified:

$$H = \frac{\dot{a}}{a} = \frac{m\phi}{\sqrt{6}} \ , \qquad \dot{\phi} = -m\sqrt{\frac{2}{3}} \ . \tag{1.3}$$

The first equation shows that if the field ϕ changes slowly, the size of the universe in this regime grows approximately as e^{Ht}, where $H = \frac{m\phi}{\sqrt{6}}$. This is the stage of inflation, which ends when the field ϕ becomes much smaller than $M_{\rm Pl} = 1$. The solution to these equations shows that after a long stage of inflation the universe initially filled with the field $\phi \gg 1$ grows exponentially [14],

$$a = a_0 \, e^{\phi^2/4} \ . \tag{1.4}$$

Thus, inflation does not require an initial state of thermal equilibrium, supercooling and tunneling from the false vacuum. It appears in the theories that can be as simple as a theory of a harmonic oscillator [15]. Only when it was realized, it became clear that inflation is not just a trick necessary to fix problems of the old big bang theory, but a generic cosmological regime.

1.2.2 Initial Conditions

But what is about the initial conditions required for chaotic inflation? Let us consider first a closed universe of initial size $l \sim 1$ (in Planck units), which emerges from the space–time foam, or from singularity, or from "nothing" in a state with the Planck density $\rho \sim 1$. Only starting from this moment, i.e. at $\rho \lesssim 1$, can we describe this domain as a *classical* universe. Thus, at this initial moment the sum of the kinetic energy density, gradient energy density, and the potential energy density is of the order unity: $\frac{1}{2}\dot{\phi}^2 + \frac{1}{2}(\partial_i\phi)^2 + V(\phi) \sim 1$ (Fig. 1.1).

We wish to emphasize, that there are no a priori constraints on the initial value of the scalar field in this domain, except for the constraint

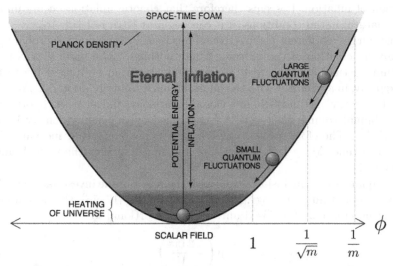

$$V(\phi) = \frac{m^2}{2}\phi^2$$

Fig. 1.1. Motion of the scalar field in the theory with $V(\phi) = \frac{m^2}{2}\phi^2$. Several different regimes are possible, depending on the value of the field ϕ. If the potential energy density of the field is greater than the Planck density $M_{\mathrm{Pl}}^4 = 1$, $\phi \gtrsim m^{-1}$, the quantum fluctuations of space–time are so strong that one cannot describe it in usual terms. Such a state is called space–time foam. At a somewhat smaller energy density (for $m \lesssim V(\phi) \lesssim 1$, $m^{-1/2} \lesssim \phi \lesssim m^{-1}$) the quantum fluctuations of space–time are small, but the quantum fluctuations of the scalar field ϕ may be large. Jumps of the scalar field due to quantum fluctuations lead to a process of eternal self-reproduction of inflationary universe which we are going to discuss later. At even smaller values of $V(\phi)$ (for $m^2 \lesssim V(\phi) \lesssim m$, $1 \lesssim \phi \lesssim m^{-1/2}$) fluctuations of the field ϕ are small; it slowly moves down as a ball in a viscous liquid. Inflation occurs for $1 \lesssim \phi \lesssim m^{-1}$. Finally, near the minimum of $V(\phi)$ (for $\phi \lesssim 1$) the scalar field rapidly oscillates, creates pairs of elementary particles, and the universe becomes hot

$\frac{1}{2}\dot{\phi}^2 + \frac{1}{2}(\partial_i\phi)^2 + V(\phi) \sim 1$. Let us consider for a moment a theory with $V(\phi) = \text{const}$. This theory is invariant under the *shift symmetry* $\phi \to \phi + c$. Therefore, in such a theory *all* initial values of the homogeneous component of the scalar field ϕ are equally probable.

The only constraint on the amplitude of the field appears if the effective potential is not constant, but grows and becomes greater than the Planck density at $\phi > \phi_{\mathrm{p}}$, where $V(\phi_{\mathrm{p}}) = 1$. This constraint implies that $\phi \lesssim \phi_{\mathrm{p}}$, but there is no reason to expect that initially ϕ must be much smaller than ϕ_{p}. This suggests that the typical initial value of the field ϕ in such a theory is $\phi \sim \phi_{\mathrm{p}}$.

Thus, we expect that typical initial conditions correspond to $\frac{1}{2}\dot{\phi}^2 \sim \frac{1}{2}(\partial_i\phi)^2 \sim V(\phi) = O(1)$. If $\frac{1}{2}\dot{\phi}^2 + \frac{1}{2}(\partial_i\phi)^2 \lesssim V(\phi)$ in the domain under consideration, then inflation begins, and then within the Planck time the terms $\frac{1}{2}\dot{\phi}^2$ and $\frac{1}{2}(\partial_i\phi)^2$ become much smaller than $V(\phi)$, which ensures continuation of inflation. It seems therefore that chaotic inflation occurs under rather natural initial conditions, if it can begin at $V(\phi) \sim 1$ [14, 16].

One can get a different perspective on this issue by studying the probability of quantum creation of the universe from "nothing." The basic idea is that quantum fluctuations can create a small universe from nothing if it can be done quickly, in agreement with the quantum uncertainty relation $\Delta E \cdot \Delta t \lesssim 1$. The total energy of scalar field in a closed inflationary universe is proportional to its minimal volume $H^{-3} \sim V^{-3/2}$ multiplied by the energy density $V(\phi)$: $E \sim V^{-1/2}$. Therefore such a universe can appear quantum mechanically within the time $\Delta t \gtrsim 1$ if $V(\phi)$ is not too much smaller than the Planck density $O(1)$.

This qualitative conclusion agrees with the result of the investigation in the context of quantum cosmology. Indeed, according to [17, 18], the probability of quantum creation of a closed universe is proportional to

$$P \sim \exp\left(-\frac{24\pi^2}{V}\right), \tag{1.5}$$

which means that the universe can be created if V is not too much smaller than the Planck density. The Euclidean approach to the quantum creation of the universe is based on the analytical continuation of the Euclidean de Sitter solution to the real time. This continuation is possible if $\dot{\phi} = 0$ at the moment of quantum creation of the universe. Thus in the simplest chaotic inflation model with $V(\phi) = \frac{m^2}{2}\phi^2$ the universe is created in a state with $V(\phi) \sim 1$, $\phi \sim m^{-1} \gg 1$ and $\dot{\phi} = 0$, which is a perfect initial condition for inflation in this model [14, 17].

One should note that there are many other attempts to evaluate the probability of initial conditions for inflation (see Chap. 5 in this volume). For example, if one interprets the square of the Hartle–Hawking wave function [19] as a probability of initial condition, one obtains a paradoxical answer $P \sim \exp(\frac{24\pi^2}{V})$, which could seem to imply that it is easier to create the universe with $V \to 0$ and with an infinitely large total energy $E \sim V^{-1/2} \to \infty$. There were many attempts to improve this anti-intuitive answer, but from my perspective these attempts were misplaced: the Hartle–Hawking wave function was derived in [19] as a wave function for the *ground state of the universe*, and therefore it describes the most probable *final* state of the universe, instead of the probability of initial conditions; see a discussion of this issue in [14, 20, 21].

Another recent attempt to study this problem was made by Gibbons and Turok [22]. They studied classical solutions describing a combined evolution of a scalar field and the scale factor of the universe, and imposed "initial conditions" not at the beginning of inflation but at its end. Since one can always reverse the direction of time in the solutions, one can always relate the conditions at the end of inflation to the conditions at its beginning. If one assumes that certain conditions at the end of inflation are equally probable, then one may conclude that the probability of initial conditions suitable for inflation must be very small [22].

From our perspective [23, 24], we have here the same paradox which is encountered in the discussion of the growth of entropy. If one starts with a well ordered system, its entropy will always grow. However, if we make a movie of this process, and play it back starting from the end of the process, then the final conditions for the original system become the initial conditions for the time-reversed system, and we will see the entropy decreasing. That is why replacing initial conditions by final conditions can be very misleading. An advantage of the inflationary regime is that it is an attractor (i.e. the most probable regime) for the family of solutions describing an expanding universe. But if one replaces initial conditions by the final conditions at the end of the process and then studies the same process back in time, the same trajectory will look like a repulsor. This is the main reason of the negative conclusion of [22].

The main problem in [22] is that the methods developed there are valid for the classical evolution of the universe, but the initial conditions for the classical evolution are determined by the processes at the quantum epoch near the singularity, where the methods of [22] are inapplicable. It is not surprising, therefore, that the results of [22] imply that initially $\dot{\phi}^2 \gg V(\phi)$. This result contradicts the results of the Euclidean approach to quantum creation of the universe [17, 18, 19] which require that initially $\dot{\phi} = 0$, see a discussion above.

As we will show in a separate publication [24], if one further develops the methods of [22], but imposes the initial conditions at the beginning of inflation, rather than at its end, one finds that inflation is most probable, in agreement with the arguments given in the first part of this section.

The discussion of initial conditions in this section was limited to the simplest versions of chaotic inflation which allow inflation at the very high energy densities, such as the models with $V \sim \phi^n$. We will return to the discussion of the problem of initial conditions in inflationary cosmology in Sects. 1.13 and 1.14, where we will analyze it in the context of more complicated inflationary models.

1.2.3 Solving the Cosmological Problems

As we will see shortly, the realistic value of the mass m is about 3×10^{-6}, in Planck units. Therefore, according to (1.4), the total amount of inflation achieved starting from $V(\phi) \sim 1$ is of the order $10^{10^{10}}$. The total duration of inflation in this model is about 10^{-30} s. When inflation ends, the scalar field ϕ begins to oscillate near the minimum of $V(\phi)$. As any rapidly oscillating classical field, it looses its energy by creating pairs of elementary particles. These particles interact with each other and come to a state of thermal equilibrium with some temperature T_{rh} [25, 26, 27, 28, 29, 30, 31]. From this time on, the universe can be described by the usual big bang theory.

The main difference between inflationary theory and the old cosmology becomes clear when one calculates the size of a typical inflationary domain at the end of inflation. The investigation of this question shows that even if the initial size of inflationary universe was as small as the Planck size $l_{\mathrm{P}} \sim 10^{-33}$ cm, after 10^{-30} s of inflation the universe acquires a huge size of $l \sim 10^{10^{10}}$ cm! This number is model-dependent, but in all realistic models the size of the universe after inflation appears to be many orders of magnitude greater than the size of the part of the universe which we can see now, $l \sim 10^{28}$ cm. This immediately solves most of the problems of the old cosmological theory [14, 15].

Our universe is almost exactly homogeneous on large scales because all inhomogeneities were exponentially stretched during inflation. The density of primordial monopoles and other undesirable "defects" becomes exponentially diluted by inflation. The universe becomes enormously large. Even if it was a closed universe of a size $\sim 10^{-33}$ cm, after inflation the distance between its "South" and "North" poles becomes many orders of magnitude greater than 10^{28} cm. We see only a tiny part of the huge cosmic balloon. That is why nobody has ever seen how parallel lines cross. That is why the universe looks so flat.

If our universe initially consisted of many domains with chaotically distributed scalar field ϕ (or if one considers different universes with different values of the field), then domains in which the scalar field was too small never inflated. The main contribution to the total volume of the universe will be given by those domains which originally contained a large scalar field ϕ. Inflation of such domains creates huge homogeneous islands out of initial chaos. (That is why I called this scenario "chaotic inflation.") Each homogeneous domain in this scenario is much greater than the size of the observable part of the universe.

1.2.4 Chaotic Inflation Versus New Inflation

The first models of chaotic inflation were based on the theories with polynomial potentials, such as $V(\phi) = \pm\frac{m^2}{2}\phi^2 + \frac{\lambda}{4}\phi^4$. But, as was emphasized in [15], the main idea of this scenario is quite generic. One should consider any particular potential $V(\phi)$, polynomial or not, with or without spontaneous symmetry breaking, and study all possible initial conditions without assuming that the universe was in a state of thermal equilibrium, and that the field ϕ was in the minimum of its effective potential from the very beginning.

This scenario strongly deviated from the standard lore of the hot big bang theory and was psychologically difficult to accept. Therefore during the first few years after the invention of chaotic inflation many authors claimed that the idea of chaotic initial conditions is unnatural, and made attempts to realize the new inflation scenario based on the theory of high-temperature phase transitions, despite numerous problems associated with it. Some authors believed that the theory must satisfy the so-called "thermal constraints" which were necessary to ensure that the minimum of the effective potential at large T should be at $\phi = 0$ [32], even though the scalar field in the models they considered was not in a state of thermal equilibrium with other particles.

The issue of thermal initial conditions played the central role in the long debate about new inflation versus chaotic inflation in the 1980s. This debate continued for many years, and a significant part of my book [14] was dedicated to it. By now the debate is over: no realistic versions of new inflation based on the theory of thermal phase transitions and supercooling have been proposed so far. Gradually it became clear that the idea of chaotic initial conditions is most general, and it is much easier to construct a consistent cosmological theory without making unnecessary assumptions about thermal equilibrium and high-temperature phase transitions in the early universe.

As a result, the corresponding terminology changed. Chaotic inflation, as defined in [15], occurs in *all* models with sufficiently flat potentials, including the potentials with a flat maximum, originally used in new inflation [33]. Now the versions of inflationary scenario with such potentials for simplicity are often called "new inflation," even though inflation begins there not as in the original new inflation scenario, but as in the chaotic inflation scenario. To avoid this terminological misunderstanding, some authors call the version of chaotic inflation scenario, where inflation occurs near the top of the scalar potential, a "hilltop inflation" [34].

1.3 Hybrid Inflation

The simplest models of inflation involve just one scalar field. However, in supergravity and string theory there are many different scalar fields, so it does make sense to study models with several different scalar fields, especially if they have some qualitatively new properties. Here we will consider one of these models, hybrid inflation [35].

The simplest version of hybrid inflation describes the theory of two scalar fields with the effective potential

$$V(\sigma, \phi) = \frac{1}{4\lambda}(M^2 - \lambda\sigma^2)^2 + \frac{m^2}{2}\phi^2 + \frac{g^2}{2}\phi^2\sigma^2 . \tag{1.6}$$

The effective mass squared of the field σ is equal to $-M^2 + g^2\phi^2$. Therefore for $\phi > \phi_c = M/g$ the only minimum of the effective potential $V(\sigma, \phi)$ is at $\sigma = 0$. The curvature of the effective potential in the σ-direction is much greater than in the ϕ-direction. Thus at the first stages of expansion of the universe the field σ rolled down to $\sigma = 0$, whereas the field ϕ could remain large for a much longer time.

At the moment when the inflaton field ϕ becomes smaller than $\phi_c = M/g$, the phase transition with the symmetry breaking occurs. The fields rapidly fall to the absolute minimum of the potential at $\phi = 0, \sigma^2 = M^2/\lambda$. If $m^2\phi_c^2 = m^2M^2/g^2 \ll M^4/\lambda$, the Hubble constant at the time of the phase transition is given by $H^2 = \frac{M^4}{12\lambda}$ (in units $M_{\text{Pl}} = 1$). If $M^2 \gg \frac{\lambda m^2}{g^2}$ and $m^2 \ll H^2$, then the universe at $\phi > \phi_c$ undergoes a stage of inflation, which abruptly ends at $\phi = \phi_c$.

Note that hybrid inflation is also a version of the chaotic inflation scenario: i am unaware of any way to realize this model in the context of the theory of high-temperature phase transitions. The main difference between this scenario and the simplest versions of the one-field chaotic inflation is in the way inflation ends. In the theory with a single field, inflation ends when the potential of this field becomes steep. In hybrid inflation, the structure of the universe depends on the way one of the fields moves, but inflation ends when the potential of the second field becomes steep. This fact allows much greater flexibility of construction of inflationary models. Several extensions of this scenario became quite popular in the context of supergravity and string cosmology, which we will discuss later.

1.4 Quantum Fluctuations and Density Perturbations

The average amplitude of inflationary perturbations generated during a typical time interval H^{-1} is given by [36, 37]

$$|\delta\phi(x)| \approx \frac{H}{2\pi} . \tag{1.7}$$

These fluctuations lead to density perturbations that later produce galaxies (see Chap. 6 in this volume). The theory of this effect is very complicated [6, 12], and it was fully understood only in the second part of the 1980s [13]. The main idea can be described as follows.

Fluctuations of the field ϕ lead to a local delay of the time of the end of inflation, $\delta t = \frac{\delta\phi}{\dot{\phi}} \sim \frac{H}{2\pi\dot{\phi}}$. Once the usual post-inflationary stage begins, the density of the universe starts to decrease as $\rho = 3H^2$, where $H \sim t^{-1}$. Therefore a local delay of expansion leads to a local density increase δ_H such that $\delta_H \sim \delta\rho/\rho \sim \delta t/t$. Combining these estimates together yields the famous result [6, 12, 13]

$$\delta_H \sim \frac{\delta\rho}{\rho} \sim \frac{H^2}{2\pi\dot{\phi}} . \tag{1.8}$$

The field ϕ during inflation changes very slowly, so the quantity $\frac{H^2}{2\pi\dot{\phi}}$ remains almost constant over an exponentially large range of wavelengths. This means that the spectrum of perturbations of the metric is flat.

A detailed calculation in our simplest chaotic inflation model of the amplitude of perturbations gives

$$\delta_H \sim \frac{m\phi^2}{5\pi\sqrt{6}} . \tag{1.9}$$

The perturbations on the scale of the horizon were produced at $\phi_{\rm H} \sim 15$ [14]. This, together with the COBE normalization $\delta_{\rm H} \sim 2{\times}10^{-5}$ gives $m \sim 3{\times}10^{-6}$, in Planck units, which is approximately equivalent to 7×10^{12} GeV. An exact value of m depends on $\phi_{\rm H}$, which in its turn depends slightly on the subsequent thermal history of the universe.

When the fluctuations of the scalar field ϕ are first produced (frozen), their wavelength is given by $H(\phi)^{-1}$. At the end of inflation, the wavelength grows by the factor of $e^{\phi^2/4}$, see (1.4). In other words, the logarithm of the wavelength l of the perturbations of metric is proportional to the value of ϕ^2 at the moment when these perturbations were produced. As a result, according to (1.9), the amplitude of the perturbations of the metric depends logarithmically on the wavelength: $\delta_{\rm H} \sim m \ln l$. A similar logarithmic dependence (with different powers of the logarithm) appears in other versions of chaotic inflation with $V \sim \phi^n$ and in the simplest versions of new inflation.

At first glance, this logarithmic deviation from scale invariance could seem inconsequential, but in a certain sense it is similar to the famous logarithmic dependence of the coupling constants in QCD, where it leads to asymptotic freedom at high energies, instead of simple scaling invariance [38, 39]. In QCD, the slow growth of the coupling constants at small momenta/large distances is responsible for nonperturbative effects resulting in quark confinement. In inflationary theory, the slow growth of the amplitude of perturbations of metric at large distances is equally important. It leads to the existence of the regime of eternal inflation and to the fractal structure of the universe on super-large scales, see Sect. 1.6.

Since the observations provide us with information about a rather limited range of l, it is often possible to parametrize the scale dependence of density perturbations by a simple power law, $\delta_{\rm H} \sim l^{(1-n_{\rm s})/2}$. An exactly flat spectrum, called Harrison–Zeldovich spectrum, would correspond to $n_{\rm s} = 1$.

The amplitude of the scalar perturbations of the metric can be characterized either by $\delta_{\rm H}$, or by a closely related quantity $\Delta_{\mathcal{R}}$ [40]. Similarly, the amplitude of tensor perturbations is given by Δ_h. Following [40, 41], one can represent these quantities as

$$\Delta_{\mathcal{R}}^2(k) = \Delta_{\mathcal{R}}^2(k_0)\left(\frac{k}{k_0}\right)^{n_{\rm s}-1}, \tag{1.10}$$

$$\Delta_h^2(k) = \Delta_h^2(k_0)\left(\frac{k}{k_0}\right)^{n_{\rm t}}, \tag{1.11}$$

where $\Delta^2(k_0)$ is a normalization constant, and k_0 is a normalization point. Here we ignored running of the indexes $n_{\rm s}$ and $n_{\rm t}$ since there is no observational evidence that it is significant.

One can also introduce the tensor/scalar ratio r, the relative amplitude of the tensor to scalar modes,

$$r \equiv \frac{\Delta_h^2(k_0)}{\Delta_{\mathcal{R}}^2(k_0)}. \tag{1.12}$$

There are three slow-roll parameters [40]

$$\epsilon = \frac{1}{2}\left(\frac{V'}{V}\right)^2, \quad \eta = \frac{V''}{V}, \quad \xi = \frac{V'V'''}{V^2}, \tag{1.13}$$

where prime denotes derivatives with respect to the field ϕ. All parameters must be smaller than one for the slow-roll approximation to be valid.

A standard slow roll analysis gives observable quantities in terms of the slow-roll parameters to first order as

$$\Delta_{\mathcal{R}}^2 = \frac{V}{24\pi^2\epsilon} = \frac{V^3}{12\pi^2(V')^2}, \tag{1.14}$$

$$n_{\rm s} - 1 = -6\epsilon + 2\eta, \tag{1.15}$$

$$r = 16\epsilon, \tag{1.16}$$

$$n_{\rm t} = -2\epsilon = -\frac{r}{8}. \tag{1.17}$$

The equation $n_{\rm t} = -r/8$ is known as the consistency relation for single-field inflation models; it becomes an inequality for multi-field inflation models. If V during inflation is sufficiently large, as in the simplest models of chaotic inflation, one may have a chance to find the tensor contribution to the CMB anisotropy. The possibility to determine $n_{\rm t}$ is less certain. The most important information which can be obtained now from the cosmological observations at present is related to (1.14) and (1.15).

Following notational conventions in [41], we use $A(k_0)$ for the scalar power spectrum amplitude, where $A(k_0)$ and $\Delta_{\mathcal{R}}^2(k_0)$ are related through

$$\Delta_{\mathcal{R}}^2(k_0) \simeq 3 \times 10^{-9} A(k_0). \tag{1.18}$$

The parameter A is often normalized at $k_0 \sim 0.05/\text{Mpc}$; its observational value is about 0.8 [41, 42, 43] (see also Chap. 6 in this volume). This leads to the observational constraint on $V(\phi)$ and on r following from the normalization of the spectrum of the large-scale density perturbations:

$$\frac{V^{3/2}}{V'} \simeq 5 \times 10^{-4}. \tag{1.19}$$

Here $V(\phi)$ should be evaluated for the value of the field ϕ which is determined by the condition that the perturbations produced at the moment when the field was equal ϕ evolve into the present time perturbations with momentum $k_0 \sim 0.05/\text{Mpc}$. In the first approximation, one can find the corresponding moment by assuming that it happened 60 e-foldings before the end of inflation. The number of e-foldings can be calculated in the slow roll approximation using the relation

$$N \simeq \int_{\phi_{\rm end}}^{\phi} \frac{V}{V'} \mathrm{d}\phi. \tag{1.20}$$

Equation (1.19) leads to the relation between r, V and H, in Planck units:

$$r \approx 3 \times 10^7 \ V \approx 10^8 \ H^2 \ . \tag{1.21}$$

Finally, recent observational data suggest [42] that

$$n_{\mathrm{s}} = 1 - 3 \left(\frac{V'}{V} \right)^2 + 2 \frac{V''}{V} = 0.95 \pm 0.016 \ , \tag{1.22}$$

for $r \ll 0.1$. These relations are very useful for comparing inflationary models with observations. In particular, the simplest versions of chaotic and new inflation predict $n_{\mathrm{s}} < 1$, whereas in hybrid inflation one may have either $n_{\mathrm{s}} < 1$ or $n_{\mathrm{s}} > 1$, depending on the model. A more accurate representation of observational constraints can be found in Sect. 1.7.

Until now we have discussed the standard mechanism of generation of perturbations of metric. However, if the model is sufficiently complicated, other mechanisms become possible. For example, one may consider a theory of two scalar fields, ϕ and σ, and assume that inflation was driven by the field ϕ, and the field σ was very light during inflation and did not contribute much to the total energy density. Therefore its quantum fluctuations also did not contribute much to the amplitude of perturbations of metric during inflation (isocurvature perturbations).

After inflation the field ϕ decays. If the products of its decay rapidly loose energy, the field σ may dominate the energy density of the universe and its perturbations suddenly become important. If, in its turn, the field σ decays, its perturbations under certain conditions can be converted into the usual adiabatic perturbations of metric. If this conversion is incomplete, one obtains a theory at odds with recent observational data [44, 45]. On the other hand, if the conversion is complete, one obtains a novel mechanism of generation of adiabatic density perturbations, which is called the curvaton mechanism [46, 47, 48, 49]. A closely related but different mechanism was also proposed in [50]. See Chap. 8 in this volume for a detailed discussion.

These mechanisms are much more complicated than the original one, but one should keep them in mind since they sometimes work in the situations where the standard one does not. Therefore they can give us an additional freedom in finding realistic models of inflationary cosmology.

1.5 Creation of Matter After Inflation: Reheating and Preheating

The theory of reheating of the universe after inflation is the most important application of the quantum theory of particle creation, since almost all matter constituting the universe was created during this process.

At the stage of inflation all energy is concentrated in a classical slowly moving inflaton field ϕ. Soon after the end of inflation this field begins to oscillate near the minimum of its effective potential. Eventually it produces many elementary particles, they interact with each other and come to a state of thermal equilibrium with some temperature T_r.

Early discussions of reheating of the universe after inflation [25] were based on the idea that the homogeneous inflaton field can be represented as a collection of the particles of the field ϕ. Each of these particles decayed independently. This process can be studied by the usual perturbative approach to particle decay. Typically, it takes thousands of oscillations of the inflaton field until it decays into usual elementary particles by this mechanism. More recently, however, it was discovered that coherent field effects such as parametric resonance can lead to the decay of the homogeneous field much faster than would have been predicted by perturbative methods, within a few dozen oscillations [26]. These coherent effects produce high energy, nonthermal fluctuations that could have significance for understanding developments in the early universe, such as baryogenesis. This early stage of rapid nonperturbative decay was called "preheating." In [27] it was found that another effect known as tachyonic preheating can lead to even faster decay than parametric resonance. This effect occurs whenever the homogeneous field rolls down a tachyonic ($V'' < 0$) region of its potential. When that occurs, a tachyonic, or spinodal instability leads to exponentially rapid growth of all long wavelength modes with $k^2 < |V''|$. This growth can often drain all of the energy from the homogeneous field within a single oscillation.

We are now in a position to classify the dominant mechanisms by which the homogeneous inflaton field decays in different classes of inflationary models. Even though all of these models, strictly speaking, belong to the general class of chaotic inflation (none of them is based on the theory of thermal initial conditions), one can break them into three classes: small field, or new inflation models [10], large field, or chaotic inflation models of the type of the model $m^2\phi^2/2$ [15], and multi-field, or hybrid models [35]. This classification is incomplete, but still rather helpful.

In the simplest versions of chaotic inflation, the stage of preheating is generally dominated by parametric resonance, although there are parameter ranges where this cannot occur [26]. In [27], it was shown that tachyonic preheating dominates the preheating phase in hybrid models of inflation. New inflation in this respect occupies an intermediate position between chaotic inflation and hybrid inflation: If spontaneous symmetry breaking in this scenario is very large, reheating occurs due to parametric resonance and perturbative decay. However, for the models with spontaneous symmetry breaking at or below the GUT scale, $\phi \ll 10^{-2}M_{\rm Pl}$, preheating occurs due to a combination of tachyonic preheating and parametric resonance. The resulting effect is very strong, so that the homogeneous mode of the inflaton field typically decays within few oscillations [28].

A detailed investigation of preheating usually requires lattice simulations, which can be achieved following [29, 30]. Note that preheating is not the last stage of reheating; it is followed by a period of turbulence [31], by a much slower perturbative decay described by the methods developed in [25], and by eventual thermalization.

1.6 Eternal Inflation

A significant step in the development of inflationary theory was the discovery of the process of self-reproduction of inflationary universe. This process was known to exist in old inflationary theory [7] and in the new one [51, 52, 53], but its significance was fully realized only after the discovery of the regime of eternal inflation in the simplest versions of the chaotic inflation scenario [54, 55]. It appears that in many inflationary models large quantum fluctuations produced during inflation may significantly increase the value of the energy density in some parts of the universe. These regions expand at a greater rate than their parent domains, and quantum fluctuations inside them lead to production of new inflationary domains which expand even faster. This leads to an eternal process of self-reproduction of the universe.

To understand the mechanism of self-reproduction one should remember that processes separated by distances l greater than H^{-1} proceed independently of one another. This is so because during exponential expansion the distance between any two objects separated by more than H^{-1} is growing with a speed exceeding the speed of light. As a result, an observer in the inflationary universe can see only the processes occurring inside the horizon of the radius H^{-1}. An important consequence of this general result is that the process of inflation in any spatial domain of radius H^{-1} occurs independently of any events outside it. In this sense any inflationary domain of initial radius exceeding H^{-1} can be considered as a separate mini-universe.

To investigate the behavior of such a mini-universe, with an account taken of quantum fluctuations, let us consider an inflationary domain of initial radius H^{-1} containing sufficiently homogeneous field with initial value $\phi \gg M_{\mathrm{Pl}}$. Equation (1.3) implies that during a typical time interval $\Delta t = H^{-1}$ the field inside this domain will be reduced by $\Delta \phi = \frac{2}{\phi}$. By comparison this expression with $|\delta \phi(x)| \approx \frac{H}{2\pi} = \frac{m\phi}{2\pi\sqrt{6}}$ one can easily see that if ϕ is much less than $\phi^* \sim \frac{5}{\sqrt{m}}$, then the decrease of the field ϕ due to its classical motion is much greater than the average amplitude of the quantum fluctuations $\delta\phi$ generated during the same time. But for $\phi \gg \phi^*$ one has $\delta\phi(x) \gg \Delta\phi$. Because the typical wavelength of the fluctuations $\delta\phi(x)$ generated during the time is H^{-1}, the whole domain after $\Delta t = H^{-1}$ effectively becomes divided into $e^3 \sim 20$ separate domains (mini-universes) of radius H^{-1}, each containing almost homogeneous field $\phi - \Delta\phi + \delta\phi$. In almost a half of these domains the field ϕ grows by $|\delta\phi(x)| - \Delta\phi \approx |\delta\phi(x)| = H/2\pi$, rather than decreases.

This means that the total volume of the universe containing the *growing* field ϕ increases 10 times. During the next time interval $\Delta t = H^{-1}$ this process repeats itself. Thus, after the two time intervals H^{-1} the total volume of the universe containing the growing scalar field increases 100 times, etc. The universe enters the eternal process of self-reproduction.

The existence of this process implies that the universe will never disappear as a whole. Some of its parts may collapse, the life in our part of the universe may perish, but there always will be some other parts of the universe where life will appear again and again, in all of its possible forms.

One should be careful, however, with the interpretation of these results. There is still an ongoing debate of whether eternal inflation is eternal only in the future or also in the past. In order to understand what is going on, let us consider any particular time-like geodesic line at the stage of inflation. One can show that for any given observer following this geodesic, the duration t_i of the stage of inflation on this geodesic will be finite. One the other hand, eternal inflation implies that if one takes all such geodesics and calculate the time t_i for each of them, then there will be no upper bound for t_i, i.e. for each time T there will exist geodesics which experience inflation for a time $t_i > T$. Even though the relative number of long geodesics can be very small, exponential expansion of space surrounding them will lead to an eternal exponential growth of the total volume of the inflationary parts of the universe.

Similarly, if one concentrates on any particular geodesic in the past time direction, one can prove that it has finite length [56], i.e. inflation in any particular point of the universe should have a beginning at some time τ_i. However, there is no reason to expect that there is an upper bound for all τ_i on all geodesics. If this upper bound does not exist, then eternal inflation is eternal not only in the future but also in the past.

In other words, there was a beginning for each part of the universe, and there will be an end for inflation at any particular point. But there will be no end for the evolution of the universe *as a whole* in the eternal inflation scenario, and at present we do not have any reason to believe that there was a single beginning of the evolution of the whole universe at some moment $t = 0$, which was traditionally associated with the big bang.

To illustrate the process of eternal inflation, we present here the results of computer simulations of evolution of a system of two scalar fields during inflation. The field ϕ is the inflaton field driving inflation; it is shown by the height of the distribution of the field $\phi(x, y)$ in a two-dimensional slice of the universe. The second field, Φ, determines the type of spontaneous symmetry breaking which may occur in the theory. We paint the surface in red, green or blue corresponding to three different minima of the potential of the field Φ. Different colors correspond to different types of spontaneous symmetry breaking, and therefore to different sets of laws of low-energy physics in different exponentially large parts of the universe.

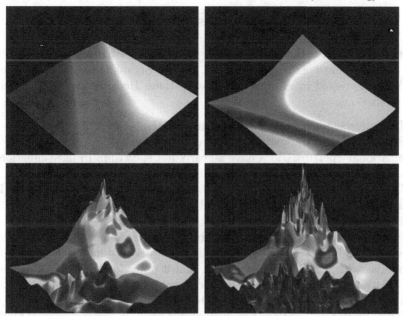

Fig. 1.2. Evolution of scalar fields ϕ and Φ during the process of self-reproduction of the universe. The height of the distribution shows the value of the field ϕ which drives inflation. The surface is painted in red (medium), green (dark) or blue (light) corresponding to three different minima of the potential of the field Φ. The laws of low-energy physics are different in the regions of different color. The peaks of the "mountains" correspond to places where quantum fluctuations bring the scalar fields back to the Planck density. Each of such places in a certain sense can be considered as a beginning of a new big bang

In the beginning of the process the whole inflationary domain is red, and the distribution of both fields is very homogeneous. Then the domain became exponentially large (but it has the same size in comoving coordinates, as shown in Fig. 1.2). Each peak of the mountains corresponds to nearly Planckian density and can be interpreted as a beginning of a new "big bang." The laws of physics are rapidly changing there, as indicated by changing colors, but they become fixed in the parts of the universe where the field ϕ becomes small. These parts correspond to valleys in Fig. 1.2. Thus quantum fluctuations of the scalar fields divide the universe into exponentially large domains with different laws of low-energy physics, and with different values of energy density.

Eternal inflation scenario was extensively studied during the last 20 years. I should mention, in particular, the discovery of the topological eternal inflation [57] and the calculation of the fractal dimension of the universe [58, 55]. The most interesting consequences of the theory of eternal inflation are related to the theory of inflationary multiverse and string theory landscape. We will discuss these subjects in Sect. 1.14.

1.7 Inflation and Observations

Inflation is not just an interesting theory that can resolve many difficult problems of the standard big bang cosmology. This theory made several predictions which can be tested by cosmological observations. Here are the most important predictions:

(1) The universe must be flat. In most models $\Omega_{\text{total}} = 1 \pm 10^{-4}$.
(2) Perturbations of the metric produced during inflation are adiabatic.
(3) Inflationary perturbations have a nearly flat spectrum. In most inflationary models the spectral index $n_{\text{s}} = 1 \pm 0.2$ ($n_{\text{s}} = 1$ means totally flat).
(4) The spectrum of inflationary perturbations should be slightly non-flat. (It is very difficult to construct a model with $n_{\text{s}} = 1$.)
(5) These perturbations are gaussian.
(6) Perturbations of the metric could be scalar, vector or tensor. Inflation mostly produces scalar perturbations, but it also produces tensor perturbations with a nearly flat spectrum, and it does *not* produce vector perturbations. There are certain relations between the properties of scalar and tensor perturbations produced by inflation.
(7) Inflationary perturbations produce specific peaks in the spectrum of CMB radiation. (For a simple pedagogical interpretation of this effect see e.g. [59]; a detailed theoretical description can be found in [60].)

It is possible to violate each of these predictions if one makes the inflationary theory sufficiently complicated. For example, it is possible to produce vector perturbations of the metric in the models where cosmic strings are produced at the end of inflation, which is the case in some versions of hybrid inflation. It is possible to have an open or closed inflationary universe, or even a small periodic inflationary universe, it is possible to have models with non-gaussian isocurvature fluctuations with a non-flat spectrum. However, it is difficult to do so, and most of the inflationary models obey the simple rules given above.

It is not easy to test all of these predictions. The major breakthrough in this direction was achieved due to the recent measurements of the CMB anisotropy. The latest results based on the WMAP experiment, in combination with the Sloan Digital Sky Survey, are consistent with predictions of the simplest inflationary models with adiabatic gaussian perturbations, with $\Omega = 1.003 \pm 0.01$, and $n_{\text{s}} = 0.95 \pm 0.016$ [42].

There are still some question marks to be examined, such as an unexpectedly small anisotropy of the CMB at large angles [41, 61] and possible correlations between low multipoles; for a recent discussion see e.g. [62, 63] and references therein (Fig. 1.3).

The observational status and interpretation of these effects is still uncertain, but if one takes these effects seriously, one may try to look for some theoretical explanations. For example, there are several ways to suppress the large angle anisotropy, see e.g. [64]. The situation with correlations between

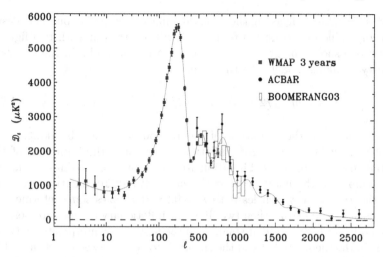

Fig. 1.3. CMB data (WMAP3, BOOMERANG03, ACBAR) versus the predictions of one of the simplest inflationary models with $\Omega = 1$ (*solid red line*), according to [43]

low multipoles requires more work. In particular, it would be interesting to study effects related to relatively light domain walls [65, 66, 67]. Another possibility is to analyze the possible effects on the CMB anisotropy which can be produced by the cosmic web structure of the perturbations in the curvaton scenario [46]. Some other possibilities are mentioned in [63]. One way or another, it is quite significant that all proposed explanations of these anomalies are based on inflationary cosmology.

One of the interesting issues to be probed by future observations is the possible existence of gravitational waves produced during inflation. The present upper bound on the tensor to scalar ratio r is not very strict, $r \lesssim 0.3$. However, new observations may either find the tensor modes or push the bound on r much further, towards $r \lesssim 10^{-2}$ or even $r \lesssim 10^{-3}$.

In the simplest monomial versions of chaotic inflation with $V \sim \phi^n$ one find the following (approximate) result: $r = 4n/N$. Here N is the number of e-folds of inflation corresponding to the wavelength equal to the present size of the observable part of our universe; typically N can be in the range of 50–60; its value depends on the mechanism of reheating. For the simplest model with $n = 2$ and $N \sim 60$ one has $r \sim 0.13 - 0.14$. On the other hand, for most of the other models, including the original version of new inflation, hybrid inflation, and many versions of string theory inflation, r is extremely small, which makes the observation of gravitational waves in such models very difficult.

One may wonder whether there are any sufficiently simple and natural models with intermediate values of r? This is an important question for those who are planning a new generation of CMB experiments. The answer to this question is positive: In the versions of chaotic inflation with potentials like

$\pm m^2\phi^2 + \lambda\phi^4$, as well as in the natural inflation scenario, one can easily obtain any value of r from 0.3 to 10^{-2}. I will illustrate it with two figures. The first one shows the graph of possible values of n_s and r in the standard symmetry breaking model with the potential

$$V = -m^2\phi^2/2 + \lambda\phi^4/4 + m^4/4\lambda = \frac{\lambda}{4}(\phi^2 - v^2)^2 , \qquad (1.23)$$

where $v = m/\sqrt{\lambda}$ is the amplitude of spontaneous symmetry breaking.

If v is very large, $v \gtrsim 10^2$, inflation occurs near the minimum of the potential, and all properties of inflation are the same as in the simplest chaotic inflation model with quadratic potential $m^2\phi^2$. If $v \ll 10$, inflation occurs as in the theory $\lambda\phi^4/4$, which leads to $r \sim 0.28$. If v takes some intermediate values, such as $v = O(10)$, then two different inflationary regimes are possible in this model: at large ϕ and at small ϕ. In the first case r interpolates between its value in the theory $\lambda\phi^4/4$ and the theory $m^2\phi^2$ (i.e. between 0.28 and 0.14). In the second case, r can take any value from 0.14 to 10^{-2}, see Fig. 1.4 [68, 69].

If one considers chaotic inflation with the potential including terms ϕ^2, ϕ^3 and ϕ^4, one can considerably alter the properties of inflationary perturbations [70]. Depending on the values of parameters, initial conditions and the required number of e-foldings N, this relatively simple class of models covers almost all parts of the area in the (r, n_s) plane allowed by the latest observational data [71], see Fig. 1.5.

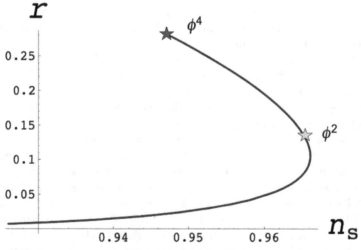

Fig. 1.4. Possible values of r and n_s in the theory $\frac{\lambda}{4}(\phi^2 - v^2)^2$ for different initial conditions and different v, for $N = 60$. In the small v limit, the model has the same predictions as the theory $\lambda\phi^4/4$. In the large v limit it has the same predictions as the theory $m^2\phi^2$. The upper branch, above the first star from below (marked as ϕ^2), corresponds to inflation which occurs while the field rolls down from large ϕ; the lower branch corresponds to the motion from $\phi = 0$

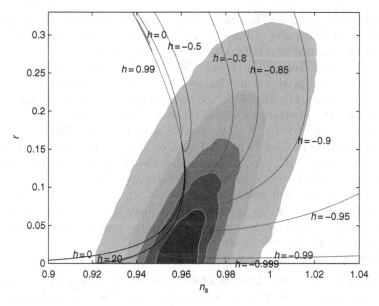

Fig. 1.5. Possible values of r and n_s for chaotic inflation with a potential including terms ϕ^2, ϕ^3 and ϕ^4 for N = 50, according to [71]. The color-filled areas correspond to 12%, 27%, 45%, 68% and 95% confidence levels according to the WMAP3 and SDSS data

Note that for all versions of the model shown in Figs. 1.4 and 1.5 the range of the cosmological evolution of the fields is $\Delta\phi > 1$, so formally these models can be called the large field models. And yet they have dramatically different properties, which do not fit into the often-used scheme dividing all models into small field models, large field models and hybrid inflation models.

1.8 Alternatives to Inflation?

The inflationary scenario is very versatile, and now, after 25 years of persistent attempts of many physicists to propose an alternative to inflation, we still do not know any other way to construct a consistent cosmological theory. Indeed, in order to compete with inflation a new theory should make similar predictions and should offer an alternative solution to many difficult cosmological problems. Let us look at these problems before starting a discussion.

(1) The homogeneity problem. Before even starting an investigation of density perturbations and structure formation, one should explain why the universe is nearly homogeneous on the horizon scale.
(2) The isotropy problem. We need to understand why all directions in the universe are similar to each other, why there is no overall rotation of the universe, etc...

(3) The horizon problem. This one is closely related to the homogeneity problem. If different parts of the universe have not been in a causal contact when the universe was born, why do they look so similar?

(4) The flatness problem. Why $\Omega \approx 1$? Why parallel lines do not intersect?

(5) The total entropy problem. The total entropy of the observable part of the universe is greater than 10^{87}. Where did this huge number come from? Note that the lifetime of a closed universe filled with hot gas with total entropy S is $S^{2/3} \times 10^{-43}$ s [14]. Thus S must be huge. Why?

(6) The total mass problem. The total mass of the observable part of the universe has mass $\sim 10^{60} M_{\mathrm{Pl}}$. Note also that the lifetime of a closed universe filled with nonrelativistic particles of total mass M is $\frac{M}{M_{\mathrm{P}}} \times 10^{-43}$ s. Thus M must be huge. But why?

(7) The structure formation problem. If we manage to explain the homogeneity of the universe, how can we explain the origin of inhomogeneities required for the large scale structure formation?

(8) The monopole problem, gravitino problem, etc.

This list is very long. That is why it was not easy to propose any alternative to inflation even before we learned that $\Omega \approx 1$, $n_{\mathrm{s}} \approx 1$, and that the perturbations responsible for galaxy formation are mostly adiabatic, in agreement with the predictions of the simplest inflationary models.

There were many attempts to propose an alternative to inflation in recent years. In general, this could be a very healthy tendency. If one of these attempts will succeed, it will be of great importance. If none of them are successful, it will be an additional demonstration of the advantages of inflationary cosmology. However, since the stakes are high, we are witnessing a growing number of premature announcements of success in developing an alternative cosmological theory (see Chap. 11 in this volume for an alternative discussion).

1.8.1 Cosmic Strings and Textures

Fifteen years ago the models of structure formation due to topological defects or textures were advertised in popular press as the models that "match the explanatory triumphs of inflation while rectifying its major failings" [72]. However, it was clear from the very beginning that these theories at best could solve only one problem (structure formation) out of the eight problems mentioned above. The true question was not whether one can replace inflation by the theory of cosmic strings/textures, but whether inflation with cosmic strings/textures is better than inflation without cosmic strings/textures. Recent observational data favor the simplest version of inflationary theory, without topological defects, or with an extremely small (few percent) admixture of the effects due to cosmic strings.

1.8.2 Pre-big Bang

An attempt to avoid the use of the standard inflationary mechanism (though still use a stage of inflation prior to the big bang) was made in the pre-big bang scenario [73]. This scenario is based on the assumption that eventually one will find a solution of the cosmological singularity problem and learn how one could transfer small perturbations of the metric through the singularity. This problem still remains unsolved, see e.g. [74]. Moreover, a detailed investigation of the homogeneity, isotropy and flatness problems in the pre-big bang scenario demonstrated that the stage of the pre-big bang inflation introduced in [73] is insufficient to solve the major cosmological problems [75].

1.8.3 Ekpyrotic/Cyclic Scenario

A similar situation emerged with the introduction of the ekpyrotic scenario [76]. The original version of this theory claimed that this scenario can solve all cosmological problems without using the stage of inflation, i.e. without a prolonged stage of an accelerated expansion of the universe, which was called in [76] "superluminal expansion." However, the original ekpyrotic scenario contained many significant errors and did not work. It is sufficient to say that instead of the big bang expected in [76], there was a big crunch [77, 78].

The ekpyrotic scenario was replaced by the cyclic scenario, which used an infinite number of periods of expansion and contraction of the universe [79]. The origin of the required scalar field potential in this model remains unclear, and the very existence of the cycles postulated in [79] have not been demonstrated. When we analyzed this scenario using the particular potential given in [79], and took into account the effect of particle production in the early universe, we found a very different cosmological regime [80, 81].

The original version of the cyclic scenario relied on the existence of an infinite number of very long stages of "superluminal expansion," i.e. inflation, in order to solve the major cosmological problems. In this sense, the original version of the cyclic scenario was not a true alternative to inflationary scenario, but its rather peculiar version. The main difference between the usual inflation and the cyclic inflation, just as in the case of topological defects and textures, was the mechanism of generation of density perturbations. However, since the theory of density perturbations in cyclic inflation requires a solution of the cosmological singularity problem [82, 83], it is difficult to say anything definite about it.

Most of the authors believe that even if the singularity problem were solved, the spectrum of perturbations in the standard version of this scenario involving only one scalar field after the singularity would be very non-flat. One may introduce more complicated versions of this scenario, involving many scalar fields. In this case, under certain assumptions about the way the universe passes through the singularity, one may find a special regime where isocurvature perturbations in one of these fields are converted into adiabatic perturbations with a nearly flat spectrum. A recent discussion of this scenario shows that this regime requires an extreme fine-tuning of initial conditions [84]. Moreover, the instability of the solutions in this regime, which was found in [84], implies that it may be very easy to switch from one regime to another under the influence of small perturbations. This may lead to a domain-like structure of the universe and large perturbations of the metric [85]. If this is the case, no fine-tuning of initial conditions could help.

One of the latest versions of the cyclic scenario attempted to avoid the long stage of accelerated expansion (low-scale inflation) and to make the universe homogeneous using some specific features of the ekpyrotic collapse [86]. The authors assumed that the universe was homogeneous prior to its collapse on the scale that becomes greater than the scale of the observable part of the universe during the next cycle. Under this assumption, they argued that the perturbations of metric produced during each subsequent cycle do not interfere with the perturbations of metric produced in the next cycle. As a result, if the universe has been homogeneous from the very beginning, it remains homogeneous on the cosmologically interesting scales in all subsequent cycles.

Is this a real solution of the homogeneity problem? The initial size of the part of the universe, which is required to be homogeneous in this scenario prior to the collapse, was many orders of magnitude greater than the Planck scale. How homogeneous should it be? If we want the inhomogeneities to be produced due to amplification of quantum perturbations, then the initial classical perturbations of the field responsible for the isocurvature perturbations must be incredibly small, smaller than its quantum fluctuations. Otherwise the initial classical inhomogeneities of this field will be amplified by the same processes that amplified its quantum fluctuations and will dominate the spectrum of perturbations after the bounce [77]. This problem is closely related to the problem mentioned above [84, 85].

Recently there was an attempt to revive the original (non-cyclic) version of the ekpyrotic scenario by involving a nonsingular bounce. This regime requires violating the null energy condition [78], which usually leads to a catastrophic vacuum instability and/or causality violation. One may hope to avoid these problems in the ghost condensate theory [87]; see a series of recent papers on this subject [88, 89, 90]. However, even the authors of the ghost condensate theory emphasize that a fully consistent version of this theory is yet to be constructed [91], and that it may be incompatible with basic gravitational principles [92].

In addition, just as the ekpyrotic scenario with the singularity [84], the new version of the ekpyrotic theory requires two fields, and a conversion of the isocurvature perturbations to adiabatic perturbations [93]. Once again, the initial state of the universe in this scenario must be extremely homogeneous: the initial classical perturbations of the field responsible for the isocurvature perturbations must be smaller than its quantum fluctuations. It does not seem possible to solve this problem without further extending this exotic model and making it a part of an even more complicated scenario.

1.8.4 String Gas Scenario

Another attempt to solve some of the cosmological problems without using inflation has been proposed by Brandenberger et al. in the context of string gas cosmology [94, 95]. The authors admitted that their model did not solve the flatness problem, so it was not a real alternative to inflation. However, they claimed that their model provided a non-inflationary mechanism of production of metric perturbations with a flat spectrum.

It would be quite interesting and important to have a new mechanism of generation of metric perturbations based on string theory. Unfortunately, a detailed analysis of the scenario proposed in [94, 95] revealed that some of its essential ingredients were either unproven or incorrect [96]. For example, the theory of generation of metric perturbations used in [94] was formulated in the Einstein frame, where the usual Einstein equations are valid. On the other hand, the bounce and the string gas cosmology were described in string frame. Then both of these results were combined without distinguishing between different frames and a proper translation from one frame to another.

If one makes all calculations carefully (ignoring other unsolved problems of this scenario), one finds that the perturbations generated in their scenario have a blue spectrum with $n = 5$, which is ruled out by cosmological observations [96]. After the conference "Inflation + 25" where this issue was actively debated, the authors of [94, 95] issued two new papers reiterating their claims [97, 98], but eventually they agreed with our conclusion expressed at this conference: the spectrum of perturbations of metric in this scenario is blue, with $n = 5$, see (43) of [99]. This rules out the models proposed in [94, 95, 97, 98]. Nevertheless, as often happens with various alternatives to inflation, some of the authors of [94, 95, 97, 98] still claim that their basic scenario remains intact and propose its further modifications [99, 100, 101].

1.8.5 Mirage Bounce

Paradoxes with the choice of frames appear in other works on bounces in cosmology as well. For example, in [102] it was claimed that one can solve all cosmological problems in the context of mirage cosmology. However, as explained in [103], in the Einstein frame in this scenario the universe does not evolve at all.

To clarify the situation without going to technical details, one may consider the following analogy. We know that all particles in our body get their masses due to spontaneous symmetry breaking in the standard model. Suppose that the Higgs field initially was out of the minimum of its potential, and experienced oscillations. During these oscillations the masses of electrons and protons also oscillated. If one measures the size of the universe in units of the (time-dependent) Compton wavelengths of the electron (which could seem to be a good idea), one would think that the scale factor of the universe oscillates (bounces) with the frequency equal to the Higgs boson mass. And yet, this "cosmological evolution" with bounces of the scale factor is an illusion, which disappears if one measures the distances in units of the Planck length $M_{\rm p}^{-1}$ (the Einstein frame).

In addition, the mechanism of generation of density perturbations used in [102] was borrowed from the paper by Hollands and Wald [104], who suggested yet another alternative mechanism of generation of metric perturbations. However, this mechanism requires investigating thermal processes at the density 90 orders of magnitude greater than the Planck density, which makes all calculations unreliable [23].

1.8.6 Bounce in Quantum Cosmology

Finally, I should mention [105], where it was argued that under certain conditions one can have a bouncing universe and produce metric perturbations with a flat spectrum in the context of quantum cosmology. However, the model of [105] does not solve the flatness and homogeneity problems. A more detailed analysis revealed that the wave function of the universe proposed in [105] makes the probability of a bounce of a large universe exponentially small [106]. The authors are working on a modification of their model, which, as they hope, will not suffer from this problem.

To conclude, at the moment it is hard to see any real alternative to inflationary cosmology, despite an active search for such alternatives. All of the proposed alternatives are based on various attempts to solve the singularity problem: one should either construct a bouncing nonsingular cosmological solution, or learn what happens to the universe when it goes through the singularity. This problem bothered cosmologists for nearly a century, so it would be great to find its solution, quite independently of the possibility to find an alternative to inflation. None of the proposed alternatives can be consistently formulated until this problem is solved.

In this respect, inflationary theory has a very important advantage: it works practically independently of the solution of the singularity problem. It can work equally well after the singularity, or after the bounce, or after the quantum creation of the universe. This fact is especially clear in the eternal inflation scenario: eternal inflation makes the processes which occurred near the big bang practically irrelevant for the subsequent evolution of the universe.

1.9 Naturalness of Chaotic Inflation

Now we will return to the discussion of various versions of inflationary theory. Most of them are based on the idea of chaotic initial conditions, which is the trademark of the chaotic inflation scenario. In the simplest versions of chaotic inflation scenario with the potentials $V \sim \phi^n$, the process of inflation occurs at $\phi > 1$, in Planck units. Meanwhile, there are many other models where inflation may occur at $\phi \ll 1$.

There are several reasons why this difference may be important. First of all, some authors argue that the generic expression for the effective potential can be cast in the form

$$V(\phi) = V_0 + \alpha\phi + \frac{m^2}{2}\phi^2 + \frac{\beta}{3}\phi^3 + \frac{\lambda}{4}\phi^4 + \sum_n \lambda_n \frac{\phi^{4+n}}{M_{\text{Pl}}^n} , \qquad (1.24)$$

and then they assume that generically $\lambda_n = O(1)$, see e.g. (128) in [107]. If this assumption were correct, one would have little control over the behavior of $V(\phi)$ at $\phi > M_{\text{Pl}}$.

Here we have written M_{Pl} explicitly, to expose the implicit assumption made in [107]. Why do we write M_{Pl} in the denominator, instead of $1000 M_{\text{Pl}}$? An intuitive reason is that quantum gravity is non-renormalizable, so one should introduce a cut-off at momenta $k \sim M_{\text{Pl}}$. This is a reasonable assumption, but it does not imply the validity of (1.24). Indeed, the constant part of the scalar field appears in the gravitational diagrams not directly, but only via its effective potential $V(\phi)$ and the masses of particles interacting with the scalar field ϕ. As a result, the terms induced by quantum gravity effects are suppressed not by factors $\frac{\phi^n}{M_{\text{Pl}}^n}$, but by factors $\frac{V}{M_{\text{Pl}}^4}$ and $\frac{m^2(\phi)}{M_{\text{Pl}}^2}$ [14]. Consequently, quantum gravity corrections to $V(\phi)$ become large not at $\phi > M_{\text{Pl}}$, as one could infer from (1.24), but only at super-Planckian energy density, or for super-Planckian masses. This justifies our use of the simplest chaotic inflation models.

The simplest way to understand this argument is to consider the case where the potential of the field ϕ is a constant, $V = V_0$. Then the theory has a *shift symmetry*, $\phi \to \phi + c$. This symmetry is not broken by perturbative quantum gravity corrections, so no such terms as $\sum_n \lambda_n \frac{\phi^{4+n}}{M_{\text{Pl}}^n}$ are generated. This symmetry may be broken by nonperturbative quantum gravity effects (wormholes? virtual black holes?), but such effects, even if they exist, can be made exponentially small [108].

On the other hand, one may still wonder whether there is any reason not to add terms like $\lambda_n \frac{\phi^{4+n}}{M_{\text{Pl}}^n}$ with $\lambda = O(1)$ to the theory. Here I will make a simple argument which may help to explain it. I am not sure whether this argument should be taken too seriously, but I find it quite amusing and unexpected.

Let us consider a theory with the potential

$$V(\phi) = V_0 + \alpha\phi + \frac{m^2}{2}\phi^2 + \lambda_n \frac{\phi^{4+n}}{M_{\text{Pl}}^n} + \frac{\xi}{2} R\phi^2 . \qquad (1.25)$$

The last term is added to increase the generality of our discussion by considering fields non-minimally coupled to gravity, including the conformal fields with $\xi = 1/6$.

Suppose first that $m^2 = \lambda_n = 0$. Then the theory can describe our ground state with a slowly changing vacuum energy only if $V_0 + \alpha\phi < 10^{-120}$, $\alpha < 10^{-120}$ [109]. This theory cannot describe inflation because α is too small to produce the required density perturbations.

Let us now add the quadratic term. Without loss of generality one can make a redefinition of the field ϕ and V_0 to remove the linear term:

$$V(\phi) = V_0 + \frac{m^2}{2}\phi^2 . \qquad (1.26)$$

This is the simplest version of chaotic inflation. The maximal value of the field ϕ in this scenario is given by the condition $\frac{m^2}{2}\phi^2 \sim 1$ (Planckian density), so the maximal amount of inflation in this model is $\sim e^{\phi^2/4} \sim e^{1/m^2}$.

If, instead, we considered a more general case with the three terms $\frac{m^2}{2}\phi^2 + \lambda_n \frac{\phi^{4+n}}{M_{\rm Pl}{}^n} + \frac{\xi}{2}R\phi^2$, the maximal amount of inflation would be

$$N < \exp\left[\min\{m^{-2}, \lambda_n^{-2/n}, \xi^{-1}\}\right] . \qquad (1.27)$$

The last constraint appears because the effective gravitational constant becomes singular at $\phi^2 \sim \xi^{-1}$.

Thus, if any of the constants $\lambda_n^{2/n}$ or ξ is greater than m^2, the total amount of inflation will be exponentially smaller than in the simplest theory $\frac{m^2}{2}\phi^2$. Therefore one could argue that if one has a possibility to choose between different inflationary theories, as in the string theory landscape, then the largest fraction of the volume of the universe will be in the parts of the multiverse with $\lambda_n^{2/n}, \xi \ll m^2$. One can easily check that for $\lambda_n^{2/n}, \xi \lesssim m^2$ the higher order terms can be ignored at the last stages of inflation, where $\phi = O(1)$. In other words, the theory behaves as purely quadratic during the last stages of inflation when the observable part of the universe was formed.

One can come to the same conclusion if one takes into account only the part of inflation at smaller values of the field ϕ, when the stage of eternal inflation is over. This suggests that the simplest version of chaotic inflation scenario is the best.

Of course, this is just an argument. Our main goal here was not to promote the model $\frac{m^2}{2}\phi^2$, but to demonstrate that the considerations of naturalness (e.g. an assumption that all λ_n should be large) depend quite crucially on the underlying assumptions. In the example given above, a very simple change of these assumptions (the emphasis on the total volume of the post-inflationary universe) was sufficient to explain the naturalness of the simplest model $\frac{m^2}{2}\phi^2$. However, the situation may become quite different if instead of the simplest theory of a scalar field combined with general relativity one starts to investigate more complicated models, such as supergravity and string theory.

1.10 Chaotic Inflation in Supergravity

In the simplest models of inflation, the field ϕ itself does not have any direct physical meaning; everything depends only on its functions such as the masses of particles and the scalar potential. However, in more complicated theories the scalar field ϕ itself may have a physical (geometrical) meaning, which may constrain the possible values of the fields during inflation. The most important example is given by $N = 1$ supergravity.

The F-term potential of the complex scalar field Φ in supergravity is given by the well-known expression (in units $M_{\rm Pl} = 1$):

$$V = e^K \left[K_{\Phi\bar{\Phi}}^{-1} |D_\Phi W|^2 - 3|W|^2 \right] . \tag{1.28}$$

Here $W(\Phi)$ is the superpotential, Φ denotes the scalar component of the superfield Φ; $D_\Phi W = \frac{\partial W}{\partial \Phi} + \frac{\partial K}{\partial \Phi} W$. The kinetic term of the scalar field is given by $K_{\Phi\bar{\Phi}} \, \partial_\mu \Phi \partial_\mu \bar{\Phi}$. The standard textbook choice of the Kähler potential corresponding to the canonically normalized fields Φ and $\bar{\Phi}$ is $K = \Phi\bar{\Phi}$, so that $K_{\Phi\bar{\Phi}} = 1$.

This immediately reveals a problem: At $\Phi > 1$ the potential is extremely steep. It blows up as $e^{|\Phi|^2}$, which makes it very difficult to realize chaotic inflation in supergravity at $\phi \equiv \sqrt{2}|\Phi| > 1$. Moreover, the problem persists even at small ϕ. If, for example, one considers the simplest case when there are many other scalar fields in the theory and the superpotential does not depend on the inflaton field ϕ, then (1.28) implies that at $\phi \ll 1$ the effective mass of the inflaton field is $m_\phi^2 = 3H^2$. This violates the condition $m_\phi^2 \ll H^2$ required for successful slow-roll inflation (the so-called η-problem).

The major progress in SUGRA inflation during the last decade was achieved in the context of the models of the hybrid inflation type, where inflation may occur at $\phi \ll 1$. Among the best models are the F-term inflation, where different contributions to the effective mass term m_ϕ^2 cancel [110], and D-term inflation [111], where the dangerous term e^K does not affect the potential in the inflaton direction. A detailed discussion of various versions of hybrid inflation in supersymmetric theories can be found in the Chaps. 3 and 4 in this volume, see also [107, 112, 113].

However, hybrid inflation occurs only on a relatively small energy scale, and many of its versions do not lead to eternal inflation. Therefore it would be nice to obtain inflation in a context of a more general class of supergravity models.

This goal seemed very difficult to achieve; it took almost 20 years to find a natural realization of the chaotic inflation model in supergravity. Kawasaki, Yamaguchi and Yanagida suggested to take the Kähler potential

$$K = \frac{1}{2}(\Phi + \bar{\Phi})^2 + X\bar{X} \tag{1.29}$$

of the fields Φ and X, with the superpotential $m\Phi X$ [114].

At first glance, this Kähler potential may seem somewhat unusual. However, it can be obtained from the standard Kähler potential $K = \Phi\bar{\Phi} + X\bar{X}$ by adding terms $\Phi^2/2 + \bar{\Phi}^2/2$, which do not give any contribution to the kinetic term of the scalar fields $K_{\Phi\bar{\Phi}} \, \partial_\mu \Phi \partial_\mu \bar{\Phi}$. In other words, the new Kähler potential, just as the old one, leads to canonical kinetic terms for the fields Φ and X, so it is as simple and legitimate as the standard textbook Kähler potential. However, instead of the U(1) symmetry with respect to rotation of the field Φ in the complex plane, the new Kähler potential has a *shift symmetry*; it does not depend on the imaginary part of the field Φ. The shift symmetry is broken only by the superpotential.

This leads to a profound change of the potential (1.28): the dangerous term e^K continues growing exponentially in the direction $(\Phi + \bar{\Phi})$, but it remains constant in the direction $(\Phi - \bar{\Phi})$. Decomposing the complex field Φ into two real scalar fields, $\Phi = \frac{1}{\sqrt{2}}(\eta + i\phi)$, one can find the resulting potential $V(\phi, \eta, X)$ for $\eta, |X| \ll 1$:

$$V = \frac{m^2}{2}\phi^2(1+\eta^2) + m^2|X|^2 \ . \tag{1.30}$$

This potential has a deep valley, with a minimum at $\eta = X = 0$. At $\eta, |X| > 1$ the potential grows up exponentially. Therefore the fields η and X rapidly fall down towards $\eta = X = 0$, after which the potential for the field ϕ becomes $V = \frac{m^2}{2}\phi^2$. This provides a very simple realization of eternal chaotic inflation scenario in supergravity [114]. This model can be extended to include theories with different power-law potentials, or models where inflation begins as in the simplest versions of chaotic inflation scenario, but ends as in new or hybrid inflation, see e.g. [115, 116].

The existence of the shift symmetry was also the basis of the natural inflation scenario [117]. The basic assumption of this scenario was that the axion field in the first approximation is massless because the flatness of the axion direction is protected by U(1) symmetry. Nonperturbative corrections lead to the axion potential $V(\phi) = V_0(1 + \cos(\phi/f_a))$. If the 'radius' of the axion potential f_a is sufficiently large, $f_a \gtrsim 3$, inflation near the top of the potential becomes possible. For much greater values of f_a one can have inflation near the minimum of the axion potential, where the potential is quadratic [118].

The natural inflation scenario was proposed back in 1990, but until now all attempts to realize this scenario in supergravity have failed. First of all, it has been difficult to find theories with large f_a. More importantly, it has been difficult to stabilize the radial part of the axion field. A possible model of natural inflation in supergravity was constructed only very recently, see Chap. 4 in this volume.

Unfortunately, we still do not know how one could incorporate the models discussed in this section in string theory. We will briefly describe some features of inflation in string theory, and refer the readers to a more detailed presentation in Chap. 4 in this volume.

1.11 Towards Inflation in String Theory

1.11.1 de Sitter Vacua in String Theory

For a long time, it had seemed rather difficult to obtain inflation in M/string theory. The main problem here was the stability of compactification of internal dimensions. For example, ignoring non-perturbative effects to be discussed below, a typical effective potential of the effective four-dimensional theory obtained by compactification in string theory of type IIB can be represented in the following form:

$$V(\varphi, \rho, \phi) \sim e^{\sqrt{2}\varphi - \sqrt{6}\rho} \; \tilde{V}(\phi) \tag{1.31}$$

Here φ and ρ are canonically normalized fields representing the dilaton field and the volume of the compactified space; ϕ stays for all other fields, including the inflaton field.

If φ and ρ were constant, then the potential $\tilde{V}(\phi)$ could drive inflation. However, this does not happen because of the steep exponent $e^{\sqrt{2}\varphi - \sqrt{6}\rho}$, which rapidly pushes the dilaton field φ to $-\infty$, and the volume modulus ρ to $+\infty$. As a result, the radius of compactification becomes infinite; instead of inflating, four-dimensional space decompactifies and becomes 10-dimensional.

Thus in order to describe inflation one should first learn how to stabilize the dilaton and the volume modulus. The dilaton stabilization was achieved in [119]. The most difficult problem was to stabilize the volume. The solution of this problem was found in [120] (KKLT construction). It consists of two steps.

First, due to a combination of effects related to the warped geometry of the compactified space and nonperturbative effects calculated directly in four-dimensional (instead of being obtained by compactification), it was possible to obtain a supersymmetric AdS minimum of the effective potential for ρ. In the original version of the KKLT scenario, it was done in the theory with the Kähler potential

$$K = -3 \log(\rho + \bar{\rho}) , \tag{1.32}$$

and with the nonperturbative superpotential of the form

$$W = W_0 + A e^{-a\rho} , \tag{1.33}$$

with $a = 2\pi/N$. The corresponding effective potential for the complex field $\rho = \sigma + i\alpha$ had a minimum at finite, moderately large values of the volume modulus field σ_0, which fixed the volume modulus in a state with a negative vacuum energy. Then an anti-$D3$ brane with the positive energy $\sim \sigma^{-2}$ was added. This addition uplifted the minimum of the potential to the state with a positive vacuum energy, see Fig. 1.6.

Instead of adding an anti-$D3$ brane, which explicitly breaks supersymmetry, one can add a D7 brane with fluxes. This results in the appearance of a D-term which has a similar dependence on ρ, but leads to spontaneous supersymmetry breaking [121]. In either case, one ends up with a metastable dS state which can decay by tunneling and formation of bubbles of 10d space with vanishing vacuum energy density. The decay rate is extremely small [120], so for all practical purposes, one obtains an exponentially expanding de Sitter space with the stabilized volume of the internal space.[2]

1.11.2 Inflation in String Theory

There are two different versions of string inflation. In the first version, which we will call modular inflation, the inflaton field is associated with one of the

---

[2] It is also possible to find de Sitter solutions in noncritical string theory [122].

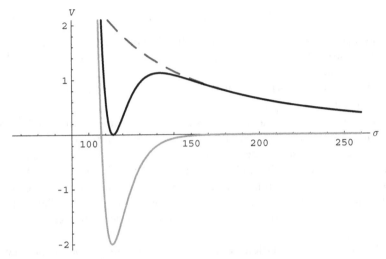

Fig. 1.6. KKLT potential as a function of $\sigma = \operatorname{Re} \rho$. The thin *green (lower) line* corresponds to the AdS stabilized potential for $W_0 = -10^{-4}$, $A = 1$, $a = 0.1$. The *dashed line* shows the additional term, which appears either due to the contribution of a $\overline{D3}$ brane or of a D7 brane. The thick *black line* shows the resulting potential with a very small but positive value of V in the minimum. The potential is shown multiplied by 10^{15}

moduli, the scalar fields which are already present in the KKLT construction. In the second version, the inflaton is related to the distance between branes moving in the compactified space. (This scenario should not be confused with inflation in the brane world scenario [123, 124]. This is a separate interesting subject, which we are not going to discuss in this chapter.)

Modular Inflation

An example of the KKLT-based modular inflation is provided by the racetrack inflation model of [125]. It uses a slightly more complicated superpotential

$$W = W_0 + Ae^{-a\rho} + Be^{-b\rho} . \qquad (1.34)$$

The potential of this theory has a saddle point as a function of the real and the complex part of the volume modulus: it has a local minimum in the direction $\operatorname{Re} \rho$, which is simultaneously a very flat maximum with respect to $\operatorname{Im} \rho$. Inflation occurs during a slow rolling of the field $\operatorname{Im} \rho$ away from this maximum (i.e. from the saddle point). The existence of this regime requires a significant fine-tuning of parameters of the superpotential. However, in the context of the string landscape scenario describing from 10^{100} to 10^{1000} different vacua (see below), this may not be such a big issue. A nice feature of this model is that it does not require adding any new branes to the original KKLT scenario, i.e. it is rather economical (Fig. 1.7.)

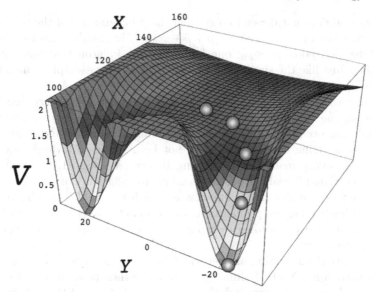

Fig. 1.7. Plot for the potential in the racetrack model (rescaled by 10^{16}). Here X stays for $\sigma = \mathrm{Re}\,\rho$ and Y stays for $\alpha = \mathrm{Im}\,\rho$. Inflation begins in a vicinity of the saddle point at $X_{\mathrm{saddle}} = 123.22$, $Y_{\mathrm{saddle}} = 0$. Units are $M_{\mathrm{Pl}} = 1$

Other interesting models of moduli inflation were developed in [126, 127, 128, 129]. An interesting property of all of these models is the existence of the regime of eternal slow-roll inflation. This property distinguishes modular inflation from the brane inflation scenario to be discussed below.

Brane Inflation

During the last few years, there were many suggestions on how to obtain hybrid inflation in string theory by considering motion of branes in the compactified space, see [130, 131] and references therein. The main problem of all of these models was the absence of stabilization of the compactified space. Once this problem was solved for dS space [120], one could try to revisit these models and develop models of brane inflation compatible with the volume stabilization.

The first idea [132] was to consider a pair of D3 and anti-D3 branes in the warped geometry studied in [120]. The role of the inflaton field ϕ in this model, which is known as the KKLMMT model, could be played by the interbrane separation. A description of this situation in terms of the effective four-dimensional supergravity involved Kähler potential

$$K = -3 \log(\rho + \bar{\rho} - k(\phi, \bar{\phi})) , \qquad (1.35)$$

where the function $k(\phi, \bar{\phi})$ for the inflaton field ϕ, at small ϕ, was taken in the simplest form $k(\phi, \bar{\phi}) = \phi\bar{\phi}$. If one makes the simplest assumption that the superpotential does not depend on ϕ, then the ϕ dependence of the potential (1.28) comes from the term $e^K = (\rho + \bar{\rho} - \phi\bar{\phi})^{-3}$. Expanding this term near the stabilization point $\rho = \rho_0$, one finds that the inflaton field has a mass $m_\phi^2 = 2H^2$. Just like the similar relation $m_\phi^2 = 3H^2$ in the simplest models of supergravity, this is not what we want for inflation.

One way to solve this problem is to consider ϕ-dependent superpotentials. By doing so, one may fine-tune m_ϕ^2 to be $O(10^{-2})H^2$ in a vicinity of the point where inflation occurs [132]. Whereas fine-tuning is certainly undesirable, in the context of string cosmology it may not be a serious drawback. Indeed, if there exist many realizations of string theory (see Sect. 1.14), then one might argue that all realizations not leading to inflation can be discarded, because they do not describe a universe in which we could live. This makes the issue of fine-tuning less problematic. Inflation in the KKLMMT model and its generalizations were studied by many authors; see Chap. 4 in this volume and references therein.

Can we avoid fine-tuning altogether? One of the possible ideas is to find theories with some kind of shift symmetry. Another possibility is to construct something like D-term inflation, where the flatness of the potential is not spoiled by the term e^K. Both of these ideas were combined together in Ref. [133] based on the model of D3/D7 inflation in string theory [134]. In this model the Kähler potential is given by

$$K = -3\log(\rho + \bar{\rho}) - \frac{1}{2}(\phi - \bar{\phi})^2 , \qquad (1.36)$$

and the superpotential depends only on ρ. The role of the inflaton field is played by the field $s = \mathrm{Re}\,\phi$, which represents the distance between the D3 and D7 branes. The shift symmetry $s \to s + c$ in this model is related to the requirement of unbroken supersymmetry of branes in a BPS state.

The effective potential with respect to the field ρ in this model coincides with the KKLT potential [120, 121]. The potential is exactly flat in the direction of the inflaton field s, until one adds a hypermultiplet of other fields $\phi_\pm$, which break this flatness due to quantum corrections and produce a logarithmic potential for the field s. The resulting potential with respect to the fields s and $\phi_\pm$ is very similar to the potential of D-term hybrid inflation [111].

During inflation, $\phi_\pm = 0$, and the field s slowly rolls down to its smaller values. When it becomes sufficiently small, the theory becomes unstable with respect to the generation of the field ϕ_+, see Fig. 1.8. The fields s and ϕ_+ roll down to the KKLT minimum, and inflation ends. For the latest developments in D3/D7 inflation see [135, 136].

All inflationary models discussed above were formulated in the context of Type IIB string theory with the KKLT stabilization. A discussion of the possibility to obtain inflation in the heterotic string theory with stable compactification can be found in [137, 138].

Finally, we should mention that making the effective potential flat is not the only way to achieve inflation. There are some models with nontrivial

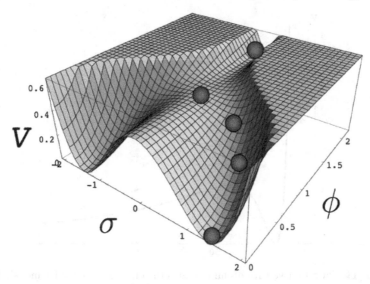

Fig. 1.8. The inflationary potential as a function of the inflaton field s and $\mathrm{Re}\,\phi_+$. In the beginning, the field s rolls along the valley $\phi_+ = 0$, and then it falls down to the KKLT minimum

kinetic terms where inflation may occur even without any potential [139]. One may also consider models with steep potentials but with anomalously large kinetic terms for the scalar fields see e.g. [140]. In application to string theory, such models, called "DBI inflation," were developed in [141].

In contrast to the moduli inflation, none of the existing versions of the brane inflation allow the slow-roll eternal inflation [142].

1.12 Scale of Inflation, the Gravitino Mass, and the Amplitude of the Gravitational Waves

So far, we did not discuss the relation of the new class of models with particle phenomenology. This relation is rather unexpected and may impose strong constraints on particle phenomenology and on inflationary models: In the simplest models based on the KKLT mechanism the Hubble constant H and the inflaton mass m_ϕ are smaller than the gravitino mass [143],

$$m_\phi \ll H \lesssim m_{3/2} \,. \tag{1.37}$$

The reason for the constraint $H \lesssim m_{3/2}$ is that the height of the barrier stabilizing the KKLT minimum is $\mathrm{O}(m_{3/2}^2)$. Adding a large vacuum energy density to the KKLT potential, which is required for inflation, may destabilize it, see Fig. 1.9. The constraint $m_\phi \ll H$ is a consequence of the slow-roll conditions.

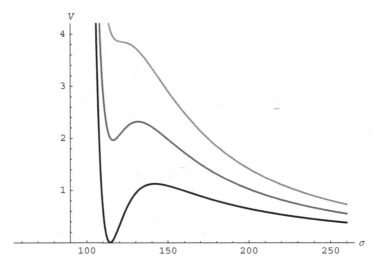

Fig. 1.9. The lowest curve with dS minimum is the one from the KKLT model. The height of the barrier in this potential is of the order $m_{3/2}^2$. The second line shows the σ-dependence of the inflaton potential. When one adds it to the theory, it always appears divided by σ^n, where in the simplest cases $n = 2$ or 3. Therefore an addition of the inflationary potential lifts up the potential at small σ. The top curve shows that when the inflation potential becomes too large, the barrier disappears, and the internal space decompactifies. This explains the origin of the constraint $H \lesssim m_{3/2}$

Therefore if one believes in the standard SUSY phenomenology with $m_{3/2} \lesssim O(1)$ TeV, one should find a realistic particle physics model where inflation occurs at a density at least 30 orders of magnitude below the Planck energy density. Such models are possible, but their parameters should be substantially different from the parameters used in all presently existing models of string theory inflation.

An interesting observational consequence of this result is that the amplitude of the gravitational waves in all string inflation models of this type should be extremely small. Indeed, according to (1.21), one has $r \approx 3 \times 10^7\ V \approx 10^8\ H^2$, which implies that

$$r \lesssim 10^8\ m_{3/2}^2 , \qquad (1.38)$$

in Planck units. In particular, for $m_{3/2} \lesssim 1$ TeV $\sim 4 \times 10^{-16}\ M_p$, which is in the range most often discussed by SUSY phenomenology, one has [144]

$$r \lesssim 10^{-24} . \qquad (1.39)$$

If CMB experiments find that $r \gtrsim 10^{-2}$, then this will imply, in the class of theories described above, that

$$m_{3/2} \gtrsim 10^{-5}\ M_p \sim 2.4 \times 10^{13}\ \text{GeV} , \qquad (1.40)$$

which is 10 orders of magnitude greater than the standard gravitino mass range discussed by particle phenomenologists.

There are several different ways to address this problem. First of all, one may try to construct realistic particle physics models with superheavy gravitinos [145, 146].

Another possibility is to consider models with the racetrack superpotential containing at least two exponents (1.34) and find parameters such that the supersymmetric minimum of the potential even before the uplifting occurs at zero energy density [143], which would mean $m_{3/2} = 0$, see Fig. 1.10. Then, by a slight change of parameters one can get the gravitino mass squared much smaller than the height of the barrier, which removes the constraint $H \lesssim m_{3/2}$.

Note, however, that in order to have $H^2 \sim V \sim 10^{-10}$ with $m_{3/2} \lesssim 1$ TeV $\sim 4 \times 10^{-16} M_{\rm p}$ in the model of [143] one would need to fine-tune the parameters of the theory with an incredible precision. This observation further strengthens the results of [147, 148], which imply that the tensor perturbations produced in all known versions of string theory inflation are undetectably small.

One could argue that since the existing versions of string theory inflation predict tensor modes with an extremely small amplitude, there is no sense to even try to detect them. From our perspective, however, the attitude should be opposite. There is a class of inflationary models that predict r in the range from 0.3 to 10^{-2}, see Sect. 1.7, so it makes a lot of sense to test this range of r even though the corresponding models have not been constructed as yet in the context of string theory.

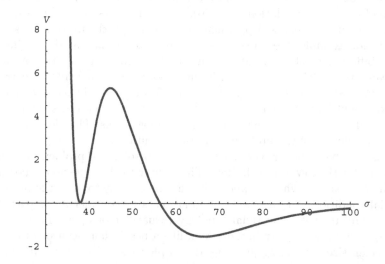

Fig. 1.10. The potential in the theory (1.34) for $A = 1$, $B = -5$, $a = 2\pi/100$, $b = 2\pi/50$, $W_0 = -0.05$. A Minkowski minimum at $V = 0$ stabilizes the volume at $\sigma_0 \approx 37$. The height of the barrier in this model is not correlated with the gravitino mass, which vanishes if the system is trapped in Minkowski vacuum. Therefore, in this model one can avoid the constraint $H \lesssim m_{3/2}$ [143]

If the tensor modes is found, the resulting situation will be similar to the situation with the discovery of the acceleration of the universe. This discovery initially puzzled string theorists, since none of the versions of string theory which existed 5 years ago could describe an accelerating universe in a stable vacuum state with a positive energy density. Eventually this problem was resolved with the development of the KKLT construction.

A possible discovery of tensor modes could lead to another constructive crisis since it may rule out many existing versions of string inflation and string phenomenology, and it may imply that the gravitino must be superheavy. Thus, investigation of gravitational waves produced during inflation may serve as a unique source of information about string theory and fundamental physics in general [144].

1.13 Initial Conditions for the Low-Scale Inflation and Topology of the Universe

One of the advantages of the simplest versions of the chaotic inflation scenario is that inflation may begin in the universe immediately after its creation at the largest possible energy density M_{Pl}^4, of a smallest possible size (Planck length), with the smallest possible mass $M \sim M_{Pl}$ and with the smallest possible entropy $S = O(1)$. This provides a true solution to the flatness, horizon, homogeneity, mass and entropy problems [14].

Meanwhile, in the new inflation scenario (more accurately, in the hilltop version of the chaotic inflation scenario), inflation occurs on the mass scale 3 orders of magnitude below M_{Pl}, when the total size of the universe was very large. If, for example, the universe is closed, its total mass at the beginning of new inflation must be greater than $10^6 M_{Pl}$, and its total entropy must be greater than 10^9. In other words, in order to explain why the entropy of the universe at present is greater than 10^{87} one should assume that it was extremely large from the very beginning. Then it becomes difficult to understand why such a large universe was homogeneous. This does not look like a real solution of the problem of initial conditions.

Thus one may wonder whether it possible to solve the problem of initial conditions for the low-scale inflation? The answer to this question is positive though perhaps somewhat unexpected: the simplest way to solve the problem of initial conditions for the low-scale inflation is to consider a compact flat or open universe with nontrivial topology (usual flat or open universes are infinite). The universe may initially look like a nearly homogeneous torus of a Planckian size containing just one or two photons or gravitons. It can be shown that such a universe continues expanding and remains homogeneous until the onset of inflation, even if inflation occurs only on a very low energy scale [149, 150, 151, 152, 153].

Consider, e.g. a flat compact universe having the topology of a torus, S_1^3,

$$ds^2 = dt^2 - a_i^2(t)\, dx_i^2 \tag{1.41}$$

with identification $x_i + 1 = x_i$ for each of the three dimensions. Suppose for simplicity that $a_1 = a_2 = a_3 = a(t)$. In this case the curvature of the universe and the Einstein equations written in terms of $a(t)$ will be the same as in the infinite flat Friedmann universe with metric $ds^2 = dt^2 - a^2(t) \, d\mathbf{x}^2$. In our notation, the scale factor $a(t)$ is equal to the size of the universe in Planck units $M_p^{-1} = 1$.

Let us assume, that at the Planck time $t_p \sim M_{Pl}^{-1} = 1$ the universe was radiation dominated, $V \ll T^4 = O(1)$. Let us also assume that at the Planck time the total size of the box was Planckian, $a(t_p) = O(1)$. In such case, the whole universe initially contained only $O(1)$ relativistic particles such as photons or gravitons, so that the total entropy of the whole universe was $O(1)$.

The size of the universe dominated by relativistic particles was growing as $a(t) \sim \sqrt{t}$, whereas the mean free path of the gravitons was growing as $H^{-1} \sim t$. If the initial size of the universe was $O(1)$, then at the time $t \gg 1$ each particle (or a gravitational perturbation of the metric) within one cosmological time would run all over the torus many times, appearing in all of its parts with nearly equal probability. This effect, called "chaotic mixing," should lead to a rapid homogenization of the universe [150, 151]. Note, that to achieve a modest degree of homogeneity required for inflation to start when the density of ordinary matter drops down, we do not even need chaotic mixing. Indeed, density perturbations do not grow in a universe dominated by ultrarelativistic particles if the size of the universe is smaller than H^{-1}. This is exactly what happens in our model. Therefore the universe should remain relatively homogeneous until the thermal energy density drops below V and inflation begins. And once it happens, the universe rapidly becomes very homogeneous.

Thus we see that in this scenario, just as in the simplest chaotic inflation scenario, inflation begins if we had a sufficiently homogeneous domain of the smallest possible size (Planck scale), with the smallest possible mass (Planck mass), and with the total entropy $O(1)$. The only additional requirement is that this domain should have identified sides, in order to make a flat or open universe compact. We see no reason to expect that the probability of formation of such domains is strongly suppressed.

One can come to a similar conclusion from a completely different point of view. The investigation of the quantum creation of a closed or an infinite open inflationary universe with $V \ll 1$ shows that this process is forbidden at the classical level, and therefore it occurs only due to tunneling. As a result, the probability of this process is exponentially suppressed [17, 18, 20]. Meanwhile, creation of the flat or open universe is possible without any need for the tunneling, and therefore there is no exponential suppression for the probability of quantum creation of a topologically nontrivial compact flat or open inflationary universe [149, 152, 153].

These results suggest that if inflation can occur only much below the Planck density, then the compact topologically nontrivial flat or open universes should be much more probable than the standard Friedmann universes described in every textbook on cosmology. This possibility is quite natural in the context of string theory, where all internal dimensions are supposed to be compact. Note, however, that if the stage of inflation is sufficiently long, it should make the observable part of the universe so large that its topology does not affect observational data.

The problem of initial conditions in string cosmology has several other interesting features. The most important one is the existence of an enormously large number of metastable de Sitter vacuum states, which makes the stage of exponential expansion of the universe almost inevitable. We will discuss this issue in the next section.

1.14 Inflationary Multiverse, String Theory Landscape and the Anthropic Principle

For many decades, people have tried to explain strange correlations between the properties of our universe, the masses of elementary particles, their coupling constants, and the fact of our existence. We know that we could not live in a five-dimensional universe, or in a universe where the electromagnetic coupling constant, or the masses of electrons and protons would be just a few times greater or smaller than their present values. These and other similar observations have formed the basis for the anthropic principle. However, for a long time many scientists believed that the universe was given to us as a single copy, and therefore speculations about these magic coincidences could not have any scientific meaning. Moreover, it would require a wild stretch of imagination and a certain degree of arrogance to assume that somebody was creating one universe after another, changing their parameters and fine-tuning their design, doing all of that for the sole purpose of making the universe suitable for our existence.

The situation changed dramatically with the invention of inflationary cosmology. It was realized that inflation may divide our universe into many exponentially large domains corresponding to different metastable vacuum states, forming a huge inflationary multiverse [52, 54, 154]. The total number of such vacuum states in string theory can be enormously large, in the range of 10^{100} or 10^{1000} [120, 155, 156, 157]. A combination of these two facts gave rise to what the experts in inflation call "the inflationary multiverse," [14, 55, 158] and string theorists call "the string theory landscape" [159].

This leads to an interesting twist in the theory of initial conditions. Let us assume first that we live in one of the many metastable de Sitter minima, say, dS_i. Eventually this dS state decays, and each of the *points* belonging to this initial state jumps to another vacuum state, which may have either a smaller vacuum energy, or a greater vacuum energy (transitions of the second type are possible because of the gravitational effects). But if the decay probability is not too large, then the total *volume* of the universe remaining in the state dS_i continues growing exponentially [9]. This is eternal inflation of the old inflation type. If the bubbles of the new phase correspond to another de Sitter space, dS_j, then some parts of the space dS_j may jump back to the state dS_i. On the other hand, if the tunneling goes to a Minkowski vacuum, such as the uncompactified 10-dimensional vacuum corresponding to the state with $\sigma \rightarrow \infty$ in Fig. 1.6, the subsequent jumps to dS states no longer occur. Similarly, if the tunneling goes to the state with a negative vacuum energy, such as the AdS vacuum in Fig. 1.10, the interior of the bubble of the new vacuum rapidly collapses. Minkowski and AdS vacua of such type are called terminal vacua, or sinks.

If initial conditions in a certain part of the universe are such that it goes directly to the sink, without an intermediate stage of inflation, then it will never return back, we will be unable to live there; so for all practical purposes such initial conditions (or such parts of the universe) can be discarded (ignoring for the moment the possibility of the resurrection of the universe after the collapse). On the other hand, if some other part of the universe goes to one of the dS states, the process of eternal inflation begins, which eventually produces an inflationary multiverse consisting of all possible dS states. This suggests that all initial conditions that allow life as we know it to exist, inevitably lead to formation of an eternal inflationary multiverse.

This scenario assumes that the vacuum transitions may bring us from any part of the string theory landscape to any other part. Here we should note that the theory of such transitions accompanied by the change of fluxes was developed for the case where dS states are not stabilized [156, 160]. A generalization of this theory for the string landscape scenario based on the KKLT mechanism of vacuum stabilization is rather nontrivial. As of now, the theory of such transitions was fully developed only for the transitions where the scalar fields change but the fluxes remain unchanged [161]. It might happen that the landscape is divided into separate totally disconnected islands, but this does not seem likely [162]. Even if the landscape is not fully transversable, one may probe all parts of the inflationary multiverse by considering the wave function of the universe corresponding to the possibility of its quantum creation in the states with different values of fluxes [163, 164].

The string theory landscape describes an incredibly large set of *discrete* parameters. However, the theory of inflationary multiverse goes even further. Some of the features of our world are determined not by the final values of the fields in the minima of their potential in the landscape, but by the dynamical, time-dependent values, which these fields were taking at different stages of the evolution of the inflationary universe. This introduces a large set of *continuous* parameters, which may take different values in different parts of the universe. For example, in the theory of dark energy, inflationary fluctuations may divide the universe into exponentially large parts with the effective value of the cosmological constant taking a continuous range of values [109]. In such models, the effective cosmological constant Λ becomes a continuous parameter. Similarly, inflationary fluctuations of the axion field make the density of dark matter a continuous parameter, which takes different values in different parts of the universe [165, 166]. Another example of a continuous parameter is the baryon asymmetry n_b/n_γ, which can take different values in different parts of the universe in the Affleck–Dine scenario of baryogenesis [167, 168].

This means that the same physical theory may yield exponentially large parts of the universe that have diverse properties. This provided the first scientific justification of the anthropic principle: We find ourselves inside a part of the universe with our kind of physical laws not because the parts with different properties are impossible or improbable, but simply because we cannot live there [52, 154].

This fact can help us understand many otherwise mysterious features of our world. The simplest example concerns the dimensionality of our universe. String theorists usually assume that the universe is 10- or 11-dimensional, so why do we live in the universe where only 4 dimensions of space–time are large? There have been many attempts to address this question, but no convincing answer has been found. This question became even more urgent after the development of the KKLT construction. Now we know that all de Sitter states, including the state in which we live now, are either unstable or metastable. They tend to decay by producing bubbles of a collapsing space, or of a 10-dimensional Minkowski space. So what is wrong about the 10-dimensional universe if it is so naturally appears in string theory?

The answer to this question was given in 1917 by Paul Ehrenfest [169]: in space–time with dimensionality $d > 4$, gravitational forces between distant bodies fall off faster than r^{-2}, and in space–time with $d < 4$, the general theory of relativity tells us that such forces are absent altogether. This rules out the existence of stable planetary systems for $d \neq 4$. A similar conclusion is valid for atoms: stable atomic systems could not exist for $d > 4$. This means that we do not need to prove that the four-dimensional space–time is a *necessary* outcome of string cosmology (in fact, it does not seem to be the case). Instead of that, we only need to make sure that the four-dimensional space–time is *possible*.

Anthropic considerations may help us to understand why the amount of dark matter is approximately five times greater than the amount of normal matter [165, 166] and why the baryon asymmetry is so small, $n_b/n_\gamma \sim 10^{-10}$ [168]. But perhaps the most famous example of this type is related to the cosmological constant problem.

Naively, one could expect the vacuum energy to be equal to the Planck density, $\rho_\Lambda \sim 1$, whereas the recent observational data show that $\rho_\Lambda \sim 10^{-120}$, in Planck units, which is approximately three times greater than the density of other matter in the universe. Why is it so small but nonzero? Why ρ_Λ constitutes is about three times greater than the density of other types of matter in the universe now? Note that long ago the density of matter was much greater than ρ_Λ, and in the future it will be much smaller.

The first anthropic solution to the cosmological constant problem in the context of inflationary cosmology was proposed in 1984 [163]. The basic assumption was that the vacuum energy density is a sum of the scalar field potential $V(\phi)$ and the energy of fluxes $V(F)$. According to [17], quantum creation of the universe is not suppressed if the universe is created at the Planck energy density, $V(\phi) + V(F) = O(1)$, in Planck units. Eventually the field ϕ rolls to its minimum at some value ϕ_0, and the vacuum energy becomes $\Lambda = V(\phi_0) + V(F)$. Since initially $V(\phi)$ and $V(F)$ could take any values with nearly equal probability, under the condition $V(\phi) + V(F) = O(1)$, we get a flat probability distribution to find a universe with a given value of the cosmological constant after inflation, $\Lambda = V(\phi_0) + V(F)$, for $\Lambda \ll 1$. The flatness of this probability distribution is crucial, because it allows us to study the probability of emergence of life for different Λ. Finally, it was argued in [163] that life as we know it is possible only for $|\Lambda| \lesssim \rho_0$, where $\rho_0 \sim 10^{-120}$ is the present energy density of the universe. This fact, in combination with inflation, which makes such universes exponentially large, provided a possible solution of the cosmological constant problem.

Shortly after that, several other anthropic solutions to the cosmological constant problem were proposed [170]. All of them were based on the assumption that life as we know it is possible only for $-\rho_0 \lesssim \rho_\Lambda \lesssim \rho_0$. This bound seemed almost self-evident to many of us at that time, and therefore in [163, 170] we concentrated on the development of the theoretical framework where the anthropic arguments could be applied to the cosmological constant.

The fact that ρ_Λ could not be much smaller than $-\rho_0$ was indeed quite obvious, since such a universe would rapidly collapse. However, the origin of the constraint $\rho_\Lambda \lesssim \rho_0$ was much less trivial. The first attempt to justify it was made in 1987 in the famous paper by Weinberg [171], but the constraint obtained there allowed the cosmological constant to be three orders of magnitude greater than its present value.

Since that time, the anthropic approach to the cosmological constant problem developed in two different directions. First of all, it became possible, under certain assumptions, to significantly strengthen the constraint on the positive cosmological constant, see e.g. [172, 173, 174, 175]. The final result of these investigations, $|\Lambda| \lesssim O(10)\, \rho_0 \sim 10^{-119}$, is very similar to the bound used in [163].

Simultaneously, new models have been developed which may allow us to put an anthropic approach to the cosmological constant problem on a firm ground. In particular, the existence of a huge number of vacuum states in string theory implies that in different parts of our universe, or in its different quantum states, the cosmological constant may take all of its possible values, from -1 to $+1$, with an increment which may be as small as 10^{-1000}. If the prior probability to be in each of these vacua does not depend strongly on Λ, one can justify the anthropic bound on Λ using the methods of [172, 173, 174, 175, 176].

However, the issue of probabilities in eternal inflation is very delicate, so one should approach anthropic arguments with some care. For example, one may try to calculate the probability to be born in a part of the universe with given properties *at a given point*. One can do this using comoving coordinates, which are not expanding during inflation [162, 177, 178, 179, 180, 181]. However, it is not obvious whether the calculation of the probabilities of physical processes at a given point, ignoring the expansion of the universe, should be used in anthropic considerations. Most of the physical entities which could be associated with "points" did not even exist before the beginning of inflation: protons did not exist, photons did not exist, galaxies did not exist. They appeared only after inflation, and their total number, and the total number of observers, is proportional to the growth of volume during inflation.

This leads to the volume-weighted [55, 182, 183, 184], or pocket-weighted [184, 185, 186] probability measures [187]. The main problem with this approach is the embarrassment of riches: the total volume of the universe occupied by any particular vacuum state, integrated over the indefinitely long history of the eternally inflating universe, is infinitely large. Thus we need to compare infinities, which is a very ambiguous task, with the answer depending on the choice of the cut-off procedure.

The volume-weighted probability measure proposed in [55] is based on the calculation of the ratio of the volumes of the parts of the universe with different properties. This is possible because if we wait long enough, eternal inflation approaches a stationary regime. Different parts of the universe expand and transform to each other. As a result, the total volumes of all parts of the universe of each particular type grow at the same rate, and the ratio of their volumes becomes time-independent [55].

This method is very good for describing the map of the inflationary multiverse, but in order to use it in anthropic considerations one should make some additional steps. According to [182], instead of calculating the ratio of volumes in different vacuum states at different densities and temperatures, we should calculate the total volume of *new* parts of the universe where life becomes possible. This ratio is related to the incoming probability current through the hypersurface of the end of inflation, or the hypersurface of a fixed density or temperature. If one uses the probability measure of [55] for anthropic considerations (which was *not* proposed in [55]), one may encounter the so-called youngness paradox [188, 189]. If one uses the prescription of [182], this paradox does not appear [21].

The results of the calculations by this method are very sensitive to the choice of the time parametrization [182, 21]. However, a recent investigation of this issue indicates that it may be possible to resolve this problem [190]. The main idea is that the parts of the universe with different properties approach the stationary regime of eternal inflation at different times. This fact was not taken into account in our earlier papers [55, 182]; the calculations of the probabilities started everywhere at the same time, even if the corresponding parts of the universe did not yet approach the stationary regime. If we start comparing the volumes of different part of the universe not at the same time after the beginning of inflation, but at the same time since the beginning of the stationarity regime, the dependence on the time parametrization disappears, at least in the simple cases where we could verify this property [190].

As we already mentioned, there are many other proposals for the calculations of probabilities in an inflationary multiverse, see e.g. [184, 186]. The results of some of these methods are not sensitive to the choice of time parametrization, but they do depend on the choice of the cut-off. A detailed discussion of this series of proposals can be found in [185, 191] and in Chap. 5 in this volume.

While discussing all of these approaches one should keep in mind yet another possibility: it is quite possible that it does not make much sense to compare infinities and talk about the probability of events that already happened. Instead of doing it, one should simply study our part of the universe, take these data as an initial input for all subsequent calculations, and study conditional probabilities for the quantities which we did not measure yet [21]. This is a standard approach used by experimentalists who continuously re-evaluate the probability of various outcomes of their future experiments on the basis of other experimental data. The non-standard part is that we should be allowed to use *all* of our observations, including our knowledge of our own properties, for the calculation of conditional probabilities.

Let us apply this limited approach to the cosmological constant problem. Twenty years ago, we already knew that our life is carbon-based, and that the amplitude of density perturbations required for the formation of galaxies was about 10^{-5}. We did not know yet what was the vacuum energy, and the prevailing idea was that we did not have much choice anyway. But with the discovery of inflation, we learned that the universe could be created differently, with different values of the cosmological constant in each of its parts created by eternal inflation. This allowed us to propose several different anthropic solutions to the cosmological constant problem based on the assumption that, for the given value of the amplitude of density perturbations and other already measured parameters, *we* cannot live in a universe with $|\Lambda| \gg 10^{-120}$. If observations would show that the cosmological constant were a million times smaller than the anthropic bound, then we would be surprised, and a theoretical explanation of this anomaly would be in order. As of now, the small value of the cosmological constant does not look too surprising, so for a while we can concentrate on solving many other problems which cannot be addressed by anthropic considerations.

Within this approach, one should not vary the constants of nature that were already known at the time when the predictions were made. In doing so, one faces the risk of repeating the old argument that the bomb does not hit the same spot twice: it is correct only until the first hit, after which the probabilities should be re-evaluated. Similarly, one should not omit the word "anthropic" from the "anthropic principle" and should not replace the investigation of the probability of *our* life with the study of life in general: we are trying to explain *our* observations rather than the possible observations made by some abstract information-processing devices. This can help us to avoid some paradoxes recently discussed in the literature [192, 193, 194].

From this discussion it should be clear that we do not really know yet which of the recently developed approaches to the theory of the inflationary multiverse is going to be more fruitful, and how far we will be able to go in this direction. One way or another, it would be very difficult to forget about what we have just learned and return to our search for the theory which unambiguously explains all parameters of our world. Now we know that some features of our part of the universe may have an unambiguous explanation, whereas some others can be purely environmental and closely correlated with our own existence.

When the inflationary theory was first proposed, its main goal was to address many problems which at that time could seem rather metaphysical: why is our universe so big? Why is it so uniform? Why parallel lines do not intersect? It took some time before we got used to the idea that the large size, flatness and uniformity of the universe should not be dismissed as trivial facts of life. Instead of that, they should be considered as observational data requiring an explanation.

Similarly, the existence of an amazingly strong correlation between our own properties and the values of many parameters of our world, such as the masses and charges of the electron and the proton, the value of the gravitational constant, the amplitude of spontaneous symmetry breaking in the electroweak theory, the value of the vacuum energy, and the dimensionality of our world, is an experimental fact requiring an explanation. A combination of the theory of inflationary multiverse and the string theory landscape provide us with a unique framework where this explanation can possibly be found.

1.15 Conclusions

Twenty five years ago, the inflationary theory looked like an exotic product of vivid scientific imagination. Some of us believed that it possessed such a great explanatory potential that it had to be correct; some others thought that it was too good to be true. Not many expected that it would be possible to verify any of its predictions in our lifetime. Thanks to the enthusiastic work of many scientists, the inflationary theory is gradually becoming a widely accepted cosmological paradigm, with many of its predictions being confirmed by observational data.

However, while the basic principles of inflationary cosmology are rather well established, many of its details keep changing with each new change of the theory of all fundamental interactions. The investigation of the inflationary multiverse and the string theory landscape force us to think about problems which sometimes go beyond the well established boundaries of physics. This makes our life difficult, sometimes quite frustrating, but also very interesting, which is perhaps the best thing that one could expect from the branch of science we have been trying to develop during the last quarter of a century.

Acknowledgments

I am grateful to the organizers of the Inflation + 25 conference and workshop for their hospitality. I would like to thank my numerous collaborators who made my work on inflation so enjoyable, especially my old friends and frequent collaborators Renata Kallosh, Lev Kofman, and Slava Mukhanov. This work was supported by the NSF grant 0244728 and by the Humboldt award.

References

1. A. D. Linde, JETP Lett. **19**, 183 (1974) [Pisma Zh. Eksp. Teor. Fiz. **19**, 320 (1974)].
2. D. A. Kirzhnits, JETP Lett. **15**, 529 (1972) [Pisma Zh. Eksp. Teor. Fiz. **15**, 745 (1972)]; D. A. Kirzhnits and A. D. Linde, Phys. Lett. B **42**, 471 (1972).
3. D. A. Kirzhnits and A. D. Linde, Annals Phys. **101**, 195 (1976).
4. A. D. Linde, Rept. Prog. Phys. **42**, 389 (1979).
5. A. A. Starobinsky, JETP Lett. **30**, 682 (1979) [Pisma Zh. Eksp. Teor. Fiz. **30**, 719 (1979)]; A. A. Starobinsky, Phys. Lett. B **91**, 99 (1980).
6. V. F. Mukhanov and G. V. Chibisov, JETP Lett. **33**, 532 (1981) [Pisma Zh. Eksp. Teor. Fiz. **33**, 549 (1981)].
7. A. H. Guth, Phys. Rev. D **23**, 347 (1981).
8. S. W. Hawking, I. G. Moss and J. M. Stewart, Phys. Rev. D **26**, 2681 (1982).
9. A. H. Guth and E. J. Weinberg, Nucl. Phys. B **212**, 321 (1983).
10. A. D. Linde, Phys. Lett. B **108**, 389 (1982); A. D. Linde, Phys. Lett. B **114**, 431 (1982); A. D. Linde, Phys. Lett. B **116**, 340 (1982); A. D. Linde, Phys. Lett. B **116**, 335 (1982).
11. A. Albrecht and P. J. Steinhardt, Phys. Rev. Lett. **48**, 1220 (1982).
12. S. W. Hawking, Phys. Lett. B **115**, 295 (1982); A. A. Starobinsky, Phys. Lett. B **117**, 175 (1982); A. H. Guth and S. Y. Pi, Phys. Rev. Lett. **49**, 1110 (1982); J. M. Bardeen, P. J. Steinhardt and M. S. Turner, Phys. Rev. D **28**, 679 (1983).
13. V. F. Mukhanov, JETP Lett. **41**, 493 (1985) [Pisma Zh. Eksp. Teor. Fiz. **41**, 402 (1985)]; V. F. Mukhanov, *Physical Foundations of Cosmology* (Cambridge University Press, Cambridge, 2005).
14. A. D. Linde, *Particle Physics and Inflationary Cosmology* (Harwood, Chur, Switzerland, 1990) `hep-th/0503203`.
15. A. D. Linde, Phys. Lett. B **129**, 177 (1983).
16. A. D. Linde, Phys. Lett. B **162**, 281 (1985).
17. A. D. Linde, Lett. Nuovo Cim. **39**, 401 (1984).
18. A. Vilenkin, Phys. Rev. D **30**, 509 (1984); A. Vilenkin, Phys. Rev. D **37**, 888 (1988); A. Vilenkin, Phys. Rev. D **39**, 1116 (1989).
19. J. B. Hartle and S. W. Hawking, Phys. Rev. D **28**, 2960 (1983).
20. A. D. Linde, Phys. Rev. D **58**, 083514 (1998), `gr-qc/9802038`.
21. A. Linde, JCAP **0701**, 022 (2007), `hep-th/0611043`.
22. G. W. Gibbons and N. Turok, `hep-th/0609095`.
23. L. Kofman, A. Linde and V. F. Mukhanov, JHEP **0210**, 057 (2002), `hep-th/0206088`.
24. L. Kofman and A. Linde, in preparation.

25. A. D. Dolgov and A. D. Linde, Phys. Lett. B **116**, 329 (1982); L. F. Abbott, E. Farhi and M. B. Wise, Phys. Lett. B **117**, 29 (1982).
26. L. Kofman, A. D. Linde and A. A. Starobinsky, Phys. Rev. Lett. **73**, 3195 (1994), hep-th/9405187; L. Kofman, A. D. Linde and A. A. Starobinsky, Phys. Rev. D **56**, 3258 (1997), hep-ph/9704452.
27. G. N. Felder, J. Garcia-Bellido, P. B. Greene, L. Kofman, A. D. Linde and I. Tkachev, Phys. Rev. Lett. **87**, 011601 (2001), hep-ph/0012142; G. N. Felder, L. Kofman and A. D. Linde, Phys. Rev. D **64**, 123517 (2001), hep-th/0106179.
28. M. Desroche, G. N. Felder, J. M. Kratochvil and A. Linde, Phys. Rev. D **71**, 103516 (2005), hep-th/0501080.
29. S. Y. Khlebnikov and I. I. Tkachev, Phys. Rev. Lett. **77**, 219 (1996), hep-ph/9603378; S. Y. Khlebnikov and I. I. Tkachev, Phys. Rev. Lett. **79**, 1607 (1997), hep-ph/9610477.
30. G. N. Felder and I. Tkachev, hep-ph/0011159.
31. G. N. Felder and L. Kofman, Phys. Rev. D **63**, 103503 (2001), hep-ph/0011160; R. Micha and I. I. Tkachev, Phys. Rev. D **70**, 043538 (2004) hep-ph/0403101; D. I. Podolsky, G. N. Felder, L. Kofman and M. Peloso, Phys. Rev. D **73**, 023501 (2006), hep-ph/0507096; G. N. Felder and L. Kofman, Phys. Rev. D **75**, 043518 (2007), hep-ph/0606256.
32. B. A. Ovrut and P. J. Steinhardt, Phys. Lett. B **133**, 161 (1983); B. A. Ovrut and P. J. Steinhardt, Phys. Rev. Lett. **53**, 732 (1984); B. A. Ovrut and P. J. Steinhardt, Phys. Rev. D **30**, 2061 (1984); B. A. Ovrut and P. J. Steinhardt, Phys. Lett. B **147**, 263 (1984).
33. A. D. Linde, Phys. Lett. B **132**, 317 (1983).
34. L. Boubekeur and D. H. Lyth, JCAP **0507**, 010 (2005), hep-ph/0502047.
35. A. D. Linde, Phys. Lett. B **259**, 38 (1991); A. D. Linde, Phys. Rev. D **49**, 748 (1994), astro-ph/9307002.
36. A. Vilenkin and L. H. Ford, Phys. Rev. D **26**, 1231 (1982).
37. A. D. Linde, Phys. Lett. B **116**, 335 (1982).
38. D. J. Gross and F. Wilczek, Phys. Rev. Lett. **30**, 1343 (1973).
39. H. D. Politzer, Phys. Rev. Lett. **30**, 1346 (1973).
40. A. R. Liddle and D. H. Lyth, *Cosmological Inflation and Large-Scale Structure* (Cambridge University Press, Cambridge, 2000).
41. H. V. Peiris et al., Astrophys. J. Suppl. **148**, 213 (2003), astro-ph/0302225. 0302225;
42. M. Tegmark et al., Phys. Rev. D **74**, 123507 (2006), astro-ph/0608632.
43. C. L. Kuo et al., arXiv:astro-ph/0611198.
44. A. D. Linde, JETP Lett. **40**, 1333 (1984) [Pisma Zh. Eksp. Teor. Fiz. **40**, 496 (1984)]; A. D. Linde, Phys. Lett. B **158**, 375 (1985); L. A. Kofman, Phys. Lett. B **173**, 400 (1986); L. A. Kofman and A. D. Linde, Nucl. Phys. B **282**, 555 (1987); A. D. Linde and D. H. Lyth, Phys. Lett. B **246**, 353 (1990).
45. S. Mollerach, Phys. Rev. D **42**, 313 (1990).
46. A. D. Linde and V. Mukhanov, Phys. Rev. D **56**, 535 (1997), astro-ph/9610219; A. Linde and V. Mukhanov, JCAP **0604**, 009 (2006), astro-ph/0511736.
47. K. Enqvist and M. S. Sloth, Nucl. Phys. B **626**, 395 (2002), hep-ph/0109214.
48. D. H. Lyth and D. Wands, Phys. Lett. B **524**, 5 (2002), hep-ph/0110002.
49. T. Moroi and T. Takahashi, Phys. Lett. B **522**, 215 (2001) [Erratum-ibid. B **539**, 303 (2002)] hep-ph/0110096.

50. G. Dvali, A. Gruzinov and M. Zaldarriaga, Phys. Rev. D **69**, 023505 (2004), astro-ph/0303591; L. Kofman, arXiv:astro-ph/0303614; A. Mazumdar and M. Postma, Phys. Lett. B **573**, 5 (2003) [Erratum-ibid. B **585**, 295 (2004)] astro-ph/0306509; F. Bernardeau, L. Kofman and J. P. Uzan, Phys. Rev. D **70**, 083004 (2004), astro-ph/0403315; D. H. Lyth, JCAP **0511**, 006 (2005), astro-ph/0510443.
51. P. J. Steinhardt, *The Very Early Universe*, ed. G. W. Gibbons, S. W. Hawking and S. Siklos (Cambridge University Press, 1983).
52. A. D. Linde, Print-82-0554, Cambridge University preprint, 1982, see http://www.stanford.edu/~alinde/1982.pdf.
53. A. Vilenkin, Phys. Rev. D **27**, 2848 (1983).
54. A. D. Linde, Phys. Lett. B **175**, 395 (1986); A. S. Goncharov, A. D. Linde and V. F. Mukhanov, Int. J. Mod. Phys. A **2**, 561 (1987).
55. A. D. Linde, D. A. Linde and A. Mezhlumian, Phys. Rev. D **49**, 1783 (1994) gr-qc/9306035.
56. A. Borde, A. H. Guth and A. Vilenkin, Phys. Rev. Lett. **90**, 151301 (2003) gr-qc/0110012.
57. A. D. Linde, Phys. Lett. B **327**, 208 (1994), astro-ph/9402031; A. Vilenkin, Phys. Rev. Lett. **72**, 3137 (1994), hep-th/9402085; A. D. Linde and D. A. Linde, Phys. Rev. D **50**, 2456 (1994), hep-th/9402115; I. Cho and A. Vilenkin, Phys. Rev. D **56**, 7621 (1997), gr-qc/9708005.
58. M. Aryal and A. Vilenkin, Phys. Lett. B **199**, 351 (1987).
59. S. Dodelson, AIP Conf. Proc. **689**, 184 (2003), hep-ph/0309057.
60. V. F. Mukhanov, Int. J. Theor. Phys. **43**, 623 (2004), astro-ph/0303072.
61. G. Efstathiou, arXiv:astro-ph/0306431; G. Efstathiou, Mon. Not. Roy. Astron. Soc. **348**, 885 (2004), astro-ph/0310207. A. Slosar and U. Seljak, Phys. Rev. D **70**, 083002 (2004), astro-ph/0404567.
62. K. Land and J. Magueijo, arXiv:astro-ph/0611518.
63. A. Rakic and D. J. Schwarz, arXiv:astro-ph/0703266.
64. C. R. Contaldi, M. Peloso, L. Kofman and A. Linde, JCAP **0307**, 002 (2003), astro-ph/0303636.
65. A. Stebbins and M. S. Turner, Astrophys. J. **339**, L13 (1989).
66. M. S. Turner, R. Watkins and L. M. Widrow, Astrophys. J. **367**, L43 (1991).
67. R. A. Battye and A. Moss, Phys. Rev. D **74**, 041301 (2006), astro-ph/0602377.
68. H. J. de Vega and N. G. Sanchez, Phys. Rev. D **74**, 063519 (2006).
69. R. Kallosh and A. Linde, arXiv:arXiv:0704.0647.
70. H. M. Hodges, G. R. Blumenthal, L. A. Kofman and J. R. Primack, Nucl. Phys. B **335**, 197 (1990).
71. C. Destri, H. J. de Vega and N. G. Sanchez, arXiv:astro-ph/0703417.
72. D. Spergel and N. Turok, Sci. Am. **266**, 52 (1992).
73. G. Veneziano, Phys. Lett. B **265**, 287 (1991); M. Gasperini and G. Veneziano, Astropart. Phys. **1**, 317 (1993), hep-th/9211021.
74. N. Kaloper, R. Madden and K. A. Olive, Nucl. Phys. B **452**, 677 (1995), hep-th/9506027; N. Kaloper, R. Madden and K. A. Olive, Phys. Lett. B **371**, 34 (1996), hep-th/9510117.
75. N. Kaloper, A. D. Linde and R. Bousso, Phys. Rev. D **59**, 043508 (1999), hep-th/9801073; A. Buonanno and T. Damour, Phys. Rev. D **64**, 043501 (2001), gr-qc/0102102.
76. J. Khoury, B. A. Ovrut, P. J. Steinhardt and N. Turok, Phys. Rev. D **64**, 123522 (2001), hep-th/0103239.

77. R. Kallosh, L. Kofman and A. D. Linde, Phys. Rev. D **64**, 123523 (2001), hep-th/0104073.
78. R. Kallosh, L. Kofman, A. D. Linde and A. A. Tseytlin, Phys. Rev. D **64**, 123524 (2001), hep-th/0106241.
79. P. J. Steinhardt and N. Turok, Phys. Rev. D **65**, 126003 (2002), hep-th/0111098.
80. G. N. Felder, A. V. Frolov, L. Kofman and A. V. Linde, Phys. Rev. D **66**, 023507 (2002), hep-th/0202017.
81. A. Linde, in *The Future of Theoretical Physics and Cosmology* (Cambridge University Press, Cambridge, 2002), p. 801, hep-th/0205259.
82. H. Liu, G. Moore and N. Seiberg, JHEP **0206**, 045 (2002), hep-th/0204168.
83. G. T. Horowitz and J. Polchinski, Phys. Rev. D **66**, 103512 (2002), hep-th/0206228.
84. K. Koyama and D. Wands, arXiv:hep-th/0703040; K. Koyama, S. Mizuno and D. Wands, arXiv:0704.1152.
85. M. Sasaki, private communication.
86. J. K. Erickson, S. Gratton, P. J. Steinhardt and N. Turok, arXiv:hep-th/0607164.
87. N. Arkani-Hamed, H. C. Cheng, M. A. Luty and S. Mukohyama, JHEP **0405**, 074 (2004), hep-th/0312099; N. Arkani-Hamed, H. C. Cheng, M. A. Luty, S. Mukohyama and T. Wiseman, JHEP **0701**, 036 (2007), hep-ph/0507120.
88. P. Creminelli, M. A. Luty, A. Nicolis and L. Senatore, JHEP **0612**, 080 (2006), hep-th/0606090.
89. E. I. Buchbinder, J. Khoury and B. A. Ovrut, arXiv:hep-th/0702154.
90. P. Creminelli and L. Senatore, arXiv:hep-th/0702165.
91. A. Adams, N. Arkani-Hamed, S. Dubovsky, A. Nicolis and R. Rattazzi, JHEP **0610**, 014 (2006), hep-th/0602178.
92. N. Arkani-Hamed, S. Dubovsky, A. Nicolis, E. Trincherini and G. Villadoro, hep-th:0704.1814.
93. J. L. Lehners, P. McFadden, N. Turok and P. J. Steinhardt, arXiv:hep-th/0702153; A. J. Tolley and D. H. Wesley, arXiv:hep-th/0703101.
94. A. Nayeri, R. H. Brandenberger and C. Vafa, Phys. Rev. Lett. **97**, 021302 (2006) hep-th/0511140.
95. R. H. Brandenberger, A. Nayeri, S. P. Patil and C. Vafa, arXiv:hep-th/0604126.
96. N. Kaloper, L. Kofman, A. Linde and V. Mukhanov, JCAP **0610**, 006 (2006) hep-th/0608200.
97. A. Nayeri, arXiv:hep-th/0607073.
98. R. H. Brandenberger, A. Nayeri, S. P. Patil and C. Vafa, arXiv:hep-th/0608121.
99. R. H. Brandenberger et al., JCAP **0611**, 009 (2006), hep-th/0608186.
100. T. Biswas, R. Brandenberger, A. Mazumdar and W. Siegel, arXiv:hep-th/0610274.
101. R. H. Brandenberger, arXiv:hep-th/0701111.
102. C. Germani, N. E. Grandi and A. Kehagias, arXiv:hep-th/0611246.
103. S. Kachru and L. McAllister, JHEP **0303**, 018 (2003), hep-th/0205209.
104. S. Hollands and R. M. Wald, Gen. Rel. Grav. **34**, 2043 (2002), gr-qc/0205058.
105. P. Peter, E. J. C. Pinho and N. Pinto-Neto, Phys. Rev. D **75**, 023516 (2007), hep-th/0610205.
106. P. Peter, E. J. C. Pinho and N. Pinto-Neto, private communication.
107. D. H. Lyth and A. Riotto, Phys. Rept. **314**, 1 (1999), hep-ph/9807278.

108. R. Kallosh, A. Linde, D. Linde and L. Susskind, Phys. Rev. **D52**, 912 (1995), hep-th/9502069.
109. A.D. Linde, in *300 Years of Gravitation*, ed. by S. W. Hawking and W. Israel (Cambridge University Press, Cambridge, 1987); J. Garriga, A. Linde and A. Vilenkin, Phys. Rev. D **69**, 063521 (2004), hep-th/0310034.
110. E. J. Copeland, A. R. Liddle, D. H. Lyth, E. D. Stewart and D. Wands, Phys. Rev. D **49**, 6410 (1994), astro-ph/9401011; G. R. Dvali, Q. Shafi and R. Schaefer, Phys. Rev. Lett. **73**, 1886 (1994), hep-ph/9406319; A. D. Linde and A. Riotto, Phys. Rev. D **56**, 1841 (1997), hep-ph/9703209.
111. P. Binetruy and G. Dvali, Phys. Lett. B **388**, 241 (1996), hep-ph/9606342; E. Halyo, Phys. Lett. B **387**, 43 (1996), hep-ph/9606423.
112. R. Kallosh and A. Linde, JCAP **0310**, 008 (2003), hep-th/0306058.
113. P. Binetruy, G. Dvali, R. Kallosh and A. Van Proeyen, Class. Quant. Grav. **21**, 3137 (2004), hep-th/0402046.
114. M. Kawasaki, M. Yamaguchi and T. Yanagida, Phys. Rev. Lett. **85**, 3572 (2000), hep-ph/0004243.
115. M. Yamaguchi and J. Yokoyama, Phys. Rev. D **63**, 043506 (2001), hep-ph/0007021; M. Yamaguchi, Phys. Rev. D **64**, 063502 (2001), hep-ph/0103045.
116. M. Yamaguchi and J. Yokoyama, Phys. Rev. D **68**, 123520 (2003), hep-ph/0307373.
117. K. Freese, J. A. Frieman and A. V. Olinto, Phys. Rev. Lett. **65**, 3233 (1990).
118. C. Savage, K. Freese and W. H. Kinney, Phys. Rev. D **74**, 123511 (2006), hep-ph/0609144.
119. S. B. Giddings, S. Kachru and J. Polchinski, Phys. Rev. **D66**, 106006 (2002), hep-th/0105097.
120. S. Kachru, R. Kallosh, A. Linde and S. P. Trivedi, Phys. Rev. D **68**, 046005 (2003), hep-th/0301240.
121. C. P. Burgess, R. Kallosh and F. Quevedo, JHEP **0310**, 056 (2003), hep-th/0309187.
122. E. Silverstein, arXiv:hep-th/0106209; A. Maloney, E. Silverstein and A. Strominger, arXiv:hep-th/0205316.
123. N. Arkani-Hamed, S. Dimopoulos and G. R. Dvali, Phys. Lett. B **429**, 263 (1998), hep-ph/9803315; I. Antoniadis, N. Arkani-Hamed, S. Dimopoulos and G. R. Dvali, Phys. Lett. B **436**, 257 (1998), hep-ph/9804398.
124. L. Randall and R. Sundrum, Phys. Rev. Lett. **83**, 3370 (1999), hep-ph/9905221.
125. J. J. Blanco-Pillado, C. P. Burgess, J. M. Cline, C. Escoda, M. Gomez-Reino, R. Kallosh, A. Linde, and F. Quevedo, JHEP **0411**, 063 (2004), hep-th/0406230.
126. J. P. Conlon and F. Quevedo, JHEP **0601**, 146 (2006), hep-th/0509012.
127. Z. Lalak, G. G. Ross and S. Sarkar, Nucl. Phys. B **766**, 1 (2007), hep-th/0503178.
128. J. J. Blanco-Pillado, C. P. Burgess, J. M. Cline, C. Escoda, M. Gomez-Reino, R. Kallosh, A. Linde, F. Quevedo,, JHEP **0609**, 002 (2006), hep-th/0603129.
129. J. R. Bond, L. Kofman, S. Prokushkin and P. M. Vaudrevange, arXiv: hep-th/0612197.
130. G. R. Dvali and S. H. H. Tye, Phys. Lett. B **450**, 72 (1999), hep-ph/9812483.
131. F. Quevedo, Class. Quant. Grav. **19**, 5721 (2002), hep-th/0210292.

132. S. Kachru, R. Kallosh, A. Linde, J. Maldacena, L. McAllister and S. P. Trivedi, JCAP **0310**, 013 (2003), hep-th/0308055.

133. J. P. Hsu, R. Kallosh and S. Prokushkin, JCAP **0312**, 009 (2003), hep-th/0311077.

134. R. Kallosh, arXiv:hep-th/0109168; C. Herdeiro, S. Hirano and R. Kallosh, JHEP **0112**, 027 (2001), hep-th/0110271; K. Dasgupta, C. Herdeiro, S. Hirano and R. Kallosh, Phys. Rev. D **65**, 126002 (2002), hep-th/0203019; J. P. Hsu and R. Kallosh, JHEP **0404**, 042 (2004), hep-th/0402047.

135. K. Dasgupta, J. P. Hsu, R. Kallosh, A. Linde and M. Zagermann, JHEP **0408**, 030 (2004), hep-th/0405247.

136. P. Chen, K. Dasgupta, K. Narayan, M. Shmakova and M. Zagermann, JHEP **0509**, 009 (2005), hep-th/0501185.

137. E. I. Buchbinder, Nucl. Phys. B **711**, 314 (2005), hep-th/0411062.

138. K. Becker, M. Becker and A. Krause, arXiv:hep-th/0501130.

139. C. Armendariz-Picon, T. Damour and V. Mukhanov, Phys. Lett. B **458**, 209 (1999), hep-th/9904075.

140. S. Dimopoulos and S. Thomas, Phys. Lett. B **573**, 13 (2003), hep-th/0307004.

141. E. Silverstein and D. Tong, Phys. Rev. D **70**, 103505 (2004), hep-th/0310221; M. Alishahiha, E. Silverstein and D. Tong, Phys. Rev. D **70**, 123505 (2004), hep-th/0404084.

142. X. Chen, S. Sarangi, S. H. Henry Tye and J. Xu, JCAP **0611**, 015 (2006), hep-th/0608082.

143. R. Kallosh and A. Linde, JHEP **0412**, 004 (2004), hep-th/0411011; J. J. Blanco-Pillado, R. Kallosh and A. Linde, JHEP **0605**, 053 (2006), hep-th/0511042.

144. R. Kallosh and A. Linde, JCAP04, 017 (2007), 0704.0647.

145. O. DeWolfe and S. B. Giddings, Phys. Rev. D **67**, 066008 (2003), hep-th/0208123.

146. N. Arkani-Hamed and S. Dimopoulos, arXiv:hep-th/0405159; N. Arkani-Hamed, S. Dimopoulos, G. F. Giudice and A. Romanino, Nucl. Phys. B **709**, 3 (2005), hep-ph/0409232.

147. D. Baumann and L. McAllister, arXiv:hep-th/0610285.

148. R. Bean, S. E. Shandera, S. H. Henry Tye and J. Xu, arXiv:hep-th/0702107.

149. Y. B. Zeldovich and A. A. Starobinsky, Sov. Astron. Lett. **10**, 135 (1984).

150. O. Heckmann and E. Schucking, in *Handbuch der Physik*, ed. S. Flugge (Springer, Berlin, 1959), Vol. 53, p. 515; G. F. Ellis, Gen. Rel. Grav. **2**, 7 (1971); J. R. Gott, Mon. Not. R. Astron. Soc. **193**, 153 (1980); C. N. Lockhart, B. Misra and I. Prigogine, Phys. Rev. **D25**, 921 (1982); H. V. Fagundes, Phys. Rev. Lett. **51**, 517 (1983).

151. N. J. Cornish, D. N. Spergel and G. D. Starkman, Phys. Rev. Lett. **77**, 215 (1996), astro-ph/9601034.

152. D. H. Coule and J. Martin, Phys. Rev. D **61**, 063501 (2000), gr-qc/9905056.

153. A. Linde, JCAP **0410**, 004 (2004), hep-th/0408164.

154. A. D. Linde, in *The Very Early Universe*, ed. G.W. Gibbons, S. W. Hawking and S. Siklos (Cambridge University Press, Cambridge, 1983), pp. 205–249, see http://www.stanford.edu/~alinde/1983.pdf.

155. W. Lerche, D. Lüst and A. N. Schellekens, Nucl. Phys. B **287**, 477 (1987).

156. R. Bousso and J. Polchinski, JHEP **0006**, 006 (2000), hep-th/0004134.

157. M. R. Douglas, JHEP **0305** 046 (2003), hep-th/0303194; F. Denef and M. R. Douglas, JHEP **0405**, 072 (2004), hep-th/0404116; M. R. Douglas and S. Kachru, arXiv:hep-th/0610102; F. Denef, M. R. Douglas and S. Kachru, arXiv:hep-th/0701050.

158. A. Linde, in *Science and Ultimate Reality: From Quantum to Cosmos*, eds. J. D. Barrow, P. C. W. Davies and C. L. Harper (Cambridge University Press, Cambridge, 2003), hep-th/0211048.

159. L. Susskind, arXiv:hep-th/0302219.

160. J. D. Brown and C. Teitelboim, Phys. Lett. B **195**, 177 (1987); J. D. Brown and C. Teitelboim, Nucl. Phys. B **297**, 787 (1988).

161. A. Ceresole, G. Dall'Agata, A. Giryavets, R. Kallòsh and A. Linde, Phys. Rev. D **74**, 086010 (2006), hep-th/0605266.

162. T. Clifton, A. Linde and N. Sivanandam, JHEP **0702**, 024 (2007), hep-th/0701083.

163. A. D. Linde, Rept. Prog. Phys. **47**, 925 (1984).

164. S. W. Hawking and T. Hertog, Phys. Rev. D **73**, 123527 (2006) hep-th/0602091.

165. A. D. Linde, Phys. Lett. B **201** (1988) 437.

166. M. Tegmark, A. Aguirre, M. Rees and F. Wilczek, Phys. Rev. D **73**, 023505 (2006), astro-ph/0511774.

167. I. Affleck and M. Dine, Nucl. Phys. B **249** (1985) 361.

168. A. D. Linde, Phys. Lett. B **160**, 243 (1985).

169. P. Ehrenfest, Proc. Amsterdam Acad. **20**, 200 (1917); P. Ehrenfest, Annalen der Physik **61**, 440 (1920).

170. A. D. Sakharov, Sov. Phys. JETP **60**, 214 (1984) [Zh. Eksp. Teor. Fiz. **87**, 375 (1984)]; T. Banks, Nucl. Phys. B **249**, 332, (1985); A. D. Linde, in *300 Years of Gravitation*, eds. S. W. Hawking and W. Israel (Cambridge University Press, Cambridge, 1987).

171. S. Weinberg, Phys. Rev. Lett. **59**, 2607 (1987).

172. H. Martel, P. R. Shapiro and S. Weinberg, Astrophys. J. **492**, 29 (1998), astro-ph/9701099.

173. J. Garriga, M. Livio and A. Vilenkin, Phys. Rev. D **61**, 023503 (2000), astro-ph/9906210; J. Garriga and A. Vilenkin, Phys. Rev. D **64**, 023517 (2001) hep-th/0011262.

174. M. Tegmark, A. Aguirre, M. Rees and F. Wilczek, Phys. Rev. D **73**, 023505 (2006) astro-ph/0511774.

175. C. H. Lineweaver and C. A. Egan, arXiv:astro-ph/0703429.

176. R. Bousso, R. Harnik, G. D. Kribs and G. Perez, arXiv:hep-th/0702115.

177. A. A. Starobinsky, in *Current Topics in Field Theory, Quantum Gravity and Strings*, Lecture Notes in Physics, eds. H. J. de Vega and N. Sanchez (Springer, Heidelberg, 1986) Vol. 206, p. 107.

178. A. S. Goncharov, A. D. Linde and V. F. Mukhanov, Int. J. Mod. Phys. A **2**, 561 (1987).

179. J. Garriga and A. Vilenkin, Phys. Rev. D **57**, 2230 (1998), astro-ph/9707292; V. Vanchurin and A. Vilenkin, Phys. Rev. D **74**, 043520 (2006), hep-th/0605015.

180. R. Bousso, Phys. Rev. Lett. **97**, 191302 (2006), hep-th/0605263.

181. D. Podolsky and K. Enqvist, arXiv:0704.0144 [hep-th].

182. J. Garcia-Bellido, A. D. Linde and D. A. Linde, Phys. Rev. D **50**, 730 (1994), astro-ph/9312039; J. Garcia-Bellido and A. D. Linde, Phys. Rev. D **51**, 429 (1995), hep-th/9408023; J. Garcia-Bellido and A. D. Linde, Phys. Rev. D **52**, 6730 (1995), gr-qc/9504022.

183. A. Vilenkin, Phys. Rev. Lett. **74**, 846 (1995), gr-qc/9406010.

184. A. Vilenkin, Phys. Rev. D **52**, 3365 (1995), gr-qc/9505031; S. Winitzki and A. Vilenkin, Phys. Rev. D **53**, 4298 (1996), gr-qc/9510054; A. Vilenkin, Phys. Rev. Lett. **81**, 5501 (1998), hep-th/9806185; V. Vanchurin, A. Vilenkin and S. Winitzki, Phys. Rev. D **61**, 083507 (2000), gr-qc/9905097; J. Garriga and A. Vilenkin, Phys. Rev. D **64**, 023507 (2001), gr-qc/0102090; R. Easther, E. A. Lim and M. R. Martin, JCAP **0603**, 016 (2006), astro-ph/0511233; A. Vilenkin, arXiv:hep-th/0602264; V. Vanchurin, arXiv:hep-th/0612215.

185. J. Garriga, D. Schwartz-Perlov, A. Vilenkin and S. Winitzki, JCAP **0601**, 017 (2006), hep-th/0509184;

186. A. Vilenkin, JHEP **0701**, 092 (2007), hep-th/0611271.

187. This family of probability measures sometimes are called "global," whereas the measures based on the comoving coordinates are called "local," see R. Bousso and B. Freivogel, arXiv:hep-th/0610132. However, this terminology, and the often repeated implication that global means acausal, are somewhat misleading. All of these measures are based on investigation of the physical evolution of a single causally connected domain of initial size $O(H^{-1})$. The early stages of the evolution of our part of the universe were influenced by the evolution of other parts of the original domain even though some of these parts at present are exponentially far away from us. One should not confuse the exponentially large particle horizon, which is relevant for understanding of the origin of our part of the universe, with the present radius of the event horizon $H^{-1} \sim 10^{28}$ cm, which is relevant for understanding of our future.

188. A. H. Guth, arXiv:astro-ph/0404546. A. H. Guth, arXiv:hep-th/0702178.

189. M. Tegmark, JCAP **0504**, 001 (2005), astro-ph/0410281.

190. A. Linde, in preparation.

191. A. Aguirre, S. Gratton and M. C. Johnson, arXiv:hep-th/0611221.

192. J. Garriga and A. Vilenkin, Prog. Theor. Phys. Suppl. **163**, 245 (2006), hep-th/0508005.

193. R. Harnik, G. D. Kribs and G. Perez, Phys. Rev. D **74**, 035006 (2006), hep-ph/0604027.

194. G. D. Starkman and R. Trotta, Phys. Rev. Lett. **97**, 201301 (2006), astro-ph/0607227.

2

Preheating After Inflation

Lev Kofman

CITA, University of Toronto, Canada
kofman@cita.utoronto.ca

Abstract. I will discuss what happens at the end point of inflation. The list of topics includes:

Preheating after inflation treated with the quantum field theory (QFT)
Recent developments in the theory of preheating
Reheating after string theory inflation, treated with the string theory
Potential observables which may be associated with (p)reheating, which are modulated cosmological fluctuations generated from preheating
Generation of gravitational waves after preheating

2.1 Generalities: Reheating the Universe

According to the inflationary scenario, the universe at early times expands quasi-exponentially in a vacuum-like state without entropy or particles.

A simple and natural realization of the vacuum-like equation of state is naturally achievable with the homogeneous scalar field $\phi(t)$ minimally coupled to gravity. Indeed, the energy-momentum tensor $T_\nu^\mu = \phi^\mu \phi_\nu - \delta_\nu^\mu \left(\frac{1}{2} \phi^\sigma \phi_\sigma - V \right)$ of a classical moving homogeneous scalar field in the potential $V(\phi)$ is simply $T_\nu^\mu = \mathrm{diag}(\epsilon, -p, -p, -p)$, where the pressure and energy are given by

$$p = \frac{1}{2} \dot\phi^2 - V \ , \quad \epsilon = \frac{1}{2} \dot\phi^2 + V \ . \tag{2.1}$$

When the potential energy dominates the kinetic energy, we have inflation with $p \approx -\epsilon$. Beginning with this simple idea, the complicated +25 years history of inflation has been about the microscopic nature of ϕ and the origin of its potential $V(\phi)$. In Fig. 2.1, I draw a chronologically ordered broad brush sketch of inflationary models for $V(\phi)$.

The theory of inflation is accompanied by the theory of the origin of particles after inflation. The details of this theory can depend on the particular

L. Kofman: *Preheating After Inflation*, Lect. Notes Phys. **738**, 55–79 (2008)
DOI 10.1007/978-3-540-74353-8_2
© Springer-Verlag Berlin Heidelberg 2008

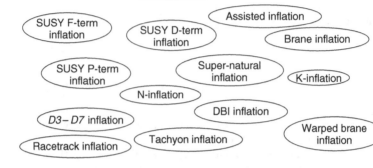

Fig. 2.1. Models of inflation

model of inflation. We will review the "particlegenesis" after inflation, in parallel with the structure of Fig. 2.1.

In the old big bang picture where the universe starts from a singularity, it was assumed that the very hot matter from the very beginning was in a state of thermal equilibrium, with the temperature gradually decreasing as the universe expands. The very early versions of inflation were embedded in the big bang picture and invoked supercooling of matter which reheats once again after inflation ends. Therefore the theory of particle creation and thermalization after inflation was dubbed as "reheating" after inflation – an anachronism of +25 years of history. In the modern version of inflation it is not necessary to postulate the hot pre-inflationary stage (although what comes "before" inflation is still not well understood).

Let us return to (2.1). During inflation, all the energy is contained in a classical, slowly moving inflaton field ϕ. Eventually the inflaton field decays and transfers all its energy to relativistic particles, to start the thermal history of the hot Friedmann universe.

The QFT of (p)reheating, i.e., the theory of particle creation from the inflaton field in an expanding universe, is a process in which quantum effects are not small, but rather generate a spectacular process where all the particles of the universe are created from the rolling classical inflaton. The theory of particle creation and the subsequent thermalization after inflation has a record of theoretical developments within QFT. The four-dimensional QFT

Lagrangian $\mathcal{L}(\phi, \chi, \psi, A_i, h_{ik}, ...)$ contains the inflaton part and other fields which give subdominant contributions to gravity. Due to the interactions of other fields with the inflaton in $\mathcal{L}$, the inflaton field decays and transfers all of its energy to relativistic particles. If the creation of particles is sufficiently slow (for instance, if the inflaton is coupled only gravitationally to the matter fields), the decay products simultaneously interact with each other and come to a state of thermal equilibrium at the reheating temperature T_r. This temperature will be the largest one in the formula for the temperature in an expanding universe

$$T = \frac{1.55}{g_*^{1/4}} \frac{\text{Mev}}{\sqrt{t\,(\text{s})}} \, , \qquad (2.2)$$

where g_* is the number of effective degrees of freedom. This gradual reheating can be treated within the framework of perturbative theory of particle creation and thermalization [1, 2, 3]. However, generically, particle production from the inflaton occurs in the non-perturbative regime. In chaotic inflationary models, soon after the end of inflation the almost homogeneous inflaton field $\phi(t)$ coherently oscillates with a very large amplitude of the order of the Planck mass M_p around the minimum of its potential. The particle production from a coherently oscillating inflaton occurs in the regime of parametric excitation [4, 5, 6]. This picture, with variation in its details, is extended to other inflationary models. For instance, in hybrid inflation (including D-term inflation), inflaton decay proceeds via a tachyonic instability of the inhomogeneous modes which accompany the symmetry breaking [7, 8]. A similar effect combined with parametric resonance is observed in the new inflationary scenario [9]. One consistent feature of preheating – non-perturbative copious particle production immediately after inflation – is that the process occurs far away from thermal equilibrium.

The transition from this stage to thermal equilibrium occurs in a few distinct stages, each one lasting much longer than the previous one. First there is the rapid preheating phase, followed by the onset of turbulent interactions between the different modes. Our understanding of this stage comes from lattice numerical simulations [10, 11] as well as from different theoretical techniques [12]. For a wide range of models, the dynamics of scalar field turbulence is largely independent of the details of inflation and preheating [13]. Finally comes thermalization, ending with equilibrium. In general, the equation of state of the universe is that of matter when it is dominated by the coherent oscillations of the inflaton field, but changes when the inflaton decays into radiation-dominated plasma [14].

Recent developments took place in string theory inflation, although the theory is still in its early stages, and it has to address the issue of the end point of inflation, i.e., (p)reheating immediately after inflation. As for QFT reheating, string theory reheating must be compatible with our thermal history. Yet, we are especially interested in the specific string theory effects during reheating.

Sooner or later in an expanding universe the stage of (p)reheating dynamically evolves into the stage of thermal equilibrium of particles, where a lot of information about the initial state will be lost. What is then an output of (p)reheating for cosmology? It is important to figure out the character of particlegenesis and thermalization in a specific model of inflation (say from Fig. 2.1) and reheat temperature T_r. Short out-of-equilibrium stages in the thermal history of the universe are responsible for the variety of observable forms of matter in the universe. The strong out-of-equilibrium character of preheating opens the possibility for crucial phenomena associated with non-equilibrium physics, including cosmological baryo/leptogenesis phase transitions, non-thermal production of heavy particles, etc. We need to know the evolution of the EOS $w(t)$ to find the connection N–$\log k$ between the number of e-folds N of inflation and the wavelength of cosmological fluctuations. The most interesting output of preheating will be its potential observables, like gravitational waves, and modulated cosmological fluctuations.

In the rest of this chapter I briefly review the basics and new developments in the theory of different types of preheating and reheating, and in the light of our current understanding, discuss the outputs of (p)reheating for cosmology.

2.2 Pair Creation by an Electric Field

Before we turn to the theory of particle creation by an oscillating inflaton, it is instructive to consider a simple prototype problem: particle creation in scalar electrodynamics (see also J. Martin's contribution in the present volume). Suppose E is the strength of the constant electric field aligned with the z-direction. Let A_μ be the four-potential of the external classical EM field, and χ be a quantum field describing massless scalar particles (of the charge e) described by the equation of motion

$$D_\mu D^\mu \chi = 0 , \tag{2.3}$$

where $D_\mu = \partial_\mu - \mathrm{i}eA_\mu$.

Next we have to choose a gauge for A_μ. It is convenient to deal with a time-dependent problem of particle creation. We can put $A_\mu = (0, 0, 0, -Et)$ Then the time-dependent part of the eigenmodes $X_k(t)\mathrm{e}^{\mathrm{i}\boldsymbol{k}\cdot\boldsymbol{x}}$ obey the equation

$$\frac{\mathrm{d}^2 X_k}{\mathrm{d}\tau^2} + \left(\kappa^2 + \tau^2\right) X_k = 0 , \tag{2.4}$$

where we introduced the dimensionless time $\tau = eEt$ and momentum $\kappa = k/eE$. The initial condition for (2.4) shall correspond to the positive frequency vacuum fluctuation $X_k(t) = \frac{1}{\sqrt{2k}}\mathrm{e}^{\mathrm{i}\kappa\tau}$. Then, (2.4) can be interpreted as the scattering of an incoming wave on an inverse parabolic potential. Far away from the apex of the potential at $\tau = 0$ where A_μ crosses zero, the solution can be written in the WKB form

$$X_k(t) = \frac{\alpha_k(t)}{\sqrt{2k}}\,e^{-ikt} + \frac{\beta_k(t)}{\sqrt{2k}}\,e^{+ikt}\,, \tag{2.5}$$

Initially $\beta_k = 0$, $\alpha_k = 1$. After scattering, β_k is no longer vanishing, and this corresponds to $n_k = |\beta_k|^2$ created particles. The analytic solution of (2.4) with the asymptotic form (2.5) is well known and gives

$$n_k = \exp\left(-\frac{\pi k^2}{eE}\right). \tag{2.6}$$

This formula is a variance of the celebrated Schwinger formula for the rate of electron–positron pair creation by a constant electric field. This scalar ED example teaches us several lessons which we shall keep in mind. Despite the constancy of the external electric field, the problem of pair creation can be written as a time-dependent problem. The instant of particles creation is associated with the time when A_μ crosses zero. Of course, the final answer (2.6) does not depend on this moment. Pair creation is a non-perturbative effect. The answer (2.6) is a non-analytic function of the coupling eE. As we will see very similar physics will take place in the process of pair creation by an oscillating inflaton field.

2.3 Linear Resonant Preheating

Consider simple chaotic inflation with the potential $V(\phi) = \frac{m^2}{2}\phi^2$. Soon after the end of inflation, the almost homogeneous inflaton field $\phi(t)$ coherently oscillates as $\phi(t) \approx \Phi(t)\sin(m_\phi t)$, with a very large initial amplitude of the order of the Planck mass $\simeq 0.1 M_{\rm Pl}$, $\Phi(t) = \frac{M_{\rm Pl}}{\sqrt{3\pi}} \cdot \frac{1}{m_\phi t}$. One can take as a toy model to describe the interaction between inflatons and radiation, i.e., other massless Bose particles χ, the Lagrangian $\mathcal{L} = -\frac{1}{2}g^2\phi^2\chi^2$. QFT of particle creation can then be constructed in the following way. Consider the Heisenberg representation of the quantum scalar field $\hat{\chi}$, with the eigenfunctions $\chi_k(t)\,e^{-i\mathbf{k}\cdot\mathbf{x}}$ where $\mathbf{k}$ is a comoving momentum. The temporal part of the eigenfunction obeys the equation

$$\ddot{\chi}_k + 3\frac{\dot{a}}{a}\dot{\chi}_k + \left(\frac{k^2}{a^2} + g^2\phi^2\right)\chi_k = 0\,, \tag{2.7}$$

with the vacuum-like initial condition $\chi_k \simeq \frac{e^{-ikt}}{\sqrt{2k}}$ in the far past. Let us seek solutions for (2.7) in the adiabatic WKB approximation form

$$a^{3/2}\chi_k(t) \equiv X_k(t) = \frac{\alpha_k(t)}{\sqrt{2\omega}}\,e^{-i\int^t \omega dt} + \frac{\beta_k(t)}{\sqrt{2\omega}}\,e^{+i\int^t \omega dt}\,, \tag{2.8}$$

where the time-dependent frequency is $\omega_k^2(t) = \frac{k^2}{a^2} + g^2\phi^2$ (neglecting small corrections $\sim H^2, \dot{H}$); initially $\beta_k(t) = 0$. The goal is to calculate particle occupation number $n_k = |\beta_k|^2$.

Equation (2.7) describes a parametric oscillator in an expanding universe. Introduce the new time variable $z = mt$ and the essential dimensionless coupling parameter $q = \frac{g^2\Phi^2}{4m^2}$. Since $\frac{m}{M_{\rm Pl}} \simeq 10^{-6}$, it is expected that $q \simeq 10^{10}g^2 \gg 1$, so the parametric resonance is broad. For large values of q, the eigenfunction $\chi_k(t)$ is changing adiabatically between the moments t_j, $j = 1, 2, 3, \cdots$, where the inflaton field is equal to zero $\phi(t_j) = 0$. The non-adiabatic changes of $\chi_k(t)$ occur only in the vicinity of t_j. Therefore, the semi-classical solution (2.8) is valid everywhere but around t_j. Thus the zeroes of the inflaton oscillations play a crucial role, very similar to the zeroes of A_μ in the scalar ED example of the previous section. Let the wave $\chi_k(t)$ have the form of the WKB solution with the pair of coefficients (α_k^j, β_k^j) before scattering at the point t_j; and the pair $(\alpha_k^{j+1}, \beta_k^{j+1})$ after scattering at t_j. The interaction term around all the points t_j is parabolic $g^2\phi^2(t) \approx g^2\Phi^2 m^2(t - t_j)^2$, and the eigenmode equation around the zero point is reduced to (2.4) of the previous section. Therefore, the outgoing amplitudes $(\alpha_k^{j+1}, \beta_k^{j+1})$ can be expressed through the incoming amplitudes (α_k^j, β_k^j) with the help of the well-known reflection and transmission amplitudes for scattering from a parabolic potential at t_j. The net result in terms of number of χ-particles $n_k^{j+1} = |\beta_k^{(j+1)}|$ created after t_j is

$$n_k^{j+1} = e^{-\pi\kappa^2} + \left(1 + 2e^{-\pi\kappa^2}\right) n_k^j - 2e^{-\pi\kappa^2/2}\sqrt{1 + e^{-\pi\kappa^2}}\sqrt{n_k^j(1 + n_k^j)}\sin\theta^j .$$
$$(2.9)$$

Here $\kappa = \frac{k^2}{a\sqrt{gm\Phi}}$ and θ^j is the phase accumulated between zeroes of $\phi(t)$.

The first term of the formula (2.9) is nothing but the variance of the Schwinger formula for the spontaneous pair creation from an external field, similar to the formula (2.6) of the previous section. The last term in (2.9) corresponds to the induced pair creation in the presence of particles created from the previous zero crossing of the inflaton. There is no analogy for this in scalar ED with a constant electric field. Finally, the second term in (2.9) is an interference between induced and spontaneous particles creation.

Formula (2.9) describes parametric resonant particle creation by an oscillating inflaton. Very quickly n_k^j becomes large and formula (2.9) can be approximated by much simpler formula

$$n_k^{j+1} = e^{2\pi\mu_k}n_k^j , \quad n_k(t) \simeq e^{\mu_k t} . \tag{2.10}$$

Here μ_k is a complex exponent, which is real for the resonant modes (positive interference) and imaginary for stable modes (negative interference). The expansion of the universe makes the phase θ_k random, and makes the resonant process stochastic, but a broad band of modes is exponentially amplified.

2.4 Non-linear Dynamics of Resonant Preheating

Due to the rapid growth of its occupation numbers the field $\chi(t, x)$ can be treated as a classical scalar field. Its appearance is described by the realization of the random gaussian field, i.e., as a superposition of standing waves with random phases and Rayleigh-distributed amplitudes. One can use many different quantities to characterize a random field, such as the spatial density of its peaks of a given height, etc. The scale of the peaks and their density depend on the characteristic scale R of the spectrum, which in our case is related to the leading resonant momentum $k_* \simeq \sqrt{gm\phi_0}a^{1/4}$ [5]. At the linear stage the phases are constant, so that the structure of the random field χ stays almost the same.

Once one field is amplified in this way, other fields that are coupled to it are themselves amplified [10, 13], so within a short time of linear preheating (of order dozens of inflaton oscillations) fluctuations of χ generate inhomogeneous fluctuations of the field ϕ. It is easy to see that fluctuations of ϕ will have a non-linear, non-gaussian character. From the equation of motion for ϕ

$$\nabla_\mu \nabla^\mu \phi + m^2 \phi^2 + g^2 \chi^2 \phi = 0 , \tag{2.11}$$

we have in Fourier space

$$\ddot{\phi}_k + 3H\dot{\phi}_k + \left(\frac{k^2}{a^2} + m^2\right)\phi_k = g^2 \phi_0(t) \int \mathrm{d}^3 q \, \chi_q \chi_{k-q}^* , \tag{2.12}$$

where we neglect the term that is third order in the fluctuations; $\phi_0(t)$ is the background oscillation. The solution of this equation with Green's functions [5] shows that ϕ fluctuations grow with twice the exponent of χ fluctuations. It also shows that the fluctuations of ϕ are non-gaussian. Sometimes this solution is interpreted as a rescattering of the particle χ_q against the condensate particle ϕ_0 at rest producing χ_{k-q} and ϕ_k, $\chi\phi_0 \to \chi\delta\phi$. However, this interpretation has significant limitations.

When the amplitudes of χ and ϕ become sufficiently large we have to deal with the fully non-linear problem. The field evolution can be well approximated using the classical equation of motion (2.11) supplemented by another equation for χ, namely

$$\nabla_\mu \nabla^\mu \chi + g^2 \phi^2 \chi = 0 . \tag{2.13}$$

Equations (2.11) and (2.13) of the non-linear preheating can be solved numerically using the LATTICEEASY program. For chaotic inflation, these results were presented in terms of the time evolution of occupation numbers $n_k(t)$ or total number density of particles $N(t)$. Figures 2.2 and 2.3 show the results of our simulations in these familiar terms of $n_k(t)$ (in combination $k^3\omega_k n_k$) and $N(t)$, as well as the evolution of the field statistics (departures from gaussianity). Here all simulation results are for model with $m = 10^{-6}M_{\mathrm{Pl}}$ (fixed by CMB normalization) and $g^2 = 2.5 \times 10^{-7}$. The size of the box was chosen as

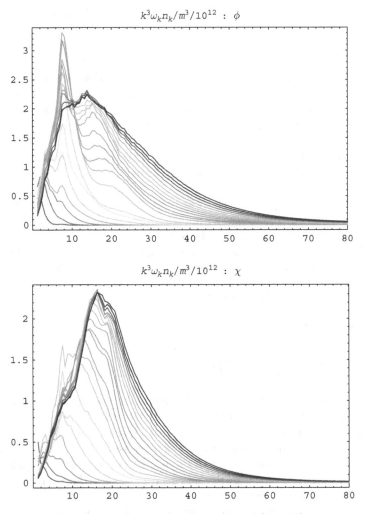

$$k^3 \omega_k n_k / m^3 / 10^{12} : \phi$$

$$k^3 \omega_k n_k / m^3 / 10^{12} : \chi$$

Fig. 2.2. Evolution of spectra of created particles

$L = 10\,m^{-1}$ and the grid contained 256^3 points. We also tried other values of g^2 and found qualitatively similar results.

Figure 2.2 shows the evolution of the spectra. The spectra show rapid growth of the occupation numbers of both fields, with a resonant peak that develops first in the infrared ($k \simeq k_*$) and then moves toward the ultraviolet as a result of rescattering. On the left panel of Fig. 2.3, one can clearly see that the occupation number of χ initially grows exponentially fast due to parametric resonance, followed by even faster growth of the ϕ field due to the interaction, in accordance with the solution of (2.12).

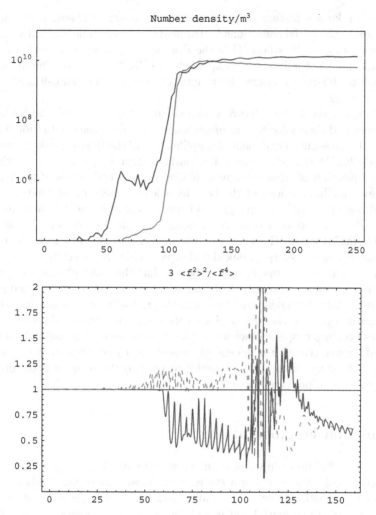

Fig. 2.3. *Left*: evolution of total number of particles. *Right*: evolution of non-gaussianity

The gaussianity of classical fields can be measured in different ways. The right panel of Fig. 2.3 shows the evolution of the ratio $\langle f^2 \rangle^2 / \langle f^4 \rangle$ (kurtosis), which is equal to unity for a gaussian field. During the linear stage of preheating, the field fluctuations form a random gaussian field, reflecting the initial quantum fluctuations that seeded them. The inhomogeneous field ϕ is generated as a non-gaussian field, in agreement with the solution of (2.12). When the fluctuation amplitude begins to get large, both fields are non-gaussian. During the later turbulent stage both fields begin to return to gaussianity.

Another known feature of preheating is the onset of chaos, when small differences in the initial conditions for the fields lead to exponentially divergent solutions: $D(t) \simeq e^{\lambda t}$, where D is the distance in phase space between the solutions and λ is the Lyapunov exponent (see [13] for details). The distance D begins to diverge exponentially exactly after the violent transition to the turbulent stage.

Let us summarize the picture which emerges when we study preheating, turbulence and thermalization in momentum space with the occupation numbers n_k. There is initial exponential amplification of the field χ, peaked around the mode k_*. At this stage the χ fluctuations form a squeezed state, which is a superposition of standing waves that make up a realization of a random gaussian field. Interactions of the two fields lead to very rapid excitation of fluctuations of ϕ, with its energy spectrum also sharply peaked around k_*. To describe generation of ϕ inhomogeneities, people use the terminology of "rescattering" of waves. However, there is a short violent stage when occupation numbers have a sharply peaked and rapidly changing spectrum. The field at this stage is non-gaussian, which signals that the wave phases are correlated. In some sense, the concept of "particles" is not very useful around that time. In the later turbulent stage when $n_k(t)$ gradually evolves and gaussianity is restored (due to the loss of phase coherency) the picture of rescattering particles becomes proper. As we will see in the next section, gaussianity is not restored for some time after the end of preheating. To understand this violent, intermediate stage, however, it is useful to turn to the reciprocal picture of field dynamics in position space [15].

2.5 Inflaton Fragmentation

The features in the occupation number spectra $n_k(t)$, namely, sharp time variations, peak at $k \sim k_*$, and strong non-gaussianity of the fields around the time of transition between preheating and turbulence suggest that we are dealing with distinct spatial features of the fields in the position space. This prompted to study the dynamics of the fields in position space.

The evolution of the fields in position space is shown in Fig. 2.4. Each frame shows the spatial profile of the fields ϕ and χ along a two-dimensional slice of the three-dimensional lattice.

The initial evolution of the fields ($t < 100$) is characterized by linear growth of fluctuations of χ. During this stage the fluctuations have the form of a superposition of standing waves with random phases, which make up a random gaussian field. The eye captures positive and negative peaks that correspond to the peaks of the initial gaussian random field χ. The peaks in this early stage correspond to the peaks of the initial gaussian random field χ. Following that phase, the oscillations of χ excite oscillations of ϕ. The first panel of Fig. (2.4) shows a typical profile near the end of this period, just as the oscillations are becoming non-linear and ϕ is becoming excited.

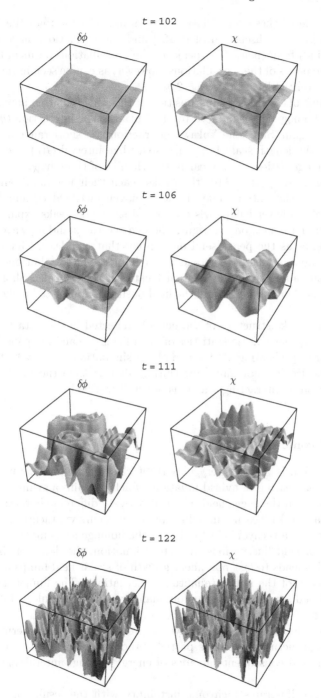

Fig. 2.4. Inflaton fragmentation via parametric resonance preheating. Values of the ϕ and χ fields in a two-dimensional slice through the lattice

The amplitude of these ϕ oscillations grows much faster than the initial χ oscillations [see our discussion of (2.12)] and the oscillations have different (and changing) frequencies. The peaks of the ϕ oscillations occur in the same places as the peaks of the χ oscillations, however, as can be seen in the bottom three panels on the left side of Fig. 2.4.

The profile of $\phi(t, \boldsymbol{x})$ is a superposition of the still oscillating homogeneous part plus inhomogeneities induced by the Yukawa-type interaction $(g^2\phi_0)\phi\chi^2$ in the Lagrangian. Since the Yukawa interaction is a short-range interaction (defined by the length scale $1/m$ for a space-like interval $G(r) \sim e^{-mr}$), induced inhomogeneities of ϕ appear in the vicinity of those in χ.

In the next stage ($t \sim 110$) the peaks reach their maximum amplitude, comparable to the initial value of the homogeneous field ϕ, and begin to spread. The two lower left panels of Fig. 2.4 show the peaks expanding and colliding. In the panels on the right, one can see the standing wave pattern loses coherence as the peaks send out ripples that collide and interfere. By $t = 124$ the fluctuations have spread throughout the lattice, but one can still see waves spreading from the original locations of the peaks. Shortly after that time all coherence is lost and the field positions appear to be like random turbulence.

The bubble-like structure of the fields is reflected in their statistics. Perhaps more surprisingly, the statistics of both fields remain non-gaussian for a long time after preheating. At the end of our simulation, at $t = 300$, the fields were still noticeably non-gaussian. During all this time the random phase approximation of interacting scalars is not well justified.

2.5.1 Tachyonic Preheating

Hybrid inflation is another very important class of inflationary models. At first glance preheating in hybrid inflation, which contains a symmetry breaking mechanism in the Higgs field sector, has a very different character than in chaotic inflation. Preheating in hybrid inflation occurs via tachyonic preheating [7], in which a tachyonic instability of the homogeneous modes drives the production of field fluctuations. In hybrid inflation, the decay of the homogeneous fields leads to fast non-linear growth of scalar field lumps associated with the peaks of the initial (quantum) fluctuations. The lumps then build up, expand and superpose in a random manner to form turbulent, interacting scalar waves.

Like parametric resonance, tachyonic preheating can be interpreted via the reciprocal picture of copious particle production far away from thermal equilibrium, and consequent cascades of energy through interacting, excited modes.

Figure 2.5 illustrates tachyonic instability with the results of numerical simulations in the model with the Higgs part of the potential $V(\phi) = \frac{1}{2}m^2\phi^2 + \frac{1}{3}\sigma\phi^3 + \frac{1}{4}\lambda\phi^4$.

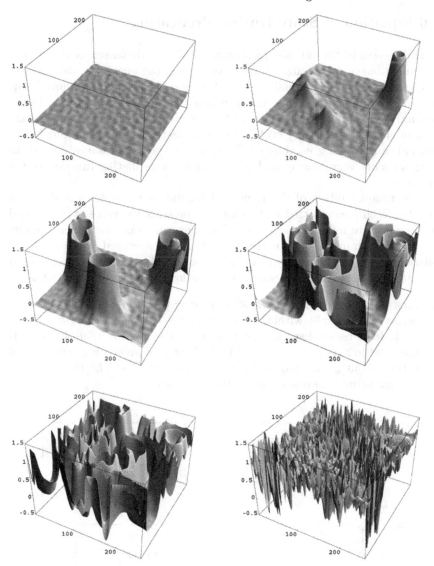

Fig. 2.5. Inflaton fragmentation via tachyonic preheating. The evolution of the field is shown at slices through the lattice

Because of the non-linear dependence of the tachyonic mass on ϕ, initial linear gaussian fluctuations quickly turn into non-linear non-gaussian field, with pronounced bubble structures.

2.6 Equation of State During Preheating

In [16] we study the out-of-equilibrium non-linear dynamics of fields after post-inflationary preheating. During preheating, the energy in the homogeneous inflaton is exponentially rapidly transferred into highly occupied out-of-equilibrium inhomogeneous modes, which subsequently evolve toward equilibrium. We compute the equation of state (EOS) during and immediately after preheating. It sharply evolves toward radiation domination long before thermal equilibrium is established. The jump of the EOS from inflaton domination occurs very quickly and its timing is an oscillating function of the couplings.

The time evolution of the EOS $w(t)$ for different couplings is shown in Fig. 2.6. Each point plotted on this figure represents the value of w averaged over a complete inflaton oscillation. This represents one of the main results of our study. Immediately after inflation, the EOS averaged over inflaton oscillations is $w = 0$. It sharply changes at the end of preheating. There are at least three important points worth emphasizing about the evolution of w.

1. First, the transition of the EOS from $w = 0$ to the value $w \sim 0.2 - 0.3$ occurs very sharply, within a time interval $\sim 10^{-36}$ s.

   Indeed, recall that the unit of time on the plots is $1/m$, where m is the inflaton mass, i.e., 10^{-37} s. The first stage of preheating is completed within about a hundred of these units, i.e., 10^{-35} s. The rise of w and gradual saturation takes roughly the same time.

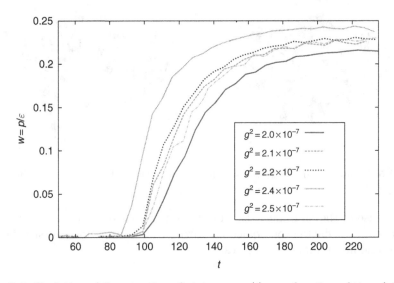

Fig. 2.6. Evolution of the equation of state $w = w(t)$ as a function of time (given in units of m^{-1}) for various couplings g^2 around $g^2 = 2 \times 10^{-7}$

2. Second, the dependence of $w(t)$ on the coupling g^2 for resonant preheating is a non-monotonic function of g^2.

    This is to say that the time during which preheating comes to an end is very weakly (logarithmically) dependent on the coupling. As seen from Fig. 2.6 the curves $w(t)$ begin to shift to the left toward an earlier end of preheating, as we vary g^2 by 5%. However, at some point the curves stop moving to the left and instead begin to return toward the right. As we change g^2 by about 25%, the cycle repeats. As we vary g^2, the function w not only shifts, but it also varies its detailed shape.

    We see that the transition time varies between $100/m$ and $150/m$. This non-monotonic behavior of the duration of preheating is explained in the theory of broad parametric resonance [5, see Sects. 6 and 9 there].

    The g^2 dependence of the EOS is the critical issue for the theory of modulated cosmological perturbations, which we will discuss in Sect. 2.8.

3. The third point is that w does not necessarily immediately go to the radiation-dominated value $1/3$.

    This is partly because immediately after preheating the light field still has a significant induced effective mass due to the interaction, and partly due to the significant residual contribution from the homogeneous inflaton [17]. Unfortunately, limitations on running longer simulations preclude us from seeing further details of the time evolution of w. However, we have a strong theoretical argument to advance the discussion further. In a model with a massive inflaton and light scalar χ even the radiation-dominated stage is transient. Indeed, sooner or later the massive inflaton particles, even if significantly under-abundant at the end of preheating, will become the dominant component, and the universe will again be matter-dominated.

2.7 Effects of Trilinear Interactions

Most studies of preheating have focused on the models with $\phi^2\chi^2$ four-legs interactions of the inflaton ϕ with another scalar field χ. A common feature of preheating is the production of a large number of inflaton quanta with non-zero momentum from re-scattering, alongside with inflatons at rest. The momenta of these relic massive inflaton particles eventually would redshift out. However, the decay of inflaton particles through four-legs $\phi\phi \to \chi\chi$ processes in an expanding universe is never complete. Thus inflaton particles later on will have a matter equation of state and come to dominate the energy density, which is not an acceptable scenario. Therefore, to avoid this, we must include in the theory of reheating interactions of the type $\phi\chi^n$, that allow the inflaton to decay completely, thus resulting in a radiation-dominated stage. Trilinear interactions are the most immediate and natural interactions of this sort.

    In [16] we investigate the effects of bosonic trilinear interactions in preheating after chaotic inflation. A trilinear interaction term allows for the complete decay of the massive inflaton particles, which is necessary for the transition to radiation domination. We found that typically the trilinear term is subdominant during early stages of preheating, but it actually amplifies parametric resonance driven by the four-legs interaction. In cases where the trilinear term does dominate during preheating, the process occurs through periodic tachyonic amplifications with resonance effects, which is so effective that preheating completes within a few inflaton oscillations. We develop an analytic theory of this process, which we call tachyonic resonance. We also study numerically the influence of trilinear interactions on the dynamics after preheating. The trilinear term eventually comes to dominate after preheating, leading to faster re-scattering and thermalization than could occur without it. Finally, we investigate the role of non-renormalizable interaction terms during preheating. We find that if they are present they generally dominate (while still in a controllable regime) in chaotic inflation models. Preheating due to these terms proceeds through a modified form of tachyonic resonance. Impact of the trilinear interaction of the inflaton can be seen in the Fig. 2.7.

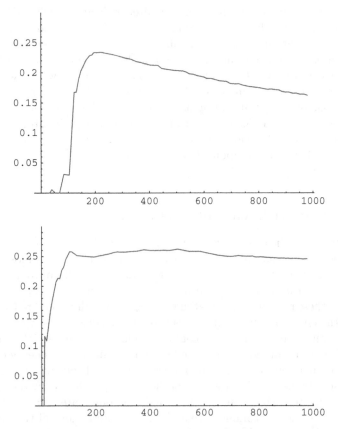

Fig. 2.7. Equation of state without (*left*) and with (*right*) three-legs interaction of inflaton

It is possible to extend analytic theory of preheating to the case of trilinear interaction $\sigma\phi\chi^2$, which describes decay of the inflaton [16]. Instead of (2.7), the temporal part of the eigenfunction χ_k obeys the equation

$$\ddot{\chi}_k + 3\frac{\dot{a}}{a}\dot{\chi}_k + \left(\frac{k^2}{a^2} + \sigma\phi\right)\chi_k = 0 \,. \tag{2.14}$$

Analytic treatment of this equation is quite different from the method of successive parabolic scatterings, because in interesting cases the wave function $\chi_k(t)$ spends significant fraction of time in the under-barrier regime. The final results for the number of created particles after j oscillations is

$$n_k^j = \exp(2jX_k)\,(2\cos\Theta_k)^{2(j-1)} \,. \tag{2.15}$$

where Θ_k is the phase and

$$X_k = 2\sqrt{2q}\, f\left(\frac{A_k}{2q}\right) \,. \tag{2.16}$$

This formula describes the effect of tachyonic parametric resonance and is complimentary to the formula (2.9).

2.8 Modulated Fluctuations from Preheating

I was searching for manifestations of extra dimensions. Superstring theory, phenomenological models with extra dimensions and other SUSY models generically predict that the coupling constants are in fact vacuum expectation values of fields like the dilaton, moduli, etc. Assuming some of these fields are light during inflation, we get generation of small classical inhomogeneities in these fields from inflation. Consequently, coupling constants inherit small inhomogeneities at scales much larger than the causal horizon in the early universe. After the moduli get pinned down to their minima, the spatial variations of coupling constants in the late time universe will be erased. However, inhomogeneities in coupling constants in the very early universe would generate modulated large-scale fluctuations in all relic species that are produced due to interactions and freezing out. Moreover (p)reheating of the inflaton field results in modulated curvature fluctuations. Even if the standard inflaton fluctuations are suppressed, in this picture we may have pure curvature cosmological fluctuations entirely generated by the modulated spatial variations of the coupling constants during preheating.

This is a completely different idea which is an alternative to the standard mechanism of generation of fluctuations. This idea was suggested in my paper "Probing string theory with modulated cosmological fluctuations" (arXiv:astro-ph/0303614) and in an independent paper by Dvali et al. in March 2003. The idea draws interest among cosmologists (about 50 citations), this June there will be a workshops in Athens, devoted to the alternative mechanisms of generation of cosmological perturbations.

I developed it further in collaboration with F. Bernardeau and J-P. Uzan. We extend this idea to the class of hybrid inflation, where the bifurcation value of the inflaton is modulated by the spatial inhomogeneities of the couplings. As a result, the symmetry breaking after inflation occurs not simultaneously in space but with the time laps in different Hubble patches inherited from the long-wavelength moduli inhomogeneities. To calculate modulated fluctuations we introduce techniques of general relativistic matching conditions for metric perturbations at the time hypersurface where the equation of state after inflation undergoes a jump, without evoking the detailed microscopic physics, as far as it justifies the jump. We apply this theory to the modulated fluctuations from the hybrid and chaotic inflations. We discuss what distinguish the modulated from the inflation-driven fluctuations, in particular, their spectral index, modification of the consistency relation and the issue of weak non-gaussianity.

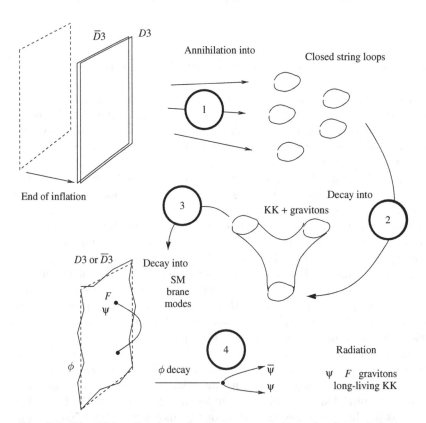

Fig. 2.8. Identifying the channels of energy cascading after brane inflation

2.9 Reheating After String Theory Inflation

In string theory realizations of inflation, the end point of inflation is often brane–anti-brane annihilation. We consider the processes of reheating of the standard model universe after brane inflation. We identify the channels of inflaton energy decay, cascading from tachyon annihilation through massive closed string loops, KK modes and brane displacement moduli to the lighter standard model particles (Fig. 2.8). Cosmological data constrain scenarios by putting stringent limits on the fraction of reheating energy deposited in gravitons and non-standard sector massive relics. We estimate the energy deposited into various light degrees of freedom in the open and closed string sectors, the timing of reheating and the reheating temperature. Production of gravitons is significantly suppressed in warped inflation. However, we predict a residual gravitational radiation background at the level $\Omega_{GW} \sim 10^{-8}$ of the present cosmological energy density. We also extend our analysis to multiple throat scenarios. A viable reheating would be possible in a single throat or in a certain subclass of multiple throat scenarios of the KKLMMT-type inflation model, but overproduction of massive KK modes poses a serious problem. The problem is quite severe if some inner manifold comes with approximate isometries (angular KK modes) or if there exists a throat of modest length other than the standard model throat, possibly associated with some hidden sector (low-lying KK modes).

2.10 Gravitational Waves from Preheating

Eventually after preheating the fields reach thermal equilibrium characterized only by the temperature. Does that mean that all traces of inflaton fragmentation history are erased? For instance, people have discussed realizations of baryogenesis at the electroweak scale via tachyonic preheating after hybrid inflation, and this process is ultimately related to the bubble-like lumps of the Higgs field that form during tachyonic preheating. Since we now see that fragmentation through bubbles can also occur in chaotic inflation, baryogenesis via out-of-equilibrium bubbles can also be extended to these models.

There is another, potentially observable consequence of the non-linear "bubble" stage of inflaton fragmentation. Lumps of the scalar fields correspond to large (order of unity) energy density inhomogeneities at the scale of those bubbles, R. Collisions of bubbles generate gravitational waves. The fraction of the total energy at the time of preheating converted into gravitational waves is significant. We estimate it is of the order of

$$\frac{\rho_{gw}}{\rho_{rad}} \simeq 0.05(RH)^2 \,, \tag{2.17}$$

where $1/H$ is the Hubble radius. This corresponds to a present-day fraction of energy density $\Omega_{GW} \sim 10^{-5}(RH)^2$. The way to understand formula (2.17) is the following: the energy converted into gravitational waves from the collision of two black holes is of the order of the black hole masses. If the mass of lumps of size R is a fraction f of a black hole of the same size, then the fraction of energy converted to gravitational waves from two lumps colliding is f. Scalar field lumps at the Hubble scale would form black holes, so in our case $f = (RH)^2$.

The present-day frequency of this gravitational radiation is

$$f \simeq \frac{M}{10^7 \text{GeV}} \text{Hz} , \qquad (2.18)$$

where $M = V^{1/4}$ is the energy scale of inflation with the potential V.

For the chaotic inflation model considered in this chapter the size of the bubbles is $R \sim$ few$/m$ and at the time they begin colliding $H \sim m/100$, so that the fraction of energy converted into gravitational waves is of the order $10^{-3} - 10^{-4}$. This figure is in agreement with the numerical calculations of gravitational wave radiation from preheating after chaotic inflation.

For chaotic inflation with M at the GUT scale the frequency (2.18) is too short and not observable. Gravitational waves continue to be generated during the turbulent stage and even during equilibrium due to thermal fluctuations, but with a smaller amplitude. It is a subject of further investigation if they can be observed. The most promising possibility for observations is, however, generation of gravity waves from low-energy hybrid inflation, where f can be much much smaller.

2.11 Looking Toward the Future

In conclusion section we will discuss the future directions and perspectives of our topics. They include scenarios of the end point of inflation, (p)reheating and beginning of the thermal equilibrium of the primordial universe. There are at least four interesting methodologies from different branches of physics to be applied. One is related to the models of inflation, since (p)reheating scenario follows from the inflationary scenario. Second is related to our understanding of the particle physics phenomenology, since it defines the interactions between particles which is vital for (p)reheating. Third is related to our deeper understanding of subtle outstanding problems of the non-equilibrium dynamics in the QFT, even at the level of simple toy models (like $\lambda\phi^4$ or $+g^2\phi^2\chi^2$). And the final aspect is about potential cosmological observables from (p)reheating which can survive "democratization" of the high-temperature thermal equilibrium of the early universe plasma, which tends to erase information about previous stages.

Potentially interesting possibilities for new (p)reheating scenarios are in the string theory inflationary models. Let me mention inflationary model based on the large-volume stabilization scheme [18]. This model involves

Kähler moduli and their partners, axions. One can think about Kähler moduli as a hole (four-cycle) in the compact Calabi–Yau manifold. Effective four-dimensional potential for Kähler moduli/axion scalar fields is suitable for inflation [19, 20]. Geometrically rolling inflaton field corresponds to the shrinking of the four-cycle (hole). The end point of inflation, depending on the parameters of the effective potential, can well be in the field theory regime as Conlon and Quevedo [19] and Bond et al. [20] assumed. However, one can have string-scale size hole at the end of inflation, where string theory effects take place [20]. Curiously enough, inflation occurs in the supergravity (i.e., field theory) regime, while it ends in the stringy regime. (P)reheating scenario in this case can be quite different from other cases.

Particle physics phenomenology provides us with the couplings between inflaton and SM and hidden sectors. The community is anticipating potential impact of the LHC results on the particle physics phenomenology, and consequently on the physics of inflation. In the mean time I will mention recent interesting model νMSM of Shaposhnikov and Tkachev [21] based on the minimal extension of the SM by three right-handed neutrinos. This model predicts parametric resonant preheating after inflation.

Thermalization in non-equilibrium QFT will be the subject of the KITP Program "Non-equilibrium Dynamics in Particle Physics and Cosmology" to be held in January–March 2008.

Potential observables from preheating are modulated cosmological fluctuations, which can be distinguished by the different consistency relations and amount of non-gaussianity; baryon asymmetry generated from preheating; and gravitational waves generated from preheating after inflation.

I also would like to mention significance of our understanding of preheating for other aspects of the inflationary theory, which at first glance are not directly related to (p)reheating.

First example. Inflationary theory is connected with the dynamics of the universe driven by a scalar field. We can study scalar field/gravity dynamics in great details using the very powerful phase portrait method (Fig.2.9), which shows the character of *all* solutions for $a(t)$ and $\phi(t)$ as trajectories in the three-dimensional phase space $\mathcal{M}^3$ with the coordinates $(H, \dot{\phi}, \phi)$ [22, 23]. How big is the fraction of inflationary solutions? For this question we need the tools of Hamiltonian dynamics to construct an invariant measure on the phase space [24]. There are complications which make the subject controversial. The physics of prior probabilities is needed to see that inflationary trajectories are typical [25].

Rigorous treatment of the scalar/gravity dynamics involves the four-dimensional phase space of the canonical variables (a, p_a, ϕ, p_ϕ) and Hamiltonian form of the Einstein equations. Since the Hamiltonian is zero, $\mathcal{H} = 0$, the phase trajectories actually reside in a three-dimensional space $\mathcal{M}^3$. One can introduce a canonical measure in the space of trajectories, $d\mu$, which is invariant under the flow of trajectories [24]. This is a local measure. If we prescribe uniform prior probability of trajectories, then we just have to compare the

Expanding universe

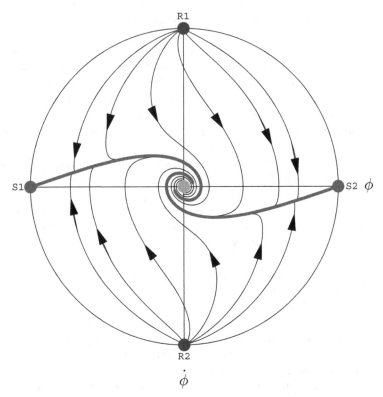

Fig. 2.9. Poincaré mapping of the phase portrait for the theory $V(\phi) = \frac{1}{2}m^2\phi^2$. The initial velocity of the scalar field rapidly decreases, which usually leads to the onset of inflation. The inflationary separatrices (the *thick red lines* to which most of the trajectories converge) are attractors when we move forward in time. At the end all trajectories are spiraling around the focus F

measure $\int_{\mathrm{inf}} d\mu$ trajectories with inflation vs $\int_{\mathrm{non\text{-}inf}} d\mu$ trajectories without inflation. It turns out that both integrals, $\int_{\mathrm{inf}} d\mu$ and $\int_{\mathrm{non\text{-}inf}} d\mu$, diverge (because the scale factor a is unbounded) and no conclusion with this prescription can be made [26]!

Consider for simplicity a flat universe with $K = 0$. In this case phase trajectories (solutions) are located at the two-dimensional hypersurface defined by the constraint equation with $K = 0$. This two-dimensional hypersurface consists of two cones which touch each other by their apex in a single point F. The upper cone corresponds to a pure expansion branch $H > 0$, while the lower cone corresponds to pure contraction $H < 0$. Consider an expanding universe (contraction is just the time reverse). All trajectories begin at the repulsion points R_1 or R_2; most of the trajectories approach the separatrices $S_1 - F$ or $S_2 - F$ and then spiral around the focus F where they end.

A recent paper [27] argued that since at the classical level scalar/gravity dynamics is reversible, one can try to judge about the measure of trajectories of interest not at the earliest stages of evolution but rather at the latest stages. In particular, if one truncates the two-dimensional integral $\int d\mu$ at some late time value of a to make it finite, the integral $\int d\mu = \int d\phi \wedge d\dot\phi$ is reduced to a one-dimensional integral $\int d\phi\, p_\phi \equiv J$. At the stage of oscillation, J is an adiabatic invariant, and together with the phase forms a pair of canonical variables (J, θ). The authors suggested measuring the priors of trajectories at the stage of scalar field oscillations, in terms of a uniform distribution of the phase θ of the oscillations. The separatrices correspond to a specific choice of the phase $\theta_*(t)$, and inflationary trajectories which converge to the separatrices have their phases crowded toward θ_*. In contrast, our prescription is to assign priors to trajectories at the very beginning around the repulsors R_1, R_2. Recent investigation [25] shows that if we repeat the trick with truncation $\int d\mu$ there, we obtain $\int d\phi\, p_\phi = J_0$ which is also conserved! (But it is not conserved in the middle of the trajectories.) We can use a uniform distribution of J_0 around singularity, and it will not map into a uniform distribution of J at the stage of the oscillations. In contrast, it gives a distribution highly peaked toward the inflationary separatric $J_*(t)$. Moreover, the modern theory of reheating after inflation does not have a long-oscillation post-inflationary stage, as we explained above, so that the prescription based on a posteriori distributions of the phase θ of oscillations after inflaton is not relevant!

Second example. Recently Shinji Mukohyama and I suggested an unusual model of inflation. This is inflation based on the fast-roll of the inflaton field. It works for the inflaton field conformally coupled to gravity. Remember, appearance of the conformal coupling was a big problem for the warped brane inflation based on the string theory model of the mobile $D3$ brane interacting with anti-brane placed on the tip of the Khlebanov–Strassler throat (KKLMMT model). In fact, it can be shown that by itself the conformal coupling is not a problem to generate inflation. This model has new features which make it very different from other models. One feature is that this inflaton with $\xi = 1/6$ (including the warped brane–anti-brane inflation) can realize the fast-roll inflation, contrary to the customary slow-roll inflation. The only feature needed for conformal inflation is the form of its potential: very shallow for the most part of the inflaton rolling and changing sharply at the end point of inflation (exactly as in the warped brane–anti-brane inflation or the hybrid models). Another feature is that fast-roll conformal inflation is a low-energy inflation (close to the least energy inflation at TeV scale). As a result it requires significantly less e-foldings N than the figure 62 typical for the GUT scale chaotic inflation. In the context of the warp geometry, as we have shown, N is directly related to the warp factor of the throat geometry of the inner manifold. Coincidentally, the same number $N \approx 37$ satisfies two different pieces of physics, one for homogeneous and isotropic universe from the low energy $1\,TeV$ inflation and another for the warped geometry mass hierarchy $M_{\mathrm{Ple}}e^{-N} \simeq TeV$. How to generate scale-free metric fluctuations in this fast-roll inflation model? Since the inflaton field is conformal, its fluctuations are not generated. Next idea, associated with modulated fluctuations, is to recall angular degrees of freedom describing relative positions of the branes in five-dimensional space orthogonal to the radial direction. It can be shown that for the fast-rolling inflaton fluctuations of the scalar fields associated with the angular coordinates have the solution corresponding to conformal scalars. Can one obtain scalar metric fluctuations in the model? The help comes from the sector of modulated cosmological fluctuations related to the scalars from MSSM or the Higgs sector. Light scalar fields from these sectors are excited during inflation, and modulate the timing hypersurface of reheating after brane–anti-brane annihilation.

References

1. A. Dolgov and A. Linde, Phys. Lett. **B116**, 329 (1982).
2. L. Abbott, E. Fahri and M. Wise, Phys. Lett. **B117**, 29 (1982).
3. A. Linde, *Particle Physics and Inflationary Cosmology* (Academic Publisher, Chur, Switzerland, 1990).
4. L. Kofman, A. Linde and A. Starobinsky, Phys. Rev. Lett. **73**, 3195 (1994).
5. L. Kofman, A. Linde and A. Starobinsky, Phys. Rev. D **D56**, 3258 (1997).
6. J. Traschen and R. Brandenberger, Phys. Rev. **D42**, 2491 (1990).
7. G. Felder et al., Phys. Rev. Lett. **87**, 011601 (2001).
8. G. Felder, L. Kofman and A. Linde, Phys. Rev. **D64**, 123517 (2001).
9. M. Desroche, G. N. Felder, J. M. Kratochvil and A. Linde, Phys. Rev. **D71**, 103516 (2005).
10. A. Khlebnikov and I. Tkachev, Phys. Rev. Lett. **77**, 219 (1996).
11. S. Y. Khlebnikov and I. I. Tkachev, Phys. Rev. Lett. **79**, 1607 (1997); T. Prokopec and T. G. Roos, Phys. Rev. **D55**, 3768 (1997).
12. J. Berges and J. Serreau, Phys. Rev. Lett. **91**, 111601 (2003); arXiv:hep-ph/0302210; arXiv:hep-ph/0410330; J. Berges, S. Borsanyi and C. Wetterich, Phys. Rev. Lett. **93**, 142002 (2004).
13. G. Felder and L. Kofman, Phys. Rev. **D63**, 0103503 (2007).
14. D. I. Podolsky, G. N. Felder, L. Kofman and M. Peloso, Phys. Rev. **D73**, 023501 (2006).
15. G. Felder and L. Kofman, Phys. Rev. **D75**, 047518 (2007).
16. J-F. Dufaux, G. N. Felder, L. Kofman, M. Peloso and D. I. Podolsky, JCAP **D75**, 123511 (2007).
17. R. Micha and I. I. Tkachev, Phys. Rev. **D70** (2004) 043538.
18. J. Conlon, F. Quevedo and K. Suruliz, JHEP **0508**, 007 (2005).
19. J. Conlon and F. Quevedo, JHEP **0601**, 146 (2006).
20. J.R. Bond, L. Kofman, S. Prolushkin and P. Vaudevange, Phys. Rev. D75: 123511 (2007), arXiv:hep-th/0612197.
21. M. Shaposhnikov and I. Tkachev, Phys. Lett. **B639**, 414 (2006).
22. V. Belinsky, L. Grishuk, I. Khalatnikov and Ya. Zel'dovich, Sov. Phys. JETP **89**, 346 (1985).

23. L. Kofman, A. Linde and A. Starobinsky, Phys. Lett. **B157**, 361 (1985).
24. G. Gibbons, S. Hawking and J. Stewart, Nucl. Phys. **B281**, 609 (1987).
25. L. Kofman and A. Linde, In preparation.
26. S. Hawking and D. Page, Nucl. Phys. **B298**, 789 (1988).
27. G. Gibbons and N. Turok, arXiv:hep-th/0609095.

3

Particle Physics Models of Inflation

David H. Lyth[1]

Physics Department, Lancaster University, Lancaster LA1 4YB, UK

Abstract. Inflation models are compared with observation on the assumption that the curvature perturbation is generated from the vacuum fluctuation of the inflaton field. The focus is on single-field models with canonical kinetic terms, classified as small- medium- and large-field according to the variation of the inflaton field while cosmological scales leave the horizon. Small-field models are constructed according to the usual paradigm for beyond Standard Model physics

3.1 Introduction

Several different types of inflation model have been proposed over the years. In this survey they are compared with observations on the assumption that the curvature perturbation is generated during inflation. The survey is based on works with my collaborators, in particular [1, 2, 3, 4].

I focus largely on the slow-roll paradigm, because it is the simplest and most widely considered possibility. It assumes that the energy density and pressure dominated by the scalar field potential V, whose value hardly varies during one Hubble time. Unless otherwise stated, we consider single-field inflation, where just one canonically normalized "inflaton" field ϕ has significant time-dependence.

In the vacuum, $V = 0$. To generate the inflationary value of V, one or more fields must be strongly displaced form the vacuum and there are two simple possibilities. In non-hybrid inflation, V is generated almost entirely by the displacement of the inflaton field from its vacuum, while in hybrid models it is generated almost entirely by the displacement of some other field χ, called the waterfall field because its eventual descent to the vacuum is supposed to be very rapid. Hybrid models are not at all artificial, being based on the concept of spontaneous symmetry breaking and restoration which is ubiquitous in early-universe cosmology.

The first slow-roll model, termed New Inflation [5] (see also [6]), was non-hybrid. It made contact with particle physics through the use of a GUT theory,

D. H. Lyth: *Particle Physics Models of Inflation*, Lect. Notes Phys. **738**, 81–118 (2008)
DOI 10.1007/978-3-540-74353-8_3   © Springer-Verlag Berlin Heidelberg 2008

but was quickly seen to generate too big a curvature perturbation [7]. Viable models using a GUT and supersymmetry were developed, including [8] what were later called hybrid inflation models. The models were rather complicated, in part because of a demand that the initial condition for observable inflation is to be set by an era of thermal equilibrium.

It was gradually recognized that prior thermal equilibrium is not necessary. A second strand of model-building, characterized by little contact with particle physics and focusing exclusively on non-hybrid models, began with the proposal of chaotic inflation [9]. Considerable attention was paid to non-Einstein gravity theories, notably the proposal of Extended Inflation [10]. In its original form that proposal is not viable if the inflaton perturbation generates the curvature perturbation [11], though it becomes viable if the curvature perturbation is generated afterward [12].

Following the formulation of a simple hybrid inflation model [13], attention went back to the connection with particle physics and supersymmetry. Almost all proposals for field theory beyond the Standard Model were considered as arenas for inflation model-building, including especially GUTs and the origin of low-energy supersymmetry breaking.

The most recent phase of model-building, beginning in about 2000, is based directly on brane world scenarios. We will consider the prediction of these kind of models without describing their string-theoretic derivation.

3.2 Beyond the Standard Model

We begin with some general ideas about the very early Universe, taking on board current thinking about what may lie beyond the Standard Model of particle physics. Guided by the desire to generate primordial perturbations from the vacuum fluctuation of scalar fields, one usually supposes that an effective four-dimensional (4D) field theory applies after the observable Universe leaves the horizon, though not necessarily with Einstein gravity.

To generate perturbations from the vacuum fluctuation we need $|aH|$ to increase with time, which is achieved by inflation defined as an era of expansion with $\ddot{a} > 0$ (repulsive gravity).[1] Perturbations would also be generated from the vacuum during an era of contraction with $\ddot{a} < 0$ The original suggestion was called the pre-Big-Bang [14]. A more recent version where the bounce corresponds to the collision of branes was called the ekpyrotic Universe [15], which was further developed to produce a cyclic Universe [16]. In these scenarios, the prediction for the perturbation depends crucially on what happens at the bounce, which is presently unclear.

Returning to the inflationary scenario, the 4D field theory which is supposed to be valid from observable inflation onwards cannot apply back to an

---

[1] As usual $a(t)$ is the scale factor of the Universe and $H \equiv \dot{a}/a$ is the Hubble parameter.

indefinitely early era. The point at which it breaks down is a matter of intense debate at present. With Einstein gravity, 4D field theory cannot be valid if the energy density exceeds the Planck scale $m_{Pl} \equiv (8\pi G)^{-1/2} = 2.4 \times 10^{18}$ GeV. This is because quantum physics and general relativity come into conflict at that scale, making it the era when classical spacetime first emerges. More generally, it is supposed that any field theory will be just an effective one, valid when relevant energy scales are below some "ultra-violet cutoff" Λ. Above the cutoff, the field theory will be replaced either by a more complete field theory, or by a completely different theory which is generally assumed to be string theory.

The measured values of the gauge couplings suggest the existence of a GUT theory, implying that field theory holds at least up to 10^{16} GeV. This has not prevented the community from considering the possibility that field theory fails at a much lower energy. The idea is that 4D spacetime would emerge as an approximation to the 10D spacetime within which string theory is supposed to hold. String theory is formulated in terms of fundamental strings (F strings), but nowadays an important role is supposed to be played by what are called D-p branes (or just D branes) with various space dimensions $0 < p \leq 9$. The electromagnetic, weak and strong forces that we experience might be confined to a particular D-3 brane, while gravity is able to penetrate to the region outside known as the bulk. An important role may be played by D strings, which are D branes with just one of our space dimensions.

3.3 The Initial Condition for Observable Inflation

The models of inflation that we are going to consider apply to at least the last 50 e-folds or so, starting with the exit from the horizon of the observable Universe. One may call this the era of observable inflation, because it is directly constrained by observation through the perturbations which it generates. Assuming Einstein gravity, observable inflation has to take place with energy density $\rho \lesssim (10^{-2} m_{Pl})^4$ or primordial gravitational would have been detected.

The era before observable inflation is not directly accessible to observation, but one may still ask about that era. In particular one may ask how the inflaton field arrives at the starting point for observable inflation. Though not compulsory, it normally is imagined that inflation begins promptly with the emergence of 4D spacetime. This is indeed desirable for two reasons. One is to prevent the observable Universe from collapsing if the density parameter Ω is initially bigger than 1 (without being fine-tuned to a value extremely close to 1). The other, which applies also to the case $\Omega < 1$, is that inflation protects an initially homogeneous region from invasion by its inhomogeneous surroundings. This is because the event horizon which represents the farthest distance that an inhomogeneity can travel, is finite during inflation. If the onset of inflation is significantly delayed, one would need either a huge initially homogeneous patch or [17] a periodic universe. In contrast, if inflation begins promptly with the emergence of 4D spacetime, the initially homogeneous region is safe provided only that it is bigger than the event horizon. For almost-exponential inflation the event horizon is of order the Hubble distance.

A simple hypothesis about the emergence of 4D spacetime was made in [9]. Working in the context of Einstein gravity, the energy density of the Universe at the Planck scale is supposed to be dominated by scalar fields, with the potential in some regions of order m_{Pl}^4 and flat enough for inflation to occur there. This setup was termed chaotic inflation, and as an example the potentials $V(\phi) \propto \phi^2$ and ϕ^4 were considered. These are generally called chaotic inflation potentials, but the proposal of [9] regarding the initial condition does not rely on a specific form for the potential. It *is* necessary though that there are regions of field space where the potential is at the Planck scale and capable of inflating. No example of such a potential has been derived from string theory.

An alternative to the chaotic inflation proposal is that inflation begins at the top of a hill in the potential, whose height is much less than m_{Pl}^4. In particular, the height could be $\lesssim (10^{16}\,\mathrm{GeV})^4$, allowing observable inflation to take place near the hilltop. This proposal is viable even if the process by which the field arrives at the hilltop is very improbable (such as the process of quantum tunnelling through a potential barrier), because inflation starting sufficiently near the hilltop gives what is called   eternal inflation [18, 19].

During eternal inflation, the volume of the inflating region grows indefinitely, and it can plausibly be argued that this indefinitely large volume outweighs any finite initial improbability. Taking into account the quantum fluctuation, it can be shown [20] that eternal inflation takes place near a hilltop provided that $|\eta| < 6$ where $\eta \equiv V''/3H^2$.

Eternal inflation near a hilltop has been called topological eternal inflation [21]. More generally, eternal inflation occurs whenever the potential over a sufficient range satisfies

$$\left(\frac{H^2}{2\pi\dot\phi_{\mathrm{class}}}\right)^2 = \frac{1}{12\pi^2}\frac{V^3}{m_{\mathrm{Pl}}^6 V'^2} > 1 \,. \tag{3.1}$$

Here $\dot\phi_{\mathrm{class}} = -V'/3H$ is the slow-roll approximation, excluding the stochastic [22] quantum fluctuation $H/2\pi$ per Hubble time. When the left hand side of (3.1) is bigger than 1, the fluctuation dominates so that it can overcome the slow-roll behaviour for an indefinitely long time during which eternal inflation occurs. In the opposite regime, the fluctuation is small and the left hand side of (3.1) becomes the spectrum of the curvature perturbation. Eternal inflation occurs with the chaotic inflation potential $V \propto \phi^p$, for sufficiently large field values [23].

Eternal inflation provides a realization of the multiverse idea, according to which all possible universes consistent with fundamental theory (nowadays, string theory) will actually exist [19, 23]. This is because eternal inflation can be of indefinitely long duration, allowing time for tunnelling to all local minimal of the scalar field potential.

3.4 Slow-roll inflation

3.4.1 Basic Equations

We will find it useful to classify the models according to the variation $\Delta\phi$ of the inflation field after the observable Universe leaves the horizon. We will call a model small-field if $\Delta\phi \ll m_{\rm Pl}$, medium-field if $\Delta\phi \sim m_{\rm Pl}$ and and large-field if $\Delta\phi \gg m_{\rm Pl}$. Hybrid inflation models are usually constructed to be of the small-field type, the idea being to make close contact with particle physics which is hardly possible for medium- and large-field models.

The inflaton field equation is

$$\ddot{\phi} + 3H\dot{\phi} + V'(\phi) = 0 \,. \tag{3.2}$$

Except near a maximum of the potential (or minimum in the case of hybrid inflation) a significant amount of inflation can hardly occur unless this equation is well-approximated by

$$3H\dot{\phi} \cong -V' \,, \tag{3.3}$$

with the energy density $3m_{\rm Pl}^2 H^2 = V + \frac{1}{2}\dot{\phi}^2$ slowly varying on the Hubble timescale:

$$\dot{H} \ll H^2 \,. \tag{3.4}$$

Equations (3.3) and (3.4) together define the slow-roll approximation, and we will use "$\cong$" to denote equalities which become exact in that approximation.

Consistency of (3.3) with the exact equation requires

$$3m_{\rm Pl}^2 H^2 \cong V \,. \tag{3.5}$$

and the flatness conditions

$$\epsilon \ll 1 \qquad |\eta| \ll 1 \,, \tag{3.6}$$

where

$$\epsilon \equiv \frac{1}{2}m_{\rm Pl}^2(V'/V)^2 \qquad \eta \equiv m_{\rm Pl}^2 V''/V \tag{3.7}$$

Requiring that successively higher derivatives of the two sides of (3.3) are equal to good accuracy gives more flatness conditions involving more slow-roll parameters. The first two are

$$|\xi^2| \ll 1, \qquad \xi^2 \equiv m_{\mathrm{Pl}}^4 \frac{V'(\mathrm{d}^3V/\mathrm{d}\phi^3)}{V^2}, \tag{3.8}$$

$$|\sigma^3| \ll 1, \qquad \sigma^3 \equiv m_{\mathrm{Pl}}^6 \frac{V'^2(\mathrm{d}^4V/\mathrm{d}\phi^4)}{V^3}. \tag{3.9}$$

The general expression is

$$|\beta_{(n)}^n| \ll 1, \qquad \beta_{(n)}^n \equiv m_{\mathrm{Pl}}^{2n} \frac{V'^{n-1}(\mathrm{d}^{n+1}V/\mathrm{d}\phi^{n+1})}{V^n}, \tag{3.10}$$

but only ξ^2 and σ^3 are ever invoked in practice.

It is obvious that these additional parameters can have either sign. The motivation for writing them as powers comes from some simple forms for V, which make $|\xi|$, $|\sigma|$ and $|\beta_{(n)}|$ at most of order η. For more general potentials one can check case-by-case how small are ξ^2 and σ^3. Usually there is at least a hierarchy

$$\eta \gg \xi^2 \gg \sigma^3 \ldots, \tag{3.11}$$

but slow-roll per se requires only that all of the slow-roll parameters are $\ll 1$ and does not require any hierarchy.

A convenient time variable is $N(t)$, the number of e-folds of expansion occurring after some initial time, given by $\mathrm{d}N = -H\mathrm{d}t$. In the slow-roll approximation

$$H' \cong -\epsilon H \tag{3.12}$$

$$\epsilon' \cong 2\epsilon(2\epsilon - \eta) \tag{3.13}$$

$$\eta' \cong 2\epsilon\eta - \xi^2, \tag{3.14}$$

$$\xi' \cong 4\epsilon\xi^2 - \eta\xi^2 - \sigma^3, \tag{3.15}$$

and so on, where a prime denotes $\mathrm{d}/\mathrm{d}N$. The first relation says that almost-exponential occurs. The second relation says that ϵ varies slowly. slow-roll does not guarantee that the other parameters are slowly varying, though this is guaranteed in the usual case that the hierarchy (3.11) holds.

The flatness conditions are obtained by successive differentiations of the slow-roll approximation. Strictly speaking, a differentiation might incur large errors so that η or higher slow-roll parameters fail to be small [compared with (3.1)]. In practice though one expects at least the first few slow-roll parameters to be small.

3.4.2 Number of e-Folds

To obtain the predictions, one needs the scale $k(\phi)$ leaving the horizon when ϕ has a given value. The number of e-folds from then until the end of slow-roll inflation at ϕ_{end} is

$$N(k) \cong m_{\mathrm{Pl}}^{-2} \int_{\phi_{\mathrm{end}}}^{\phi} \left(\frac{V}{V'}\right) \mathrm{d}\phi = m_{\mathrm{Pl}}^{-1} \left| \int_{\phi_{\mathrm{end}}}^{\phi} \frac{\mathrm{d}\phi}{\sqrt{2\epsilon(\phi)}} \right|. \tag{3.16}$$

For definiteness we will evaluate the predictions for the biggest cosmological scale $k = a_0 H_0$, where the subscript 0 denotes the present epoch, and denote $N(a_0 H_0)$ simply by N. The prediction for any other scale can be obtained using

$$N(k) = N - \ln(k/H_0) \equiv N - \Delta N(k) . \qquad (3.17)$$

Taking the shortest cosmological scale to be the one enclosing mass $M = 10^4 M_\odot$, those scales span a range $\Delta N = 14$.

The value of N depends on the evolution of the scale factor after inflation. With the maximum inflation scale $V^{1/4} = 10^{16}$ GeV and radiation domination from inflation onwards, $N = 61$. Delaying reheating until $T \sim$ MeV, with matter domination before that, reduces this by 14. With the maximum inflation scale it is therefore reasonable to adopt as an estimate

$$N = 54 \pm 7 , \qquad (3.18)$$

Reducing the inflation scale reduces N by $\ln(V^{1/4}/10^{16} \text{ GeV})$, and the lowest scale usually considered is 10^{10} GeV or so, reducing the above central value to 40.

Based on this discussion it seems fair to say that the fractional uncertainty in N is likely to be at most of order 20%. As we shall see, the corresponding uncertainties in the predictions are of the same order in a wide range of models. On the other hand, a very low inflation scale and/or Thermal Inflation [24] could reduce N by an indefinite amount. The only absolute constraint is $N > 14$, required so that perturbations are generated on all cosmological scales. Also, a long era of domination by the kinetic term of a scalar field (kination), corresponding to $P = \rho$, could increase the estimate [25] by up to 14. Taking all of that on board the maximum range would be $14 < N < 75$.

In non-hybrid models, ϵ usually increases with time and inflation ends when one of the flatness conditions fails, after which ϕ goes to its vacuum expectation value (vev). From its definition, ϵ increasing with time corresponds to $\ln V$ being concave-downward. In this case, the value of ϕ_* obtained from (3.16) will typically be insensitive to ϕ_{end}, making the model more predictive.

In some hybrid models, ϵ decreases with time ($\ln V$ concave-upward), and inflation ends only when the waterfall field is destabilized. In other hybrid inflation models though, ϵ increases with time ($\ln V$ concave-downward), and *slow-roll* inflation may end before the waterfall field is destabilized through the failure of one of the flatness conditions. If that happens, a few more e-folds of inflation can take place while the inflaton oscillates about its vev (locked inflation [26]), until the amplitude of the oscillation becomes low enough to destabilize the waterfall field.

3.4.3 Predictions

The vacuum fluctuation of the inflaton generates a practically gaussian perturbation, with spectrum $\mathcal{P}_\phi(k) = (H_k/2\pi)^2$ where the subscript k indicates horizon exit $k = aH$. This perturbation generates a time-independent curvature perturbation with spectrum [7]

$$\mathcal{P}_\zeta(k) = \frac{1}{24\pi^2 m_{\rm Pl}^4} \frac{V_k}{\epsilon_k} . \qquad (3.19)$$

The error in this estimate will come from the error in $\mathcal{P}_\phi$ and the slow-roll approximation. Both are expected to give a small fractional error, of order $\max\{\epsilon, \eta\}$. Differentiating with respect to $\ln k$ to get the spectral index may incur a fractional error $\gtrsim 1$ if η is rapidly varying [27], but that is not the case in the usual models. Differentiating (3.19) using (3.12) and (3.13) give the spectral tilt;

$$n - 1 \equiv \frac{d \ln \mathcal{P}_\zeta}{d \ln k} = 2\eta_k - 6\epsilon_k . \qquad (3.20)$$

If in addition $d\eta/dN$ (equivalently, ξ^2) is slowly varying this may be differentiated again to obtain the running,

$$\frac{dn}{d \ln k} = -16\epsilon\eta + 24\epsilon^2 + 2\xi^2 . \qquad (3.21)$$

Observable inflation can take place near a maximum or minimum of the potential even with the flatness condition $|\eta| \ll 1$ mildly violated to become $|\eta| \sim 1$ (so-called fast-roll inflation [28], though note that $\dot\phi$ is still small making H almost constant).[2] This quite natural possibility would give tilt $|n - 1| \simeq 1$, which is also quite compatible with the original arguments of Harrison [29] and Zel'dovich [30] for $n \sim 1$ and all known environmental arguments. The very small tilt now observed is not required by any general consideration, and a large tilt $n - 1 \sim -0.3$ had previously been considered as a serious possibility to make a critical-density CDM model more viable [11].

During inflation, the vacuum fluctuation generates a primordial tensor perturbation, setting the initial amplitude for gravitational waves which oscillate after horizon entry. The spectrum $\mathcal{P}_T$ of this perturbation is conveniently specified by the tensor fraction $r \equiv \mathcal{P}_T/\mathcal{P}_\zeta$. In the slow-roll approximation [11],[3]

$$r = 16\epsilon = -8n_T , \qquad (3.22)$$

where $n_T \equiv d \ln \mathcal{P}_T/d \ln k$. The second relation has become known as the consistency condition, and its violation would show that the curvature perturbation is not generated by a single-field slow-roll inflation.

Using the observed value for the spectrum of the curvature perturbation, the tensor fraction is given by

$$r = \left(\frac{V^{1/4}}{3.3 \times 10^{16}\,{\rm GeV}} \right)^4 . \qquad (3.23)$$

---

[2] Very close to a maximum is the regime of eternal inflation, which presumably precedes fast- or slow-roll inflation.

[3] The definition of r in this reference was slightly different.

The tensor fraction can also be related to $\Delta\phi$. Suppose that slow-roll persists to almost the end of inflation and that $\ln V$ is concave-downward throughout. Then $|V/V'|$ is continuously increasing, and (3.16) gives $2\epsilon < N^{-2}(\Delta\phi/m_{\rm Pl})^2$. This can be written [3, 31]

$$16\epsilon = r < 0.003 \left(\frac{50}{N}\right)^2 \left(\frac{\Delta\phi}{m_{\rm Pl}}\right)^2. \qquad (3.24)$$

Now suppose instead that slow-roll persists to the end of inflation, without any requirement on the shape of the potential. As a consequence of slow-roll, ϵ varies little during one Hubble time and there are only 50 or so Hubble times. It follows that one may expect ϵ to be at least roughly constant, in which case the right hand side of (3.24) provides at least a rough estimate of the actual value of $\Delta\phi$.

Finally, let us adopt the most conservative possible position and consider just the change $\Delta\phi_4$ during the four e-folds after the observable Universe leaves the horizon, that being the era when an observable tensor perturbation may actually be generated. Then it is certainly safe to assume that ϵ has negligible variation, leading to the quite firm estimate

$$r \simeq \frac{1}{2}\left(\frac{\Delta\phi_4}{m_{\rm Pl}}\right)^2. \qquad (3.25)$$

In Fig. 3.1, the r–n plane is divided into three regions, according to whether V and $\ln V$ are concave-upward or concave-downward while cosmological scales leave the horizon. Figure 3.2 repeats the plot in the $\ln r$–n plane.

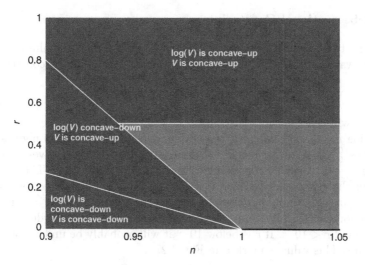

Fig. 3.1. The r–n plane is divided into three regions, according to whether V and $\ln V$ are concave-upward or concave-downward while cosmological scales leave the horizon.

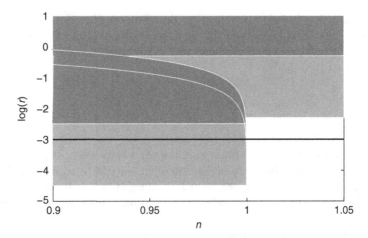

Fig. 3.2. The plot of Figure 1 is repeated in the $\ln r$-n plane

If the concave-upward -downward behaviour persists till the end of slow-roll inflation, the right-hand region is inhabited exclusively by hybrid inflation models, since otherwise inflation would never end. With that assumption, (3.24) and (3.25) imply that the lightly shaded region of the figures is excluded if $\Delta\phi > 0.1m_{\rm Pl}$, and that the heavily shaded region region is excluded if $\Delta\phi > m_{\rm Pl}$. [In the right-hand region, corresponding to concave-upward $\ln V$, we used (3.25) with $\Delta\phi_4 = \Delta\phi$; the actual bound will be tighter since in reality $\Delta\phi_4 < \Delta\phi$.]

3.4.4 Observational Constraints

According to observation [32] value of the spectrum $\mathcal{P}_\zeta$ has the almost scale-invariant value $(5 \times 10^{-5})^2$, with negligible error. This gives the constraint

$$V^{1/4}/\epsilon^{1/4} = 0.027m_{\rm Pl} = 6.6 \times 10^{16}\,{\rm GeV}, \qquad (3.26)$$

which we will call the cmb constraint.

Setting $r = 0$ and taking n to be scale-independent, observation gives [32] $n \simeq 0.948^{+0.015}_{-0.018}$. Allowing r and a scale-independent $dn/d\ln k$ gives a higher n and $n' \simeq -0.10 \pm 0.05$, consistent with no running at 2σ level. The allowed region in the r–n plane is shown in Fig. 3.3. (This is a corrected version of the figure in [32], kindly supplied by the authors). The bound $r = 0$ is seen to apply for $r \ll 0.1$.[4] Within a few years there will be either a detection of r or a bound $r < 10^{-2}$. If r is below 10^{-3} it will probably be undetectable by any means. This value is marked in Fig. 3.2.

---

[4] The 1-σ limit with r set equal to zero is tighter than the limit read off from setting $r = 0$ in the r–n plot, because the joint probability distribution is non-gaussian.

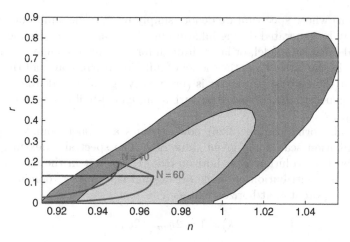

Fig. 3.3. The *closed areas* show the regions allowed by observation at 66 and 95% confidence levels. The *curved lines* are the Natural Inflation predictions for $N = 20$ and $N = 75$, and the *horizontal lines* are the corresponding multi-field Chaotic Inflation predictions. The junction of each pair of lines corresponds to single-field Chaotic Inflation

From all this, we see that small- and medium-field generally give $r \lesssim 10^{-2}$. This means that the predicted tensor fraction is unlikely to be observed. It also means that the prediction for the spectral tilt can be taken as simply $n - 1 = 2\eta$; to reproduce the observed negative tilt the potential of a small- or medium-field model should be concave-downwards while cosmological scales leave the horizon.

3.4.5 Beyond the Standard Paradigm

Throughout we have adopted the standard paradigm, whereby the curvature perturbation ζ is generated by the inflaton perturbation in a single-field slow-roll inflation model. In general there will exist other light fields, each possessing a perturbation with the nearly flat spectrum $(H/2\pi)^2$, any one of which might be responsible for the curvature perturbation.

The predictions in this more general scenario are best calculated through the δN formalism [33, 34, 35, 36, 37]. As our main focus is on the standard paradigm we just give some basic results without derivation. It is convenient to use at horizon exit a field basis $\{\phi, \sigma_i\}$, where ϕ points along the inflaton trajectory and the σ_i ($i = 2 \ldots M$) are orthogonal. The perturbation $\delta\phi$ then generates the same time-independent curvature perturbation as in the single-field case, whose spectrum we denote by $\mathcal{P}_{\zeta_\phi}$. The orthogonal perturbations give no contribution to the curvature at horizon exit, but one or more of them may generate an additional contribution later which may be dominant by the time that the curvature perturbation settles down to the final time-independent value (obtaining as cosmological scales start to approach the horizon) whose spectrum we denote simply by $\mathcal{P}_\zeta$. The additional contribution may be generated during inflation in which case we are dealing with a multi-field inflation model, or later through for example the curvaton mechanism [38]. In the latter case, the model of inflation is irrelevant; all that matters is that the Hubble parameter is slowly varying. Liberated from the constraint to generate the curvature perturbation, model-building becomes much easier [39].

The cmb normalization (3.26) now becomes an upper bound, implying a lower inflation scale for a given value of ϵ. The spectral index in general depends on the evolution after horizon exit [1, 34], but in the most natural case that the contribution of single orthogonal field $\sigma \equiv \sigma_i$ dominates it is given by the potential at horizon exit as

$$n(k) - 1 = 2\eta_{\sigma\sigma} - 2\epsilon , \tag{3.27}$$

where $\eta_{\sigma\sigma} = \partial^2 V/\partial\sigma^2$. (The case that two contributions are comparable may arise by accident, or in special models where ϕ and an orthogonal field are related such as the one involving axion physics which is described in [40].)

Since the tensor perturbation depends only on H the tensor fraction r is reduced;

$$r = 16\epsilon \frac{\mathcal{P}_{\zeta\phi}}{\mathcal{P}_\zeta} < 16\epsilon . \tag{3.28}$$

It is negligible if an orthogonal contribution dominates.

We did not mention non-gaussianity. According to the standard paradigm, the non-gaussianity is [41] about 100 times smaller than the level that can be detected from the CMB anisotropy (and/or galaxy surveys) though it has recently been claimed [42] that a measurement from the 21-cm anisotropy might be possible. In contrast, non-standard paradigms may easily generate non-gaussianity at an observable level; in particular the curvaton and inhomogeneous reheating scenarios are expected to generate non-gaussianity at a level that is at least marginally observable through the cmb. If non-gaussianity is observed we will be dealing with functions (of rotationally invariant scalars formed from the wave-vectors that define the bispectrum, trispectrum etc.) as opposed to numbers, which will provide powerful information about the origin of the curvature perturbation.

All of this assumes slow-roll inflation. That possibility is compatible with the simultaneous detection of a tensor perturbation *and* non-gaussianity only if some orthogonal field can generate the non-gaussianity without being dominant (a highly constrained scenario [43]). The main alternative to slow-roll inflaton seems to be inflation with non-quadratic kinetic terms, called k-inflation [44], of which special forms are the brane world DBI inflation scenario [45] and ghost inflation [46].

3.5 Modular Inflation

We begin our survey of inflation models with the most plausible medium-field model, which goes by the name of modular inflation. This is a non-hybrid model in which the inflaton is a modulus. It was suggested a long time ago [47], and its possible realization in the context of brane worlds is under investigation at present.

Moduli may play other roles too in the early Universe, and we describe their properties before getting to the inflation model. For the present purpose a modulus may be defined as a field with a potential of the form

$$V = V_0 f\left(\frac{\phi}{m_{\rm Pl}}\right) , \qquad (3.29)$$

This is supposed to hold in the range $0 < \phi \lesssim m_{\rm Pl}$, with the function $f(x)$ and its low derivatives of order 1 at a generic point. At the vev, where f and f' vanish, the mass-squared $m^2 \equiv V''$ is typically of order $V_0/m_{\rm Pl}^2$. If the potential has a maximum, it will typically be located at a distance of order $m_{\rm Pl}$ from the vev with the tachyonic mass-squared V'' typically of order $-m^2$.

Fields with this property are expected (though not inevitable) in a field theory derived from string theory. Usually the field theory is taken to be supersymmetric though moduli are expected anyway. Moduli are usually supposed to have interactions of only gravitational strength, corresponding to a lifetime $\Gamma \sim m^3/m_{\rm Pl}^2$. Alternatively though, a modulus may have interactions of ordinary strength, in particular gauge interactions. The fixed point of the symmetry group is then called a point of enhanced symmetry. Such a point might correspond to either the vev or to a maximum of the potential. It may even be possible for both of these to be points of enhanced symmetry, involving different symmetry groups.

Moduli may affect cosmology in several ways. Usually they are considered in the context of supersymmetry, and the simplest expectation for the mass is then $m \sim$ TeV, corresponding to what we may call light moduli. A light modulus is typically displaced strongly from its vev during inflation, by an amount which puts its subsequent oscillation and gravitational-strength decay into conflict with nucleosynthesis. To avoid this "moduli problem" one may suppose that all moduli are heavy, or that there is Thermal Inflation [24].

Now we turn to modular inflation. It is usually supposed to take place near a maximum or saddle-point of the potential, with just one modulus ϕ varying significantly. As many moduli typically exist, that may not be easy to arrange. Supposing that it happens let us set $\phi = 0$ at the maximum and consider the power series for the potential. The generic expectation would be for the quadratic term alone to provide at least a crude approximation to the potential in the slow-roll regime, corresponding to

$$V(\phi) = V_0 \left(1 + \frac{1}{2}\eta_0 \frac{\phi^2}{m_{\rm Pl}^2}\right) \qquad (3.30)$$

But this requires [from (3.29)] roughly $\eta_0 \sim -1$ which gives spectral tilt $n - 1 \sim -1$ in contradiction with observation. To provide a modular inflation model one suppresses the quadratic term, either by means of a symmetry [48] or more usually by fine tuning (see for instance [49]).

If the suppressed quadratic term is still required to dominate while cosmological scales leave the horizon, one obtains the scale-independent prediction $n = 1 + 2\eta_0$ which can agree with observation by choice of η_0. This prediction is scale-independent which might in the future allow it to be distinguished from other predictions for n. Of course, one has to invoke additional terms to end inflation, presumably at a value $\phi_{\text{end}} \sim m_{\text{Pl}}$. The tensor fraction is

$$r = 2 \left(\frac{\phi_{\text{end}}}{m_{\text{Pl}}} \right)^2 (1 - n)^2 e^{-N(1-n)} \sim 10^{-3.5} \left(\frac{\phi_{\text{end}}}{m_{\text{Pl}}} \right)^2 . \tag{3.31}$$

Taking $\phi_{\text{end}} \sim m_{\text{Pl}}$ gives the result shown in Fig. 3.10. The tensor fraction is unobservable, but corresponds to a high normalization scale $V^{1/4} \sim 10^{15}\,\text{GeV}$, meaning that we are not dealing with a light modulus.

It is more reasonable to suppose that the suppressed quadratic term is negligible. Then, as a rough approximation it may be reasonable to write

$$V \simeq V_0 \left[1 - \left(\frac{\phi}{\mu} \right)^p \right] , \tag{3.32}$$

with $p \gtrsim 3$ (not necessarily an integer) and $\mu \sim m_{\text{Pl}}$.

If this approximation holds for some reasonable length of time after cosmological scales leave the horizon it gives

$$\phi_*^{p-2} = \left[p(p - 2)\mu^{-p} N m_{\text{Pl}}^2 \right]^{-1} , \tag{3.33}$$

(independently of ϕ_{end}) and

$$n - 1 = -\frac{2}{N} \left(\frac{p - 1}{p - 2} \right) . \tag{3.34}$$

For the range $3 < p < \infty$ with $N = 50$ we get $0.92 < n < 0.96$. The cmb normalization corresponds to a tensor fraction

$$r \simeq \frac{0.001}{(p - 2)^4} \left(\frac{\mu}{m_{\text{Pl}}} \right)^{\frac{p}{2p-4}} \left(\frac{50}{N} \right)^{\frac{2(p-1)}{p-2}} . \tag{3.35}$$

This is shown in Fig. 3.10 with $\mu = m_{\text{Pl}}$. Again, the tensor fraction is too small to detect but still corresponds to a high energy scale $V^{1/4} \sim 10^{15}\,\text{GeV}$. These estimates agree to rough order of magnitude with results obtained numerically using potentials derived from string theory (see for instance [49]).

3.6 Small-Field Models

A range of small-field models has been proposed. Before describing them we make some general remarks, followed by a very basic treatment of supersymmetry which is invoked in most small-field models.

The motivation for small-field models comes from ideas about what is likely to lie beyond the Standard Model of particle physics. Choosing the origin as the fixed point of the relevant symmetries, the tree-level potential will have a power series expansion,

$$V(\phi) = V_0 \pm \frac{1}{2}m^2\phi^2 + M\phi^3 + \frac{1}{4}\lambda\phi^4 + \sum_{d=5}^{\infty} \lambda_d m_{\mathrm{Pl}}^4 \left(\frac{\phi}{m_{\mathrm{Pl}}}\right)^d \qquad (3.36)$$

The lower-order terms of (3.36), which do not involve m_{Pl}, are renormalizable terms (corresponding to a renormalizable quantum field theory). The higher order terms, which disappear in the limit $m_{\mathrm{Pl}} \to \infty$, are non-renormalizable terms. We are taking m^2 positive and as indicated the quadratic term might have either sign. The other renormalizable terms will usually be positive, but the non-renormalizable terms might have either sign.

According to a widely held view, non-renormalizable terms of arbitrarily high order are expected, with magnitudes big enough to place this expansion out of control at $\phi \gtrsim m_{\mathrm{Pl}}$. The typical expectation is $|\lambda_d| \sim 1$ if m_{Pl} is the ultra-violet cutoff and $|\lambda_d| \sim (m_{\mathrm{Pl}}/\Lambda)^d$ (the latter corresponding to the replacement $m_{\mathrm{Pl}} \to \Lambda$) if the cutoff Λ is smaller. This view is part of a more general one, according to which the lagrangian of a field theory ought to contain all terms that are allowed by the symmetries, with coefficients typically of order 1 in units of the ultra-violet cutoff (see for instance [50]).

If the field theory is replaced by a more complete field theory above the cutoff, the λ_d can be calculated and will be of the advertised order of magnitude if ϕ has unsuppressed interactions. But if instead it is replaced by string theory above the cutoff, then estimates of λ_d should come from string theory. Such estimates are at present not available, except for moduli.[5] In general then, one is free to accept or not the usual view about non-renormalizable terms.[6]

Following [1], let us see what sort of conditions the terms in (3.36) must satisfy, to achieve inflation in the small-field regime $\phi \ll m_{\mathrm{Pl}}$. We discount the possibility of extremely accurate cancellations between different terms.

---

[5] In the case of moduli (3.29) implies a strong suppression of the couplings. However, the inflaton in a small-field model is not usually supposed to be a modulus because the origin in small-field models is usually taken to be the fixed point of the symmetries of some unsuppressed interactions, which would make the origin a point of enhanced symmetry for the modulus.

[6] This is less true if supergravity is invoked because the non-renormalizable terms are then present and out of control for generic choices of the functions defining the theory. But one can still make special choices to avoid the problem.

This means that the constant term has to dominate, and that we require the addition of any one other term to respect the flatness condition $|\eta| \ll 1$, the other flatness conditions then being automatic.[7]

We shall not consider the cubic term, which usually is forbidden by a symmetry. For the other terms $|\eta| \ll 1$ is equivalent to

$$m^2 \ll \frac{V_0}{m_{\mathrm{Pl}}^2} \simeq 3H_*^2 \tag{3.37}$$

$$\lambda \ll \frac{V_0}{m_{\mathrm{Pl}}^4} \frac{m_{\mathrm{Pl}}^2}{\phi^2} \tag{3.38}$$

$$\lambda_d \ll \frac{V_0}{m_{\mathrm{Pl}}^4} \left(\frac{m_{\mathrm{Pl}}^2}{\phi^2} \right)^{\frac{d-2}{2}}. \tag{3.39}$$

One might think that the second and third conditions can always be satisfied by making ϕ small enough, but this is not correct because there is a lower limit on the variation of ϕ. Indeed, during just the ten or so e-folds while cosmological scales leave the horizon (3.16) and (3.26) require ϕ to change by at least $10^4 V^{1/2}/m_{\mathrm{Pl}}$ and ϕ cannot be smaller than that on all such scales. We conclude that

$$\lambda \lesssim 10^{-8} \tag{3.40}$$

$$\lambda_d \lesssim 10^{-8} \left(\frac{10^{16}\,\mathrm{GeV}}{V_0^{1/4}} \right)^{2(d-4)}. \tag{3.41}$$

The first condition requires λ to be very small, and the second condition requires at least the first few λ_d to be very small unless the inflation scale is well below $10^{16}\,\mathrm{GeV}$. Supersymmetry can ensure these conditions, either by itself or combined with an internal symmetry. Alternatively one can invoke just an internal symmetry corresponding to $\phi \to \phi +$ const, making ϕ a PNGB, though as we remark later that is not so easy to arrange as one might think.

Finally, we recall that for a generic field in an effective field theory, m_{Pl} in (3.36) might be replaced by an ultra-violet cutoff $\Lambda_{\mathrm{UV}} < m_{\mathrm{Pl}}$, arising either because heavy fields have been integrated out, or because large extra dimensions come into play. One hopes that such a thing does not happen for the inflaton field, because it would make it more difficult to satisfy the flatness conditions [53]. Fortunately, the presence of large extra dimensions does not

---

[7] In a supersymmetric theory one instead consider A-term inflation [51, 52]. Dropping the constant term V_0, one can choose a flat direction (say in the space of the MSSM scalars) in which the leading non-renormalizable term in the superpotential generates an A-term. Then a fine-tuned match between three terms in the potential can give $V' = V'' = 0$ for a particular field value. Inflation can then take place near that value and naturally reproduce the cmb normalization. By a suitable choice of the fine-tuning it can also reproduce the observed spectral index, though it can also give any value in the slow-roll range $0 \lesssim n \lesssim 2$ [52].

in itself prevent m_{Pl} from being the effective cutoff for at least some of the fields.

3.7 Supersymmetry: General Features

Field theory beyond the Standard Model is usually required to possess supersymmetry. Supersymmetry [54] is an extension of Lorentz invariance. Its outstanding prediction is that each fermion should have bosonic superpartners, and vice versa, with identical mass and couplings in the limit of unbroken supersymmetry. Supersymmetry has to be broken in our Universe.

Supersymmetry is usually taken to be a local symmetry, and is then called supergravity because it automatically incorporates gravity.[8] In that case the breaking is spontaneous. In many situations, global supersymmetry is used with the expectation that it will provide a good approximation to supergravity. In that case the breaking can be spontaneous and/or explicit.

We shall deal with the simplest version of supersymmetry, known as $N = 1$ supersymmetry, which alone seems able to provide a viable extension of the Standard Model. Here, each spin-half field is paired with either a complex spin-zero field (making a chiral supermultiplet), or else with a gauge boson field (making a gauge supermultiplet). With supergravity, the graviton (spin two) comes with a gravitino (spin 3/2). With spontaneously broken global supersymmetry there is instead a spin 1/2 goldstino.

One motivation for supersymmetry concerns the mass of the Higgs particle, given by the vev of $\partial^2 V/\partial\phi^2$ where ϕ is the Higgs field. The function V that we have up till now being calling simply the potential is only an effective one, and not the "bare" potential entering into the lagrangian which defines the field theory. Interactions of the scalar fields with themselves and each other change the bare potential into an effective potential. We will be concerned with perturbative quantum effects represented by Feynman diagrams. If we including just tree-level (no-loop) diagrams, the effective potential is still given by the power series (3.36) with different (renormalized) values for the coefficients in the series. Loop corrections give further renormalization of the coefficients, which is our immediate concern. (They also give the potential logarithmic terms that have to be added to the power series, which we come to later.)

The point now is that the loop "correction" in a generic field theory will be large, driving the physical mass up to a value of order the ultra-violet cutoff. As the latter is usually supposed to be many orders of magnitude above the physical Higgs mass, one must in the absence of supersymmetry fine-tune the bare mass so that it almost exactly cancels the loop correction. To protect the Higgs mass from this fine tuning, one needs to keep the loop correction under

---

[8] Some brane world scenarios explicitly break local supersymmetry which means there is actually explicitly broken global supersymmetry.

control by means of a symmetry which would make it zero in the unbroken limit. The best symmetry for doing that job in the case of the Higgs field is supersymmetry.[9]

In a supersymmetric extension of the Standard Model, each particle species must come with a superpartner. It turns out that at least two Higgs fields are then needed. Keeping just two, one arrives at the Minimal Supersymmetric Standard Model (MSSM), which is a globally supersymmetric theory with canonically normalized fields. The partners of the quarks and leptons are called squarks and sleptons, those of the Higgs fields are called higgsinos, and those of the gauge fields are called gauginos.

Unbroken supersymmetry would require that each Standard Model particle has the same mass as its partner. This is not observed, which means that the global supersymmetry possessed by the MSSM must be broken in the present vacuum. To agree with observation it turns out that the breaking has to be explicit as opposed to spontaneous. To ensure that supersymmetry continues to do its job of stabilizing the potential against loop corrections, the breaking must be of a special kind called soft breaking. Soft supersymmetry breaking has to give slepton and squark masses very roughly of order $100\,\text{GeV}$. They cannot be much smaller or they would have been observed, and they cannot be much more bigger if supersymmetry is to do its job of stabilizing the Higgs mass.

Softly broken supersymmetry explains with high accuracy the observed ratio of the three gauge couplings (determining the strengths of the strong, weak and electromagnetic interactions) on the hypothesis that there is a GUT. This feature is actually preserved if one allows the squarks and sleptons to be extremely heavy (hence not observable), a proposal known as Split Supersymmetry.

The LHC will soon determine the nature of the fundamental interactions immediately beyond the Standard Model, and may or may not find evidence for supersymmetry. In the latter case we will know that supersymmetry is too badly broken to be relevant for the Standard Model. It might still be relevant in the early Universe and in particular during inflation, but there is no doubt that increased emphasis will then be placed on non-supersymmetric inflation models. A good candidate for non-supersymmetric inflation would be modular inflation. Alternatively, one might make the inflaton a PNGB, or just accept extreme fine tuning.

---

[9] If a symmetry other than supersymmetry were to be used, the Higgs field ϕ would become a PNGB corresponding to a shift symmetry $\phi \to \phi + \text{const}$. It is difficult for a shift symmetry to protect the Higgs mass, because the symmetry will be broken by the strong couplings that the Higgs is known to possess. This problem can be overcome by what is called the Little Higgs mechanism but the resulting schemes are complicated especially if the ultra-violet cutoff is supposed to be many orders of magnitude bigger than the observed mass.

3.8 Supersymmetry: Form of the Potential

In a supergravity theory, the potential is a function of the complex scalar fields, of the form

$$V(\phi_i) = V_+(\phi_i) - 3m_{\rm Pl}^2 m_{3/2}^2(\phi_i) . \tag{3.42}$$

The first term is positive, and spontaneously breaks supersymmetry.

In the vacuum, $m_{3/2}(\phi_i)$ becomes the gravitino mass which we denote simply be $m_{3/2}$. Let us denote the vev of the first term by $M_{\rm S}^4$. The near-cancellation of the two terms in the vacuum is unexplained (the cosmological constant problem). The explicitly broken global supersymmetry seen in the MSSM sector is supposed to be obtained from the full potential as an approximation. To achieve this the spontaneous breaking must take place in some "hidden sector" with some "messenger" sector communicating (mediating) between the hidden sector and the MSSM sector. The value of $M_{\rm S}$ required to give squark and slepton masses of order $100\,{\rm GeV}$ depends on the strength of the mediation. Let us characterize it by $M_{\rm mess}$, with $100\,{\rm GeV} = M_{\rm S}^2/M_{\rm mess}$. Gravitational-strength mediation ("gravity mediation") corresponds to $M_{\rm mess} \sim m_{\rm Pl}$ and the biggest reasonable range is $10^4\,{\rm GeV} \lesssim M_{\rm mess} \lesssim 10^{12}$.[10] The corresponding gravitino mass is between $1\,{\rm eV}$ and $10^6\,{\rm GeV}$.

Coming to inflation, supersymmetry stabilizes the potential against loop corrections just as in the MSSM Higgs case. Also, the small λ required in the tree-level potential can be obtained quite naturally. One generally assumes that the first term of (3.42) dominates since there is no reason to expect a fine cancellation. Assuming that supersymmetry in the early Universe is broken at least as strongly as in the vacuum, this requires $V \gtrsim M_{\rm S}^4$. Partly for that reason, very low-scale inflation is difficult to achieve.

Now we come to what has been called the η problem. The supergravity potential can be written as the sum of two terms, called the F term and the D term. In most inflation models V comes from the F term. Then, each scalar field typically has mass-squared at least of order $m^2 \gtrsim V/m_{\rm Pl}^2 = 3H^2$. For the inflaton this is in mild conflict with the slow-roll requirement $|\eta| \ll 1$ [55, 56, 57, 58].

Even if we allow the curvature perturbation to be generated after inflation, say in the curvaton model, we still need $m^2 \ll V/m_{\rm Pl}^2$ for the curvaton field. In that case there may be a problem even after inflation, because a generic supergravity theory still gives each scalar field an effective mass at least of order H [59] except during radiation domination [60], which will tend to drive each field to its unperturbed value and kill the curvature perturbation.

---

[10] The upper limit corresponds to anomaly mediation, which is gravity mediation suppressed by a loop factor. The lower limit is an interpretation of $M_{\rm S} \gg 100\,{\rm GeV}$, required so that the hidden sector is indeed hidden.

Returning to the standard scenario for generating the curvature perturbation, we typically need $|\eta| \sim 0.01$ to generate the observed spectral tilt. This represents an order 1% fine-tuning which is not too severe. What is perhaps more serious is that the η problem calls into question the validity of any model which is formulated within the context of global supersymmetry. It is easy to ensure $|\eta| \ll 1$ in such a theory, but having done that the supergravity correction may still be big and completely alter the model. In a typical global supersymmetry model though, the same is true of other types of correction as well.

3.9 One-Loop Correction

Loop corrections add a logarithmic term to the effective potential. In the direction of any field ϕ, the one-loop correction is

$$\Delta V(\phi) = \sum_i \frac{\pm \mathcal{N}_i}{64\pi^2} M_i^4(\phi) \ln \left[\frac{M_i^2(\phi)}{Q^2} \right] . \tag{3.43}$$

This is called the Coleman–Weinberg potential. The sum goes over all particle species, with the plus/minus sign for bosons/fermions, and $\mathcal{N}_i$ the number of spin states. The quantity $M_i^2(\phi)$ is the effective mass-squared of the species, in the presence of the constant ϕ field. For a scalar, $M_i^2 = \partial^2 V/\partial \phi_i^2$, which is valid for ϕ itself as well as other scalars.

The quantity Q is called the renormalization scale. If the loop correction were calculated to all orders, the potential would be independent of Q. In a given situation, Q should be set equal to a typical energy scale so as to minimize the size of the loop correction and its accompanying error. Focusing on the inflaton potential, we should set Q equal to a typical value of ϕ (one within the range which corresponds to horizon exit for cosmological scales). That having being done, the *magnitude* of ΔV will typically be negligible, but its derivatives may easily be significant.

If supersymmetry were unbroken, each spin-1/2 field would have a scalar- or gauge field partner with the same mass and couplings, causing the loop correction to vanish. In reality supersymmetry is broken. To see how things work out, let us consider the loop correction from a chiral supermultiplet, consisting of a spin-1/2 particle with a scalar partner. The partner is a complex field $\psi = (\psi_1 + i\psi_2)/\sqrt{2}$, whose real components ψ_i have true masses m_i. If there is an interaction $\frac{1}{2}\lambda'\phi^2|\psi^2|$, this gives $M_i^2 = m_i^2 + \frac{1}{2}\lambda'\phi^2$ ($i = 1, 2$). (We use the prime to distinguish this coupling from the self-coupling λ in the tree-level potential (3.36) of the inflaton.) The spin-1/2 field typically has true mass $m_f = 0$, and its interaction with ϕ generates an effective mass-squared $M_f^2(\phi) = \frac{1}{2}\lambda'\phi^2$. (This result is not affected by either spontaneous or soft supersymmetry breaking.) When ϕ is much bigger than m_i, the loop correction is therefore

$$\Delta V \simeq \frac{1}{32\pi^2} \left[\sum_{i=1,2} \left(m_i^2 + \frac{1}{2}\lambda'\phi^2 \right)^2 - 2 \left(\frac{1}{2}\lambda'\phi^2 \right)^2 \right] \ln\frac{\phi}{Q} . \tag{3.44}$$

The coefficient of ϕ^4 vanishes by virtue of the supersymmetry. For the other terms, we will consider two cases. Suppose first that global supersymmetry is spontaneously broken during inflation. Then it turns out that typically $m_1^2 = -m_2^2$, causing the coefficient of ϕ^2 in (3.44) to vanish. This leaves

$$\Delta V \simeq \frac{m_1^4}{32\pi^2} \ln\frac{\phi}{Q} . \tag{3.45}$$

In this case the derivatives of ΔV are independent of Q, making its choice irrelevant as the magnitude of ΔV is negligible.

Now suppose instead that global supersymmetry is explicitly (softly) broken during inflation, the coefficient of ϕ^2 in 3.44 does not vanish, but instead typically dominates the constant term. Adding the loop correction to the mass term of the tree-level potential gives

$$\Delta V = \frac{1}{2} \left[m^2 + \frac{\lambda'}{32\pi^2} \left(m_1^2 + m_2^2 \right) \ln\frac{\phi}{Q} \right] \phi^2 . \tag{3.46}$$

This expression is valid over a limited range of ϕ, if Q set equal to a value of ϕ within that range. If a large range of ϕ is under consideration, it should be replaced by an expression of the form

$$\Delta V = \frac{1}{2}m^2(\phi)\,\phi^2. \tag{3.47}$$

The "running mass" $m^2(\phi)$ is calculated from what are called renormalization group equations (RGEs).

The above discussion involved the loop correction due to a chiral super-multiplet. Couplings involving chiral super multiplets, such as λ', are called Yukawa couplings and they can be very small. We could instead have discussed the loop correction due to a gauge supermultiplet, consisting of a spin-1/2 field whose partner is a gauge field. The couplings involving gauge super multiplets are called gauge couplings and denoted usually by g. They are not expected to be very small. The loop correction from a gauge supermultiplet is essentially of the above form, with λ' replaced by g.

Finally, if there is no supersymmetry, the loop correction typically desta-bilizes the tree-level potential, and in particular it gives to the mass of each scalar field a contribution which is typically of order the ultra-violet cutoff. To obtain an acceptable potential, and in particular acceptable masses, one has to invoke a fine-tuned cancellation between the loop correction and the tree-level potential. Considering just the contribution from the spin-1/2 part of 3.44, and adding it to the self-coupling of ϕ, one has

$$\Delta V = \frac{1}{4}\left[\lambda - \left(\frac{\lambda'}{4\pi}\right)^2 \ln\frac{\phi}{Q}\right]\phi^4. \tag{3.48}$$

As with the mass, the RGE's give a more accurate result, corresponding to $\Delta V = \frac{1}{4}\lambda(\phi)\phi^4$ with a running coupling $\lambda(\phi)$.

3.10 Small-Field Models: Moving Away from the Origin

In this section we consider small-field potentials with the shape shown in Fig. 3.4. We begin with non-hybrid models, taking the origin as the fixed point of the symmetries. Then the minimum of the potential corresponds to a nonzero vev, and the potential vanishes there. Such models are usually called New Inflation models, since that was the name given to the first viable slow-roll model which happened to be of that kind.

The situation for New Inflation is similar to the one we discussed for modular inflation. Keeping the quadratic term alone cannot be a good approximation throughout inflation. Assuming that the quadratic term is already negligible when cosmological scales leave the horizon, the approximation (3.32) seems reasonable, with $p \gtrsim 3$ and now $\mu \ll m_{\mathrm{Pl}}$. With this approximation the spectral tilt is given by (3.34). The tensor fraction is given by (3.35) with $\mu \ll m_{\mathrm{Pl}}$ making it absolutely negligible, and allowing an inflation scale far below $10^{15}\,\mathrm{GeV}$.

The original New Inflation model corresponded to $p = 4$;

$$V \simeq V_0 - \frac{1}{4}\lambda\phi^4 + \cdots. \tag{3.49}$$

To be precise, the inflaton was supposed to be the GUT Higgs, taken to be practically massless, whose Mexican–Hat potential was generated by a running coupling coming from the non-supersymmetric Coleman–Weinberg potential. The cmb normalization now requires $\lambda = 3 \times 10^{-13}(50/N)^3$. This ruled out the model in its original form, because λ was the GUT gauge coupling with known magnitude of order 10^{-1}. A viable version of the model was obtained

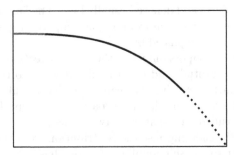

Fig. 3.4. Modular, new, inverted hybrid, mutated hybrid

[61] by declaring that the inflaton is a gauge singlet, making λ a Yukawa coupling whose value can be chosen at will.

Instead of invoking the approximation (3.32), we might suppose that the quadratic term dominates while cosmological scales leave the horizon but a higher term dominates soon afterward. The simplest potential of this kind is

$$V = V_0 - \frac{1}{2}m^2\phi^2 - \frac{1}{4}\lambda\phi^4 + \cdots . \tag{3.50}$$

A supersymmetric realization of this case making close contact with particle physics is given in [62] (see also [1]), which is very fine-tuned if the inflaton is required to generate the curvature perturbation. There is also a non-supersymmetric realization invoking a Little Higgs mechanism [63, 64], making ϕ a PNGB with a periodic potential. The prediction for this model is the same as for (3.30), with the difference that ϕ_{end} will be far below m_{Pl} making the inflation scale far below 10^{16} GeV.

Turning to hybrid inflation, the simplest possibility is inverted hybrid inflation [65] where the origin remains the fixed point of symmetries, and one simply reverses the sign of m^2, m_ψ^2 and λ' in the usual hybrid inflation potential [vord below]. The negative sign of λ' is difficult to arrange especially in a supersymmetric model, and severe fine-tuning is also required [66].

Instead one can make ϕ a PNGB so that it has a periodic potential [63, 64, 67]. The shift symmetry is broken both by the potential $V(\phi)$ and by the coupling of ϕ to the waterfall field. The inflationary trajectory does not pass through the fixed point of the symmetries, and taking the origin to be a maximum of the potential is just an arbitrary choice. Instead of making ϕ a true PNGB, one can arrange that at least it is effectively one during inflation, in the sense that the potential then becomes flat in some well-defined limit [57, 58, 68]. For both types of model it seems possible for the magnitude of the spectrum and the spectral tilt to be in agreement with observation by suitable choice of parameters. The inflation scale can be many orders of magnitude below 10^{15} GeV.

3.11 Moving Toward the Origin; Power-Law Potential

In this section we consider potentials of the form illustrated in Fig. 3.5, of either the small-field or medium-field type. We begin with potentials that can be approximated by (3.32) with $p < 0$. Such potentials give the prediction (3.34) for the spectral index and (3.35) for the tensor fraction.

With $p = -4$, (3.32) has been derived in a brane world scenario, where $\mu \sim m_{\text{Pl}}$ is allowed corresponding to a medium-field model [69]. This is a hybrid inflation model, with the usual potential schematically of the form

$$V(\phi, \chi) = V(\phi) + \frac{1}{2}m^2\phi^2 - \frac{1}{2}m_\chi^2\chi^2 + \frac{1}{2}\lambda'\chi^2\phi^2 + \frac{1}{4}\lambda\chi^4 . \tag{3.51}$$

At $\phi > \phi_c \equiv m_\chi/\sqrt{\lambda'}$ the waterfall field is driven to zero, leaving $V(\phi)$ given by (3.32). The unusually form of $V(\phi)$ here arises because the inflaton field ϕ corresponds to the distance between branes attracted towards each other. Inflation in this model ends when the branes coalesce.

Colliding brane inflation has the usual η problem, in that the potential is expected to have a term $\frac{1}{2}m^2\phi^2$ with $m^2 \sim H^2$. But the brane world scenario can motivate a non-canonical normalization of a specific form, leading to what is called DBI inflation which can take place even with $m^2 \sim H^2$. We shall not present the results for that case.

At the end of this brane world inflation, F and D strings are typically produced. At present it is not clear how that affects the viability of the model, because the evolution of the string network has not been reliably calculated.

The potential (3.32) with various values of p had been derived earlier in the context of ordinary field theory, with $\mu \ll m_{Pl}$ corresponding to a small-field model. The mechanism, referred to as mutated [70] or smooth [71] hybrid inflation, is the following. The waterfall field is not fixed during inflation, but instead adjusts to continually minimize the potential. The effective potential is then $V(\phi, \xi(\phi))$, and for simplicity the ϕ-dependence at fixed χ is taken to be negligible. In this way [65] one can obtain any $p < 0$ (not necessarily integral) as well as $p > 1$. Taking negative p, the upper bound on r (evaluated by setting $\Delta\phi < m_{Pl}$) is shown in Fig. 3.10.

This is a good place to mention another potential of the kind shown in Fig. 3.5;

$$V \simeq V_0 \left[1 - \exp\left(-q\frac{\phi}{m_{Pl}}\right)\right], \qquad (3.52)$$

with q of order 1. It occurs if inflation takes place in field space where the kinetic function has a pole, irrespective of the form of the potential [58], with model-dependent values of q such as $q = 1$ or $\sqrt{2}$. It can also be obtained by transforming R^2 gravity or scalar-tensor gravity to the Einstein frame, giving $q = \sqrt{2/3}$. Notice that these modified-gravity theories should not be used in

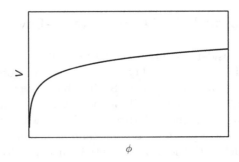

Fig. 3.5. F- and D-term inflation, colliding brane, mutated hybrid

conjunction with the standard supergravity potential, because that potential is evaluated in the Einstein frame.

The potential is supposed to apply in the regime where V_0 dominates, which is $\phi \gtrsim m_{\mathrm{Pl}}$. Inflation ends at $\phi_{\mathrm{end}} \sim m_{\mathrm{Pl}}$, and when cosmological scales leave the horizon, we have $\phi \simeq \ln(q^2 N) m_{\mathrm{Pl}}/q$ and

$$n \simeq 1 + 2\eta = 1 - \frac{2}{N}. \tag{3.53}$$

The predicted cmb normalization (for $q = 1$ and $N = 50$) is shown in Fig. 3.10 as a cross.

3.12 F and D Term Inflation

Now we suppose that the potential is dominated by the loop correction, in a model invoking spontaneously broken global supersymmetry. We focus initially on the case that the supergravity correction is negligible, asking later whether that is reasonable in specific models. In the regime $\phi \gg \phi_c$ the potential is then given by (3.45), while in the limit $\phi \to \phi_c$ it vanishes [because $M_i(\phi)$ in (3.43) vanishes]. The mass-squared in (3.45) is proportional to some coupling g which controls the strength of the spontaneous supersymmetry breaking. The potential during inflation is therefore of the form

$$V(\phi) = V_0 \left(1 + \frac{g^2}{8\pi^2} f(\phi) \ln \frac{\phi}{Q} \right), \tag{3.54}$$

where $f = 1$ for $\phi \gg \phi_c$ and $f \to 0$ as $\phi \to \phi_c$. The potential has the form shown in Fig. 3.5.

For $\phi \gg \phi_c$,

$$\eta = -\frac{g^2}{8\pi^2} \frac{m_{\mathrm{Pl}}^2}{\phi^2} = -\epsilon \frac{m_{\mathrm{Pl}}}{\phi}. \tag{3.55}$$

Consider first the regime

$$g^2 \gg 8\pi^2 \phi_c^2/m_{\mathrm{Pl}}^2. \tag{3.56}$$

slow-roll inflation ends at $\phi_{\mathrm{end}} = 2gm_{\mathrm{Pl}}^2/4\pi \gg \phi_c$, because $\eta = 1$ there. After *slow-roll* inflation ends, ϕ oscillates about $\langle \phi \rangle = 0$. A few e-folds [of order $\ln(\phi_{\mathrm{end}}/\phi_c)$] of "locked" inflation then occur, until the amplitude falls below ϕ_c.

The integral (3.16) is dominated by the limit ϕ giving

$$\phi \simeq \sqrt{\frac{N}{4\pi^2}} gm_{\mathrm{Pl}}. \tag{3.57}$$

To be in the desired regime $\phi \ll m_{\mathrm{Pl}}$ we need $g \ll 1$ which might be in conflict with (3.56). Proceeding anyway one finds $n = 1 - 1/N \simeq 0.98$, and the cmb normalization $r = 0.0011(50/N)g^2$. This prediction (with $N = 50$) is shown as a star in Fig. 3.10.

All this is with g in the regime (3.56). If we decrease g smoothly to reach the opposite regime $g^2 \ll 8\pi^2\phi_c^2/m_{Pl}^2$, $\phi(N)$ approaches ϕ_c, the cmb normalization decreases and n approaches 1 [72].

Two versions of this model exist in the literature, referred to generally as F-term [57, 58, 73] and D-term [74, 75] inflation.[11] In both cases, the starting point is a simple global supersymmetry theory with canonical kinetic terms, giving the hybrid inflation potential (3.51) with $V(\phi)$ perfectly flat.

In the F-term case, g is a Yukawa coupling, which can be chosen to be small yielding a small-field model. The cmb normalization fixes the vev of the waterfall field, as $\Lambda \simeq 6 \times 10^{15}\,\mathrm{GeV}$. Identifying the waterfall field(s) as a subset of the GUT Higgs fields motivates this value. Turning that around, the GUT model predicts roughly the observed magnitude for the spectrum of the curvature perturbation.

As we are dealing with an F term, the η problem exists; we expect $V \simeq \pm m^2\phi^2$ with $m^2 \sim H^2$. To have a viable model m^2 needs to be tuned down by a factor of order 0.01 but there is no reason why it should be negligible. The case of positive m^2 has been investigated in [76] and negative m^2 in [3]. The latter case gives an attractive model because it corresponds to hilltop inflation as in Fig. 3.9. After eternal inflation near the hilltop, the field can roll in the negative ϕ direction. After redefining the origin and reversing the sign of ϕ we recover the small-field model considered in Sect. 3.10. Taking the case (3.56), the spectral index and the height of the potential have been calculated, and are lower than in the original model.

In the D-term case, g is a gauge coupling which presumably cannot be small. The vev of the waterfall field has the same cmb normalization as in the F-term case. This vev is expected to be of order the string scale, relating D term inflation directly to string theory.

There is no η problem for the D-term model, but the tree-level potential $V(\phi)$ is still not expected to be flat because we are dealing with a medium-field model where non-renormalizable terms are out of control. There is no particular reason to think that the tree-level $V(\phi)$ will be quadratic, but one may adopt the quadratic form as a parameterization. The case of positive mass-squared was considered in [77, 78], and negative mass-squared in [3, 79]. As in the F term case, it gives an attractive inflation model and with the height of the potential and the spectral index both lower than with the original model.

In both the F and D-term models, the inflationary energy scale without a tree-level potential is $V \simeq g^2\Lambda^4$. Cosmic strings are generically produced

---

[11] The supergravity potential can be written as the sum of an F term and a D term. With the D term one is driven more or less inevitably to this type of model, but many other possibilities exist with the F term.

with tension $\mu \sim V^{1/2}$, and the cmb constraint $\mu^{1/2} \lesssim 10^{15} \, \mathrm{GeV}$ imposes restrictions on the parameter space.

3.13 Tree-Level Hybrid Inflation

All of the models considered so far can give a spectral index which is consistent with observation at the time of writing, provided that N is not too far below the expected value $\simeq 50$). Now we turn to small- and medium-field models which at least in their simplest form are ruled out by their prediction for the spectral index (as always, on the assumption that the inflaton perturbation generates the curvature perturbation).

Any small- or medium-field model with a concave-upward potential is ruled out. Such models are of the hybrid type, unless the potential becomes concave-downward after cosmological scales leave the horizon. Taking the fixed point as the origin of symmetries, we distinguish between potentials with positive slope as in Fig. 3.6, and with negative slope as in Fig. 3.8.

A negative slope can arise from non-perturbative quantum effects [80]. More usually, one finds models with positive slope as in Fig. 3.6, coming from a tree-level hybrid inflation model with (say) a quadratic potential. The

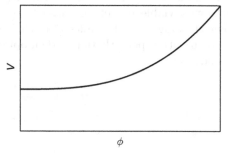

Fig. 3.6. Tree-level hybrid

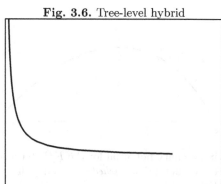

Fig. 3.7. Dynamical supersymmetry breaking

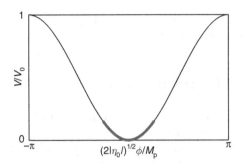

Fig. 3.8. Natural/chaotic inflation

potential including the waterfall field χ is [13] of the form (3.51) with $V(\phi) = \frac{1}{2}m^2\phi^2$.

A well-motivated tree-level hybrid inflation model, called Supernatural Inflation by its authors [81], uses softly broken global supersymmetry. The waterfall field is, in our nomenclature, a light modulus [82]. In contrast with most models of inflation, the inflationary scale is low corresponding to $V_0^{1/4} \sim M_S \sim 10^{10}\,\mathrm{GeV}$, the idea being that there is gravity-mediated supersymmetry breaking both during inflation and in the vacuum, the only difference in the former case being that the last term of (3.42) has not yet kicked in order to achieve a viable model the masses m_χ and m are taken to be respectively somewhat bigger and smaller than their generic values of order H_*. The observed curvature perturbation is then obtained with λ' just a few orders of magnitude below 1.

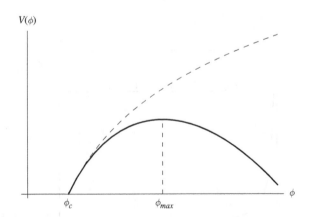

Fig. 3.9. Sketch of the inflationary potential for the F/D-term scenario in its simplest form, without a tree-level potential (*dashed line*) and with a concave-downward tree-level potential (*continuous line*)

The origin $\chi = 0$ is taken by the authors to be, in our nomenclature, a point of enhanced symmetry. The relevant symmetries cannot be those of the Standard Model because $\langle \chi \rangle \sim m_{\text{Pl}}$. After inflation the waterfall field oscillates about its vev, but it is supposed to decay into SM particles before nucleosynthesis so that it presents no moduli problem. This makes the vev another point of enhanced symmetry, the symmetries now being those of the Standard Model [82].

As with practically all inflation models, the inflaton is invoked just to give inflation and is not part of any extension of the Standard Model that has been proposed for other purposes. Models similar in spirit have been proposed (beginning with [83]) that are based on extensions of the Standard Model that serve other purposes too. They have an even lower inflation scale, corresponding to a mediation strength stronger than gravitational. They invoke fine tunings, which may however be reasonable within the context of string theory and branes. They can give either ordinary or inverted hybrid inflation, but in both cases the spectral tilt is practically zero in contradiction with observation. To avoid this problem though, it seems possible to generate the curvature perturbation during preheating [84].

In considering tree-level hybrid inflation, one has to remember that the coupling of the inflaton to the waterfall field generates a calculable loop correction to the potential, which can be concave-downward and rescue the model. This still leaves a large region of parameter space in which the one-loop correction from this source is negligible [85], though in some part of that space one should still worry about the two-loop correction [81]. In any case the coupling of the inflaton to fields other than the waterfall field can also generate a concave-downward loop correction. We consider this possibility next, in the context of the running-mass model.

A different possibility for generating a concave-downward potential would be to include the leading non-renormalizable term with a negative sign, generating a maximum as we discussed already for F- and D-term inflation. The possibility has not been investigated at the time of writing.

3.14 Running Mass Models

The loop correction with soft supersymmetry breaking generates a running mass. If the mass belongs to the inflaton we have a running-mass inflation model. The usual model [86] starts with the Supernatural inflation model that we mentioned earlier. At $\phi = m_{\text{Pl}}$, the running mass $m^2(\phi)$ is supposed to be of order V_0/m_{Pl}^2, which is the minimum value in a generic supergravity theory. The inflaton is supposed to have couplings (gauge, or maybe Yukawa) that are not too small, and it is supposed that $m^2(\phi)$ passes through zero before it stops running. The running associated with a given loop will stop when ϕ falls below the mass of the particle in the loop.

The potential near $m^2(\phi) = 0$ is flat enough to support inflation. To see this, we can use (3.46) which is valid over any small range of ϕ and will therefore be valid around the minimum. It can be written in the form

$$V = V_0 \left[1 + \frac{1}{2} \eta_0 \frac{\phi^2}{m_{\text{Pl}}^2} \left(\ln \frac{\phi}{\phi_*} - \frac{1}{2} \right) \right], \tag{3.58}$$

which leads to

$$m_{\text{Pl}} \frac{V'}{V_0} = \eta_0 \frac{\phi}{m_{\text{Pl}}} \ln \frac{\phi}{\phi_*}. \tag{3.59}$$

The potential has a maximum or minimum at $\phi = \phi_*$, at which $\eta = \eta_0$, and near which

$$\eta = \eta_0 \left(1 + \ln \frac{\phi_*}{\phi} \right). \tag{3.60}$$

A maximum is favoured theoretically, because a minimum requires a hybrid inflation model with ϕ_c tuned to be near the minimum.

To estimate $|\eta_0|$, we can make the crude approximation that (3.60) is valid at $\phi \sim m_{\text{Pl}}$, where $|\eta|$ is supposed to be of order 1. Then

$$|\eta_0| \sim 1/\ln(m_{\text{Pl}}/\phi_*). \tag{3.61}$$

This will give $|\eta_0| \ll 1$ if ϕ_* is exponentially below m_{Pl}, and with the reasonable requirement $\phi_* \gtrsim 100\,\text{GeV}$ it gives something like $|\eta_0| \sim 10^{-1}$. For a generic value of $\phi(N)$ this corresponds to $|n - 1| \sim 0.1$ which is outside the observational bound. One can satisfy current observation by choosing the parameters so that $\phi(N) = \phi_*$ around the middle of the cosmological range of scales, corresponding to the spectrum having a maximum at that point [87]. The running of the spectral index at that point is $dn/d\ln k \simeq -2\eta_0^2$, and we are requiring $|\eta_0| \sim 10^{-1}$. This is allowed by present observations, though it will soon be ruled out or confirmed.

To see whether the condition $\phi(N) \simeq \phi_*$ is reasonable, as well as to calculate the cmb normalization, we need

$$N(\phi) = -\frac{1}{|\eta_0|} \ln \left(\ln \frac{\phi_{\text{end}}}{\phi_*} \ln \frac{\phi_*}{\phi} \right). \tag{3.62}$$

If slow-roll inflation ends at $|\eta| \sim 1$, and (3.60) is still roughly valid there, $|\eta_0| \ln(\phi_*/\phi_{\text{end}}) \sim 1$ and (3.62) requires roughly $|\eta_0| \simeq \exp(-N|\eta_0|)$ which is more or less compatible with $|\eta_0| \sim 0.1$, and also more or less satisfies the cmb normalization with $V_0^{1/4} \sim 10^{-10}\,\text{GeV}$.

A running mass has also been considered in the context of a two-field modular inflation model [20, 88]. The two real fields are components of a complex field Φ. The maximum of the tree-level potential, chosen as $\Phi = 0$, represents a point of enhanced symmetry, and its height is $V_0^{1/4} \sim 10^{10}\,\text{GeV}$ corresponding to gravity-mediated supersymmetry breaking. Writing $\Phi \equiv |\Phi|e^{i\theta}$, the potential depends on both θ and $|\Phi|$. The tree-level negative mass-squared defined at the origin is supposed to have the generic value corresponding to $|\eta_0| \sim 1$, but interactions cause the mass to run. This turns the maximum into a crater, and it makes the potential very flat at the rim so that inflation can take place there.

There is a family of trajectories characterized by the initial value of θ. The curvature perturbation in this two-field model was calculated from the δN formalism. Near a special value of θ, chosen as zero, θ can be chosen to reproduce the cmb normalization is reproduced with $V_0^{1/4} \sim M_S \sim 10^{10}$ GeV. It seems to be possible to reproduce the observed spectral index by choice of parameters.

3.15 Large-Field Models

Now we turn to large-field models. They give a significant tensor perturbation $r \sim 10^{-2}$, which will be observed or ruled out in the near future.

The field variation cannot actually be extremely large, because (3.16) requires $\Delta\phi/m_{\text{Pl}} < \sqrt{2\epsilon_{\text{max}}}N \ll 50$. Two kinds of potential have been considered. One [9] is the Chaotic Inflation potential $V \propto \phi^p$ with p an even integer. The slow-roll parameters are

$$\epsilon = \frac{p^2}{2}\frac{m_{\text{Pl}}^2}{\phi^2}, \qquad \eta = p(p-1)\frac{m_{\text{Pl}}^2}{\phi^2}. \tag{3.63}$$

Inflation ends at $\phi_{\text{end}} \simeq p m_{\text{Pl}}$ When cosmological scales leave the horizon, we find from (3.16) that $\phi = \sqrt{2Np}\, m_{\text{Pl}}$, giving

$$n - 1 = -\frac{2+p}{2N} = -\frac{2+p}{100}, \qquad r = \frac{4p}{N} = 0.08p. \tag{3.64}$$

Current observational constraints practically rule out the case $p \geq 4$. Future observation will rule out or support the remaining case $p = 2$. The cmb normalization for $V = \frac{1}{2}m^2\phi^2$ is $m = 1.8 \times 10^{13}$ GeV, and for $V = \frac{1}{4}\lambda\phi^4$ it is $\lambda = 7 \times 10^{-14}$. If the curvature perturbation is not generated by the inflaton, these become upper bounds, and there is no spectral index constraint.

Another simple possibility is to use a sinusoidal potential

$$V = \frac{1}{2}V_0\left[1 + \cos\left(\frac{\sqrt{2|\eta_0|}\phi}{m_{\text{Pl}}}\right)\right]. \tag{3.65}$$

Here, the origin has been taken to be the maximum of the potential, and $\eta_0 < 0$ is the value of η there. This was called Natural Inflation by its authors [89]. The vev is at $\langle\phi\rangle = -\pi m_{\text{Pl}}/\sqrt{2|\eta_0|}$.

With this potential $\phi(N)$ is given by

$$\sin\left(\sqrt{\frac{|\eta_0|}{2}}\frac{\phi}{m_{\text{Pl}}}\right) = \sqrt{\frac{1}{1+|\eta_0|}}\, e^{-N|\eta_0|}, \tag{3.66}$$

leading to

$$\epsilon = \frac{1}{2N} \frac{2N|\eta_0|}{e^{2N|\eta_0|} - 1}, \qquad \eta = \epsilon - |\eta_0|. \qquad (3.67)$$

The maximum is at $\phi = 0$, and eternal inflation can take place there providing the initial condition for observable inflation. But if $N|\eta_0| \ll 1$, observable inflation itself will not begin until the potential is near the minimum, corresponds to the "chaotic inflation" potential $V = \frac{1}{2}m^2\phi^2$. The prediction in the r–n plane is shown in Figs. 3.3 and 3.10. We see that the current bound on n requires $r \gtrsim 10^{-2}$. This means that Natural Inflation will eventually be confirmed or ruled out, though it may turn out to be indistinguishable from chaotic inflation.

Large-field models are difficult to understand within the generally accepted rules for constructing field theories beyond the Standard Model, whereby the higher order terms in the expansion (3.36) are under control only for $\phi \ll m_{Pl}$. Some possibilities do exist though.

First, the inflationary trajectory may lie in the space of many fields, corresponding say $\phi = \sum_{i=1}^{N} a_i \phi_i / \sqrt{\sum a_i^2}$. Then, with say all a_i equal, we can have $\phi \gg m_{Pl}$ with each $\phi_i \ll m_{Pl}$. This was called Assisted Inflation by its authors [90]. At first sight one might think that the proposal lacks content, since a rotation of the field basis can always make ϕ one of the fields. The point though is that the field theory may select a particular basis, as the one in which the power series (3.36) is expected to be relevant. It has been argued [91] that this will be the case if each ϕ_i has a sinusoidal potential, leading to what they called N-flation. Then, if inflation takes place near the minimum of the potential one can have ϕ^2 chaotic inflation even though the proportionality $V \propto \phi^2$ does not persist up to the Planck scale.

A second possibility is for the inflationary trajectory may wind many times around the fixed point of the symmetries, at a distance $\lesssim m_{Pl}$ from that point. Something like this has been suggested in the context of string theory [92], giving a sinusoidal potential corresponding to Natural Inflation. Finally, it may be possible to evade the general rule that (3.36) is out of control at $\phi \gg m_{Pl}$, if the field theory is derived from a special higher-dimensional setup. This is the idea of Gauge Inflation [64, 93, 94], where the inflaton is the fifth component of a gauge field living in a 5D theory, which becomes a PNGB in the 4D theory. This again can give a sinusoidal potential. None of these proposals allows V to increase continually up to the Planck scale, in the spirit of the Chaotic Inflation proposal.

3.16 Warm Inflation

In all of the inflation models mentioned so far, energy loss by the inflation field ϕ is assumed to be negligible on the grounds that ϕ changes only slowly with time. Including this energy loss will give an equation of the form

$$\ddot{\phi} + (3H + \Gamma)\dot{\phi} + V' = 0 \,, \tag{3.68}$$

where Γ is some time-dependent quantity. The warm inflation model [95] assumes that Γ is significant, or even dominant ($\Gamma \gg H$).

The extent to which warm inflation is possible was investigated in the GUT hybrid inflation model [96] using an earlier calculation of the energy loss [97]. It does not occur in the original GUT hybrid model but apparently can occur if the inflaton has a suitable interaction with a spin-half particle. The curvature perturbation in warm inflation receives a contribution from the thermal fluctuation, which dominates the contribution of the vacuum fluctuation if Γ is dominant.

3.17 Present Status and Outlook

Figure 3.10 summarizes most of the predictions that we have been discussing, always assuming that the inflaton perturbation generates the curvature perturbation. (Recall that the alternative was considered in Sect. 3.4.5.)

Consider first small- and medium-field models. For these models the tilt is directly related to the curvature of the potential, $n - 1 = 2\eta$. As a result, the recently-observation negative tilt has had a dramatic effect, ruling out whole classes of otherwise attractive models. These include the original tree-level hybrid inflation model, in particular those rather well-motivated versions which

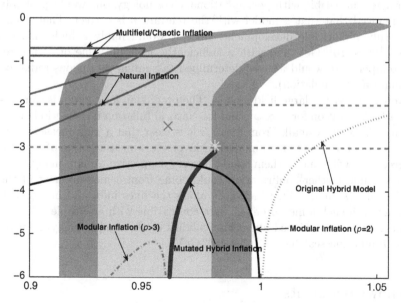

Fig. 3.10. The shaded regions are the allowed by observation as in Fig. 3.3, and the predictions are described in the text. Planned observation will detect r or give a limit $r < 10^{-2}$, and $r < 10^{-3}$ will probably never be observed

invoke during inflation the vacuum supersymmetry-breaking mechanism. The running-mass variant of tree-level hybrid inflation is not yet ruled out, but it will be if the observational bound on the running of n gets much tighter.

Among simple single-field slow-roll models, the ones that agree with observation are modular inflation, and hybrid inflation with a concave-downward potential. The latter can be achieved by what are usually termed simply F- and D-term inflation, involving the loop correction generated by spontaneously broken global supersymmetry. They can also be achieved by mutated hybrid inflation.

All of these simple models give (exactly or as what should be a reasonable approximation) a distinctive prediction for the scale-dependence of the tilt, of the form

$$n - 1 = -\left(\frac{p-1}{p-2}\right)\frac{2}{N(k)} \ . \tag{3.69}$$

This gives the scale-dependence (running)

$$\frac{1}{2}\frac{dn}{d\ln k} = -\left(\frac{p-2}{p-1}\right)\left(\frac{n-1}{2}\right)^2 \ . \tag{3.70}$$

Several years down the line it might be possible to measure this level of running, for instance through a measurement of the 21-cm anisotropy. A confirmation of the above prediction would select within observational uncertainty values for both N and p. If the former were in the relatively narrow range compatible with post-inflationary cosmology, one would probably be convinced that that a model with the relevant p is correct. That would be a truly remarkable development, since it would imply a high inflation scale $V^{1/4} \sim 10^{15}$ GeV and with a sufficiently accurate value of N the reheat temperature would also be determined (assuming continuous radiation domination after inflation).

Now consider the large-field models. The prediction for r and n is compatible with observation for $V \propto \phi^2$, and for Natural Inflation if the period of the potential is not too small. From Fig. 3.3 it is clear that a joint measurement of r and n can rule out these models. Conversely, a measurement of r and n in agreement with one of them would be very suggestive. Again, many years down the line further confirmation could come from a measurement of the running of $n(k)$ and $r(k)$, which goes along the lines indicated in Fig. 3.3. And, again, if such a measurement were compatible with a sensible value for N one would be convinced about the validity of the model, implying again the high inflation scale now $V^{1/4} \sim 10^{16}$ GeV.

Acknowledgements

I thank my collaborators, and in particular Lotfi Boubekeur and Laila Alabidi who supplied the figures. The research is supported by PPARC grant

PPA/G/S/2003/00076. DHL is supported by PP/D000394/1 and by EU grants MRTN-CT-2004-503369 and MRTN-CT-2006-035863.

References

1. D. H. Lyth and A. Riotto, Phys. Rept. **314**, 1 (1999).
2. A. R. Liddle and D. H. Lyth, *Cosmological Inflation and Large Scale Structure*, (CUP, Cambridge, 2000).
3. L. Boubekeur and D. H. Lyth, JCAP **0507**, 010 (2005).
4. L. Alabidi and D. H. Lyth, JCAP **0605**, 016 (2006).
5. A. D. Linde, Phys. Lett. B **108**, 389 (1982); A. Albrecht and P. J. Steinhardt, Phys. Rev. Lett. **48**, 1220 (1982).
6. S. W. Hawking and I. G. Moss, Phys. Lett. B **110**, 35 (1982).
7. A. H. Guth and S. Y. Pi, Phys. Rev. Lett. **49**, 1110 (1982). S. W. Hawking, Phys. Lett. B **115**, 295 (1982). A. A. Starobinsky, Phys. Lett. B **117**, 175 (1982). J. M. Bardeen, P. J. Steinhardt and M. S. Turner, Phys. Rev. D **28**, 679 (1983).
8. B. Ovrut and P. J. Steinhardt, Phys. Rev. Lett. **53**, 732 (1984); K. Enqvist D. V. Nanopoulos, Nucl. Phys. B **252**, 508 (1985).
9. A. D. Linde, Phys. Lett. B **129**, 177 (1983).
10. D. La and P. J. Steinhardt, Phys. Rev. Lett. **62**, 376 (1989).
11. A. R. Liddle and D. H. Lyth, Phys. Lett. B **291**, 391 (1992); A. R. Liddle and D. H. Lyth, Phys. Rept. **231**, 1 (1993).
12. F. Di Marco and A. Notari, Phys. Rev. D **73**, 063514 (2006).
13. A. D. Linde, Phys. Rev. D **49**, 748 (1994).
14. M. Gasperini and G. Veneziano, Mod. Phys. Lett. A **8**, 3701 (1993).
15. J. Khoury, B. A. Ovrut, P. J. Steinhardt and N. Turok, Phys. Rev. D **64**, 123522 (2001).
16. P. J. Steinhardt and N. Turok, Phys. Rev. D **65**, 126003 (2002).
17. A. Linde, JCAP**0410**, 004, `hep-th/0408164` (2004).
18. P. J. Steinhardt, 1982 Princeton preprint UPR-0198T; A. Vilenkin, Phys. Rev. D **27**, 2848 (1983).
19. A.D. Linde, Print-82-0554, Cambridge University preprint, 1982 (http://www.stanford.edu/ alinde/1982.pdf).
20. K. Kadota and E. D. Stewart, JHEP **0307**, 013 (2003).
21. A. D. Linde, Phys. Lett. B **327**, 208, 1994b; A. Vilenkin, Phys. Rev. Lett. **72**, 3137 (1994).
22. A. A. Starobinsky, in *Lecture Notes in Physics*, Vol. 242, eds. H. J. de Vega and N. Sanchez (Springer, Berlin, 1986).
23. A. D. Linde, Phys. Lett. B **175**, 395 (1986).
24. G. Lazarides, C. Panagiotakopoulos and Q. Shafi, Phys. Rev. Lett. **56**, 557 (1986). D. H. Lyth and E. D. Stewart, Phys. Rev. Lett. **75**, 201 (1995) D. H. Lyth and E. D. Stewart, Phys. Rev. D **53**, 1784 (1996).
25. A. R. Liddle and S. M. Leach, Phys. Rev. D **68**, 103503 (2003).
26. G. Dvali and S. Kachru, arXiv:hep-th/0309095; K. Dimopoulos and M. Axenides, JCAP **0506**, 008 (2005).
27. J. O. Gong and E. D. Stewart, Phys. Lett. B **510**, 1 (2001).
28. A. Linde, JHEP **0111**, 052 (2001).
29. R. Harrison, Phys. Rev. D **1**, 2726 (1970).

30. Ya. B. Zel'dovich, Astron. Astrophys. **5**, 84 (1970).
31. D. H. Lyth, Phys. Rev. Lett. **78**, 1861 (1997).
32. D. N. Spergel et al., arXiv:astro-ph/0603449.
33. A. A. Starobinsky, JETP Lett. **42**, 152 (1985) [Pisma Zh. Eksp. Teor. Fiz. **42**, 124 (1985)].
34. M. Sasaki and E. D. Stewart, Prog. Theor. Phys. **95**, 71 (1996).
35. D. H. Lyth, K. A. Malik and M. Sasaki, JCAP **0505**, 004 (2005).
36. D. H. Lyth and Y. Rodriguez, Phys. Rev. Lett. **95** (2005) 121302.
37. M. Sasaki and T. Tanaka, Prog. Theor. Phys. **99**, 763 (1998).
38. S. Mollerach, Phys. Rev. D **42**, 313 (1990); A. D. Linde and V. Mukhanov, Phys. Rev. D **56**, 535 (1997); D. H. Lyth and D. Wands, Phys. Lett. B **524**, 5 (2002); T. Moroi and T. Takahashi, Phys. Lett. B **522**, 215 (2001) [Erratum-ibid. B **539**, 303 (2002)]; D. H. Lyth, C. Ungarelli and D. Wands, Phys. Rev. D **67**, 023503 (2003).
39. K. Dimopoulos and D. H. Lyth, Phys. Rev. D **69**, 123509 (2004).
40. K. Dimopoulos, G. Lazarides, D. Lyth and R. Ruiz de Austri, JHEP **0305**, 057 (2003).
41. J. Maldacena, JHEP **0305**, 013 (2003).
42. A. Cooray, arXiv:astro-ph/0610257.
43. L. Boubekeur and D. H. Lyth, Phys. Rev. D **73**, 021301 (2006).
44. C. Armendariz-Picon, T. Damour and V. F. Mukhanov, Phys. Lett. B **458**, 209 (1999); J. Garriga and V. F. Mukhanov, Phys. Lett. B **458**, 219 (1999).
45. M. Alishahiha, E. Silverstein and D. Tong, Phys. Rev. D **70**, 123505 (2004);
46. N. Arkani-Hamed, P. Creminelli, S. Mukohyama and M. Zaldarriaga, JCAP **0404**, 001 (2004).
47. P. Binetruy and M. K. Gaillard, Phys. Rev. D **34**, 3069 (1986).
48. G. G. Ross and S. Sarkar, Nucl. Phys. B **461**, 597 (1996).
49. J. J. Blanco-Pillado et al., JHEP **0411**, 063 (2004); Z. Lalak, G. G. Ross and S. Sarkar, arXiv:hep-th/0503178; J. J. Blanco-Pillado et al., Inflating in a better racetrack, JHEP **0609**, 002 (2006).
50. S. Weinberg, *The Quantum Theory of Fields. Vol. 1: Foundations*, (Sect. 12.3) (Cambridge University Press, 2000).
51. R. Allahverdi, K. Enqvist, J. Garcia-Bellido and A. Mazumdar, arXiv:hep-ph/0605035; D. H. Lyth, arXiv:hep-ph/0605283; R. Allahverdi, K. Enqvist, J. Garcia-Bellido, A. Jokinen and A. Mazumdar, arXiv:hep-ph/0610134; R. Allahverdi, A. Kusenko and A. Mazumdar, arXiv:hep-ph/0608138; R. Allahverdi, A. Jokinen and A. Mazumdar, arXiv:hep-ph/0610243.
52. J. C. Bueno Sanchez, K. Dimopoulos and D. H. Lyth, arXiv:hep-ph/0608299.
53. D. H. Lyth, Phys. Lett. B **419**, 57 (1998).
54. S. Weinberg, *The Quantum Theory of Fields. Vol. 3* , (Cambridge University Press 2000).
55. B. Ovrut and P. J. Steinhardt, Phys. Lett. B **133**, 161 (1983).
56. G. D. Coughlan, R. Holman, P. Ramond and G. G. Ross, Phys. Lett. B **140**, (1984); M. Dine, W. Fischler and D. Nemeschansky, Phys. Lett. B **136**, 169 (1984).
57. E. J. Copeland, A. R. Liddle, D. H. Lyth, E. D. Stewart and D. Wands, Phys. Rev. D **49**, 6410 (1994).
58. E. D. Stewart, Phys. Rev. D **51**, 6847 (1995).
59. M. Dine, L. Randall and S. Thomas, Phys. Rev. Lett. **75**, 398 (1995).

60. D. H. Lyth and T. Moroi, JHEP **0405**, 004 (2004).
61. Q. Shafi and A. Vilenkin, Phys. Rev. Lett. **52**, 691 (1984); S. Y. Pi, Phys. Rev. Lett. **52**, 1725 (1984).
62. M. Dine and A. Riotto, Phys. Rev. Lett. **79**, 2632 (1997).
63. N. Arkani-Hamed, H. C. Cheng, P. Creminelli and L. Randall, JCAP **0307**, 003 (2003).
64. D. E. Kaplan and N. J. Weiner, JCAP **0402**, 005 (2004).
65. D. H. Lyth and E. D. Stewart, Phys. Rev. D **54**, 7186 (1996).
66. S. F. King and J. Sanderson, Phys. Lett. **B412**, 19 (1997).
67. J. D. Cohn and E. D. Stewart, Phys. Lett. B **475**, 231 (2000).
68. M. K. Gaillard, D. H. Lyth and H. Murayama, Phys. Rev. D **58**, 123505 (1998).
69. S. H. Henry Tye, arXiv:hep-th/0610221.
70. E. D. Stewart, Phys. Lett. B **345**, 414 (1995).
71. G. Lazarides and C. Panagiotakopoulos, Phys. Rev. D **52**, 559 (1995).
72. B. Kyae and Q. Shafi, arXiv:hep-ph/0504044.
73. G. R. Dvali, Q. Shafi and R. K. Schaefer, Phys. Rev. Lett. **73**, 1886 (1994).
74. E. D. Stewart, Phys. Rev. D **51**, 6847 (1995).
75. P. Binetruy and G. R. Dvali, D-term inflation, Phys. Lett. B **388**, 241 (1996). E. Halyo, Phys. Lett. B **387**, 43 (1996).
76. C. Panagiotakopoulos, Phys. Lett. B **402**, 257 (1997); C. Panagiotakopoulos, Phys. Rev. D **55**, 7335 (1997).
77. D. H. Lyth, Phys. Lett. B **419**, 57 (1998).
78. C. F. Kolda and J. March-Russell, Phys. Rev. D **60**, 023504 (1999).
79. C. M. Lin and J. McDonald, Phys. Rev. D **74**, 063510 (2006).
80. W. H. Kinney and A. Riotto, Astropart. Phys. **10**, 387 (1999).
81. L. Randall, M. Soljacic and A. H. Guth, Nucl. Phys. B **472**, 377 (1996).
82. M. Berkooz, M. Dine and T. Volansky, Phys. Rev. D **71**, 103502 (2005).
83. M. Bastero-Gil and S. F. King, Phys. Lett. B **423**, 27 (1998).
84. M. Bastero-Gil, V. Di Clemente and S. F. King, Phys. Rev. D **70**, 023501 (2004).
85. D. H. Lyth, Phys. Lett. B **466**, 85 (1999).
86. E. D. Stewart, Phys. Lett. B **391**, 34 (1997); E. D. Stewart, Phys. Rev. D **56**, 2019 (1997).
87. L. Covi, D. H. Lyth and A. Melchiorri, Phys. Rev. D **67**, 043507 (2003); L. Covi, D. H. Lyth, A. Melchiorri and C. J. Odman, Phys. Rev. D **70**, 123521 (2004).
88. K. Kadota and E. D. Stewart, JHEP **0312**, 008 (2003); D. h. Jeong, K. Kadota, W. I. Park and E. D. Stewart, JHEP **0411**, 046 (2004); K. Kadota and E. D. Stewart, arXiv:hep-ph/0409330.
89. K. Freese, J. A. Frieman and A. V. Olinto, Phys. Rev. Lett. **65**, 3233 (1990); F. C. Adams, J. R. Bond, K. Freese, J. A. Frieman and A. V. Olinto, Phys. Rev. D **47**, 426 (1993).
90. A. R. Liddle, A. Mazumdar and F. E. Schunck, Phys. Rev. D **58**, 061301 (1998).
91. S. Dimopoulos, S. Kachru, J. McGreevy and J. G. Wacker, arXiv:hep-th/0507205; R. Easther and L. McAllister, arXiv:hep-th/0512102.
92. J. E. Kim, H. P. Nilles and M. Peloso, JCAP **0501**, 005 (2005).
93. N. Arkani-Hamed, H. C. Cheng, P. Creminelli and L. Randall, Phys. Rev. Lett. **90**, 221302 (2003).
94. F. Paccetti Correia, M. G. Schmidt and Z. Tavartkiladze, arXiv:hep-th/0504083.
95. A. Berera and L. Z. Fang, Phys. Rev. Lett. **74**, 1912 (1995).

96. L. M. H. Hall and I. G. Moss, Phys. Rev. D **71**, 023514, (2005) hep-ph/0408323; M. Bastero-Gil and A. Berera, Phys. Rev. D **71**, 063515 (2005); M. Bastero-Gil and A. Berera, Phys. Rev. D **72**, 103526 (2005); M. Bastero-Gil and A. Berera, arXiv:hep-ph/0610343.

97. A. Berera and R. Ramos, Phys. Rev. **D71**, 023513 (2005).

4

Inflation in String Theory

Renata Kallosh

Department of Physics, Stanford University, Stanford, CA 94305
kallosh@stanford.edu

Abstract. In this chapter we describe the recent progress achieved in the construction of inflationary models in the context of string theory with flux compactification and moduli stabilization. We also discuss a possibility to test string theory through cosmological observations.

4.1 Introduction

This chapter addresses some problems of string theory in explaining the cosmological observations and some recent progress made in the construction of inflationary models in the context of flux compactification and moduli stabilization.[1] In view of the available precision observational data supporting inflationary cosmology, as well as the new data expected to come in a few years from now, we will also discuss some possibilities to test string theory through cosmological observations.

It is important to find out how the observational cosmology can probe string theory, since our universe is an ultimate laboratory of fundamental physics. High-energy accelerators will probe the scale of energies way below GUT scales. Cosmology and astrophysics are the major sources of data in the gravitational sector of the fundamental physics (above GUT, near Planck scale).

One can argue that M/string theory is fundamental: it has sectors with perturbatively finite quantum gravity. It includes supersymmetry and supergravity and has a potential to describe the standard model of particle physics and beyond. It selects $d = 10$ critical string theory and $d = 11$ M-theory. These two dimensions are also maximal dimensions for supergravity, $d = 10$ for chiral supergravity and $d = 11$ for the non-chiral one. These theories are almost unique. And in any case, it is the best and most advanced theory

---

[1] Recent reviews on flux compactification and moduli stabilization can be found in [1, 2, 3] and on inflation in string theory in [4, 5].

R. Kallosh: *Inflation in String Theory*, Lect. Notes Phys. **738**, 119–156 (2008)
DOI 10.1007/978-3-540-74353-8_4   © Springer-Verlag Berlin Heidelberg 2008

beyond standard model that we have now. But does it have any falsifiable predictions for cosmology?

To confront observational cosmology, one usually assumes the existence of some effective four-dimensional $\mathcal{N} = 1$ supergravity based on flux compactification and moduli stabilization, derivable from superstring theory. In this context, string theory has already provided a possible explanation for the dark energy of the universe via an effective cosmological constant of the metastable de Sitter vacua [6]. The most recent analysis of data on dark energy [7] confirms the consistency of the cosmological ΛCDM concordance model with the simplest form of dark energy, the cosmological constant. These data on dark energy in [7] are taken from supernovae, gamma ray bursts, acoustic oscillations, nucleosynthesis, large-scale structure, and the Hubble constant. The idea of the landscape of string vacua [6, 8, 9] supports the possibility of an anthropic explanation of the observable value of the cosmological constant.

Several models of inflation have been derived since 2003 in the compactified string theory with the so-called KKLT scenario of moduli stabilization [6]. Prior to this recent progress, string theory had a major problem of runaway moduli. Many interesting ideas were suggested, but the runaway moduli did not allow to have any type of internally consistent cosmology, see for example [10, 11, 12].

The first string inflation model based on the KKLT construction, with all moduli stabilized at the exit from inflation, is the brane–anti-brane annihilation scenario in the warped geometry, the KKLMMT model [13]. This model belongs to a general class of brane inflation models [10, 14] where the inflaton field, whose evolution drives inflation, is associated with the relative position of branes in the compactified space. Another class of string inflation models, which we will discuss later, modular inflation, does not consider brane dynamics. It assumes that the inflaton is one of the many moduli fields present in the KKLT construction.

In the new models the inflaton field is the only field (or some combination of fields) which is not stabilized before the exit from inflation. Each of these models relies on particular assumptions. Some of these models make clear predictions for observables and are therefore falsifiable by data. Some other models are more speculative and need more work before they can give definite predictions. There is an issue of fine-tuning and the problem of identifying stringy quantum corrections, which requires much deeper understanding.

The future developments in string cosmology and our attitude towards various models of inflation may depend strongly on several crucial pieces of information, which may become available during the next few years. Here is the list of the most important observables, which may shift the interest from one class of models of inflation to another.[2]

---

[2] See also Chap. 3 which contains a review of the models of inflation constructed during the last 25 years.

(a) A precision measurement of the tilt of the spectrum of scalar perturbations, n_s, which provides the measure of the violation of the scale invariance, $n_s - 1$.

The current value of n_s, which takes into account the WMAP3 results, is close to 0.95, if one ignores a possible contribution of the gravitational waves from inflation and from cosmic strings [15, 16]. This value is below the WMAP1 value, which was about 0.98. As an example of potential importance of future clarification of the value of spectral index we may refer to **D**-term inflation in supergravity [17, 18] and their string theory version, $D3/D7$ brane inflation [19, 20]. These models naturally have $n_s = 0.98$ and no gravitational waves. This was the perfect value for WMAP1, but it may be on a high side for WMAP3. On the other hand some models of modular inflation in string theory [21, 22, 23, 24, 25] with $n_s \sim 0.95-0.96$ did not originally look so good, but became much more attractive with WMAP3. New data on n_s will provide a powerful selection tool of valid models of inflation.

(b) A possible discovery of primordial gravitational waves from inflation, i.e., the measurement of the tensor to scalar ratio $r = T/S$.

The current limit is given by $r < 0.3$. A new series of observations may possibly test the models with $r \gtrsim 10^{-3}$. The simplest model of chaotic inflation [26] with $m^2\phi^2$ potential predicts $r \approx 0.15$, with an analogous prediction for the chaotic inflation in supergravity [27]. This level is expected to be reached during the next few years, particularly with Planck and dedicated polarization experiments, such as BICEP (down to $r \gtrsim 5 \times 10^{-2}$), Spider (down to $r \gtrsim 10^{-2}$) and others, perhaps all the way down to $r > 10^{-3}$. It has been clarified recently in [28] and in [29] that all known models of brane inflation, including the DBI inflation model [30, 31], do not lead to a prediction of an observable r.[3] The hope remains that the new brane inflation models with tensors may be constructed.

The model of assisted inflation [32, 33] and related to it the proposal of N-flation model of string theory [34, 35] are basically reducible to a chaotic inflation with the corresponding level of observable gravity waves. We will discuss below to which extent such models can be actually derived from string theory. Other models of string inflation typically predict $r < 10^{-3}$, which would make tensor perturbations almost impossible to detect.

The discovery/non-discovery of tensor fluctuations would be crucial for the selection of inflationary models. A discovery of gravitational waves with $r \sim 10^{-1}-10^{-3}$ would make it very important to understand whether inflationary models predicting large r can be derived from string theory. It would eliminate a majority of other models of brane inflation and/or modular inflation, which predict a non-detectable level of gravitational waves.

---

[3] We are grateful to D. Baumann, R. Bean, D. Lyth, L. McAllister, and H. Tye for the discussion of this issue.

(c) A possible discovery of cosmic strings produced by the end of inflation.[4]

It has been recognized recently that the discovery of cosmic strings produced by the end of inflation may be one of the most compelling potential observational windows into physics at the string scale [4, 13, 37, 38]. The main point here is that the current CMB experimental bound[5] on the tension of cosmic strings, $G\mu \leq 2 \times 10^{-7}$ [40, 41] is difficult to achieve generically and simultaneously predict the existence of light cosmic strings satisfying the bound. If however, the signal from such light cosmic strings are discovered via the B-polarization signal due to vector modes, the preferred class of models of inflation in string theory may be associated with warped throat geometry as in various versions of the KKLMMT model [4, 5, 13]. This is the basic class of models with a natural suppression mechanism for the tension of cosmic strings due to the position along the warped throat in the fifth dimension, which changes the energy scale of the four-dimensional physics. Another mechanism of production of cosmic strings, satisfying the observational bound on the tension, has been suggested in strongly coupled heterotic M-theory [42].

Other observables, e.g., the non-gaussianity may also become important in future (see Chap. 9 and [43]).

In Sect. 4.2 of this paper we discuss the relations between cosmology and particle physics phenomenology and in Sect. 4.3 we discuss the impact of string theory/supergravity on some issues in cosmology. In particular, we describe some interesting cosmological models based on $\mathcal{N} = 1$ supergravity models, which have not yet been implemented in string theory. In Sect. 4.4 we discuss some brane inflation models. In Sect. 4.5 the set of modular inflation models, which do not require the presence of branes, is presented. In Sect. 4.7 we discuss the N-flation/assisted inflation models. Finally, in Sect. 4.8 we focus on possible fundamental reasons for the flatness of the inflaton potentials from the perspective of string theory.

4.2 Cosmology and Particle Physics Phenomenology

For a long time we did not have any string theory interpretation of the acceleration of the universe. This problem was resolved in 2003 with the invention of the KKLT scenario [6] and its generalizations. By construction, the moduli are first stabilized in some anti-de Sitter space with a negative cosmological constant (CC). The relevant Kähler potential and superpotential in the simplest case are

$$K = -3\ln(T + \bar{T}) , \qquad W = W_0 + Ae^{-aT} . \tag{4.1}$$

---

[4] A detailed discussion of this topic can be found in Chap. 10 [36].

[5] A stronger bound $G\mu \leq 1.5 \times 10^{-8}$ has been recently claimed from the Parkes Pulsar Timing Array project [39]. The full project is expected to be able to either detect gravity waves from the cosmic strings or reduce the limit to $G\mu \leq 5 \times 10^{-12}$.

Here W_0 is the superpotential originating from fluxes stabilizing the axion–dilaton and complex structure moduli. The exponential term comes from gaugino condensation or wrapped brane instantons. This scenario requires in addition some mechanism of uplifting of the AdS vacua to a de Sitter space with a positive CC of the form $\delta V = \frac{C}{(T+\bar{T})^n}$. In all known cases this procedure always leads to metastable de Sitter vacua, see Fig. 4.1 for the simplest case of the original KKLT model. In addition to the dS minimum at some finite value of the volume modulus $\sigma = \frac{T+\bar{T}}{2}$, there is always a Dine–Seiberg Minkowski vacuum corresponding to an infinite 10-dimensional space with an infinite volume of the compactified space, $\sigma \rightarrow \infty$. The lifetime of metastable dS vacua usually is much greater than the lifetime of the universe.

There are numerous ways to find flux vacua in string theory, with all possible values of the cosmological constant. This is known as the landscape of string vacua [6, 8, 9]. The concept of the landscape has already changed quite a few settings in particle physics phenomenology. The first and most striking example is that of the split supersymmetry [44] where the new ideas of supersymmetry breaking were consistently realized without a requirement that supersymmetry has to protect the smallness of the Higgs mass.

New ideas of particle phenomenology in the context of supergravity and moduli stabilization were developed in [3, 45, 46, 47, 48, 49, 50, 51, 52], leading to a set of new predictions for the spectrum of particles to be detected in the future.

Recent progress in dS vacuum stabilization in string theory has influenced particle phenomenology by demonstrating that metastable vacua are quite

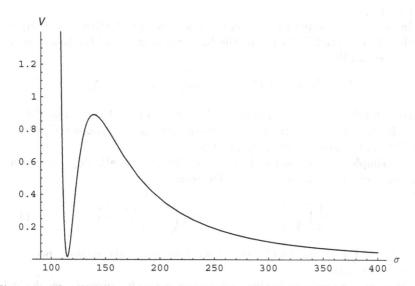

Fig. 4.1. The KKLT potential as a function of the volume of extra dimensions $\sigma = \frac{T+\bar{T}}{2}$

legitimate. This triggered a significant new trend in supersymmetric model building, starting with [53]. The long-standing prejudice, that the models of dynamical supersymmetry breaking must have no supersymmetric vacua, is abandoned. New metastable positive energy vacua with the lifetime longer than the age of the universe were found in supersymmetric gauge models. Models with metastable vacua represent an interesting and valid alternative approach to particle phenomenology.

4.3 String Theory Inspired Supergravity Models and Cosmology

Traditionally cosmological models of inflation use a single scalar field with a canonical kinetic term of the form $L_{\text{kin}} = \frac{1}{2} \partial_\mu \phi \partial_\nu \phi \, g^{\mu\nu}$ with some particular self-interaction, like $\frac{1}{2} m^2 \phi^2$ or $\frac{1}{4} \lambda \phi^4$.

In supersymmetric models one cannot have single scalars fields. Scalars come in pairs, they are always complex in supersymmetry. For example, axion and dilaton, or axion and radial modulus. Generically, in supergravity and string theory there is a multi-dimensional moduli space, the scalar fields ϕ^i, $\bar{\phi}^{\bar{i}}$ playing the role of complex coordinates in Kähler geometry.

$$L_{\text{kin}} = G_{i\bar{\imath}}(\phi, \bar{\phi}) \partial_\mu \phi^i \partial_\nu \bar{\phi}^{\bar{\imath}} \, g^{\mu\nu} \,, \qquad G_{i\bar{\imath}} \equiv \partial_i \partial_{\bar{\imath}} K(\phi, \bar{\phi}) \,. \qquad (4.2)$$

The metric in the moduli space $G_{i\bar{\imath}}(\phi, \bar{\phi})$ is derived from the Kähler potential $K(\phi^i, \bar{\phi}^{\bar{\imath}})$, as shown above. An effective inflaton may be a particular direction in moduli space.

In cases where string theory can be defined by an effective $\mathcal{N} = 1$ supergravity the potential is defined by the Kähler potential and the holomorphic superpotential $W(\phi)$:

$$V(\phi, \bar{\phi}) = F_{\text{F}} + V_{\text{D}} = e^K(|DW|^2 - 3|W|^2) + V_{\text{D}} \,. \qquad (4.3)$$

Here the total potential consists of the **F**-term and the **D**-term. The **F**-term potential V_{F} depends on the Kähler potential and the superpotential whereas the **D**-term is related to gauge symmetries.

The simple expressions for the one-field slow-roll inflationary parameters must be generalized in these models, for example,

$$\epsilon = \frac{1}{2} \left(\frac{V'}{V} \right)^2 \quad \Rightarrow \quad \epsilon = \left(\frac{G^{i\bar{\imath}} \partial_i V \partial_{\bar{\imath}} V}{V^2} \right) \,, \qquad (4.4)$$

where $G^{i\bar{\imath}}$ is the inverse to $G_{i\bar{\imath}}$, the Green function in the moduli space, see [54] for more details.

One more comment is due here on the distinctive features of cosmological models in string theory. The scalar fields often have geometrical meaning: distance between branes, size of internal dimensions, size of supersymmetric cycles on which branes can be wrapped. The axion fields originate from some form of fields and therefore they are paired into a complex field with particular moduli. In simple examples, which often appear in string theory,

$$K = -c\ln(\Phi + \bar{\Phi}) \qquad (4.5)$$

with some constant c. The origin of the logarithm and shift symmetry (independence on $\Phi - \bar{\Phi}$) in (4.5) will be explained in Sect. 4.4.2. For the total volume–axion $c = 3$, which results in no-scale supergravity. For the dilaton–axion $c = 1$, etc. The kinetic term for axion and its partner is given by

$$L_{\text{kin}} = c\frac{\partial_\mu\Phi\partial_\nu\bar{\Phi}g^{\mu\nu}}{(\Phi + \bar{\Phi})^2} = \frac{1}{2}[(\partial\phi)^2 + \mathrm{e}^{-2\sqrt{2/c}\,\phi}(\partial a)^2] . \qquad (4.6)$$

Here we take $\Phi = \mathrm{e}^{\sqrt{2/c}\,\phi} + \mathrm{i}\sqrt{2/c}\,a$ so that the modulus field ϕ has a canonical kinetic term. However, the axion a is coupled to the modulus ϕ and this coupling cannot be removed unless ϕ is fixed to a constant value. Typically it is difficult to separate the evolution of the axion and dilaton fields. Both of them are evolving and both are stabilized only at the exit from inflation, as we will show later.

4.3.1 Supergravity Models: Examples of Chaotic and Axion Valley Inflation

In the effective $\mathcal{N} = 1$ supergravity any choice of a Kähler potential and a holomorphic superpotential provides a valid theory. However, at present only specific versions of the effective $\mathcal{N} = 1$ $d = 4$ supergravity have been derived from a consistent string theory. String theory, in principle, offers a better understanding of quantum corrections. In practice, it remains a major challenge to identify the quantum corrections in the context of non-perturbative compactified string theory with fluxes. As far as we know now, string theory strongly limits the choice of the effective $\mathcal{N} = 1$ $d = 4$ supergravity models associated with string theory comparative to generic $\mathcal{N} = 1$ $d = 4$ supergravity.

The idea that the shift symmetry may help to protect flat directions of the potential was proposed long time ago [11, 55, 56]. In the string inspired models it was mostly in the context of Kähler potentials $K = -c\ln(\Phi + \bar{\Phi})$, or some generic, non-specified Kähler potentials. At that time it was not known how to stabilize the runaway moduli in such models. As we will see, the simple Kähler-shift symmetric KKLT model, against the naive shift symmetry expectations, does not have axionic flat directions near the minimum of the potential.

To achieve inflation based on the KKLT models we need to take a more complicated Kähler potential together with some additional ingredients, see Sect. 4.5 on modular inflation. We present below two examples of $\mathcal{N} = 1$ supergravity models for inflation (which have not been realized in string theory yet), which may explain flat directions for the inflaton field due to shift symmetry with Kähler potentials of the form $K = \frac{1}{2}(\Phi + \bar{\Phi})^2$. In this section, it is explained that such Kähler potentials with the shift symmetry do appear in some versions of string theory (see also Sect. 4.4.2). Here, after presenting the examples, we identify some remaining problems which appear when one tries to implement the simplest versions of chaotic inflation or natural inflation in string theory.

Steep Axions in the KKLT-Type Models with the Kähler Shift Symmetry

The generic property of KKLT models is that the exponential terms in the superpotential, which stabilize the volume modulus, simultaneously stabilize the axion field. The masses of these two fields are not much different, as one can easily see in many models, starting from the simplest KKLT model with $K = -3\ln(T + \bar{T})$, $W = W_0 + Ae^{-aT}$. The Kähler potential possesses the shift symmetry $T \to T + i\delta$ with real δ. With $T = \sigma + i\alpha$ the kinetic term for scalars and the potential $V(\sigma, \alpha)$ are:

$$L_{\text{kin}} = \frac{3}{4\sigma^2}[(\partial\sigma)^2 + (\partial\alpha)^2] \,,$$

$$V = \frac{aAe^{-2a\sigma}\left(A(3 + a\sigma) + 3e^{a\sigma}W_0\cos[a\alpha]\right)}{6\sigma^2} + \frac{D}{\sigma^3} \,.$$

Near the minimum of the potential at σ_0, α_0 the canonical fields are $\sqrt{3/2}\,\sigma/\sigma_0$ and $\sqrt{3/2}\,\alpha/\sigma_0$. Therefore the curvatures in the axion–volume directions plotted in σ, α variables are practically the same for canonically normalized fields. There is no significant flatness in the axion direction comparative to the volume modulus direction near the minimum of the potential.

The Supergravity Version of Chaotic Inflation

The model proposed in [27] is based on the Kähler potential with shift symmetry for the inflaton field and with the holomorphic superpotential, which breaks this symmetry,

$$K = \frac{1}{2}(\Phi + \bar{\Phi})^2 + X\bar{X} \,, \qquad W = m\Phi X \,. \tag{4.7}$$

The Kähler potential does not depend on the inflaton $\varphi = -i(\Phi - \bar{\Phi})/\sqrt{2}$, but the superpotential does depend on it. This model has a very steep potential with respect to all fields except φ. In the φ-direction it has a very simple potential $\frac{1}{2}m^2\varphi^2$. Thus this model is the supergravity version of the chaotic inflation model for a single scalar field with the potential $\frac{1}{2}m^2\varphi^2$.

As of now we do not know whether one can derive the supergravity chaotic inflation model [27] from string theory. The reason is that the Kähler potentials with shift symmetry for the closed string moduli usually have the form $K = -c\ln(\Phi + \bar{\Phi})$ and therefore would lead to the runaway behavior of the potential of the form $V \sim \frac{1}{(\Phi+\bar{\Phi})^c}$. This is very different from $V \sim e^{\frac{1}{2}(\Phi+\bar{\Phi})^2}$, which is an important feature of the model in [27]. Such type of Kähler potentials $K = \frac{1}{2}(\Phi+\bar{\Phi})^2$ with shift symmetry have been studied for the open string moduli [58, 59, 60] (as we will explain later for $D3/D7$ brane system). However, these fields have restricted range since they correspond to the distance between branes. It would be very interesting to find a valid regime of string theory capable of reproducing the supergravity version of the chaotic inflation [27], or its generalization, as an effective $\mathcal{N} = 1$ supergravity. It would be particularly important if both supersymmetry and gravitational waves were discovered.

The Axion Valley Model (Natural Inflation in Supergravity)

The natural inflation PNGB model for the pseudo-Nambu–Goldstone boson [61, 62] is based on a potential of the form $\Lambda^4(1 \pm \cos(\phi/f))$ with $f \geq 0.7 m_{Pl}$ ($m_{Pl} \equiv G^{-1/2} = 1.22 \times 10^{19}\,\text{GeV}$) and $\Lambda \sim M_{GUT}$. In [62] an attempt was made to derive this potential from string theory with an axion–dilaton field S. The canonical axion field was identified with $\sim \frac{\text{Im}S}{\text{Re}S}$. A closely related idea that the axions in string/M-theory may play the role of dark energy was proposed in [63]. To obtain natural inflation or axion dark energy in supergravity it was necessary to stabilize the dilaton, ReS, and to keep an almost flat potential for the axion, ImS. Until now, this goal was not achieved. Similarly, no realization of natural inflation or of the axion dark energy was proposed in string theory.

However, as we now show, it is indeed possible to develop a consistent realization of natural inflation in supergravity. We consider the KKLT model with all fields fixed at their minima, and add to it a field Φ with a shift-symmetric Kähler potential and a non-perturbative superpotential which breaks the shift symmetry of the Kähler potential:

$$K = \frac{1}{4}(\Phi + \bar{\Phi})^2 , \qquad W = w_0 + Be^{-b\Phi} , \qquad (4.8)$$

with[6]

$$V_\Phi = e^K(|DW|^2 - 3|W|^2) = V_1(x) - V_2(x)\cos(b\beta) . \qquad (4.9)$$

Here $\Phi = x + i\beta$ and

$$V_1(x) = e^{x(-2b+x)}B^2(-3 + 2(x-b)^2 + e^{2bx}(-3 + 2x^2)w_0^2 , \qquad (4.10)$$

$$V_2 = 2Be^{bx}w_0(3 + 2bx - 2x^2) . \qquad (4.11)$$

The presence of the KKLT model is to modify the potential constructed from (4.8) in two aspects. It rescales the overall value of the Φ field potential and adds to it a positive constant. The effective uplifting can make the potential at the minimum of Φ close to zero (from the positive side). This rescaling can

---

[6] One can also use two exponents and/or other more complicated version of the model.

be absorbed by an effective rescaling of w_0 and B. Thus we have a model with the canonical kinetic term for both x and β and the following potential

$$g^{-1/2}L = \frac{1}{2}[(\partial x)^2 + (\partial \beta)^2] - V(x, \beta) , \qquad (4.12)$$

where the axion valley potential is

$$V(x, \beta) = V_1(x) - V_2(x)\cos(b\beta) - V_0, \qquad V_0 = V_1(x_0) - V_2(x_0)\cos(b\beta_0) , \qquad (4.13)$$

and x_0, β_0 is the point where the potential has a minimum so that the potential vanishes at the minimum. $V_1(x), V_2(x)$ are given in (4.10) and (4.11).

If the minimum lies at $\beta_0 = 0$ the potential at stabilized x at the point x_0 takes the form of the natural PNGB model potential:

$$V = V_2(x_0)(1 - \cos(b\beta)) . \qquad (4.14)$$

Our goal is to make the potential in (4.12) for the x field steep and the potential for the β field very flat. This is indeed possible, as different from the KKLT model (4.1). In the KKLT model, the potential is equally steep for the volume modulus and the axion near the minimum of the potential, see Fig. 4.2. Meanwhile, in the axion valley model, e.g., for $B = 1$, $b = 0.05$, $w_0 = 10^{-4}$, we find that the potential in the x-direction is steep, and we have a nice nearly flat valley for the axion β, which may play the role of the inflaton field, see Fig. 4.3 where the potential is also multiplied by 10^3.

To make this model compatible with the WMAP3 data, we may put the system to the minimum in x at $x = x_0$ and use the values for the parameters suggested in [64] for the potential $V = \Lambda^4(1 - \cos(\phi/f))$. We need $V_2(x_0) = \Lambda^4$ with Λ at the GUT scale and our parameter $1/b$ corresponds to $\sqrt{8\pi}f$ in [64]. We have to take into account that in a supergravity setting we are working in units where $M = M_{Pl} \equiv m_{Pl}/\sqrt{8\pi} = 1$, $m_{Pl} = 1.22 \times 10^{19}$ GeV.

There are two limiting cases to consider. In the first case, $0.7m_{Pl} \leq f \ll 5m_{Pl}$ ($0.04 \ll b \leq 0.28$), inflation takes place near the maximum of the potential, as in the new inflation scenario. In the second case, $f \geq 5m_{Pl}$ ($b \leq 0.04$), the potential is very flat at the minimum and the model is close to the simplest chaotic inflation scenario with a quadratic potential. In this regime, for $x_0 < 1$, the COBE/WMAP normalization of inflationary perturbations implies that $w_0 B b^2 \sim 1.5 \times 10^{-12}$. Clearly, such parameters are possible from the point of view of supergravity, particularly with an account of the rescaling mentioned above when combining this model with the KKLT potential for uplifting.

To the best of our knowledge, the axion valley model (4.8) proposed above provides the first explicit realization of the natural inflation in supergravity. It does realize the standard lore that the shift symmetry of the Kähler potential may protect a nearly flat axion potential. It gives a simple example of such a model, where the partner of the axion is stabilized and the total potential has a stable dS minimum.

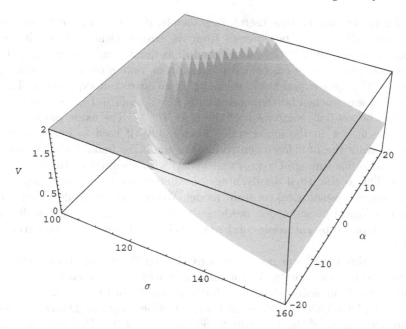

Fig. 4.2. The funnel-type potential of the KKLT model depending on the volume σ and the axion α from [57]. Figure 4.1 presents a slice of this potential in the σ direction at the minimum for the axion. The potential in the axion direction is as steep as in the volume modulus direction

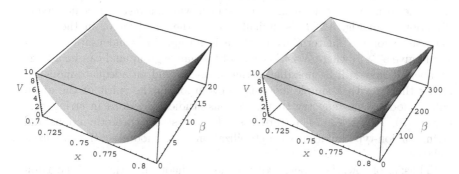

Fig. 4.3. Axion valley potential (4.12) and (4.13). On the *left figure*, there is a view on the axion valley. There is a sharp minimum for x and a very shallow minimum for β. The β-direction is practically flat for β from 0 to 20 (in Planck units), whereas in the x-direction the potential evolves significantly when x changes by 0.1. On the *right figure*, the potential is plotted for β from 0 to 300. The plot shows the periodicity in the axion variable β. Both β and x have canonical kinetic terms

This model may become useful, in principle, if dark energy in future experiments will prove to be different from the cosmological constant. It has been shown recently [65] that scalar moduli as quintessence in supergravity either behave as a pure cosmological constant or violate constraints coming from the fifth force experiments, such as Cassini spacecraft experiment. This conclusion was reached for the models with $K = -c \ln(T + \bar{T})$, and was valid only for the "radial" component of the field, $T + \bar{T}$. The axion moduli like $T - \bar{T}$ or $\Phi - \bar{\Phi}$ in the axion valley model (4.8), if used as quintessence, seem to avoid the fifth force constraint. On the other hand, for such models a double fine-tuning of $V_1(x_0)$ and $V_2(x_0)$ with accuracy of 10^{-120} is required. This is the usual disadvantage of quintessence models as compared to the simple cosmological constant model. We remind the readers that for the cosmological constant one needs a single fine-tuning, which, hopefully, can be addressed by anthropic considerations in the framework of the stringy landscape.

So far this model has not been derived from the string theory, but it is available in supergravity. The main reason why it is not easy to get an axion valley from string theory is that one has to justify the Kähler potential in (4.8). This type of Kähler was identified, e.g., in $D3/D7$ model via an expansion of the logarithmic potential in (4.19). However, in this particular situation it is not easy to argue that the superpotential in (4.8) can be used. In the same model with the $K3 \times \frac{T^2}{\mathbb{Z}_2}$ compactification one can use the Kähler–Hodge manifold $\frac{SO(2,n)}{SO(2) \times SO(n)}$ with the Kähler potential $K = \ln[(x_0 + \bar{x}_0)^2 - \sum_{i=1}^{i=n}(x_i + \bar{x}_i)^2]$ [66]. This Kähler potential is perfectly suitable for our purpose after expansion near the minimum for $x_0 + \bar{x}_0$. However, in this situation the exponential terms in the superpotential originate from the instantons and not from the gaugino condensation [67, 68, 69]. They have the form $e^{-2\pi n x_i}$ with integer n which will not result in a flat axion direction as in Fig. 4.3. It is difficult to get the small factor in the exponent since for the quaternions x_i there is no gaugino condensation which would give $e^{-\frac{2\pi}{N} x_i}$ where N is the rank of the gauge group and can be a large number. It may still be possible in the future to find an adequate model in string theory with a simple axion-type evolution in the spirit of the axion valley model. One may try to use various studies of axions in string theory in [70, 71, 72, 73, 74] toward cosmology of string theoretic axions and stringy axion-type inflation with stabilization of all moduli at the exit from inflation.

Thus, as of now, the supergravity version of chaotic inflation and the axion valley inflation models have not been implemented in string theory. In what follows we will describe some interesting models of inflation in string theory and comment on their attractive features and problems.

4.4 Brane Inflation in String Theory

4.4.1 KKLMMT-Like Models of Inflation in the Warped Throat Geometry

The first inflationary scenario in string theory with compactification of extra dimensions and stabilized moduli was proposed in the KKLMMT paper [13]. This model is a stringy development of the previously introduced concept of brane inflation [10, 14]. It is based on the KKLT mechanism of moduli stabilization, so that at the exit from inflation all moduli are stabilized. Inflation occurs as in the hybrid inflation scenario [75], the distance between mobile $D3$ and $\overline{D3}$ branes playing the role of the inflaton field. Hybrid inflation ends due to the brane–anti-brane annihilation. Both inflation and the brane–anti-brane annihilation take place in the warped throat geometry [76, 77]. This model without fine-tuning has a so-called η-problem [78]: the inflaton mass-squared is large, $m^2 \sim H^2$, and inflation is short. In effective supergravity this can be seen as coming from the Kähler potential of the form

$$K = -3\ln\left[(T + \bar{T}) - \Phi\bar{\Phi}\right] , \qquad (4.15)$$

where the distance between $D3$ and $\overline{D3}$ branes is related to Φ and $(T + \bar{T})$ is related to the volume of the compactification. At some fixed value of $T + \bar{T}$ given by $(T + \bar{T})_0$ there is a standard Kähler potential $K = \phi\bar{\phi}$ for the redefined distance field $\phi = \frac{\sqrt{3}\Phi}{\sqrt{(T+\bar{T})_0}}$, which comes from the expansion near the minimum of the volume:

$$K = -3\ln[(T + \bar{T}) - \Phi\bar{\Phi}] \approx -3\ln[(T + \bar{T})_0] + \phi\bar{\phi} + ... \qquad (4.16)$$

The inflaton potential for the field ϕ has a form $e^K V_W$ where $V_W = (|DW|^2 - 3|W|^2)$. With $e^K = e^{\phi\bar{\phi}}$, the η-problem is due to the e^K part of the potential. Because of this term, the second derivative of the potential is of order H^2 instead of $\sim 10^{-2}H^2$ as required by the flatness of the spectrum of inflationary perturbations. One of the possibilities to improve the situation is to use a contribution from the superpotential which cancels that from e^K. It is possible to avoid the η-problem in the KKLMMT class of models via some fine-tuning by accounting for stringy quantum corrections [79] which lead to some modifications of the superpotential. The most recent detailed investigation of these and relevant issues was performed in [80, 81]. In particular, the structure of the superpotential of the form

$$W = W_0 + A(\phi)e^{-aT} \qquad (4.17)$$

was clarified. Here the pre-exponential factor $A(\phi)$ acquires the dependence on the position of the $D3$ brane in the throat. However, the complete and predictive phenomenology of these models is still to be worked out. An extended review of this class of models and their generalizations was presented recently in [4, 5].

From the perspective of the observational points explained above, this class of models of $D3$–$\overline{D3}$ brane annihilation in the warped throat geometry has the following features:

(a) The value of the tilt of the spectrum n_s is not unambiguously predictable and depends on the details of the model, like the choice of the fine-tuning and/or the choice of the modification of the original model.

(b) In all versions of this model the level of primordial gravitational waves from inflation is predicted to be extremely small.

(c) Here is the most attractive feature of inflationary models in this class: They have $U(1)$ symmetry, and the corresponding cosmic strings can be easily produced by the end of inflation. Such strings can appear in many versions of the hybrid inflation scenario, which may lead to cosmological problems. However, in the KKLMMT model, the tension of these strings can be easily controlled [13] by the warp factor in the throat geometry, $e^{-\frac{2\pi K}{3g_s M}}$ [76, 77]. By a proper choice of the integer fluxes K and M one can make the string tension and string contribution to the perturbations of density rather small. But under certain conditions these strings may provide a detectable contribution to gravitational waves [38, 82]. They may also contribute to the CMB polarization. If the B-type polarization in CMB will be detected as coming from the vector modes generated from cosmic strings [40, 41, 83], this class of models will become particularly attractive. A discovery of very light cosmic strings will force us to look for a better understanding of the theoretical problems in this class of inflationary models in string theory.

4.4.2 On Shift Symmetry in $D3/D7$ Brane Inflation

The $D3/D7$ brane inflation model [19, 20, 84] is a stringy version of the **D**-term inflation in supergravity. This model has a number of interesting features. It relies on one of the most theoretically advanced examples of stabilization of all moduli of M-theory on $K3 \times K3$ manifold and type IIB string theory on $K3 \times \frac{T^2}{\mathbb{Z}_2}$ orientifold [66, 67, 68, 69, 85]. The model has an approximate shift symmetry [58, 59, 60], which results in the required flatness of the inflaton potential. The original **D**-term inflation [17, 18] has a nearly flat potential since the F-term is vanishing. However, in $D3/D7$ brane inflation model the F-term potential is required for the volume stabilization. Therefore the shift symmetry of the Kähler potential is important for the effective supergravity model of inflation in which the volume of the extra dimensions is stabilized. The Kähler potential is

$$K = -3\ln\left[(T + \bar{T}) - \frac{1}{2}(\Phi + \bar{\Phi})^2\right] , \qquad (4.18)$$

where the two-dimensional distance between $D3$ and $D7$ branes is related to $(\Phi + \bar{\Phi})$ and $(\Phi - \bar{\Phi})$ fields and the volume is related to $T + \bar{T}$. The Kähler potential does not depend on the $(\Phi - \bar{\Phi})$ field, which is therefore an inflaton field. [7] As before, $(T + \bar{T})$ is related to the volume of the compactification. After the volume stabilization at $T + \bar{T} = (T + \bar{T})_0$, the Kähler potential takes the following form:

$$K = -3\ln[(T + \bar{T}) - \frac{1}{2}(\Phi + \bar{\Phi})^2] \approx -\ln[(T + \bar{T})_0^3] + \frac{1}{2}(\phi + \bar{\phi})^2 + \dots \quad (4.19)$$

This tree-level Kähler potential does not depend on the inflaton field $(\phi - \bar{\phi})$ under certain geometric conditions. This eliminates the η-problem in the inflaton direction if all quantum corrections are small. An example of such geometric conditions is the case of the $K3 \times \frac{T^2}{\mathbb{Z}_2}$ manifold, where the Kähler potential of the form closely related to the one in (4.18) was derived in [66] from the $\mathcal{N} = 2$ supergravity structure, where

$$K = -\ln\left[i(\bar{X}^\Lambda F_\Lambda - X^\Lambda \bar{F}_\Lambda\right] . \quad (4.20)$$

Here the holomorphic prepotential $\mathcal{F}$ depends on coordinates $X^\Lambda = (X^0, X^A)$ of the Kähler manifold and $F_\Lambda \equiv \partial_\Lambda \mathcal{F}$. This Kähler potential is invariant under symplectic transformations associated with duality symmetry in string theory:

$$\begin{pmatrix} X \\ F \end{pmatrix}' = \begin{pmatrix} A & B \\ C & D \end{pmatrix} \begin{pmatrix} X \\ F \end{pmatrix} . \quad (4.21)$$

In the case of the cubic prepotential,

$$\mathcal{F} = \frac{C_{ABC}}{3!} \frac{X^A X^B X^C}{X^0} , \qquad \Lambda = (0, A) , \quad (4.22)$$

the formula (4.20) always leads to a Kähler potential with a manifest shift symmetry

$$K = -\ln\left[i\frac{C_{ABC}}{3!}(z - \bar{z})^A(z - \bar{z})^B(z - \bar{z})^C\right] , \quad (4.23)$$

where $z^A = \frac{X^A}{X^0}$ are the special coordinates of the Kähler manifold. The symplectic invariance of the Kähler potential in $\mathcal{N} = 2$ supergravity plays a fundamental role for the attractor mechanism and the computation of the entropy of stringy BPS black holes [87].

It was explained in [60] that the shift symmetry of $D3/D7$ model is a subgroup of the duality symmetry of string theory associated with the cubic superpotential $\mathcal{F} = s(tu - \frac{1}{2}x^2)$ where s is the axion–dilaton, t is the volume of the $K3$ manifold, u is the complex structure of the $\frac{T^2}{\mathbb{Z}_2}$ manifold and x is

---

[7] Inflationary models in supergravity with shift symmetric Kähler potential analogous to (4.18) were studied in [86].

the $D3$ brane position. The formulae (4.18) and (4.19) for the Kähler potential are somewhat simplified version of the exact expression. The purpose of the simplification is to allow an easy comparison with the analogous Kähler potential without shift symmetry in (4.15) and (4.16).

An assumption that stringy quantum corrections do not break badly the tree-level shift symmetry, which is broken only softly by Coleman–Weinberg type logarithmic corrections to the potential, replaces the fine-tuning which is often required for the flatness of the inflationary potential in other models. These logarithmic corrections are present due to Fayet–Iliopoulos terms (magnetic fluxes on $D7$ brane). Note that the Kähler potential (4.18) and (4.19) has a shift symmetry

$$\phi \to \phi + i\delta , \qquad \delta = \bar{\delta} \tag{4.24}$$

of the same type as the chaotic inflation supergravity model [27].

Let us compare it to the models of the KKLMMT type. Stringy corrections of the type derived in [79, 80] must be significant in models [13] to provide the solution of the η problem via the superpotential dependence on the inflaton: these corrections must cancel the $m^2 \sim H^2$ contribution from the Kähler potential. In the case of the $D3/D7$ model with the term $\frac{1}{2}(\phi + \bar{\phi})^2$ in the Kähler potential, there is no η-problem in the $(\phi - \bar{\phi})$ direction if the stringy corrections of the type studied in [79, 80] are small. An interesting feature of such corrections is that their value may depend on the stabilized values of other moduli, e.g., some complex structure moduli. The choice of fluxes stabilizing the complex structure moduli may therefore help to control these corrections and to make them large or small, depending on the model.

D-term inflation is closely related to theoretical issues of D-term uplifting [88] of AdS vacua to dS vacua in the KKLT-type construction. In string theory the relevant models use fluxes on $D7$ branes, which upon volume stabilization reduce to Fayet–Iliopoulos (FI) terms in supergravity. In presence of dark energy/positive CC the significance of FI terms in supergravity/string cosmology is very important. This has led to recent progress toward a better understanding of the FI terms [89, 90] in supergravity/string theory.

Some issues with consistent constructions of D-term uplifting with FI terms for de Sitter vacua were only recently clarified [48, 51, 52, 91, 92, 93, 94, 95, 96]. Here one should note that in string theory there are no constant FI terms, only field-dependent D-terms! They become "constant" FI-terms after moduli stabilization in effective supergravity. The D-term uplifting and stringy D-terms inflation have been studied recently [81, 97, 98] and one can expect more investigations of these issues in future. Also new ideas on F-term uplifting of AdS vacua to dS vacua with positive energy have been proposed recently [99, 100, 101]. Note that in supergravity it is easy to find AdS vacua with negative CC. The uplifting mechanisms (D-term or F-term) are designed to convert the AdS supergravity vacua into those with positive CC. Therefore the studies of these uplifting mechanisms in a context of effective supergravity may provide us with an answer to a *profound issue: where does the positive energy of the universe come from?*

To the extent to which $D3/D7$ brane inflationary model is reducible to the D-term inflation, the situation with three major observational possibilities is the following:

(a) A generic value of the tilt of the spectrum is $n_s \approx 0.98$, which is a bit higher than the current number relevant to models without gravitational waves from inflation. Therefore here the future precision data on n_s will be important. It will also require an effort to suppress n_s, if necessary, as in [102, 103], toward smaller values in a way motivated by the theory. It would be interesting to find out what kind of string theory quantum corrections may help in this direction.

(b) The model predicts an undetectable level of primordial gravitational waves from inflation.

(c) D-term inflation, as well as its stringy version $D3/D7$ brane inflationary model, have $U(1)$ symmetry and the corresponding cosmic strings are generically produced. These strings are heavy unless a special effort is made to get rid of them [84, 89, 103, 104, 105, 106, 107]. It does not seem possible to produce very light cosmic strings which may eventually be detected via the B-type polarization in CMB from the vector modes.

4.4.3 DBI Inflation

The idea of DBI inflation [30, 31] is to consider a relativistic motion of the mobile D3 brane. The action of the brane is therefore considered in the proper Born–Infeld form without expansion that keeps only the second derivative terms. There is a limit on the maximal velocity which is required to make the action consistent. All this leads to an unusual and interesting model of inflation, which, in principle, may predict some non-gaussianity and significant gravitational waves. The actual predictions seem to depend on specific assumptions on the geometry in which the brane is embedded, etc. As an example of such assumptions, we refer to [108], where it was shown that to match the data, $D3$ must move close to the tip of the warped throat. In models with 60 e-folds in the KS throat [76], the non-gaussianity was shown to exceed the current bounds. However, in other geometries things may be different and have to be studied separately. One of the interesting features of this model is that some minimal level of non-gaussianity is always expected: this makes the model falsifiable by future data.[8]

---

[8] Most recent analysis of this model in [28] and, particularly in [29], indicates that the observable r may not be possible in all developed brane inflation models including the DBI model.

4.4.4 Assisted M5 Brane Inflation

We would like to comment here on one more model of inflation in the heterotic M-theory with multiple moving M5 branes [109]. This model is an attempt to find a version of assisted inflation [32, 33] in M-theory where a large amount of branes may help to realize an effectively flat potential to assist the inflation which is not possible for a single brane. The problem of the heterotic M-theory is that the stabilization of the orbifold length, the volume of compactification, and other moduli require a regime of a strong coupling. However, it is interesting that if we take the prediction of the phenomenological part of the model regarding a spectral index we find the following change from WMAP1 to WMAP3 data. For $n_s \approx 0.98$ the number of M5 branes required for the assistance effect is about 89. For the new value of the index $n_s \approx 0.95$ one needs 66 branes. This model gives an example of the situation when the new data can be easily accommodated within the same model.

4.4.5 A Remark on Conceptional Issues in Brane Inflation

All models of brane inflation starting with [14] have an interesting and novel feature with regard to previously known models: a possibility of an interpretation of the four-dimensional scalars as distance between branes, as excitations of strings stretched between branes, etc. This advantage, however, is somewhat difficult to realize in a clear and consistent way in the context of a compactified internal space.

The conceptual problems of brane inflation in general reflect the difficulty of describing the action of branes, or a probe brane, in a background of compactified internal space. This is a reflection of the major problem of string theory where the description of the open string sector and D-branes is not clearly formulated in the framework of the closed string theory.

More work will be required here to clarify the geometry of extra dimensions in the presence of moving extended objects and the relevant effective supergravity in four dimensions.

4.5 Modular Inflation in String Theory

The models of the so-called modular inflation, i.e., inflation with moduli fields corresponding to the closed string sector, are conceptually simpler than the models of brane inflation. Several interesting models belonging to this class have been derived lately; we are going to discuss them below.

One could wonder why we started with brane inflation in our investigation of the KKLT-based inflationary models? The answer is simple: we have been unable to identify any flat directions suitable for inflation in the original KKLT models. However, perhaps, some simple axion-type inflation models are available in string theory with stabilized moduli. In Sect. 4.3.1 of this paper, we have proposed an explicit $\mathcal{N} = 1$ supergravity model which has an axion valley with a stabilized scalar direction and a flat axion direction, see Fig. 4.3. To the best of our knowledge, this is the first explicit realization of natural inflation in supergravity. However, so far we have been unable to find any such models in string theory.

We therefore focus here on the existing modular inflation models derived from string theory: the racetrack inflation [21, 22], better racetrack inflation [24], large volume Kähler inflation [23], and its generalized version, roulette inflation [25]. Unlike the brane inflation models discussed in the previous section, all of these modular inflation models allow eternal slow-roll inflation.

The spectral index is $n_s \approx 0.95$ for the racetrack models; $n_s \approx 0.96$ for the large volume Kähler inflation. In all cases the amplitude of tensor perturbations is extremely small, and there are no cosmic strings. If the Planck satellite confirms n_s at this level and does not detect gravity waves and cosmic strings, we will have to pay serious attention to these models.

Their conceptual simplicity comparatively to the brane inflation models is due to the fact that one can use the framework of closed string theory only (i.e., without open strings) and relate it to the effective $\mathcal{N} = 1$ $d = 4$ supergravity. One has to deduce the structure of the Kähler potential and the superpotential and find out the parameters which provide inflation. This is where the fine-tuning enters.

A toy model of racetrack inflation [21] is based on one complex modulus and has five parameters with significant fine-tuning. The gaugino-induced racetrack inflation [22] has two complex moduli, nine parameters with less fine-tuning, and has most features as in [21]. More realistic models of string theory compactification, the "better racetrack models" [24], start with two complex moduli and in case of five parameters require a significant fine-tuning. Large volume Kähler inflation models [23, 25] require three or more moduli, start with 11 parameters and need less fine-tuning.

4.5.1 Racetrack Inflation as Eternal Topological Inflation

Slow-roll inflation is realized if a scalar potential $V(\phi)$ is positive in a region where the following conditions are satisfied:

$$\epsilon \equiv \frac{1}{2} \left(\frac{V'}{V} \right)^2 \ll 1 \qquad \eta \equiv \frac{V''}{V} \ll 1 . \qquad (4.25)$$

Primes refer to derivatives with respect to the scalar field, which is assumed to be canonically normalized. Satisfying these conditions is not easy for typical potentials since the inflationary region has to be very flat. Furthermore, after finding such a region we are usually faced with the issue of initial conditions: Why should the field ϕ start in the particular slow-roll domain?

For the simplest chaotic inflation models of the type of $\frac{m^2}{2}\phi^2$ this problem can be easily resolved, see Chap. 1 and [110]. The problem of initial conditions in the theories where inflation is possible only at the densities much smaller than the Planck density is much more complicated; for a possible solution see, e.g., [111]. In any case, one can always argue that even if the probability of proper initial conditions for inflation is strongly suppressed, the possibility to have eternal inflation infinitely rewards those domains where inflation occurs. In other words, one may argue that the problem of initial conditions in the theories where eternal inflation is possible becomes largely irrelevant (see Chap. 5).

Eternal inflation [112, 113] is not an automatic property of all inflationary models. Many versions of the hybrid inflation scenario, including some of the versions used recently for the implementation of inflation in string theory, do not have this important property. Fortunately, inflation is eternal in all models where it occurs near the flat top of the scalar potential. Moreover, in this case eternal inflation occurs even at the classical level, due to the eternal expansion of topological defects [114, 115, 116].

The racetrack inflation [21] gives an example of eternal topological inflation within string theory moduli space, generalizing the KKLT scenario. The model differs from the simplest KKLT case only in the form assumed for the nonperturbative superpotential, which is taken to have the modified racetrack form

$$K = -3\ln(T + \bar{T}), \qquad W = W_0 + A\,\mathrm{e}^{-aT} + B\,\mathrm{e}^{-bT}. \qquad (4.26)$$

Such superpotential would be obtained through gaugino condensation in a theory with a product gauge group. The constant term W_0 results from fluxes and represents the effective superpotential as a function of all the fields that have been fixed already, such as the dilaton and complex structure moduli. As in the simplest KKLT model, the scalar potential is a sum of two parts $V = V_F + \delta V$. The first term comes from the standard $\mathcal{N} = 1$ supergravity formula for the F-term potential in (4.3). The uplifting potential, δV, is taken in the form $\delta V = \frac{E}{X^2}$ and $T \equiv X + \mathrm{i}Y$. The total potential is

$$\begin{aligned}
V &= \frac{E}{X^\alpha} + \frac{\mathrm{e}^{-aX}}{6X^2}\left[aA^2\,(aX + 3)\;\mathrm{e}^{-aX} + 3W_0aA\cos(aY)\right] + \\
&\quad + \frac{\mathrm{e}^{-bX}}{6X^2}\left[bB^2\,(bX + 3)\;\mathrm{e}^{-bX} + 3W_0bB\cos(bY)\right] + \\
&\quad + \frac{\mathrm{e}^{-(a+b)X}}{6X^2}\left[AB\,(2abX + 3a + 3b)\cos((a - b)Y)\right]
\end{aligned} \qquad (4.27)$$

Notice that, to the order to which we are working, the Kähler potential depends only on X and not on Y. For fields rolling slowly in the Y direction this feature helps to address the η-problem of F-term inflation. All dependence on the axion field Y is via $\cos(aY)$, $\cos(bY)$, and $\cos((a - b)Y)$ as the result of the shift symmetry of the Kähler potential, which is broken by the axion dependence in the exponents in the superpotential.

This potential has several de Sitter (or anti-de Sitter) minima, depending on the values of the parameters A, a, B, b, W_0, E. In general it has a very rich structure, due in part to the competition of the different periodicities of the Y-dependent terms. In particular, $a - b$ can be very small, as in standard racetrack models, since we can choose $a = 2\pi/M$, $b = 2\pi/N$ with $N \sim M$ and both large integers. Notice that in the limit $(a - b) \to 0$ and $W_0 \to 0$, the Y direction becomes exactly flat. We can then tune these parameters (and AB) in order to obtain flat regions suitable for inflation.

The values of the parameters of this potential are: $A = \frac{1}{50}, B = -\frac{35}{1000}, a = \frac{2\pi}{100}, b = \frac{2\pi}{90}, W_0 = -\frac{1}{25000}$. Note that the potential is periodic with period 900, i.e., there is a set of two degenerate minima at every $Y = 900n$ where $n = 0, 1, 2, ...,$ etc., as shown in Fig. 4.4.

The shape of the potential is very sensitive to the values of the parameters. Figure 4.5 illustrates a region of the scalar potential for which inflation is possible. With these values the two minima seen in Fig. 4.5 occur for field values $X_{\mathrm{min}} = 96.130$ and $Y_{\mathrm{min}} = \pm 22.146$, and the inflationary saddle point lies at $X_{\mathrm{saddle}} = 123.22$ and $Y_{\mathrm{saddle}} = 0$. It is crucial that this model contains two degenerate minima since this guarantees the existence of causally disconnected regions of space which are in different vacua. These regions necessarily have a domain wall between them where the field is near the saddle point and thus eternal inflation is taking place, provided that the slow-roll conditions are satisfied there. It is then inevitable to have regions close to the saddle in which inflation occurs, with a sufficiently large duration to explain our flat and homogeneous universe.

The racetrack model as well as many other string inflation models has some interesting scaling properties, e.g.,

$$a \to a/\lambda, \quad b \to b/\lambda, \quad E \to \lambda^2 E, \tag{4.28}$$

$$A \to \lambda^{3/2} A, \quad B \to \lambda^{3/2} B, \quad W_0 \to \lambda^{3/2} W_0. \tag{4.29}$$

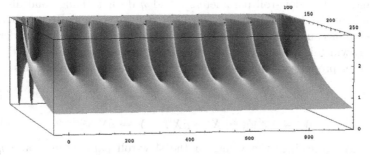

Fig. 4.4. A plot for the racetrack potential (rescaled by 10^{16}). Inflation begins in a vicinity of any of the saddle points. Units are $M_{\mathrm{pl}} = 1$. As one can see, the potential is periodic in the axion direction, but it is very much different from the potential of natural inflation: there is no axion valley here

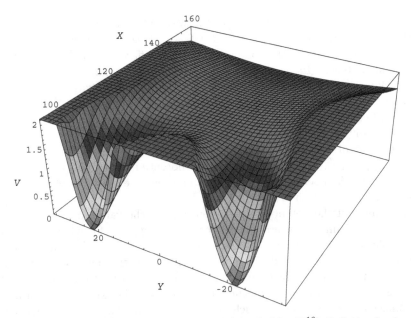

Fig. 4.5. Plot for a racetrack-type potential (rescaled by 10^{16}). Inflation begins in a vicinity of the saddle point and ends up in one of the two minima, depending on initial conditions. Note that near the minima the potential has a KKLT-type funnel shape where the curvature in volume and axion direction is of the same scale for canonical variables $X/X_{\min}$ and $Y/X_{\min}$

Under all these rescalings the potential does not change under the condition that the fields also rescale as

$$X \to \lambda X, \quad Y \to \lambda Y, \tag{4.30}$$

in which case the location of the extrema also rescale. One can verify that the values of the slow-roll parameters ϵ and η do not change and also the amplitude of the density perturbations $\frac{\delta\rho}{\rho}$ remains the same. It is important to take into account that the kinetic term in this model is invariant under the rescaling, which is not the case for canonically normalized fields.

Another property of this model is given by the following rescalings

$$a \to a/\mu, \quad b \to b/\mu, \quad E \to E/\mu, \tag{4.31}$$

$$V \to \mu^{-3}V, \quad X \to \mu X, \quad Y \to \mu Y. \tag{4.32}$$

Under these rescalings the values of the slow-roll parameters ϵ and η do not change; however, the amplitude of the density perturbations $\frac{\delta\rho}{\rho}$ scales as $\mu^{-3/2}$.

These two types of rescalings allow to generate many other models from the known ones, in particular to change the positions of the minima or, if one is interested in eternal inflation, one can easily change $\frac{\delta\rho}{\rho}$ keeping the potential flat.

One can study the slow-roll inflation, by examining field motion near the saddle point which occurs between the two minima identified above. At the saddle point the potential has a maximum in the Y direction and a minimum in the X direction, so the initial motion of a slowly rolling scalar field is in the Y direction.

To compute observable quantities for the CMB we numerically evolve the scalar field starting close to the saddle point, and let the fields evolve according to the cosmological evolution equations for non-canonically normalized scalar fields. The results of the numerical evolution confirm that the inflaton is primarily in the axionic direction Y at the very beginning of inflation, as must be the case since Y is the unstable direction at the saddle point. Eventually both the axion Y and the volume modulus X reach the absolute minimum. The spectral index is found to be $n_{\rm s} \approx 0.95$ in the COBE region of the spectrum.

4.5.2 Better Racetrack Inflation Model

The KKLT model with stabilization of just one complex Kähler modulus is a simplified toy model of the generic stabilization phenomenon with many moduli. One of the simplest more realistic models of this kind has two Kähler moduli. It is based on an explicit compactification of type IIB string theory: the orientifold of degree 18 hypersurface $\mathbb{P}^4_{[1,1,1,6,9]}$, an elliptically fibered Calabi–Yau over $\mathbb{P}^2$. The stabilization of moduli in this model was performed in [117] where it was also shown how $D3$ instantons generate a non-perturbative superpotential, thus providing an explicit realization of the KKLT scenario.

The model is a Calabi–Yau threefold with the number of Kähler moduli $h^{1,1} = 2$ and the number of complex structure moduli $h^{2,1} = 272$. The 272 parameter prepotential for this model is not known. However, one can restrict ourselves to the slice of the complex structure moduli space which is fixed under the action of the discrete symmetry $\Gamma \equiv \mathbb{Z}_6 \times \mathbb{Z}_{18}$. This allows to reduce the moduli space of the complex Calabi–Yau structures to just two parameters, since the slice is two-dimensional. This restricted model was studied intensely in string theory. The remaining 270 moduli are required to vanish to support this symmetry. The defining equation for the Calabi–Yau two-parameter subspace of the total moduli space is

$$f = x_1^{18} + x_2^{18} + x_3^{18} + x_4^3 + x_5^2 - 18\psi x_1 x_2 x_3 x_4 x_5 - 3\phi x_1^6 x_2^6 x_3^6 \,. \qquad (4.33)$$

The axion–dilaton and all complex structure moduli are stabilized by fluxes. The remaining two Kähler moduli are stabilized by a non-perturbative superpotential. For this model we identify situations for which a linear combination of the axionic parts of the two Kähler moduli acts as an inflaton. As in the previous racetrack scenario, inflation begins at a saddle point of the scalar potential and proceeds as an eternal topological inflation.

The Kähler geometry of the two Kähler moduli $h^{1,1} = 2$ was specified in [117]. We denote them by $\tau_{1,2} = X_{1,2} + iY_{1,2}$. These moduli correspond geometrically to the complexified volumes of the divisors (or four cycles) D_4 and D_5, and give rise to the gauge couplings for the field theories on the $D7$ branes which wrap these cycles. For this manifold the Kähler potential is given by

$$K = -2\ln R = -2\ln\left[\frac{1}{36}((\tau_2 + \bar{\tau}_2)^{3/2} - (\tau_1 + \bar{\tau}_1)^{3/2})\right], \qquad (4.34)$$

where R denotes the volume of the underlying Calabi–Yau space. The flat directions of the potential are lifted by $D3$ instantons, which generate the following non-perturbative superpotential:

$$W = W_0 + A\,e^{-a\tau_1} + B\,e^{-b\tau_2}. \qquad (4.35)$$

Given these expressions for K and W, the scalar potential takes the following form:

$$
\begin{aligned}
V_F + \delta V = {} & \frac{216}{(X_2^{3/2} - X_1^{3/2})^2}\Big\{B^2b(bX_2^2 + 2bX_1^{3/2}X_2^{1/2} + 3X_2)e^{-2bX_2} \\
& + A^2a(3X_1 + 2aX_2^{3/2}X_1^{1/2} + aX_1^2)e^{-2aX_1} \\
& + 3Bb\,W_0X_2e^{-bX_2}\cos(bY_2) + 3Aa\,W_0X_1e^{-aX_1}\cos(aY_1) \\
& + 3ABe^{-aX_1-bX_2}(aX_1 + bX_2 + 2abX_1X_2)\cos(-aY_1 + bY_2))\Big\} \\
& + \frac{D}{(X_2^{3/2} - X_1^{3/2})^2}
\end{aligned}
\qquad (4.36)
$$

where the last term is the uplifting term δV. Notice that this potential is parity invariant, $(X_i, Y_i) \to (X_i, -Y_i)$, with Y_i being pseudo-scalars. It is also invariant under the two discrete shifts, $Y_1 \to Y_1 + 2\pi m_1/a$ and $Y_2 \to Y_2 + 2\pi m_2/b$, where the m_i are arbitrary integers. There is also the approximate $U_R(1)$ R-symmetry, $a\,\delta Y_1 = b\,\delta Y_2 = \Delta$, which becomes exact in the limit $W_0 \to 0$.

We next ask whether slow-roll evolution is possible with this superpotential and Kähler potential. Searching the parameter space, we are able to find choices for which the scalar potential behaves similarly to the original racetrack inflation potential. Starting at the saddle point, since only one of the four real directions is unstable, we have sufficient freedom to make this direction flat enough to give rise to successful inflation.

Our goal was to find a set of parameters required for inflation with the COBE normalization of power spectrum. These examples are not particularly easy to find. The example with $P(k_0) = 4 \times 10^{-10}$ and $n_s = 0.95$, has the following parameters: $W_0 = 5.22666 \times 10^{-6}$, $A = 0.56$, $B = 7.46666 \times 10^{-5}$, $a = 2\pi/40$, $b = 2\pi/258$, $D = 6.21019 \times 10^{-9}$. With these choices of the parameters the minimum described above is located at $X_1 = 98.75839$, $X_2 = 171.06117$,

$Y_1 = 0, Y_2 = 129$, corresponding to a volume $R = 99$ in string units, which is large enough to trust the effective field theory treatment we use.

It is difficult to plot the potential since it is a function of four variables. Here we will only show the behavior of this potential as a function of the axion variables Y_1, Y_2 at the minimum of the radial variables X_1, X_2, and the potential as a function of the radial variables X_1, X_2 at the minimum of the angular variables Y_1, Y_2. Figure 4.6 illustrates the behavior of the potential near the minimum of the potential.

We have checked that the eigenvalues of the Hessian (mass^2) matrix are all positive, verifying that it is indeed a local minimum. The value of the masses for the moduli at this minimum turn out to be of order $10^{-6} - 10^{-7}$ in Planck units. Inflation occurs near the saddle point located at $X_1 = 108.96194, X_2 = 217.68875, Y_1 = 20, Y_2 = 129$. At this point the mass matrix has three positive eigenvalues and one negative one in the direction of $(\delta X_1, \delta X_2, \delta Y_1, \delta Y_2) = (0, 0, -0.6546, 0.7560)$, corresponding to a purely axion direction. This is the initial direction of the slow-roll away from the saddle point toward the non-trivial minimum described above.

The value of the effective potential at the saddle point is $V \sim 3.35 \times 10^{-16}$ in Planck units, so that the scale of inflation is $V^{1/4} = 3.25 \times 10^{14}$ GeV. This is a rather small scale. The ratio of tensor to scalar perturbations in this scenario is very small, $r \ll 1$, so the gravitational waves produced in this scenario will be very hard to observe.

To find the slow-roll parameter η at the saddle point (recall that $\epsilon = 0$ automatically at a saddle point), as well as to compute the inflationary trajectories, we must use the generalized definitions of the slow-roll parameters.

This leads to $n_s \approx 0.95$ and a long period of inflation, 980 e-foldings after the end of eternal inflation.

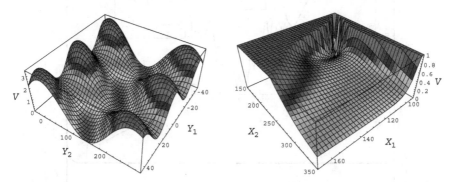

Fig. 4.6. On the *left* there is a potential as a function of the axion variables Y_1, Y_2 at the minimum of the radial variables X_1, X_2, in units 10^{-15} of the Planck density. On the *right* there is a potential as a function of the radial variables X_1, X_2 at the minimum of the angular variables Y_1, Y_2, in units 10^{-14} of the Planck density

We have computed the power spectrum for the model under consideration by first numerically evolving the full set of field equations, which can be
efficiently written in the form

$$\frac{d\phi_i}{dN} = \frac{1}{H}\dot{\phi}_i(\pi_i)$$

$$\frac{d\pi_i}{dN} = -3\pi_i - \frac{1}{H}\frac{\partial}{\partial\phi_i}\left(V(\phi_i) - L_{\text{kin}}\right) \tag{4.37}$$

Here $\mathcal{L}_{\text{kin}}$ is defined in (4.2) and N is the number of e-foldings starting from
the beginning of inflation, $\pi_i = \partial\mathcal{L}_{\text{kin}}/\partial\dot{\phi}_i$ are the canonical momenta, and the
time derivatives $\dot{\phi}_i$ are regarded as functions of π_i. We use initial conditions
where the field starts from rest along the unstable direction, close enough to
the saddle point to give more than 60 e-foldings of inflation. In fact our starting
point corresponds to the boundary of the eternally inflating region around the
saddle point. An example of the inflationary trajectories for all the fields is
shown in Figs. 4.7 and 4.8. The choice of the satisfactory parameters for this
model is not unique, just as in the racetrack scenario [21]. There is a rescaling
of parameters which does not alter the inflationary dynamics or the height
of the potential; it rescales the fields, but leaves the slow-roll parameters and
the amplitude of density perturbations invariant. There is also a second set
of rescalings, which does rescale the potential and the amplitude of density
fluctuations.

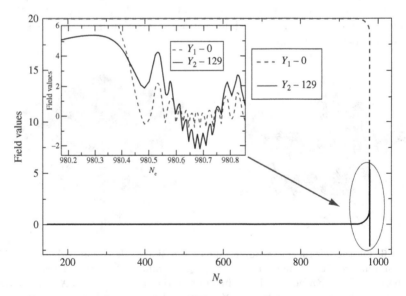

Fig. 4.7. Evolution of the axionic fields Y_1, Y_2 during inflation. We plot the values
of $Y_i - Y_i^{\min}$. *Inset* shows oscillations around the minimum at the end of inflation

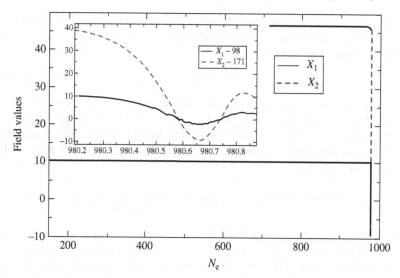

Fig. 4.8. Evolution of the X_1, X_2 directions during the inflationary period. We plot the values of $X_i - X_i^{\min}$. *Inset* shows oscillations around the minimum at the end of inflation

Evaluating the spectral index at 55 e-foldings before the end of inflation gives the spectral properties relevant for the CMB. For $W_0 = 5.227 \times 10^{-6}$ the spectral index reaches its largest value

$$n_s \approx 0.95 \, . \tag{4.38}$$

This is the same value that was found in the original racetrack model. The value of W_0 has to be tuned at the level of a percent to keep the spectral index from decreasing into a range of phenomenologically disfavored values.

In the KKLT model, with the superpotential containing only one exponent for the volume modulus, one could not have inflation without adding moving branes. In the original racetrack inflation scenario [21] it was possible to find the first working inflationary model without adding any new branes to the KKLT vacuum stabilization scenario. In the better racetrack model inflation is achieved in a theory with two moduli fields, without introducing the standard racetrack potentials with two exponential terms for each of them. This suggests that by increasing the number of moduli fields and/or the number of non-perturbative contributions to the superpotential, inflation may become easier to achieve and with less fine-tuning.

4.5.3 Gaugino Condensation Induced Racetrack Models of Inflation

An interesting development of racetrack models of inflation was suggested in [22]. They did not use the flux contribution to the superpotential, i.e., they chose $W_0 = 0$. They have assumed that the total volume modulus is fixed and focused their attention on the dilaton–axion field and one more complex field with the canonical Kähler potential,

$$K = -\ln(S + \bar{S}) + \chi\bar{\chi} . \tag{4.39}$$

The choice of the non-perturbative superpotential for these two fields was made on the basis of some prior studies of the effects of the gaugino condensation where the superpotential may have, in addition to the exponential dependence on the dilaton–axion, a specific dependence on other moduli. In their particular model

$$W = \chi^p A N_1 M^3 \mathrm{e}^{-S/N_1} + \chi^{p'} B N_2 M^3 \mathrm{e}^{-S/N_2} \left(\frac{M^2}{(\alpha + \beta\chi)^2}\right)^\gamma . \tag{4.40}$$

Thus the model has two complex moduli, $S = s + i\phi$ and $\chi = x e^{i\theta}$, and depends on nine parameters. The specific choice was made for a good inflationary model: $A = 1.5, B = 8.2, N_1 = 10, N_2 = 9$. The additional interactions/parameters, which were absent in previous models, are $p = p' = 0.5$, $\alpha = 1, \beta = 2.3$, and finally $\gamma = 10^{-4}$. The model does not seem to require a significant fine-tuning. However, the smallness of γ is important as the mass-squared of θ is proportional to γ. The qualitative picture of this model is very close to the racetrack models described above in a sense that the potential has a saddle point with the flat θ-axion direction, where the eternal topological inflation may take place. Inflation ends with an exit stage, when the system reaches the minimum of the potential and all moduli are stabilized.

4.5.4 Inflation in Models with Large Volume of Compactification

Another interesting class of inflationary models [23, 25] has been developed on the basis of the so-called large volume compactification [118], where the value of the W_0 in the superpotential is not small, as in simplest KKLT models, and therefore the α' correction to the Kähler potential plays a significant role. An example of a successful model is codified in the Kähler potential

$$K = -2\ln\left[\alpha\left((T_1 + \bar{T}_1)^{3/2} - \lambda_2(T_2 + \bar{T}_2)^{3/2} - \lambda_3(T_3 + \bar{T}_3)^{3/2}\right) + \frac{\xi}{2}\right], \tag{4.41}$$

and the superpotential

$$W = W_0 + \sum_{i=1}^{3} A_i e^{a_1 T_i} . \tag{4.42}$$

Here the term $\xi/2$ is due to the α' corrections to the Kähler potential. The model has 11 parameters, to compare with five parameters in the previously discussed racetrack and better racetrack models. It offers therefore more possibilities to look for inflationary slopes. The choice made in [23], as well as in roulette case [25], has been first of all to find the conditions on the parameters of the model which allow to stabilize the moduli T_1 and T_2 in a way that they actually do not participate in inflation; only the third modulus T_3 is driving inflation. In [23] the axion $\mathrm{Im}T_3$ of the third modulus T_3 is also frozen at its minimum, and the inflaton is given by $\mathrm{Re}T_3$. The evolution of this field takes place in a nice τ-trough.

In a more general case considered in [25], two complex moduli are fixed, as before, but in the third modulus, $T_3 = \tau + i\theta$, both the volume modulus τ and the axion θ are not fixed. As the result, there are many inflationary trajectories in the landscape of this model. The potential is exponentially flat in the τ-direction and has a periodic structure in the θ-direction.

The cosmological evolution of the complex field T_3 has been evaluated numerically using the "SuperCosmology" code [57] designed for models with generic moduli space metric and arbitrary number of complex fields with any Kähler potential and superpotential. The trajectories depend on initial conditions in the τ, θ plane. The randomness of (τ, θ) initial conditions allows for a large ensemble of trajectories. Features in the ensemble of histories include roulette trajectories, with long-lasting inflation in the direction of the rolling axion, enhanced in number of e-foldings over those restricted to lie in the τ-trough. The asymptotic flatness of the potential makes possible an eternal slow-roll inflation. A wide variety of potentials and inflaton trajectories agree with the cosmic microwave background and large-scale structure data.

The ABC for this class of models is simple: the value of the n_s is predicted to be 0.96 in agreement with current data, no observable gravitational waves, no cosmic strings.

4.6 N-flation/Assisted Inflation

At present N-flation [34, 35] is the only inflationary model studied in the context of string theory that may result in an effective chaotic inflation with a significant level of gravitational waves. It is close to the ideas of assisted inflation proposed earlier in [32, 33]. The main idea of assisted inflation is that each field feels not only the downward force from its own potential but also the collective frictional force from all fields. Therefore slow-roll is easier to achieve and the individual fields do not have to exceed Planck scale vevs.

The equations of motion for a set of scalar fields in generic situation with the moduli space metric (in real notation with $L_{\mathrm{kin}} = \frac{1}{2}G_{ij}(\phi)\partial_\mu\phi^i\partial^\mu\phi^j$) are [54, 56]:

$$\ddot{\phi}^i + \Gamma^i_{jk}(\phi)\dot{\phi}^j\dot{\phi}^k + 3H\dot{\phi}^i + G^{ij}(\phi)\partial_j V = 0 . \tag{4.43}$$

Here $G^{ij}(\phi)$ is the inverse metric of the moduli space and $\Gamma^i_{jk}(\phi)$ are the Christoffel symbols in the moduli space. If $G_{ij} = \delta_{ij}$ and $V = \sum_i V_i(\phi_i)$, i.e., if the metric of the moduli space is flat and if the potential is a sum of the potentials of the individual fields, the assistance effect becomes clear:

$$\ddot{\phi}^i + 3H\dot{\phi}^i + \partial_i V_i = 0 \,, \qquad H^2 = \frac{\sum_i V_i}{3M_{\mathrm{Pl}}^2} \,. \qquad (4.44)$$

Each field responds to its own potential (there is no summation in the term $\partial_i V_i$ above), but the friction via the Hubble parameter comes from all fields and can be significantly stronger than in the case without assistance.

Thus in the general case of moduli space metric we need to identify in string theory the situations when the simplified (4.44) gives a good approximation for the actual complicated dynamics given by (4.43).

An interesting attempt to do so was made in [34]. The model requires a large number of axions, $N \sim 240\,(M_{\mathrm{Pl}}/f)^2$, where f is the generic axion decay constant. For $f \approx 10^{-1}M_{\mathrm{Pl}}$, one should have $N \approx 10^4$. String theory may provide such a large number of axions, there are known examples of up to 10^5 axions. If f is a smaller fraction of M_{Pl} the number of required axions grows.

One may wonder whether the phenomenological assumptions made in [34, 35] can be justified in the known framework of compactified string theory. The main assumption is that in the effective supergravity model with numerous complex moduli,

$$t_n = \frac{\phi_n}{f_n} + \mathrm{i}M^2 R_n^2 \,, \qquad (4.45)$$

all moduli R_n^2 quickly go to their minima. Then only the axions $\frac{\phi_n}{f_n}$ remain to drive inflation. The reason for this assumption is that the Kähler potential depends only on the volume modulus of all two cycles, $R_n^2 = -\frac{\mathrm{i}}{2M^2}(t_n - \bar{t}_n)$, but it does not depend on the axions $\frac{\phi_n}{f_n} = \frac{1}{2}(t_n + \bar{t}_n)$, so one could expect that the axion direction in the first approximation remains flat. Let us examine this assumption more carefully.

The Kähler potential is given by the same formula as that in (4.23), which is an exact expression for $\mathcal{N} = 2$ supergravities with cubic prepotentials. If the superpotential does not depend on t_n, the potential has a runaway dependence on the moduli R_n^2:

$$V \sim \mathrm{e}^K = \frac{1}{\frac{\mathrm{i}C^{lmn}}{3!}(t_l - \bar{t}_l)(t_m - \bar{t}_m)(t_n - \bar{t}_n)} \,. \qquad (4.46)$$

Here the Calabi–Yau intersection numbers C^{lmn} can be positive as well as negative. In this approximation, the potential is flat in the axion directions, but the vacuum is unstable.

The instanton contribution to the superpotential,

$$W = \sum_n w_n \mathrm{e}^{2\pi t_n} \,, \qquad (4.47)$$

is supposed to stabilize the volume moduli R_n^2 quickly, whereas the axions are expected to slowly reach their minima. When all volume moduli are fixed, the moduli space metric is flat up to some constant rescalings. This assumption makes it easy to establish that the total potential with stabilized volume moduli is reduced to the *sum of potentials of each axion*, since the problem is reduced to the global SUSY potential

$$V \approx \sum_n V_n \approx \sum_n |\partial_n W|^2 \,. \tag{4.48}$$

Thus both conditions specified above, the flatness of the moduli space metric and the decoupling of the potentials, are supposed to be satisfied [34, 35]. This leads to the uncoupled set of light massive axions, which makes the assistance effect easily possible in accordance with (4.44).

Can the assumption that *volume moduli stabilized quickly and axions slow-roll for a long time* be justified in known models of string theory which stabilize both types of complex fields?

The study of the potentials for 10^4 fields is difficult to carry out. However, we have some experience with exactly the same type of KKLT potentials in the case of one or two complex moduli, which we described in the previous sections of this chapter. The generic property of this class of models is that the same exponential terms in the superpotential that stabilize the volume moduli simultaneously stabilize the axions. Therefore the masses of these two fields are not much different. One can easily see it in the simplest model of the KKLT potential, Fig. 4.2.

We can now look at more complicated examples. The racetrack potential has a complicated profile, shown in Figs. 4.4 and 4.5. Near the saddle point the axion has a flat maximum, the volume is at the local minimum. Therefore the significant part of inflation proceeds via the slow-roll of the axion till the motion of the axion kicks the volume modulus out of the local minimum so that both fields reach the absolute minimum in the waterfall-type evolution. However, one can easily see in Fig. 4.5 that near the minimum there is no significant flatness in the axion direction comparative to the volume modulus direction, same as in the simplest KKLT potential in Fig. 4.2.

In the better racetrack model the potential near the saddle point for the two axions and another one for the two volume moduli are plotted in Fig. 4.6. The trajectories for the two axions and the volume moduli are shown in Figs. 4.7 and 4.8. Here again, near the saddle point there is a flat maximum in the direction which is a combination of two axions. However, near the absolute minimum all directions are steep.

A related general observation was recently made in the models with a large volume of compactification where various trajectories with different initial conditions were studied. It was stressed in [25] that it was easy to find the trajectories evolving only in the volume-τ direction, however, so far no inflationary trajectories in the axion-θ direction have been found in this model.

Thus all known string theory models in which the stabilization of the scalar and the pseudo-scalar fields is possible due to the exponential terms in the superpotential do indeed have a cos-type dependence on the axion field. However, comparing with the potentials in (4.27) and (4.36), one can see that they are quite different from the simplified PNGB (Pseudo-Nambu–Goldstone Boson) potential $V = \Lambda^4[1 - \cos(\phi/f)]$ with the fixed radial field. In the case of two axions Y_1, Y_2 and two volume moduli X_1, X_2 in (4.36) there is a term with $\cos(aY_1)$, $\cos(bY_2)$, as well as a mixing term $\cos(bY_2 - aY_1)$. More importantly, the dependence on X_1, X_2 is rather involved, which leads to a complicated dynamics, in general.

It may be useful to compare the actual potential (4.36) with the simplified potential depending on two axions in [119], where it was argued that the models with two axions may lead to an assistance effect. However, the main assumption in [119] is that the potential of the form $V = \Lambda_1^4[1 - \cos(a\phi_1)] + \Lambda_2^4[1 - \cos(b\phi_1 + c\phi_2)]$ can be derived from string theory with the superpotential (4.35). This assumption is problematic at this stage because of the volume stabilization issue. Both axions and volume moduli undergo a dynamical evolution according to the potentials in (4.27) and (4.36) and the use of the pure multi-axion potential is not valid for the known models derived from string theory.

The axion valley model proposed in Sect. 4.3.1, see Fig. 4.3, would support the ideas in [119], and it is valid in supergravity. But this model is still to be derived from string theory.

It may still be possible to justify the N-flation model in string theory. For example, one can try to find a string theory realization of the axion valley model described in Sect. 4.3.1, or of some other model of this kind where the basic assumptions made in [34, 35] are satisfied.

Another possibility is to start with initial conditions where all volume moduli lie very close to the minimum of the potential and axions are away from the minimum. If the motion of all axions leads to a significant friction, because of the assisted inflation, one may hope that the volume moduli will stay near their minima, and the regime of N-flation model will be valid. This is not a very attractive proposition since it requires a lot of additional fine-tuning of the initial conditions.

Finally, it may happen that even if one considers a simultaneous motion of all interacting fields, the axions and the radial moduli, the simple fact that there are many of them may be sufficient for the existence of the assisted inflation regime. This possibility requires a more detailed investigation.

Thus, the assisted models of inflation in string theory require more work, just as all other models described in this talk. Assisted inflation in string theory will become a particularly important issue if the gravitational waves from inflation are detected.

4.7 Discussion

String cosmology has several different but closely related goals: to find inflationary models based on string theory, to identify some of their predictions which may be related to the specifically stringy nature of the inflationary models, and, by doing so, to test string theory by comparing its predictions with observations.

In the discussion of inflationary models in string theory and supergravity we have described the shift symmetry which may under certain conditions explain the required flatness of the inflaton direction of the potential. Examples include chaotic inflation in a supergravity model [27] and the axion valley model (supergravity version of the natural inflation) proposed here in Sect. 4.3.1. In the axion valley model shown in Fig. 4.3 the scalar is heavy and quickly stabilizes, whereas the pseudo-scalar remains very light before it reaches the minimum of the potential. This is in contrast with the KKLT model where both the scalar and the pseudo-scalar (the volume modulus and the axion) have approximately the same curvature of the potential near the minimum, as shown in Fig. 4.2. The supergravity version of chaotic inflation and the axion valley model use the shift symmetry of the Kähler potential in a very effective way. Therefore it would be most interesting to find string theory versions of these models.

An inflationary $D3/D7$ brane model [19, 20, 84] discussed in Sect. 4.4.2 is also based on shift symmetry slightly broken by quantum corrections [58, 59, 60]. In $D3/D7$ brane model the shift symmetry originates from the $\mathcal{N} = 2$ supergravity structure of the Kähler potential shown in (4.20) and (4.23). In $\mathcal{N} = 2$ models the Kähler potential[9] is given by $K = -\ln\left[\mathrm{i}\frac{C_{ABC}}{3!}(z - \bar{z})^A(z - \bar{z})^B(z - \bar{z})^C\right]$ and the shift symmetry is generic. Here C_{ABC} are the intersection numbers of the compactification manifold. The $\mathcal{N} = 2$ supersymmetry of these models may be broken by fluxes and/or nonperturbative corrections, but the total potential may still enjoy a slightly broken shift symmetry under the transformation of the inflaton field $z \rightarrow z + \delta$, with real parameter δ. In the racetrack models in Sects. 4.5.1 and 4.5.2 the Kähler potentials also exhibit the shift symmetry, since they are given by the $\mathcal{N} = 2$ formula, see for example (4.34). The shift symmetry of the total potential is, however, strongly broken near the minimum and slightly broken near the saddle point.

In other models which we have discussed here the shift symmetry is not relevant for the flatness of the inflaton potential. It is not present in the KKLMMT model [13], which therefore in most cases relies on a special dependence on the inflaton field in the superpotential to cancel the one in the Kähler potential. Such a cancellation requires fine-tuning, which may be possible in the context of a huge stringy landscape.

In the inflationary models with extra large volume of compactification [23, 25] the dynamics is mostly based on the very flat radial modulus direction. There is no shift symmetry of the Kähler potential and of the total scalar potential in the inflaton direction $T_3 + \bar{T}_3$ on which the Kähler potential

---

[9] It is interesting that the theory of $\mathcal{N} = 2$ supersymmetric black holes in string theory and the attractor mechanism stabilizing the moduli near the black hole horizon are both based on precisely the same class of Kähler potentials, see for example (30) in [87].

depends strongly. The flatness is achieved via some set of hierarchy relations between the parameters of the model with three complex fields. The flatness in the axion direction (the remnant of the shift symmetry of the Kähler potential in (4.41)) does help to increase the number of e-foldings [25], but it is not a major feature of the model.

We also discussed the string theory N-flation model [34, 35] of assisted inflation [32, 33]. We have pointed out that the assumption in [34, 35] that only the axions define the dynamics due to the shift symmetry of the Kähler potential may be valid in the context of the axion valley supergravity model presented in Sect. 4.3.1 here, but not in the currently known constructions of string theory. We explained, in particular, why this assumption is not valid in the simplest KKLT model of moduli stabilization.

Nevertheless, it might be possible to find an assisted inflation regime in string theory by investigating a combined dynamics of all string theory moduli. This, as well as finding a string theory generalization of the supergravity versions of chaotic and natural inflation discussed above, would be particularly important if the primordial gravitational waves from inflation were detected.

In this chapter we have described a representative subclass of string inflation models, but nevertheless the list of the models which we have discussed remains certainly incomplete. Moreover, one should clearly understand that the whole subject is relatively new, and it is difficult to make any far-reaching conclusions at this stage. For example, the time span between the invention of the simplest chaotic inflation model $m^2\phi^2$ [26] and its supergravity version [27] is about 17 years. Similarly, the time span between the invention of the natural inflation [61] and its implementation in supergravity (see Sect. 4.3.1) is also about 17 years. Meanwhile the KKLT mechanism of vacuum stabilization has been proposed only 4 years ago, and therefore the development of the cosmological models based on string theory with a stabilized vacuum has begun only very recently.

The search of new models of string theory inflation should go in parallel with the further development of string theory. One may try to find some new theoretical structures which will lead to interesting inflationary models in string theory. This is one of the important challenges for string theory and cosmology. The string theory community is well aware of these challenges, see for example [2, 3, 120, 121], and one may hope that a better understanding of the core string theory may lead to better models for cosmology.

Under certain conditions, the new developments may allow us to test string theory by current and future precision data in cosmology. The conditions include a reliable derivation of inflationary models from string theory. These models should have distinct predictions for observables like the spectral index n_s, the level of gravitational waves, and the abundance of light cosmic strings. If these conditions are satisfied, one may hope that in few years from now, when the precision data will become available and the derivation of the inflationary models will be refined, it will be possible to test the string theory assumptions underlying the derivation of the corresponding inflationary models.

Acknowledgments

I am grateful to the organizers and participants of "Inflation+25" for the most stimulating atmosphere of the conference. I thank all my collaborators on the projects described in the talk and L. McAllister, T. Banks, D. Baumann, R. Bean, S. Dimopoulos, S. Kachru, L. Kofman, A. Linde, D. Lyth, L. McAllister, E. Silverstein, H. Tye and J. Wacker for the discussions of various topics in this talk. This work was supported by the NSF grant 0244728 and by the A. von Humboldt award.

References

1. M. Grana, Phys. Rep. **423**, 91–158 (2006), hep-th/0509003.
2. M. R. Douglas and S. Kachru (2006), hep-th/0610102.
3. R. Blumenhagen, B. Kors, D. Lust and S. Stieberger (2006), hep-th/0610327.
4. S. H. Henry Tye (2006), hep-th/0610221.
5. J. M. Cline (2006), hep-th/0612129.
6. S. Kachru, R. Kallosh, A. Linde and S. P. Trivedi, Phys. Rev. **D68**, 046005 (2003), hep-th/0301240.
7. E. L. Wright (2007), astro-ph/0701584.
8. R. Bousso and J. Polchinski, JHEP **06**, 006 (2000), hep-th/0004134.
9. L. Susskind (2003), hep-th/0302219.
10. F. Quevedo, Class. Quant. Grav. **19**, 5721–5779 (2002), hep-th/0210292.
11. T. Banks, M. Berkooz, S. H. Shenker, G. W. Moore and P. J. Steinhardt, Phys. Rev. **D52**, 3548–3562 (1995), hep-th/9503114.
12. P. Binetruy and M. K. Gaillard, Phys. Rev. **D34**, 3069–3083 (1986).
13. S. Kachru, R. Kallosh, A. Linde, J. M. Maldacena, L. McAllister and S. P. Trivedi, JCAP **0310**, 013 (2003), hep-th/0308055.
14. G. R. Dvali and S. H. H. Tye, Phys. Lett. **B450**, 72–82 (1999), hep-ph/9812483.
15. D. N. Spergel et al. (2006), astro-ph/0603449.
16. M. Tegmark et al., Phys. Rev. **D74**, 123507 (2006), astro-ph/0608632.
17. P. Binetruy and G. R. Dvali, Phys. Lett. **B388**, 241–246 (1996), hep-ph/9606342.
18. E. Halyo, Phys. Lett. **B387**, 43–47 (1996), hep-ph/9606423.
19. K. Dasgupta, C. Herdeiro, S. Hirano and R. Kallosh, Phys. Rev. **D65**, 126002 (2002), hep-th/0203019.
20. F. Koyama, Y. Tachikawa and T. Watari, Phys. Rev. **D69**, 106001 (2004), hep-th/0311191.
21. J. J. Blanco-Pillado et al., JHEP **11**, 063 (2004), hep-th/0406230.
22. Z. Lalak, G. G. Ross and S. Sarkar (2005), hep-th/0503178.

23. J. P. Conlon and F. Quevedo, JHEP **01**, 146 (2006), hep-th/0509012.
24. J. J. Blanco-Pillado et al., JHEP **09**, 002 (2006), hep-th/0603129.
25. J. R. Bond, L. Kofman, S. Prokushkin and P. M. Vaudrevange (2006), hep-th/0612197.
26. A. D. Linde, Phys. Lett. **B129**, 177–181 (1983).
27. M. Kawasaki, M. Yamaguchi and T. Yanagida, Phys. Rev. Lett. **85**, 3572–3575 (2000), hep-ph/0004243.
28. D. Baumann and L. McAllister (2006), hep-th/0610285.
29. R. Bean, S. E. Shandera, S. H. Henry Tye and J. Xu (2007), hep-th/0702107.
30. E. Silverstein and D. Tong, Phys. Rev. **D70**, 103505 (2004), hep-th/0310221.
31. M. Alishahiha, E. Silverstein and D. Tong, Phys. Rev. **D70**, 123505 (2004), hep-th/0404084.
32. A. R. Liddle, A. Mazumdar and F. E. Schunck, Phys. Rev. **D58**, 061301 (1998), astro-ph/9804177.
33. P. Kanti and K. A. Olive, Phys. Lett. **B464**, 192–198 (1999), hep-ph/9906331.
34. S. Dimopoulos, S. Kachru, J. McGreevy and J. G. Wacker (2005), hep-th/0507205.
35. R. Easther and L. McAllister, JCAP **0605**, 018 (2006), hep-th/0512102.
36. M. Sakellariadou (2007), hep-th/0702003.
37. E. J. Copeland, R. C. Myers and J. Polchinski, JHEP **06**, 013 (2004), hep-th/0312067.
38. J. Polchinski (2004), hep-th/0412244.
39. F. A. Jenet et al. (2006), astro-ph/0609013.
40. L. Pogosian, I. Wasserman and M. Wyman (2006), astro-ph/0604141.
41. U. Seljak and A. Slosar, Phys. Rev. **D74**, 063523 (2006), astro-ph/0604143.
42. K. Becker, M. Becker and A. Krause, Phys. Rev. **D74**, 045023 (2006), hep-th/0510066.
43. P. Creminelli, L. Senatore, M. Zaldarriaga and M. Tegmark (2006), astro-ph/0610600.
44. N. Arkani-Hamed and S. Dimopoulos, JHEP **06**, 073 (2005), hep-th/0405159.
45. J. F. G. Cascales, M. P. Garcia del Moral, F. Quevedo and A. M. Uranga, JHEP **02**, 031 (2004), hep-th/0312051.
46. P. G. Camara, L. E. Ibanez and A. M. Uranga, Nucl. Phys. **B708**, 268–316 (2005), hep-th/0408036.
47. D. Lust, S. Reffert and S. Stieberger, Nucl. Phys. **B727**, 264–300 (2005), hep-th/0410074.
48. K. Choi, A. Falkowski, H. P. Nilles and M. Olechowski, Nucl. Phys. **B718**, 113–133 (2005), hep-th/0503216.
49. J. P. Conlon, F. Quevedo and K. Suruliz, JHEP **08**, 007 (2005), hep-th/0505076.
50. O. Lebedev, V. Lowen, Y. Mambrini, H. P. Nilles and M. Ratz (2006), hep-ph/0612035.
51. B. S. Acharya, K. Bobkov, G. L. Kane, P. Kumar and J. Shao (2007), hep-th/0701034.
52. D. Cremades, M. P. G. del Moral, F. Quevedo and K. Suruliz (2007), hep-th/0701154.
53. K. Intriligator, N. Seiberg and D. Shih, JHEP **04**, 021 (2006), hep-th/0602239.
54. M. Sasaki and E. D. Stewart, Prog. Theor. Phys. **95**, 71–78 (1996), astro-ph/9507001.

55. P. Binetruy and M. K. Gaillard, Phys. Lett. **B195**, 382 (1987).
56. M. K. Gaillard, H. Murayama and K. A. Olive, Phys. Lett. **B355**, 71–77 (1995), hep-ph/9504307.
57. R. Kallosh and S. Prokushkin (2004), hep-th/0403060.
58. J. P. Hsu, R. Kallosh and S. Prokushkin, JCAP **0312**, 009 (2003), hep-th/0311077.
59. H. Firouzjahi and S. H. H. Tye, Phys. Lett. **B584**, 147–154 (2004), hep-th/0312020.
60. J. P. Hsu and R. Kallosh, JHEP **04**, 042 (2004), hep-th/0402047.
61. K. Freese, J. A. Frieman and A. V. Olinto, Phys. Rev. Lett. **65**, 3233–3236 (1990).
62. F. C. Adams, J. R. Bond, K. Freese, J. A. Frieman and A. V. Olinto, Phys. Rev. **D47**, 426–455 (1993), hep-ph/9207245.
63. K. Choi, Phys. Rev. **D62**, 043509 (2000), hep-ph/9902292.
64. C. Savage, K. Freese and W. H. Kinney, Phys. Rev. **D74**, 123511 (2006), hep-ph/0609144.
65. P. Brax and J. Martin (2006), hep-th/0612208.
66. C. Angelantonj, R. D'Auria, S. Ferrara and M. Trigiante, Phys. Lett. **B583**, 331–337 (2004), hep-th/0312019.
67. R. Kallosh, A.-K. Kashani-Poor and A. Tomasiello, JHEP **06**, 069 (2005), hep-th/0503138.
68. N. Saulina, Nucl. Phys. **B720**, 203–210 (2005), hep-th/0503125.
69. P. S. Aspinwall and R. Kallosh, JHEP **10**, 001 (2005), hep-th/0506014.
70. T. Banks and M. Dine, Nucl. Phys. **B505**, 445–460 (1997), hep-th/9608197.
71. T. Banks, M. Dine and M. Graesser, Phys. Rev. **D68**, 075011 (2003), hep-ph/0210256.
72. T. Banks, M. Dine, P. J. Fox and E. Gorbatov, JCAP **0306**, 001 (2003), hep-th/0303252.
73. P. Svrcek and E. Witten, JHEP **06**, 051 (2006), hep-th/0605206.
74. K. Choi and K. S. Jeong, JHEP **01**, 103 (2007), hep-th/0611279.
75. A. D. Linde, Phys. Rev. **D49**, 748–754 (1994), astro-ph/9307002.
76. I. R. Klebanov and M. J. Strassler, JHEP **08**, 052 (2000), hep-th/0007191.
77. S. B. Giddings, S. Kachru and J. Polchinski, Phys. Rev. **D66**, 106006 (2002), hep-th/0105097.
78. E. J. Copeland, A. R. Liddle, D. H. Lyth, E. D. Stewart and D. Wands, Phys. Rev. **D49**, 6410–6433 (1994), astro-ph/9401011.
79. M. Berg, M. Haack and B. Kors, Phys. Rev. **D71**, 026005 (2005), hep-th/0404087.
80. D. Baumann et al., JHEP **11**, 031 (2006), hep-th/0607050.
81. C. P. Burgess, J. M. Cline, K. Dasgupta and H. Firouzjahi (2006), hep-th/0610320.
82. T. Damour and A. Vilenkin, Phys. Rev. **D71**, 063510 (2005), hep-th/0410222.
83. E. Jeong and G. F. Smoot (2006), astro-ph/0612706.
84. K. Dasgupta, J. P. Hsu, R. Kallosh, A. Linde and M. Zagermann, JHEP **08**, 030 (2004), hep-th/0405247.
85. P. K. Tripathy and S. P. Trivedi, JHEP **03**, 028 (2003), hep-th/0301139.
86. P. Brax and J. Martin, Phys. Rev. **D72**, 023518 (2005), hep-th/0504168.
87. S. Ferrara, R. Kallosh and A. Strominger, Phys. Rev. **D52**, 5412–5416 (1995), hep-th/9508072.

88. C. P. Burgess, R. Kallosh and F. Quevedo, JHEP **10**, 056 (2003), hep-th/0309187.
89. P. Binetruy, G. Dvali, R. Kallosh and A. Van Proeyen, Class. Quant. Grav. **21**, 3137–3170 (2004), hep-th/0402046.
90. H. Elvang, D. Z. Freedman and B. Kors, JHEP **11**, 068 (2006), hep-th/0606012.
91. G. Villadoro and F. Zwirner, Phys. Rev. Lett. **95**, 231602 (2005), hep-th/0508167.
92. A. Achucarro, B. de Carlos, J. A. Casas and L. Doplicher, JHEP **06**, 014 (2006), hep-th/0601190.
93. S. L. Parameswaran and A. Westphal, JHEP **10**, 079 (2006), hep-th/0602253.
94. K. Choi and K. S. Jeong, JHEP **08**, 007 (2006), hep-th/0605108.
95. E. Dudas and Y. Mambrini, JHEP **10**, 044 (2006), hep-th/0607077.
96. M. Haack, D. Krefl, D. Lust, A. Van Proeyen and M. Zagermann (2006), hep-th/0609211.
97. P. Brax et al. (2006), hep-th/0610195.
98. J. Rocher and M. Sakellariadou, JCAP **0611**, 001 (2006), hep-th/0607226.
99. E. Dudas, C. Papineau and S. Pokorski (2006), hep-th/0610297.
100. H. Abe, T. Higaki, T. Kobayashi and Y. Omura (2006), hep-th/0611024.
101. R. Kallosh and A. Linde (2006), hep-th/0611183.
102. M. Bastero-Gil, S. F. King and Q. Shafi (2006), hep-ph/0604198.
103. C.-M. Lin and J. McDonald, Phys. Rev. **D74**, 063510 (2006), hep-ph/0604245.
104. R. Kallosh and A. Linde, JCAP **0310**, 008 (2003), hep-th/0306058.
105. R. Kallosh (2001), hep-th/0109168.
106. J. Urrestilla, A. Achucarro and A. C. Davis, Phys. Rev. Lett. **92**, 251302 (2004), hep-th/0402032.
107. R. Jeannerot and M. Postma, JCAP **0607**, 012 (2006), hep-th/0604216.
108. S. Kecskemeti, J. Maiden, G. Shiu and B. Underwood, JHEP **09**, 076 (2006), hep-th/0605189.
109. K. Becker, M. Becker and A. Krause, Nucl. Phys. **B715**, 349–371 (2005), hep-th/0501130.
110. A. D. Linde, *Particle Physics and Inflationary Cosmology* (Harwood, Chur, Switzerland, 1990, 362 p.), hep-th/0503203.
111. A. Linde, JCAP **0410**, 004 (2004), hep-th/0408164.
112. A. Vilenkin, Phys. Rev. **D27**, 2848 (1983).
113. A. D. Linde, Phys. Lett. **B175**, 395–400 (1986).
114. A. D. Linde, Phys. Lett. **B327**, 208–213 (1994), astro-ph/9402031.
115. A. Vilenkin, Phys. Rev. Lett. **72**, 3137–3140 (1994), hep-th/9402085.
116. A. D. Linde and D. A. Linde, Phys. Rev. **D50**, 2456–2468 (1994), hep-th/9402115.
117. F. Denef, M. R. Douglas and B. Florea, JHEP **06**, 034 (2004), hep-th/0404257.
118. V. Balasubramanian, P. Berglund, J. P. Conlon and F. Quevedo, JHEP **03**, 007 (2005), hep-th/0502058.
119. J. E. Kim, H. P. Nilles and M. Peloso, JCAP **0501**, 005 (2005), hep-ph/0409138.
120. H. Ooguri and C. Vafa (2006), hep-th/0605264.
121. K. Becker, M. Becker and J. H. Schwarz, *String Theory and M-Theory. A Modern Introduction* (Cambridge University Press, 2007, 739 p.).

5

Predictions in Eternal Inflation

Sergei Winitzki

Department of Physics, LMU Munich, Germany
serge@theorie.physik.uni-muenchen.de

Abstract. In generic models of cosmological inflation, quantum fluctuations strongly influence the spacetime metric and produce infinitely many regions where the end of inflation (reheating) is delayed until arbitrarily late times. The geometry of the resulting spacetime is highly inhomogeneous on scales of many Hubble sizes. The recently developed string-theoretic picture of the "landscape" presents a similar structure, where an infinite number of de Sitter, flat, and anti-de Sitter universes are nucleated via quantum tunneling. Since observers on the Earth have no information about their location within the eternally inflating universe, the main question in this context is to obtain statistical predictions for quantities observed at a random location. I describe the problems arising within this statistical framework, such as the need for a volume cutoff and the dependence of cutoff schemes on time slicing and on the initial conditions. After reviewing different approaches and mathematical techniques developed in the past two decades for studying these issues, I discuss the existing proposals for extracting predictions and give examples of their applications

5.1 Eternal Inflation

The general idea of eternally inflating spacetime was first introduced and developed in the 1980s [1, 2, 3, 4] in the context of slow-roll inflation. Let us begin by reviewing the main features of eternal inflation, following these early works.

A prototypical model contains a minimally coupled scalar field ϕ (the "inflaton") with an effective potential $V(\phi)$ that is sufficiently flat in some range of ϕ. When the field ϕ has values in this range, the spacetime is approximately de Sitter with the Hubble rate

$$\frac{\dot{a}}{a} = \sqrt{\frac{8\pi}{3}V(\phi)} \equiv H(\phi) . \tag{5.1}$$

(We work in units where $G = c = \hbar = 1$.) The value of H remains approximately constant on timescales of several Hubble times ($\Delta t \gtrsim H^{-1}$), while

S. Winitzki: *Predictions in Eternal Inflation*, Lect. Notes Phys. **738**, 157–191 (2008)
DOI 10.1007/978-3-540-74353-8_5
© Springer-Verlag Berlin Heidelberg 2008

the field ϕ follows the slow-roll trajectory $\phi_{sr}(t)$. Quantum fluctuations of the scalar field ϕ in de Sitter background grow linearly with time [5, 6, 7],

$$\langle \hat{\phi}^2(t + \Delta t) \rangle - \langle \hat{\phi}^2(t) \rangle = \frac{H^3}{4\pi^2} \Delta t , \qquad (5.2)$$

at least for time intervals Δt of order several H^{-1}. Due to the quasi-exponential expansion of spacetime during inflation, Fourier modes of the field ϕ are quickly stretched to super-Hubble length scales. However, quantum fluctuations with super-Hubble wavelengths cannot maintain quantum coherence and become essentially classical [6, 7, 8, 9, 10]; this issue is discussed in more detail in Sect. 5.1.2. The resulting field evolution $\phi(t)$ can be visualized [1, 8, 11] as a Brownian motion with a "random jump" of typical step size $\Delta\phi \sim H/(2\pi)$ during a time interval $\Delta t \sim H^{-1}$, superimposed onto the deterministic slow-roll trajectory $\phi_{sr}(t)$. A statistical description of this "random walk"-type evolution $\phi(t)$ is reviewed in Sect. 5.2.1.

The "jumps" at points separated in space by many Hubble distances are essentially uncorrelated; this is another manifestation of the well-known "no-hair" property of de Sitter space [12, 13, 14]. Thus the field ϕ becomes extremely inhomogeneous on large (super-horizon) scales after many Hubble times. Moreover, in the semi-classical picture it is assumed [2] that the local expansion rate $\dot{a}/a \equiv H(\phi)$ tracks the local value of the field $\phi(t, \mathbf{x})$ according to the Einstein equation (5.1). Here $a(t, \mathbf{x})$ is the scale factor function which varies with $\mathbf{x}$ only on super-Hubble scales, $a(t, \mathbf{x})\Delta x \gtrsim H^{-1}$. Hence, the spacetime metric can be visualized as having a slowly varying, "locally de Sitter" form (with spatially flat coordinates $\mathbf{x}$),

$$g_{\mu\nu}dx^\mu dx^\nu = dt^2 - a^2(t, \mathbf{x})d\mathbf{x}^2 . \qquad (5.3)$$

The deterministic trajectory $\phi_{sr}(t)$ eventually reaches a (model-dependent) value ϕ_* signifying the end of the slow-roll inflationary regime and the beginning of the reheating epoch (thermalization). Since the random walk process will lead the value of ϕ away from $\phi = \phi_*$ in some regions, reheating will not begin everywhere at the same time. Moreover, regions where ϕ remains in the inflationary range will typically expand faster than regions near the end of inflation where $V(\phi)$ becomes small. Therefore, a delay of the onset of reheating will be rewarded by additional expansion of the proper 3-volume, thus generating more regions that are still inflating. This feature is called "self-reproduction" of the inflationary spacetime [3]. Since each Hubble-size region evolves independently of other such regions, one may visualize the spacetime as an ensemble of inflating Hubble-size domains (Fig. 5.1).

The process of self-reproduction will never result in a global reheating if the probability of jumping away from $\phi = \phi_*$ and the corresponding additional volume expansion factors are sufficiently large. The corresponding quantitative conditions and their realization in typical models of inflation are reviewed in Sect. 5.3.1. Under these conditions, the process of self-reproduction of inflating regions continues forever. At the same time, every

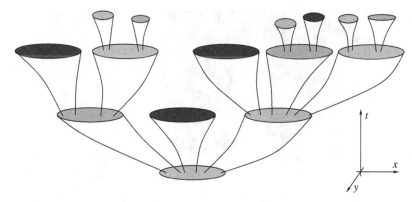

Fig. 5.1. A qualitative diagram of self-reproduction during inflation. *Shaded space-like domains* represent Hubble-size regions with different values of the inflation field ϕ. The time step is of order H^{-1}. *Dark-colored shades* are regions undergoing reheating ($\phi = \phi_*$); *lighter-colored shades* are regions where inflation continues. On average, the number of inflating regions grows with time

given comoving worldline (except for a set of measure zero; see Sect. 5.3.1) will sooner or later reach the value $\phi = \phi_*$ and enter the reheating epoch. The resulting situation is known as "eternal inflation" [3]. More precisely, the term "eternal inflation" means future-eternal self-reproduction of inflating regions [15].[1] To emphasize the fact that self-reproduction is due to random fluctuations of a field, one refers to this scenario as "eternal inflation of random walk type." Below we use the terms "eternal self-reproduction" and "eternal inflation" interchangeably.

Observers like us may appear only in regions where reheating already took place. Hence, it is useful to consider the locus of all reheating events in the entire spacetime; in the present example, it is the set of spacetime points x there $\phi(x) = \phi_*$. This locus is called the *reheating surface* and is a non-compact, spacelike three-dimensional hypersurface [16, 18]. It is important to realize that a finite, initially inflating 3-volume of space may give rise to a reheating surface having an infinite 3-volume, and even to infinitely many causally disconnected pieces of the reheating surface, each having an infinite 3-volume (see Fig. 5.2). This feature of eternal inflation is at the root of several technical and conceptual difficulties, as will be discussed below.

Everywhere along the reheating surface, the reheating process is expected to provide appropriate initial conditions for the standard "hot big bang" cosmological evolution, including nucleosynthesis and structure formation. In other words, the reheating surface may be visualized as the locus of the "hot big bang" events in the spacetime. It is thus natural to view the reheating

---

[1] It is worth emphasizing that the term "eternal inflation" refers to future-eternity of inflation in the sense described above, but does not imply past-eternity. In fact, inflationary spacetimes are generically *not* past-eternal [16, 17].

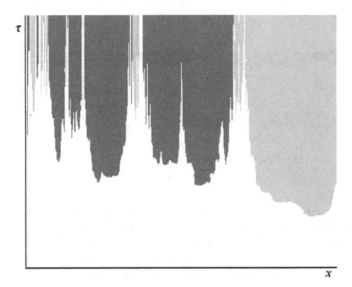

Fig. 5.2. A 1+1-dimensional slice of the spacetime structure in an eternally inflating universe (numerical simulation in [19]). *Shades of different colors* represent different, causally disconnected regions where reheating took place. The reheating surface is the *line* separating the white (inflating) domain and the shaded domains

surface as the initial equal-time surface for astrophysical observations in the post-inflationary epoch. Note that the observationally relevant range of the primordial spectrum of density fluctuations is generated only during the last 60 e-foldings of inflation. Hence, the duration of the inflationary epoch that preceded reheating is not directly measurable beyond the last 60 e-foldings; the total number of e-foldings can vary along the reheating surface and can be in principle arbitrarily large.[2]

The phenomenon of eternal inflation is also found in multi-field models of inflation [22, 23], as well as in scenarios based on Brans–Dicke theory [24, 25, 26], topological inflation [27, 28], braneworld inflation [29], "recycling universe" [30] and the string theory landscape [31]. In some of these models, quantum tunneling processes may generate "bubbles" of a different phase of the vacuum (see Sect. 5.2.5 for more details). Bubbles will be created randomly at various places and times, with a fixed rate per unit 4-volume. In the interior of some bubbles, additional inflation may take place, followed by a new reheating surface. The interior structure of such bubbles is sketched in Fig. 5.3.

---

[2] For instance, it was shown that holographic considerations do not place any bounds on the total number of e-foldings during inflation [20]. For recent attempts to limit the number of e-foldings using a different approach, see, e.g., [21]. Note also that the effects of "random jumps" are negligible during the last 60 e-foldings of inflation, since the produced perturbations must be of order 10^{-5} according to observations.

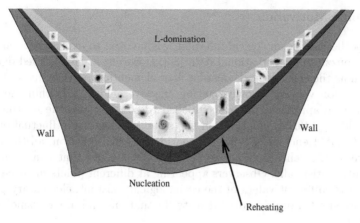

Fig. 5.3. A spacetime diagram of a bubble interior. The infinite, spacelike reheating surface is shown in *darker shade.* Galaxy formation is possible within the spacetime region indicated.

The nucleation event and the formation of bubble walls is followed by a period of additional inflation, which terminates by reheating. Standard cosmological evolution and structure formation eventually give way to a Λ-dominated universe. Infinitely many galaxies and possible civilizations may appear within a thin spacelike slab running along the interior reheating surface. This reheating surface appears to interior observers as an infinite, spacelike hypersurface [32]. For this reason, such bubbles are called "pocket universes," while the spacetime is called a "multiverse." (Generally, the term "pocket universe" refers to a non-compact, connected component of the reheating surface [33].)

In scenarios of this type, each bubble is causally disconnected from most other bubbles.[3] Hence, bubble nucleation events may generate infinitely many statistically inequivalent, causally disconnected patches of the reheating surface, every patch giving rise to a possibly infinite number of galaxies and observers. This feature significantly complicates the task of extracting physical predictions from these models. This class of models is referred to as "eternal inflation of tunneling type."

In the following subsections, I discuss the motivation for studying eternal inflation as well as physical justifications for adopting the effective stochastic picture. Different techniques developed for describing eternal inflation are reviewed in Sect. 5.2. Section 5.3 contains an overview of methods for extracting predictions and a discussion of the accompanying "measure problem."

---

[3] Collisions between bubbles are rare [34]; however, effects of bubble collisions are observable in principle [35].

5.1.1 Some Motivation

The hypothesis of cosmological inflation was invoked to explain several outstanding puzzles in observational data [36]. However, some observed quantities (such as the cosmological constant Λ or elementary particle masses) may be expectation values of slowly varying effective fields χ_a. Within the phenomenological approach, we are compelled to consider also the fluctuations of the fields χ_a during inflation, on the same footing as the fluctuations of the inflaton ϕ. Hence, in a generic scenario of eternal inflation, all the fields χ_a arrive at the reheating surface $\phi = \phi_*$ with values that can be determined only statistically. Observers appearing at different points in space may thus measure different values of the cosmological constant, elementary particle masses, spectra of primordial density fluctuations and other cosmological parameters.

It is important to note that inhomogeneities in observable quantities are created on scales far exceeding the Hubble horizon scale. Such inhomogeneities are not directly accessible to astrophysical experiments. Nevertheless, the study of the global structure of eternally inflating spacetime is not merely of academic interest. Fundamental questions regarding the cosmological singularities, the beginning of the universe and of its ultimate fate, as well as the issue of the cosmological initial conditions, all depend on the knowledge of the global structure of the spacetime as predicted by the theory, whether or not this global structure is directly observable (see, e.g., [37, 38]). In other words, the fact that some theories predict eternal inflation influences our assessment of the viability of these theories. In particular, the problem of initial conditions for inflation [39] is significantly alleviated when eternal inflation is present. For instance, it was noted early on that the presence of eternal self-reproduction in the "chaotic" inflationary scenario [40] essentially removes the need for the fine-tuning of the initial conditions [3, 41]. More recently, constraints on initial conditions were studied in the context of self-reproduction in models of quintessence [42] and k-inflation [43].

Since the values of the observable parameters χ_a are random, it is natural to ask for the probability distribution of χ_a that would be measured by a randomly chosen observer. Understandably, this question has been the main theme of much of the work on eternal inflation. Obtaining an answer to this question promises to establish a more direct contact between scenarios of eternal inflation and experiment. For instance, if the probability distribution for the cosmological constant Λ were peaked near the experimentally observed, puzzlingly small value (see, e.g., [44] for a review of the cosmological constant problem), the smallness of Λ would be explained as due to observer selection effects rather than to fundamental physics. Considerations of this sort necessarily involve some anthropic reasoning; however, the relevant assumptions are minimal. The basic goal of theoretical cosmology is to select physical theories of the early universe that are most compatible with astrophysical observations, including the observation of our existence. It appears reasonable to assume that the civilization of Planet Earth evolved near a randomly chosen star compatible with the development of life, within a randomly chosen galaxy where such stars exist. Many models of inflation generically include eternal inflation and hence predict the formation of infinitely many galaxies where civilizations like ours may develop. It is then also reasonable to assume that our civilization is typical among all the civilizations that evolved in galaxies formed at any time in the universe. This assumption is called the "principle of mediocrity" [18].

To use the "principle of mediocrity" for extracting statistical predictions from a model of eternal inflation, one proceeds as follows [18, 45]. In the example with the fields χ_a described above, the question is to determine the probability distribution for the values of χ_a that a random observer will measure. Presumably, the values of the fields χ_a do not directly influence the emergence of intelligent life on planets, although they may affect the efficiency of structure formation or nucleosynthesis. Therefore, we may assume a fixed, χ_a-dependent mean number of civilizations $\nu_{\mathrm{civ}}(\chi_a)$ per galaxy and proceed to ask for the probability distribution $P_G(\chi_a)$ of χ_a near a randomly chosen galaxy. The observed probability distribution of χ_a will then be

$$P(\chi_a) = P_G(\chi_a)\nu_{\mathrm{civ}}(\chi_a) \,. \tag{5.4}$$

One may use the standard "hot big bang" cosmology to determine the average number $\nu_G(\chi_a)$ of suitable galaxies per unit volume in a region where reheating occurred with given values of χ_a; in any case, this task does not appear to pose difficulties of principle. Then the computation of $P_G(\chi_a)$ is reduced to determining the volume-weighted probability distribution $\mathcal{V}(\chi_a)$ for the fields χ_a within a randomly chosen 3-volume along the reheating surface. The probability distribution of χ_a will be expressed as

$$P(\chi_a) = \mathcal{V}(\chi_a)\nu_G(\chi_a)\nu_{\mathrm{civ}}(\chi_a) \,. \tag{5.5}$$

However, defining $\mathcal{V}(\chi_a)$ turns out to be far from straightforward since the reheating surface in eternal inflation is an infinite 3-surface with a complicated geometry and topology. The lack of a natural, unambiguous, unbiased measure on the infinite reheating surface is known as the "measure problem" in eternal inflation. Existing approaches and measure prescriptions are discussed in Sect. 5.3, where two main alternatives (the "volume-based" and "worldline-based" measures) are presented. In Sect. 5.3.2 I give arguments in favor of using the volume-based measure for computing the probability distribution of values χ_a measured by a random observer. The volume-based measure has been applied to obtain statistical predictions for the gravitational constant in Brans–Dicke theories [24, 25], cosmological constant (dark energy) [46, 47, 48, 49, 50, 51], particle physics parameters [52, 53, 54] and the amplitude of primordial density perturbations [48, 51, 55, 56].

The issue of statistical predictions has recently come to the fore in conjunction with the discovery of the string theory landscape. According to various estimates, one expects to have between 10^{500} and 10^{1500} possible vacuum states of string theory [31, 57, 58, 59, 60]. The string vacua differ in the geometry of spacetime compactification and have different values of the effective cosmological constant (or "dark energy" density). Transitions between vacua may happen via the well-known Coleman–deLuccia tunneling mechanism [32]. Once the dark energy dominates in a given region, the spacetime becomes locally de Sitter. Then the tunneling process will create infinitely many disconnected "daughter" bubbles of other vacua. Observers like us may appear within any of the habitable bubbles. Since the fundamental theory does not specify a single "preferred" vacuum, the probability distribution of vacua remains to be determined as found by a randomly chosen observer. The "volume-based" and "worldline-based" measures can be extended to scenarios with multiple bubbles, as discussed in more detail in Sect. 5.3.4. Some recent results obtained using these measures are reported in [61, 62].

5.1.2 Physical Justifications of the Semi-classical Picture

The standard framework of inflationary cosmology asserts that vacuum quantum fluctuations with super-horizon wavelengths become classical inhomogeneities of the field ϕ. The calculations of cosmological density perturbations generated during inflation [7, 11, 63, 64, 65, 66, 67] also assume that a "classicalization" of quantum fluctuations takes place via the same mechanism. In the calculations, the statistical average $\langle \delta\phi^2 \rangle$ of classical fluctuations on super-Hubble scales is simply set equal to the quantum expectation value $\langle 0| \hat{\phi}^2 |0\rangle$ in a suitable vacuum state. While this approach is widely accepted in the cosmology literature, a growing body of research is devoted to the analysis of the quantum-to-classical transition during inflation (see, e.g., [68] for an early review). Since a detailed analysis would be beyond the scope of the present text, I merely outline the main ideas and arguments relevant to this issue.

A standard phenomenological explanation of the "classicalization" of the perturbations is as follows. For simplicity, let us restrict our attention to a slow-roll inflationary scenario with one scalar field ϕ. In the slow-roll regime, one can approximately regard ϕ as a massless scalar field in de Sitter background spacetime [4]. Due to the exponentially fast expansion of de Sitter spacetime, super-horizon Fourier modes of the field ϕ are in squeezed quantum states with exponentially large ($\sim e^{Ht}$) squeezing parameters [69, 70, 71, 72, 73, 74]. Such highly squeezed states have a macroscopically large uncertainty in the field value ϕ and thus quickly decohere due to interactions with gravity and with other fields. The resulting mixed state is effectively equivalent to a statistical ensemble with a Gaussian-distributed value of ϕ. Therefore one may compute the statistical average $\langle \delta\phi^2 \rangle$ as the quantum expectation value $\langle 0| \hat{\phi}^2 |0\rangle$ and interpret the fluctuation $\delta\phi$ as a classical "noise." A heuristic description of the "classicalization" [4] is that the quantum commutators of the creation and annihilation operators of the field modes, $[\hat{a}, \hat{a}^\dagger] = 1$, are much smaller than the expectation values $\langle a^\dagger a \rangle \gg 1$ and are thus negligible.

A related issue is the backreaction of fluctuations of the scalar field ϕ on the metric.[4] According to the standard theory (see, e.g., [67, 83] for reviews), the perturbations of the metric arising due to fluctuations of ϕ are described by an auxiliary scalar field (sometimes called the "Sasaki–Mukhanov variable") in a fixed de Sitter background. Thus, the "classicalization" effect should apply equally to the fluctuations of ϕ and to the induced metric perturbations. At the same time, these metric perturbations can be viewed, in an appropriate coordinate system, as fluctuations of the local expansion rate $H(\phi)$ due to local fluctuations of ϕ [4, 7, 84]. Thus one arrives at the picture of a "locally de Sitter" spacetime with the metric (5.3), where the Hubble rate $\dot{a}/a = H(\phi)$ fluctuates on super-horizon length scales and locally follows the value of ϕ via the classical Einstein equation (5.1).

The picture as outlined is phenomenological and does not provide a description of the quantum-to-classical transition in the metric perturbations at the level of field theory. For instance, a fluctuation of ϕ leading to a local increase of $H(\phi)$ necessarily violates the null energy condition [85, 86, 87]. The cosmological implications of such "semi-classical" fluctuations (see, e.g., the scenario of "island cosmology" [88, 89, 90]) cannot be understood in detail within the framework of the phenomenological picture.

A more fundamental approach to describing the quantum-to-classical transition of perturbations was developed using non-equilibrium quantum field theory and the influence functional formalism [91, 92, 93, 94]. In this approach, decoherence of a pure quantum state of ϕ into a mixed state is entirely due to the self-interaction of the field ϕ. In particular, it is predicted that no decoherence would occur for a free field with $V(\phi) = m^2\phi^2/2$. This result is at variance with the accepted paradigm of "classicalization" as outlined above. If the source of the "noise" is the coupling between different perturbation modes of ϕ, the typical amplitude of the "noise" will be second order in the perturbation. This is several orders of magnitude smaller than the amplitude of "noise" found in the standard approach. Accordingly, it is claimed [95, 96] that the magnitude of cosmological perturbations generated by inflation is several orders of magnitude smaller than the results currently accepted as standard, and that the shape of the perturbation spectrum depends on the details of the process of "classicalization" [97]. Thus, the results obtained via the influence functional techniques do not appear to reproduce the phenomenological picture of "classicalization" as outlined above. This mismatch emphasizes the need for a deeper understanding of the nature of the quantum-to-classical transition for cosmological perturbations.

Finally, let us mention a different line of work which supports the "classicalization" picture. In [8, 98, 99, 100, 101, 102], calculations of (renormalized) expectation values such as $\langle \hat{\phi}^2 \rangle$, $\langle \hat{\phi}^4 \rangle$, etc., were performed for field operators $\hat{\phi}$

---

[4] The backreaction effects of the long-wavelength fluctuations of a scalar field during inflation have been investigated extensively (see, e.g., [75, 76, 77, 78, 79, 80, 81, 82]).

in a fixed de Sitter background. The results were compared with the statistical averages

$$\int P(\phi,t)\phi^2 d\phi, \quad \int P(\phi,t)\phi^4 d\phi, \quad \text{etc.}, \tag{5.6}$$

where the distribution $P(\phi,t)$ describes the "random walk" of the field ϕ in the Fokker–Planck approach (see Sect. 5.2.1). It was shown that the leading late-time asymptotics of the quantum expectation values coincide with the corresponding statistical averages (5.6). These results appear to validate the "random walk" approach, albeit in a limited context (in the absence of backreaction).

5.2 Stochastic Approach to Inflation

The *stochastic approach to inflation* is a semi-classical, statistical description of the spacetime resulting from quantum fluctuations of the inflation field(s) and their backreaction on the metric [1, 2, 103, 104, 105, 106, 107, 108, 109, 110, 111]. In this description, the spacetime remains everywhere classical but its geometry is determined by a stochastic process. In the next subsections I review the main tools used in the stochastic approach for calculations in the context of random walk type, slow-roll inflation. Models involving tunneling-type eternal inflation are considered in Sect. 5.2.5.

5.2.1 Random Walk-Type Eternal Inflation

The important features of random walk-type eternal inflation can be understood by considering a simple slow-roll inflationary model with a single scalar field ϕ and a potential $V(\phi)$. The slow-roll evolution equation is

$$\dot\phi = -\frac{1}{3H}\frac{dV}{d\phi} = -\frac{1}{4\pi}\frac{dH}{d\phi} \equiv v(\phi), \tag{5.7}$$

where $H(\phi)$ is defined by (5.1) and $v(\phi)$ is a model-dependent function describing the "velocity" $\dot\phi$ of the deterministic evolution of the field ϕ. The slow-roll trajectory $\phi_{\rm sr}(t)$, which is a solution of (5.7), is an attractor [112, 113] for trajectories starting with a wide range of initial conditions.[5]

As discussed in Sect. 5.1.2, the super-horizon modes of the field ϕ are assumed to undergo a rapid quantum-to-classical transition. Therefore one regards the spatial average of ϕ on scales of several H^{-1} as a *classical* field variable. The spatial averaging can be described with help of a suitable window function,

$$\langle\phi(\mathbf{x})\rangle \equiv \int W(\mathbf{x}-\mathbf{y})\phi(\mathbf{y})d^3\mathbf{y}. \tag{5.8}$$

---

[5] See [43] for a precise definition of an attractor trajectory in the context of inflation.

It is implied that the window function $W(\mathbf{x})$ decays quickly on physical distances $a\,|\mathbf{x}|$ of order several H^{-1}. From now on, let us denote the volume-averaged field simply by ϕ (no other field ϕ will be used).

As discussed above, the influence of quantum fluctuations leads to random "jumps" superimposed on top of the deterministic evolution of the volume-averaged field $\phi(t,\mathbf{x})$. This may be described by a Langevin equation of the form [2]

$$\dot{\phi}(t,\mathbf{x}) = v(\phi) + N(t,\mathbf{x}) ,\tag{5.9}$$

where $N(t,\mathbf{x})$ stands for "noise" and is assumed to be a Gaussian random function with known correlator [2, 114, 115, 116]

$$\langle N(t,\mathbf{x})N(\tilde{t},\tilde{\mathbf{x}})\rangle = C(t,\tilde{t},|\mathbf{x}-\tilde{\mathbf{x}}|\,;\phi) .\tag{5.10}$$

An explicit form of the correlator C depends on the specific window function W used for averaging the field ϕ on Hubble scales [116]. However, the window function W is merely a phenomenological device used in lieu of a complete *ab initio* derivation of the stochastic inflation picture. One expects, therefore, that results of calculations should be robust with respect to the choice of W. In other words, any uncertainty due to the choice of the window function must be regarded as an imprecision inherent in the method. For instance, a robust result in this sense is an exponentially fast decay of correlations on timescales $\Delta t \gtrsim H^{-1}$,

$$C(t,\tilde{t},|\mathbf{x}-\tilde{\mathbf{x}}|\,;\phi) \propto \exp\left(-2H(\phi)\,|t-\tilde{t}|\right) ,\tag{5.11}$$

which holds for a wide class of window functions [116].

For the purposes of the present consideration, we only need to track the evolution of $\phi(t,\mathbf{x})$ along a single comoving worldline $\mathbf{x} = \text{const.}$ Thus, we will not need an explicit form of $C(t,\tilde{t},|\mathbf{x}-\tilde{\mathbf{x}}|\,;\phi)$ but merely the value at coincident points $t = \tilde{t}$, $\mathbf{x} = \tilde{\mathbf{x}}$, which is computed in the slow-roll inflationary scenario as [2]

$$C(t,t,0;\phi) = \frac{H^2(\phi)}{4\pi^2} .\tag{5.12}$$

[This represents the fluctuation (5.2) accumulated during one Hubble time, $\Delta t = H^{-1}$.] Due to the property (5.11), one may neglect correlations on time scales $\Delta t \gtrsim H^{-1}$ in the "noise" field.[6] Thus, the evolution of ϕ on time scales $\Delta t \gtrsim H^{-1}$ can be described by a finite-difference form of the Langevin equation (5.9),

$$\phi(t+\Delta t) - \phi(t) = v(\phi)\Delta t + \sqrt{2D(\phi)\Delta t}\,\xi(t) ,\tag{5.13}$$

where

$$D(\phi) \equiv \frac{H^3(\phi)}{8\pi^2}\tag{5.14}$$

---

[6] Taking these correlations into account leads to a picture of "color noise" [117, 118]. In what follows, we only consider the simpler picture of "white noise" as an approximation adequate for the issues at hand.

and ξ is a normalized random variable representing "white noise,"

$$\langle \xi \rangle = 0, \quad \langle \xi^2 \rangle = 1, \tag{5.15}$$

$$\langle \xi(t)\xi(t + \Delta t) \rangle = 0 \quad \text{for} \quad \Delta t \gtrsim H^{-1}. \tag{5.16}$$

Equation (5.13) is interpreted as describing a Brownian motion $\phi(t)$ with the systematic "drift" $v(\phi)$ and the "diffusion coefficient" $D(\phi)$. In a typical slow-roll inflationary scenario, there will be a range of ϕ where the noise dominates over the deterministic drift,

$$v(\phi)\Delta t \ll \sqrt{2D(\phi)\Delta t}, \quad \Delta t \equiv H^{-1} . \tag{5.17}$$

Such a range of ϕ is called the "diffusion-dominated regime." For ϕ near the end of inflation, the amplitude of the noise is very small, and so the opposite inequality holds. This is the "deterministic regime" where the random jumps can be neglected and the field ϕ follows the slow-roll trajectory.

5.2.2 Fokker–Planck Equations

A useful description of the statistical properties of $\phi(t)$ is furnished by the probability density $P(\phi, t)\mathrm{d}\phi$ of having a value ϕ at time t. As in the case of the Langevin equation, the values $\phi(t)$ are measured along a single, randomly chosen comoving worldline $\mathbf{x} = \text{const}$. The probability distribution $P(\phi, t)$ satisfies the Fokker–Planck (FP) equation whose standard derivation we omit [119, 120],

$$\partial_t P = \partial_\phi \left[-v(\phi)P + \partial_\phi \left(D(\phi)P \right) \right] . \tag{5.18}$$

The coefficients $v(\phi)$ and $D(\phi)$ are in general model-dependent and need to be calculated in each particular scenario. These calculations require only the knowledge of the slow-roll trajectory and the mode functions of the quantized scalar perturbations. For ordinary slow-roll inflation with an effective potential $V(\phi)$, the results are well-known expressions (5.7) and (5.14). The corresponding expressions for models of k-inflation were derived in [43] using the relevant quantum theory of perturbations [121].

It is well known that there exists a "factor ordering" ambiguity in translating the Langevin equation into the FP equation if the amplitude of the "noise" depends on the position. Specifically, the factor $D(\phi)$ in (5.13) may be replaced by $D(\phi + \theta\Delta t)$, where $0 < \theta < 1$ is an arbitrary constant. With $\theta \neq 0$, the term $\partial_{\phi\phi} (DP)$ in (5.18) will be replaced by a different ordering of the factors,

$$\partial_{\phi\phi} (DP) \rightarrow \partial_\phi \left[D^\theta \partial_\phi \left(D^{1-\theta} P \right) \right] . \tag{5.19}$$

Popular choices $\theta = 0$ and $\theta = 1/2$ are called the Ito and the Stratonovich factor ordering, respectively. Motivated by the considerations of [122], we choose $\theta = 0$ as shown in (5.13) and (5.18). Given the phenomenological nature of the Langevin equation (5.13), one expects that any ambiguity due to the choice of θ represents an imprecision inherent in the stochastic approach. This imprecision is typically of order $H^2 \ll 1$ [123].

The quantity $P(\phi, t)$ may be also interpreted as the fraction of the *comoving* volume (i.e., coordinate volume $d^3\mathbf{x}$) occupied by the field value ϕ at time t. Another important characteristic is the volume-weighted distribution $P_V(\phi, t)d\phi$, which is defined as the *proper* 3-volume (as opposed to the comoving volume) of regions having the value ϕ at time t. (To avoid considering infinite volumes, one may restrict one's attention to a finite comoving domain in the universe and normalize $P_V(\phi, t)$ to unit volume at some initial time $t = t_0$.) The volume distribution satisfies a modified FP equation [4, 106, 108],

$$\partial_t P_V = \partial_\phi \left[-v(\phi)P_V + \partial_\phi \left(D(\phi)P_V \right) \right] + 3H(\phi)P_V \,, \qquad (5.20)$$

which differs from (5.18) by the term $3HP_V$ that describes the exponential growth of 3-volume in inflating regions.[7]

Presently we consider scenarios with a single scalar field; however, the formalism of FP equations can be straightforwardly extended to multi-field models (see, e.g., [122]). For instance, the FP equation for a two-field model is

$$\partial_t P = \partial_{\phi\phi} \left(DP \right) + \partial_{\chi\chi} \left(DP \right) - \partial_\phi \left(v_\phi P \right) - \partial_\chi \left(v_\chi P \right) \,, \qquad (5.21)$$

where $D(\phi, \chi)$, $v_\phi(\phi, \chi)$ and $v_\chi(\phi, \chi)$ are appropriate coefficients.

5.2.3 Methods of Solution

In principle, one can solve the FP equations forward in time by a numerical method, starting from a given initial distribution at $t = t_0$. To specify the solution uniquely, the FP equations must be supplemented by boundary conditions at both ends of the inflating range of ϕ [110, 111]. At the reheating boundary ($\phi = \phi_*$), one imposes the "exit" boundary conditions,

$$\partial_\phi \left[D(\phi)P \right]_{\phi=\phi_*} = 0, \quad \partial_\phi \left[D(\phi)P_V \right]_{\phi=\phi_*} = 0 \,. \qquad (5.22)$$

These boundary conditions express the fact that random jumps are very small at the end of inflation and cannot move the value of ϕ away from $\phi = \phi_*$. If the potential $V(\phi)$ reaches Planck energy scales at some $\phi = \phi_{\max}$ (this happens generally in "chaotic"-type inflationary scenarios with unbounded potentials), the semi-classical picture of spacetime breaks down for regions with $\phi \sim \phi_{\max}$. Hence, a boundary condition must be imposed also at $\phi = \phi_{\max}$. For instance, one can use the absorbing boundary condition,

$$P(\phi_{\max}) = 0 \,, \qquad (5.23)$$

---

[7] A more formal derivation of (5.20) as well as details of the interpretation of the distributions P and P_V in terms of ensembles of worldlines can be found in [124].

which means that Planck-energy regions with $\phi = \phi_{\max}$ disappear from consideration [110, 111].

Once the boundary conditions are specified, one may write the general solution of the FP equation (5.18) as

$$P(\phi, t) = \sum_\lambda C_\lambda P^{(\lambda)}(\phi)\, e^{\lambda t}\,, \qquad (5.24)$$

where the sum is performed over all the eigenvalues λ of the differential operator

$$\hat{L}P \equiv \partial_\phi \left\{ -v(\phi)P + \partial_\phi\left[D(\phi)P\right]\right\}\,, \qquad (5.25)$$

and the corresponding eigenfunctions $P^{(\lambda)}$ are defined by

$$\hat{L}P^{(\lambda)}(\phi) = \lambda P^{(\lambda)}(\phi)\,. \qquad (5.26)$$

The constants C_λ can be expressed through the initial distribution $P(\phi, t_0)$.

By an appropriate change of variables $\phi \to z$, $P(\phi) \to F(z)$, the operator $\hat{L}$ may be brought into a manifestly self-adjoint form [104, 106, 107, 108, 123, 125],

$$\hat{L} \to \frac{\mathrm{d}^2}{\mathrm{d}z^2} + U(z)\,. \qquad (5.27)$$

Then one can show that all the eigenvalues λ of $\hat{L}$ are non-positive; in particular, the (algebraically) largest eigenvalue $\lambda_{\max} \equiv -\gamma < 0$ is non-degenerate and the corresponding eigenfunction $P^{(\lambda_{\max})}(\phi)$ is positive everywhere [43, 123]. Hence, this eigenfunction describes the late-time asymptotic of the distribution $P(\phi, t)$,

$$P(\phi, t) \propto P^{(\lambda_{\max})}(\phi)\, e^{-\gamma t}\,. \qquad (5.28)$$

The distribution $P^{(\lambda_{\max})}(\phi)$ is the "stationary" distribution of ϕ per comoving volume at late times. The exponential decay of the distribution $P(\phi, t)$ means that at late times most of the comoving volume (except for an exponentially small fraction) has finished inflation and entered reheating.

Similarly, one can represent the general solution of (5.20) by

$$P_V(\phi, t) = \sum_{\tilde{\lambda}} C_{\tilde{\lambda}} P_V^{(\tilde{\lambda})}(\phi) e^{\tilde{\lambda} t}\,, \qquad (5.29)$$

where

$$[\hat{L} + 3H(\phi)]P^{(\tilde{\lambda})}(\phi) = \tilde{\lambda} P^{(\tilde{\lambda})}(\phi)\,. \qquad (5.30)$$

By the same method as for the operator $\hat{L}$, it is possible to show that the spectrum of eigenvalues $\tilde{\lambda}$ of the operator $\hat{L} + 3H(\phi)$ is bounded from above and that the largest eigenvalue $\tilde{\lambda}_{\max} \equiv \tilde{\gamma}$ admits a non-degenerate, everywhere positive eigenfunction $P^{(\tilde{\gamma})}(\phi)$. However, the largest eigenvalue $\tilde{\gamma}$ may be either positive or negative. If $\tilde{\gamma} > 0$, the late-time behavior of $P_V(\phi, t)$ is

$$P_V(\phi, t) \propto P^{(\tilde{\gamma})}(\phi)e^{\tilde{\gamma}t} , \tag{5.31}$$

which means that the total proper volume of all the inflating regions grows with time. This is the behavior expected in eternal inflation: the number of independently inflating domains increases without limit. Thus, the condition $\tilde{\gamma} > 0$ is the criterion for the presence of eternal self-reproduction of inflating domains. The corresponding distribution $P^{(\tilde{\gamma})}(\phi)$ is called the "stationary" distribution [26, 110, 111, 126].

If $\tilde{\gamma} \leq 0$, eternal inflation does not occur and the entire space almost surely (i.e., with probability 1) enters the reheating epoch at a finite time.

If the potential $V(\phi)$ is of "new" inflationary type [6, 127, 128, 129, 130] and has a global maximum at say $\phi = \phi_0$, the eigenvalues γ and $\tilde{\gamma}$ can be estimated (under the usual slow-roll assumptions on V) as [123]

$$\gamma \approx \frac{V''(\phi_0)}{8\pi V(\phi_0)} H(\phi_0) < 0, \quad \tilde{\gamma} \approx 3H(\phi_0) > 0 . \tag{5.32}$$

Therefore, eternal inflation is generic in the "new" inflationary scenario.

Let us comment on the possibility of obtaining solutions $P(\phi, t)$ in practice. With the potential $V(\phi) = \lambda\phi^4$, the full time-dependent FP equation (5.18) can be solved analytically via a non-linear change of variable $\phi \rightarrow \phi^{-2}$ [125, 131, 132]. This exact solution, as well as an approximate solution $P(\phi, t)$ for a general potential, can also be obtained using the saddle-point evaluation of a path-integral expression for $P(\phi, t)$ [133]. In some cases the eigenvalue equation $\hat{L}P^{(\lambda)} = \lambda P^{(\lambda)}$ may be reduced to an exactly solvable Schrödinger equation. These cases include potentials of the form $V(\phi) = \lambda e^{\mu\phi}$, $V(\phi) = \lambda\phi^{-2}$, $V(\phi) = \lambda \cosh^{-2}(\mu\phi)$; see, e.g., [123] for other examples.

A general approximate method for determining $P(\phi, t)$ for arbitrary potentials [134, 135, 136] consists of a perturbative expansion,

$$\phi(t) = \phi_0(t) + \delta\phi_1(t) + \delta\phi_2(t) + \cdots , \tag{5.33}$$

applied directly to the Langevin equation. The result is (at the lowest order) a Gaussian approximation with a time-dependent mean and variance [134],

$$P(\phi, t) \approx \frac{1}{\sqrt{2\pi\sigma^2(t)}} \exp\left\{ -\frac{[\phi - \phi_0(t)]^2}{2\sigma^2(t)} \right\} , \tag{5.34}$$

$$\sigma^2(t) \equiv \frac{H'^2(\phi_{\rm sr})}{\pi} \int_{\phi_{\rm sr}}^{\phi_{\rm in}} \frac{H^3}{H'^3}d\phi, \tag{5.35}$$

$$\phi_0(t) \equiv \phi_{\rm sr}(t) + \frac{H''}{2H'}\sigma^2(t) + \frac{H'}{4\pi}\left(\frac{H_{\rm in}^3}{H_{\rm in}'^2} - \frac{H^3}{H'^2} \right), \tag{5.36}$$

where $\phi_{\rm sr}(t)$ is the slow-roll trajectory and $\phi_{\rm in}$ is the initial value of ϕ. While methods based on the Langevin equation do not take into account boundary conditions or volume weighting effects, the formula (5.34) provides an adequate approximation to the distribution $P(\phi, t)$ in a useful range of ϕ and t [136].

5.2.4 Gauge Dependence Issues

An important feature of the FP equations is their dependence on the choice of the time variable. One can consider a replacement of the form

$$t \to \tau, \quad d\tau \equiv T(\phi)dt , \tag{5.37}$$

understood in the sense of integrating along comoving worldlines $\mathbf{x} = \text{const}$, where $T(\phi) > 0$ is an arbitrary function of the field. For instance, a possible choice is $T(\phi) \equiv H(\phi)$, which makes the new time variable dimensionless,

$$\tau = \int H dt = \ln a . \tag{5.38}$$

This time variable is called "scale factor time" or "e-folding time" since it measures the number of e-foldings along a comoving worldline.

The distributions $P(\phi, \tau)$ and $P_V(\phi, \tau)$ are defined as before, except for considering the 3-volumes along hypersurfaces of equal τ. These distributions satisfy FP equations similar to (5.18)–(5.20). With the replacement (5.37), the coefficients of the new FP equations are modified as follows [123],

$$D(\phi) \to \frac{D(\phi)}{T(\phi)}, \quad v(\phi) \to \frac{v(\phi)}{T(\phi)} , \tag{5.39}$$

while the "growth" term $3HP_V$ in (5.20) is replaced by $3HT^{-1}P_V$. The change in the coefficients may significantly alter the qualitative behavior of the solutions of the FP equations. For instance, stationary distributions defined through the proper time t and the e-folding time $\tau = \ln a$ were found to have radically different behaviors [18, 111, 126]. This sensitivity to the choice of the "time gauge" τ is unavoidable since hypersurfaces of equal τ may preferentially select regions with certain properties. For instance, most of the proper volume in equal-t hypersurfaces is filled with regions that have gained expansion by remaining near the top of the potential $V(\phi)$, while hypersurfaces of equal scale factor will under-represent those regions. Thus, a statement such as "most of the volume in the universe has values of ϕ with high $V(\phi)$" is largely gauge dependent.

In the early works on eternal inflation [25, 110, 111, 126], the late-time asymptotic distribution of volume $P_V^{(\bar{\gamma})}(\phi)$ along hypersurfaces of equal proper time [see (5.31)] was interpreted as the stationary distribution of field values in the universe. However, the high sensitivity of this distribution to the choice of the time variable makes this interpretation unsatisfactory. Also, it was noted [137] that equal proper-time volume distributions predict an unacceptably small probability for the currently observed CMB temperature. The reason for this result is the extreme bias of the proper-time gauge toward over-representing regions where reheating occurred very recently [18, 138]. One might ask whether hypersurfaces of equal scale factor or some other choice of time gauge would provide less biased answers. However, it turns out [124] that there exists no *a priori* choice of the time gauge τ that provides unbiased equal-τ probability distributions for all potentials $V(\phi)$ in models of slow-roll inflation (see Sect. 5.3.3 for details).

Although the FP equations necessarily involve a dependence on gauge, they do provide a useful statistical picture of the distribution of fields in the universe. The FP techniques can also be used for deriving several gauge-independent results. For instance, the presence of eternal inflation is a gauge-independent statement (see also Sect. 5.3.1): if the largest eigenvalue $\tilde{\gamma}$ is positive in one gauge of the form (5.37), then $\tilde{\gamma} > 0$ in every other gauge [139]. Using the FP approach, one can also compute the fractal dimension of the inflating domain [139, 140] and the probability of exiting inflation through a particular point ϕ_* of the reheating boundary in the configuration space (in case there exists more than one such point).

The exit probability can be determined as follows [43, 122]. Let us assume for simplicity that there are two possible exit points ϕ_* and ϕ_E, and that the initial distribution is concentrated at $\phi = \phi_0$, i.e.,

$$P(\phi, t = 0) = \delta(\phi - \phi_0) , \tag{5.40}$$

where $\phi_E < \phi_0 < \phi_*$. The probability of exiting inflation through $\phi = \phi_E$ during a time interval $[t, t + dt]$ is

$$dp_{exit}(\phi_E) = -v(\phi_E)P(\phi_E, t)dt \tag{5.41}$$

[note that $v(\phi_E) < 0$]. Hence, the total probability of exiting through $\phi = \phi_E$ at any time is

$$p_{exit}(\phi_E) = \int_0^\infty dp_{exit}(\phi_E) = -v(\phi_E) \int_0^\infty P(\phi_E, t)dt . \tag{5.42}$$

Introducing an auxiliary function $F(\phi)$ as

$$F(\phi) \equiv -v(\phi) \int_0^\infty P(\phi, t)dt , \tag{5.43}$$

one can show that $F(\phi)$ satisfies the gauge-invariant equation,

$$\partial_\phi \left[\partial_\phi \left(\frac{D}{v} F \right) - F \right] = \delta(\phi - \phi_0) . \tag{5.44}$$

This is in accord with the fact that $p_{exit}(\phi_E) = F(\phi_E)$ is a gauge-invariant quantity. Equation (5.44) with the boundary conditions

$$F(\phi_*) = 0, \quad \partial_\phi \left(\frac{D}{v} F \right) \bigg|_{\phi=\phi_E} = 0 , \tag{5.45}$$

can be straightforwardly integrated and yields explicit expressions for the exit probability $p_{exit}(\phi_E)$ as a function of the initial value ϕ_0 [43]. The exit probability $p_{exit}(\phi_*)$ can be determined similarly.

5.2.5 Self-Reproduction of Tunneling Type

Until now, we considered eternal self-reproduction due to random walk of a scalar field. Another important class of models includes self-reproduction due to bubble nucleation.[8] Such scenarios of eternal inflation were studied in [34, 141, 142, 143, 144, 145].

In a locally de Sitter universe dominated by dark energy, nucleation of bubbles of false vacuum may occur due to tunneling [14, 32, 146, 147]. Since the bubble nucleation rate κ per unit 4-volume is very small [32, 148],

$$\kappa = \mathrm{O}(1) H^{-4} \exp\left(-S_{\mathrm{I}} - \frac{\pi}{H^2}\right) , \qquad (5.46)$$

where S_{I} is the instanton action and H is the Hubble constant of the de Sitter background, bubbles will generically not merge into a single false-vacuum domain [34]. Hence, infinitely many bubbles will be nucleated at different places and times. The resulting "daughter" bubbles may again contain an asymptotically de Sitter, infinite universe, which again gives rise to infinitely many "grand-daughter" bubbles. This picture of eternal self-reproduction was called the "recycling universe" [30]. Some (or all) of the created bubbles may support a period of additional inflation followed by reheating, as shown in Fig. 5.3.

In the model of [30], there were only two vacua which could tunnel into each other. A more recently developed paradigm of "string theory landscape" [31] involves a very large number of metastable vacua, corresponding to local minima of an effective potential in field space. The value of the potential at each minimum is the effective value of the cosmological constant Λ in the corresponding vacuum. Figure 5.4 shows a phenomenologist's view of the "landscape." Vacua with $\Lambda \leq 0$ do not allow any further tunneling[9] and are called "terminal" vacua [153], while vacua with $\Lambda > 0$ are called "recyclable" since they can tunnel to other vacua with $\Lambda > 0$ or $\Lambda \leq 0$. Bubbles of recyclable vacua will give rise to infinitely many nested "daughter" bubbles. A conformal diagram of the resulting spacetime is outlined in Fig. 5.5. Of course, only a finite number of bubbles can be drawn; the bubbles actually form a fractal structure in a conformal diagram [154].

A statistical description of the "recycling" spacetime can be obtained [30, 153] by considering a single comoving worldline $\mathbf{x} = $ const that passes through different bubbles at different times. (It is implied that the worldline is randomly chosen from an ensemble of infinitely many such worldlines passing through different points $\mathbf{x}$.) Let the index $\alpha = 1, ..., N$ label all the available types of bubbles. For calculations, it is convenient to use the e-folding time $\tau \equiv \ln a$. We are interested in the probability $f_\alpha(\tau)$ of passing through

---

[8] Both processes may be combined in a single scenario [30], but we shall consider them separately for clarity.

[9] Asymptotically flat $\Lambda = 0$ vacua cannot support tunneling [149, 150, 151]; vacua with $\Lambda < 0$ will quickly collapse to a "big crunch" singularity [32, 152].

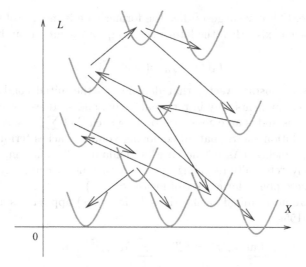

Fig. 5.4. A schematic representation of the "landscape of string theory," consisting of a large number of local minima of an effective potential. The variable X collectively denotes various fields and Λ is the effective cosmological constant. *Arrows show possible tunneling transitions between vacua*

a bubble of type α at time τ. This probability distribution is normalized by $\sum_\alpha f_\alpha = 1$; the quantity $f_\alpha(\tau)$ can be also visualized as the fraction of the comoving volume occupied by bubbles of type α at time τ. Denoting by $\kappa_{\alpha\beta}$ the nucleation rate for bubbles of type α within bubbles of type β [computed according to (5.46)], we write the "master equation" describing the evolution of $f_\alpha(\tau)$,

$$\frac{\mathrm{d}f_\beta}{\mathrm{d}\tau} = \sum_\alpha \left(-\kappa_{\alpha\beta} f_\beta + \kappa_{\beta\alpha} f_\alpha \right) \equiv \sum_\alpha M_{\beta\alpha} f_\alpha, \qquad (5.47)$$

where we introduced the auxiliary matrix $M_{\alpha\beta}$. Given a set of initial conditions $f_\alpha(0)$, one can evolve $f_\alpha(\tau)$ according to (5.47).

To proceed further, one may now distinguish the following two cases: either terminal vacua exist (some β such that $\kappa_{\alpha\beta} = 0$ for all α) or all the vacua

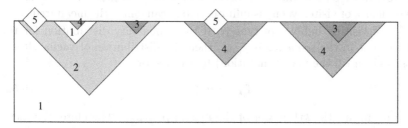

Fig. 5.5. A conformal diagram of the spacetime where self-reproduction occurs via bubble nucleation. Regions labeled *"5"* are asymptotically flat ($\Lambda = 0$)

are recyclable. (Theory suggests that the former case is more probable [59].) If terminal vacua exist, then the late-time asymptotic solution can be written as [153]

$$f_\alpha(\tau) \approx f_\alpha^{(0)} + s_\alpha e^{-q\tau}, \tag{5.48}$$

where $f_\alpha^{(0)}$ is a constant vector that depends on the initial conditions and has non-zero components only in terminal vacua, and s_α does not depend on initial conditions and is an eigenvector of $M_{\alpha\beta}$ such that $\sum_\alpha M_{\beta\alpha} s_\alpha = -q s_\beta$, $q > 0$. This solution shows that all comoving volume reaches terminal vacua exponentially quickly. (As in the case of random walk inflation, there are infinitely many "eternally recycling" points $\mathbf{x}$ that never enter any terminal vacua, but these points form a set of measure zero.)

If there are no terminal vacua, the solution $f_\alpha(\tau)$ approaches a constant distribution [155],

$$\lim_{\tau \to \infty} f_\alpha(\tau) \approx f_\alpha^{(0)}, \quad \sum_\beta M_{\beta\alpha} f_\alpha^{(0)} = 0, \tag{5.49}$$

$$f_\alpha^{(0)} = H_\alpha^4 \exp\left(\frac{\pi}{H_\alpha^2}\right). \tag{5.50}$$

In this case, the quantities $f_\alpha^{(0)}$ are independent of initial conditions and are interpreted as the fractions of time spent by the comoving worldline in bubbles of type α.

One may adopt another approach and ignore the duration of time spent by the worldline within each bubble. Thus, one describes only the *sequence* of the bubbles encountered along a randomly chosen worldline [62, 155, 156]. If the worldline is initially in a bubble of type α, then the probability $\mu_{\beta\alpha}$ of entering the bubble of type β as the next bubble in the sequence after α is

$$\mu_{\beta\alpha} = \frac{\kappa_{\beta\alpha}}{\sum_\gamma \kappa_{\gamma\alpha}}. \tag{5.51}$$

(For terminal vacua α, we have $\kappa_{\gamma\alpha} = 0$ and so we may define $\mu_{\beta\alpha} = 0$ for convenience.) Once again we consider landscapes without terminal vacua separately from terminal landscapes. If there are no terminal vacua, then the matrix $\mu_{\alpha\beta}$ is normalized, $\sum_\beta \mu_{\beta\alpha} = 1$, and is thus a stochastic matrix [157] describing a Markov process of choosing the next visited vacuum. The sequence of visited vacua is infinite, so one can define the mean frequency $f_\alpha^{(\text{mean})}$ of visiting bubbles of type α. If the probability distribution for the first element in the sequence is $f_{(0)\alpha}$, then the distribution of vacua after k steps is given (in the matrix notation) by the vector

$$\mathbf{f}_{(k)} = \boldsymbol{\mu}^k \mathbf{f}_{(0)}, \tag{5.52}$$

where $\boldsymbol{\mu}^k$ means the kth power of the matrix $\boldsymbol{\mu} \equiv \mu_{\alpha\beta}$. Therefore, the mean frequency of visiting a vacuum α is computed as an average of $f_{(k)\alpha}$ over n consecutive steps in the limit of large n:

$$\mathbf{f}^{(\text{mean})} = \lim_{n\to\infty} \frac{1}{n} \sum_{k=1}^{n} \mathbf{f}_{(k)} = \lim_{n\to\infty} \frac{1}{n} \sum_{k=1}^{n} \mu^k \mathbf{f}_{(0)}. \tag{5.53}$$

[It is proved in the theory of Markov processes that the limit $f_\alpha^{(\text{mean})}$ given by (5.53) almost surely coincides with the mean frequency of visiting the state α; see, e.g., [158, Chap. 5, Theorem 2.1], and [159, Theorem 3.5.9.].] It turns out that the distribution $\mathbf{f}^{(\text{mean})}$ is independent of the initial state $\mathbf{f}_{(0)}$ and coincides with the distribution (5.49) found in the continuous-time description [155].

If there exist terminal vacua, then almost all sequences will have a finite length. The distribution of vacua in a randomly chosen sequence is still well-defined and can be computed using (5.53) without the normalizing factor $1/n$,

$$\mathbf{f}^{(\text{mean})} = (1 - \mu)^{-1} \mu \mathbf{f}_{(0)}, \tag{5.54}$$

but now the resulting distribution depends on the initial state $\mathbf{f}_{(0)}$ [62, 156].

5.3 Predictions and Measure Issues

As discussed in Sect. 5.1.1, a compelling question in the context of eternal inflation is how to make statistical predictions of observed parameters. One begins by determining whether eternal inflation is present in a given model.

5.3.1 Presence of Eternal Inflation

Since the presence of eternal self-reproduction in models of tunneling type is generically certain (unless the nucleation rate for terminal vacua is unusually high), in this section we restrict our attention to eternal inflation of the random walk type.

The hallmark of eternal inflation is the unbounded growth the total number of independent inflating regions. The total proper 3-volume of the inflating domain also grows without bound at late times, at least when computed along hypersurfaces of equal proper time or equal scale factor. However, the proper 3-volume is a gauge-dependent quantity, and one may construct time gauges where the 3-volume decreases with time even in an everywhere expanding universe [124]. The 3-volume of an *arbitrary* family of equal-time hypersurfaces cannot be used as a criterion for the presence of eternal inflation. However, a weaker criterion is sufficient: eternal inflation is present if (and only if) there *exists* a choice of time slicing with an unbounded growth of the 3-volume of inflating domains [124]. Equivalently, eternal inflation is present if a finite co-moving volume gives rise to infinite physical volume [137]. Thus, the presence or absence of eternal inflation is a gauge-independent statement. One may, of course, use a particular gauge (such as the proper time or e-folding time) for calculations, as long as the result is known to be gauge-independent. It can be shown that eternal inflation is present if and only if the 3-volume grows in the e-folding time slicing [124].

The presence of eternal inflation has been analyzed in many specific scenarios. For instance, eternal inflation is generic in "chaotic" [3, 41, 160] and "new" [8] inflationary models. It is normally sufficient to establish the existence of a "diffusion-dominated" regime, that is, a range of ϕ where the typical amplitude $\delta\phi \sim H$ of "jumps" is larger than the typical change of the field, $\dot{\phi}\Delta t$, during one Hubble time $\Delta t = H^{-1}$. For models of scalar field inflation, the condition is

$$H^2 \gg H' . \tag{5.55}$$

Such a range of ϕ is present in most slow-roll models of inflation. (For an example of an inflationary scenario where eternal inflation is generically *not* present, see [161].) A strict formal criterion for the presence of eternal inflation is the positivity of the largest eigenvalue $\tilde{\gamma}$ of the operator $\hat{L}+3H(\phi)$, as defined in Sect. 5.2.3.

The causal structure of the eternally inflating spacetime and the topology of the reheating surface can be visualized using the construction of "eternal comoving points" [139]. These are comoving worldlines $\mathbf{x} = $ const that forever remain within the inflating domain and never enter the reheating epoch. These worldlines correspond to places where the reheating surface reaches $t = \infty$ in a spacetime diagram (see Fig. 5.2). It was shown in [139] using topological arguments that the presence of inflating domains at arbitrarily late times entails the existence of infinitely many such "eternal points." The set of all eternal points within a given three-dimensional spacelike slice is a measure of zero fractal set. The fractal dimension of this set can be understood as the fractal dimension of the inflating domain [124, 139, 140] and is invariant under any *smooth* coordinate transformations in the spacetime.

The existence of eternal points can be used as another invariant criterion for the presence of eternal inflation. The probability $X(\phi)$ of having an eternal point in an initial Hubble-size region with field value ϕ can be found as the solution of a gauge-invariant, non-linear diffusion equation [139]

$$D\partial_{\phi\phi}X + v\partial_\phi X - 3H(1-X)\ln(1-X) = 0 , \tag{5.56}$$

with zero boundary conditions. Eternal inflation is present if there exists a non-trivial solution $X(\phi) \not\equiv 0$ of this equation.

5.3.2 Observer-Based Measure in Eternal Inflation

In theories where observable parameters χ_a are distributed randomly, one would like to predict the values of χ_a most likely to be observed by a random (or "typical") observer. More generally, one looks for the probability distribution $P(\chi_a)$ of observing the values χ_a. As discussed in Sect. 5.1.1, considerations of this type necessarily involve some form of the "principle of mediocrity" [18]. On a more formal level, one needs to construct an ensemble of the "possible observers" and to define a probability measure on this ensemble. In inflationary cosmology, observers appear only along the reheating surface. If eternal inflation is present, the reheating surface contains infinitely many causally disconnected and (possibly) statistically inequivalent domains. The principal difficulties in the probabilistic approach are due to a lack of a natural definition of measure on such surfaces.[10]

Existing proposals for an observer-based measure fall in two major classes, which may be designated as "volume-based" vs. "worldline-based." The difference between these classes is in the approach taken to construct the ensemble of observers. In the "volume" approach [18, 19, 45, 111, 153, 163], the ensemble contains every observer appearing in the universe, at any time or place. In the "worldline" approach [156, 164, 165, 166], the ensemble consists of observers appearing near a single, randomly selected comoving worldline $\mathbf{x} = \text{const}$ (a timelike geodesic could also be used). If the ensemble contains infinitely many observers (this is typically the case for volume-based ensembles), a regularization is needed to obtain specific probability distributions. Finding and applying suitable regularization procedures is a separate technical issue explored in Sect. 5.3.3 and 5.3.4. I begin with a general discussion of these measure prescriptions.

A number of previously considered "volume-based" measure proposals were found to be lacking in one aspect or another [18, 111, 123, 164, 167, 168], the most vexing problem being the dependence on the choice of the time gauge [110, 111, 137]. The requirement of time gauge independence is sufficiently important to reject any measure proposal that suffers from the gauge ambiguity. A prescription manifestly free from gauge dependence is the "spherical cutoff" measure [45]. This prescription provides unambiguous predictions for models of random walk-type eternal inflation if the reheating condition $\phi = \phi_*$ corresponds to a topologically compact and connected locus in the field space $\{\phi, \chi_a\}$ (see Sect. 5.3.3). For models where the reheating condition is met at several disconnected loci in field space (tunneling-type eternal inflation belongs to this class), one can use the recently proposed prescription of "comoving cutoff" [153, 163]. Since no other volume-based prescriptions are currently considered viable, we refer to the mentioned spherical cutoff/comoving cutoff prescriptions simply as the "volume-based measure."

Similarly, existing measure proposals of the "worldline" type appear to converge essentially to a single prescription [62, 156] (however, see [169, 170] for the most recent developments). We refer to this prescription as the "worldline-based measure."

The main difference between the worldline-based and volume-based measures is in their dependence on the initial state. When considering the volume-based measure, one starts from a finite initial spacelike 3-volume. (Final results

---

[10] To avoid confusion, let us note that the recent work [162] proposes a measure in the phase space of trajectories rather than an observer-based measure in the sense discussed here.

are insensitive to the choice of this 3-surface in spacetime or to its geometry.) The initial state consists of the initial values of the fields χ_a within the initial volume, and possibly a label α corresponding to the type of the initial bubble. When considering the worldline-based measure, one assumes knowledge of these data at the initial point of the worldline.[11] It turns out that the volume-based measure always yields results that are independent of the initial conditions. This agrees with the concept of the "stationarity" of a self-reproducing universe [110, 111, 126]; the universe forgets the initial state in the course of eternal self-reproduction. In contrast, probabilities obtained using the worldline-based measure always depend on the initial state (except for the case of a "non-terminal" landscape, i.e., a landscape scenario without terminal vacua). A theory of initial conditions is necessary to obtain a specific prediction from the worldline-based measure.

At this time, there is no consensus as to which of the two measures is the physically relevant one. The present author is inclined to regard the two measures as reasonable answers to two differently posed questions. The first question is to determine the probability distribution for observed values of χ_a, given that the observer is randomly chosen from all the observers present in the entire spacetime. Since we have no knowledge as to the total duration of inflation in our past or the total number of bubble nucleations preceding the most recent one, it appears reasonable to include in the ensemble all the observers that will ever appear anywhere in the spacetime. The answer to the first question is thus provided by the volume-based measure.

The second question is posed in a rather different manner. In the context of tunneling-type eternal inflation, upon discovering the type of our bubble we may wish to leave a message to a future civilization that may arise in our future after an unspecified number of nested bubble nucleations. The analogous situation in the context of random walk-type inflation is a hypothetical observer located within an inflating region of spacetime who wishes to communicate with future civilizations that will eventually appear when inflation is over. The only available means of communication is leaving information on paper in a sealed box. The message might contain the probability distribution $P(\chi_a)$ for parameters χ_a that we expect the future civilization to observe. In this case, the initial state is known at the time of writing the message. It is clear that the box can be discovered only by future observers near its worldline. It is then natural to choose the ensemble of observers appearing along this worldline. Starting from the known initial state, one would then compute $P(\chi_a)$ according to the worldline-based measure.

Calculations using the worldline-based measure usually do not require regularization (except for the case of non-terminal landscape) because the

---

[11] Naturally, it is assumed that the initial state is in the self-reproduction regime. For random walk-type models, the initial 3-volume is undergoing inflation rather than reheating and for tunneling models, the initial volume is not situated within a terminal bubble.

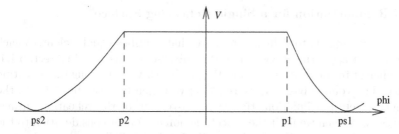

Fig. 5.6. Illustrative inflationary potential with a flat self-reproduction regime $\phi_1 < \phi < \phi_2$ and deterministic regimes $\phi_*^{(1)} < \phi < \phi_1$ and $\phi_2 < \phi < \phi_*^{(2)}$

worldline-based ensemble of observers is almost surely finite [156]. For instance, in random walk-type models the worldline-based measure predicts the exit probability distribution p_{exit}, which can be computed by solving a suitable differential equation [see (5.44)]. However, the ensemble used in the volume-based measure is infinite and requires a regularization. Known regularization methods are reviewed in Sect. 5.3.3 and 5.3.4.

A simple toy model [18] where the predictions of the volume-based measure can be obtained analytically is a slow-roll scenario with a potential shown in Fig. 5.6. The potential is flat in the range $\phi_1 < \phi < \phi_2$ where the evolution is diffusion-dominated, while the evolution of regions with $\phi > \phi_2$ or $\phi < \phi_1$ is completely deterministic (fluctuation-free). It is assumed that the diffusion-dominated range $\phi_1 < \phi < \phi_2$ is sufficiently wide to cause eternal self-reproduction of inflating regions. There are two thermalization points, $\phi = \phi_*^{(1)}$ and $\phi = \phi_*^{(2)}$, which may be associated to different types of true vacuum and thus to different observed values of cosmological parameters. The question is to compare the volumes $\mathcal{V}_1$ and $\mathcal{V}_2$ of regions thermalized into these two vacua. Since there is an infinite volume thermalized into either vacuum, one looks for the volume ratio $\mathcal{V}_1/\mathcal{V}_2$.

The potential is symmetric in the range $\phi_1 < \phi < \phi_2$, so it is natural to assume that Hubble-size regions exiting the self-reproduction regime at $\phi = \phi_1$ and at $\phi = \phi_2$ are equally abundant. Since the evolution within the ranges $\phi_*^{(1)} < \phi < \phi_1$ and $\phi_2 < \phi < \phi_*^{(2)}$ is deterministic, the regions exiting the self-reproduction regime at $\phi = \phi_1$ or $\phi = \phi_2$ will be expanded by fixed amounts of e-foldings, which we may denote N_1 and N_2, respectively,

$$N_j = -4\pi \int_{\phi_j}^{\phi_*^{(j)}} \frac{H(\phi)}{H'(\phi)} \mathrm{d}\phi . \tag{5.57}$$

Therefore the volume of regions thermalized at $\phi = \phi_*^{(j)}$, where $j = 1, 2$, will be increased by the factors $\exp(3N_j)$. Hence, the volume ratio is

$$\frac{\mathcal{V}_1}{\mathcal{V}_2} = \exp\left(3N_1 - 3N_2\right). \tag{5.58}$$

5.3.3 Regularization for a Single Reheating Surface

The task at hand is to define a measure that ascribes equal weight to each observer ever appearing anywhere in the universe. As discussed in Sect. 5.1.1, it is sufficient to construct a measure $\mathcal{V}(\chi_a)$ of the 3-volume along the reheating surface. The volume-based measure $P(\chi_a)$ will then be given by (5.5). In the presence of eternal inflation, the proper 3-volume of the reheating surface diverges even when we limit the spacetime domain under consideration to the comoving future of a finite initial spacelike 3-volume. Therefore, the reheating surface needs to be regularized.

In this section we consider the case when the reheating condition is met at a topologically compact and connected locus in the configuration space $\{\phi, \chi_a\}$. In this case, every connected component of the reheating surface in spacetime will contain all the possible values of the fields χ_a, and all such connected pieces are statistically equivalent. Hence, it suffices to consider a single connected piece of the reheating surface. A situation of this type is illustrated in Fig. 5.7. A sketch of the random walk in configuration space is shown in Fig. 5.8.

A simple regularization scheme is known as the "equal-time cutoff." One considers the part of the reheating surface formed before a fixed time $t_{\max}$; that part is finite as long as $t_{\max}$ is finite. Then one can compute the distribution of the quantities of interest within that part of the reheating surface. Subsequently, one takes the limit $t_{\max} \to \infty$. The resulting

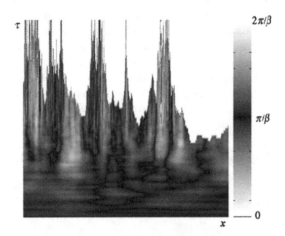

Fig. 5.7. A 1+1-dimensional slice of spacetime in a two-field inflationary model (numerical simulation in [19]). *Shades* denote different values of the field χ, which takes values in the periodically identified interval $[0, 2\pi/\beta]$. The *white region* represents the thermalized domain. The boundary of the thermalized domain is the reheating surface (cf. Fig. 5.2), which contains all the possible values of the field χ

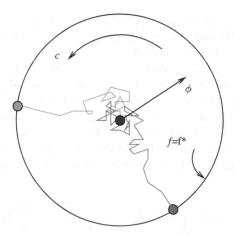

Fig. 5.8. A random walk in configuration space for a two-field inflationary model considered in [19]. The center of the field space is a diffusion-dominated regime. The reheating condition, $\phi = \phi_*$, selects a compact and connected region (a circle) in configuration space. The problem is to determine the volume-weighted probability distribution for the values of χ at reheating

distribution can be found from the solution $P_V^{(\tilde{\gamma})}(\phi)$ of the "stationary" FP equation,

$$[\hat{L} + 3H(\phi)]P_V^{(\tilde{\gamma})} = \tilde{\gamma}P_V^{(\tilde{\gamma})} , \qquad (5.59)$$

with the largest eigenvalue $\tilde{\gamma}$ (see Sect. 5.2.3). However, both the eigenvalue $\tilde{\gamma}$ and the distribution $P_V^{(\tilde{\gamma})}(\phi)$ depend rather sensitively on the choice of the equal-time hypersurfaces. Since there appears to be no preferred choice of the cutoff hypersurfaces in spacetime, the equal-time cutoff cannot serve as an unbiased measure. Also, it was shown in [124] that the unbiased result (5.58) cannot be obtained via an equal-time cutoff with any choice of the time gauge.

The "spherical cutoff" measure prescription [45] regularizes the reheating surface in a different way. A finite region within the reheating surface is selected as a spherical region of radius R around a randomly chosen center point. (Since the reheating 3-surface is spacelike, the distance between points can be calculated as the length of the shortest path within the reheating 3-surface.) Then the distribution of the quantities of interest is computed within the spherical region. Subsequently, the limit $R \to \infty$ is evaluated. Since every portion of the reheating surface is statistically the same, the results are independent of the choice of the center point. The spherical cutoff is gauge-invariant since it is formulated entirely in terms of the intrinsic properties of the reheating surface.

While the spherical cutoff prescription successfully solves the problem of regularization, there is no universally applicable analytic formula for the resulting distribution. Application of the spherical cutoff to general models of random walk inflation requires a direct numerical simulation of the stochastic field dynamics in the inflationary spacetime. Such simulations were reported in [19, 111, 140, 171] and used the Langevin equation (5.9) with a specific stochastic ansatz for the noise field $N(t, \mathbf{x})$.

Apart from numerical simulations, the results of the spherical cutoff method may be obtained analytically in a certain class of models [19]. One such case is a multi-field model where the potential $V(\phi, \chi_a)$ is independent of χ_a within the range of ϕ where the "diffusion" in ϕ dominates. Then the distribution in χ_a is flat when field ϕ exits the regime of self-reproduction and resumes the deterministic slow-roll evolution. One can derive a gauge-invariant Fokker–Planck equation for the volume distribution $P_V(\phi, \chi_a)$, using ϕ as the time variable [19],

$$\partial_\phi P_V = \partial_{\chi\chi} \left(\frac{D}{v_\phi} P_V \right) - \partial_\chi \left(\frac{v_\chi}{v_\phi} P_V \right) + \frac{3H}{v_\phi} P_V, \qquad (5.60)$$

where D, v_ϕ and v_χ are the coefficients of the FP equation (5.21). This equation is valid for the range of ϕ where the evolution of ϕ is free of fluctuations. By solving (5.60), one can calculate the volume-based distribution of χ_a predicted by the spherical cutoff method as $P_V(\phi_*, \chi)$. Note that the mentioned restriction on the potential $V(\phi, \chi_a)$ is important. In general, the field ϕ cannot be used as the time variable since the surfaces of constant ϕ are not everywhere spacelike due to large fluctuations of ϕ in the diffusion-dominated regime.

5.3.4 Regularization for Multiple Types of Reheating Surfaces

Let us begin by considering a simpler example: an inflationary scenario with an *asymmetric* slow-roll potential $V(\phi)$ having two minima $\phi_*^{(1)}$, $\phi_*^{(2)}$. This scenario has two possibilities for thermalization, possibly differing in the observable parameters χ_a. More generally, one may consider a scenario with n different minima of the potential, possibly representing n distinct reheating scenarios. It is important that the minima $\phi_*^{(j)}$, $j = 1, ..., n$ are topologically disconnected *in the configuration space*. This precludes the possibility that different minima are reached within one connected component of the reheating hypersurface in spacetime. Additionally, the fields χ_a may fluctuate across each connected component in a way that depends on the minimum j. Thus, the distribution $P(\chi_a; j)$ of the fields χ_a at each connected component of the reheating surface may depend on j. In other words, the different components of the reheating surface may be statistically inequivalent with respect to the distribution of χ_a on them. To use the volume-based measure for making predictions in such models, one needs a regularization method that is applicable to situations with a large number of disconnected and statistically inequivalent components of the reheating hypersurface.

In such situations, the spherical cutoff prescription (see Sect. 5.3.3) yields only the distribution of χ_a across one connected component, since the sphere of a finite radius R will never reach any other components of the reheating surface. Therefore, the spherical cutoff needs to be supplemented by a "weighting" prescription, which would assign weights p_j to the minimum labeled j. In scenarios of tunneling type, such as the string theory landscape, observers may find themselves in bubbles of types $\alpha = 1, ..., N$. Again, a weighting prescription is needed to determine the probabilities p_α of being in a bubble of type α.

Two different weighting prescriptions have been formulated, using the volume-based [153, 163] and the worldline-based approach [156], respectively. The first prescription is called the "comoving cutoff" while the second the "worldline" or the "holographic" cutoff. For clarity, we illustrate these weighting prescriptions on models of tunneling type where infinitely many nested bubbles of types $\alpha = 1, ...N$ are created and where some bubbles are "terminal," i.e., contain no "daughter" bubbles. In the volume-based approach, each bubble receives equal weight in the ensemble; in the worldline-based approach, only bubbles intersected by a selected worldline are counted and given equal weight. Let us now examine these two prescriptions in more detail.

Since the set of all bubbles is infinite, one needs to perform a "regularization," that is, one needs to select a very large but finite subset of bubbles. The weight p_α will be calculated as the fraction of bubbles of type α within the selected subset; then the number of bubbles in the subset will be taken to infinity. Technically, the two prescriptions differ in the details of the regularization. The difference between the two prescriptions can be understood pictorially (Fig. 5.9). In a spacetime diagram, one draws a finite number of timelike comoving geodesic worldlines emitted from an initial 3-surface toward the future. (It can be shown that the results are independent of the choice of these lines, as long as that choice is uncorrelated with the bubble nucleation process [163].) Each of these lines will intersect only a finite number of bubbles, since the final state of any worldline is (almost surely) a terminal bubble. The subset of bubbles needed for the regularization procedure is defined as the set of all bubbles intersected by at least one line. At this point, the volume-based approach assigns equal weight to each bubble in the subset, while the worldline-based approach assigns equal weight to each bubble along each worldline. As a result, the volume-based measure counts each bubble in the subset only once, while the worldline-based measure counts each bubble as many times as it is intersected by some worldlines. After determining the weights p_α by counting the bubbles as described, one increases the number of worldlines to infinity and evaluates the limit values of p_α.

It is clear that the volume-based measure represents the counting of bubbles in the entire universe, and it is appropriate that each bubble is being counted only once. On the other hand, the worldline-based measure counts bubbles occurring along a single worldline, ignoring the bubbles produced in other parts of the universe and introducing an unavoidable bias due to the initial conditions at the starting point of the worldline.

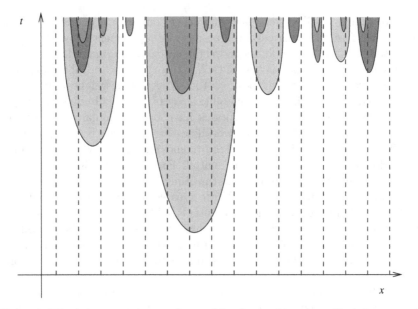

Fig. 5.9. Weighting prescriptions for models of tunneling type. *Shaded regions* are bubbles of different types. *Dashed vertical lines* represent randomly chosen comoving geodesics used to define a finite subset of bubbles

Explicit formulas for p_α were derived for a tunneling-type scenario (with terminal vacua) in the volume-based approach [153],

$$p_\alpha = \sum_\beta H_\beta^q \kappa_{\alpha\beta} s_\beta \, , \tag{5.61}$$

where $\kappa_{\alpha\beta}$ is the matrix of nucleation rates, q and s_α are the quantities defined by (5.48), and H_β is the Hubble parameter in bubbles of type β. The expressions for p_α obtained from the worldline-based measure are given by (5.54). As we have noted before, the volume-based measure assigns weights p_α that are independent of initial conditions, while the weights obtained from the worldline-based measure depend sensitively on the type of bubble where the counting begins.

In the case of a non-terminal landscape, both the volume-based and the worldline-based measures give identical results for p_α, which coincide with the mean frequency (5.49) of visiting a bubble of type α [155, 156].

With the weighting prescriptions just described, the volume-based and the worldline-based measure proposals can be considered complete. In other words, we have two alternative prescriptions that can be applied (in principle) to arbitrary models of random walk or tunneling-type eternal inflation. Further research is needed to reach a definite conclusion concerning the viability of these measure prescriptions.

Acknowledgments

The author is grateful to Gabriel Lopes Cardoso, Matthew Johnson, and Andrei Linde for useful discussions, and Andrei Barvinsky and Alex Vilenkin for comments on the manuscript.

References

1. A. Vilenkin, Phys. Rev. **D27**, 2848 (1983).
2. A. A. Starobinsky, in *Current Topics in Field Theory, Quantum Gravity and Strings, Lecture Notes in Physics* 206, eds H. J. de Vega and N. Sanchez (Springer Verlag), p. 107.
3. A. D. Linde, Phys. Lett. **B175**, 395–400 (1986).
4. A. S. Goncharov, A. D. Linde and V. F. Mukhanov, Int. J. Mod. Phys. **A2**, 561–591 (1987).
5. A. Vilenkin and L. H. Ford, Phys. Rev. **D26**, 1231 (1982).
6. A. D. Linde, Phys. Lett. **B116**, 335 (1982).
7. A. A. Starobinsky, Phys. Lett. **B117**, 175–178 (1982).
8. A. Vilenkin, Nucl. Phys. **B226**, 527 (1983).
9. R. H. Brandenberger, Nucl. Phys. **B245**, 328 (1984).
10. A. H. Guth and S.-Y. Pi, Phys. Rev. **D32**, 1899–1920 (1985).
11. J. M. Bardeen, P. J. Steinhardt and M. S. Turner, Phys. Rev. **D28**, 679 (1983).
12. G. W. Gibbons and S. W. Hawking, Phys. Rev. **D15**, 2738–2751 (1977).
13. S. W. Hawking and I. G. Moss, Phys. Lett. **B110**, 35 (1982).
14. S. W. Hawking and I. G. Moss, Nucl. Phys. **B224**, 180 (1983).
15. A. Vilenkin, Eternal inflation and chaotic terminology, `gr-qc/0409055`.
16. A. Borde and A. Vilenkin, Phys. Rev. Lett. **72**, 3305–3309 (1994).
17. A. Borde, A. H. Guth and A. Vilenkin, Phys. Rev. Lett. **90**, 151301 (2003).
18. A. Vilenkin, Phys. Rev. **D52**, 3365–3374 (1995).
19. V. Vanchurin, A. Vilenkin and S. Winitzki, Phys. Rev. **D61**, 083507 (2000).
20. D. A. Lowe and D. Marolf, Phys. Rev. **D70**, 026001 (2004).
21. C.-H. Wu, K.-W. Ng and L. H. Ford, Constraints on the duration of inflationary expansion from quantum stress tensor fluctuations, `gr-qc/0608002`.
22. A. D. Linde, Phys. Lett. **B259**, 38–47 (1991).
23. A. D. Linde, Phys. Rev. **D49**, 748–754 (1994).
24. J. Garcia-Bellido, A. D. Linde and D. A. Linde, Phys. Rev. **D50**, 730–750 (1994).
25. J. Garcia-Bellido, Nucl. Phys. **B423**, 221–242 (1994).
26. J. Garcia-Bellido and D. Wands, Phys. Rev. **D52**, 5636–5642 (1995).
27. A. D. Linde, Phys. Lett. **B327**, 208–213 (1994).
28. A. Vilenkin, Phys. Rev. Lett. **72**, 3137–3140 (1994).
29. K. E. Kunze, Phys. Lett. **B587**, 1–6 (2004).
30. J. Garriga and A. Vilenkin, Phys. Rev. **D57**, 2230–2244 (1998).
31. L. Susskind, The anthropic landscape of string theory, `hep-th/0302219`.
32. S. R. Coleman and F. De Luccia, Phys. Rev. **D21**, 3305 (1980).
33. A. H. Guth, Eternal inflation, Ann. N.Y. Acad. Sci. **950**, 66–82 (2001).
34. A. H. Guth and E. J. Weinberg, Nucl. Phys. **B212**, 321 (1983).

35. J. Garriga, A. H. Guth and A. Vilenkin, Eternal inflation, bubble collisions, and the persistence of memory, hep-th/0612242.
36. A. H. Guth, Phys. Rev. **D23**, 347–356 (1981).
37. J. Garriga, V. F. Mukhanov, K. D. Olum and A. Vilenkin, Int. J. Theor. Phys. **39**, 1887–1900 (2000).
38. J. Garriga and A. Vilenkin, Phys. Rev. **D64**, 043511 (2001).
39. D. S. Goldwirth and T. Piran, Phys. Rept. **214**, 223–291 (1992).
40. A. D. Linde, Phys. Lett. **B129**, 177–181 (1983).
41. A. D. Linde, Mod. Phys. Lett. **A1**, 81 (1986).
42. J. Martin and M. A. Musso, Phys. Rev. **D71**, 063514 (2005).
43. F. Helmer and S. Winitzki, Phys. Rev. **D74**, 063528 (2006).
44. S. M. Carroll, Living Rev. Rel. **4**, 1 (2001).
45. A. Vilenkin, Phys. Rev. Lett. **81**, 5501–5504 (1998).
46. J. Garriga, T. Tanaka and A. Vilenkin, Phys. Rev. **D60**, 023501 (1999).
47. J. Garriga and A. Vilenkin, Phys. Rev. **D61**, 083502 (2000).
48. J. Garriga, M. Livio and A. Vilenkin, Phys. Rev. **D61**, 023503 (2000).
49. J. Garriga and A. Vilenkin, Phys. Rev. **D67**, 043503 (2003).
50. J. Garriga, A. Linde and A. Vilenkin, Phys. Rev. **D69**, 063521 (2004).
51. J. Garriga and A. Vilenkin, Prog. Theor. Phys. Suppl. **163**, 245–257 (2006).
52. M. Tegmark, A. Vilenkin and L. Pogosian, Phys. Rev. **D71**, 103523 (2005).
53. M. Tegmark, A. Aguirre, M. Rees and F. Wilczek, Phys. Rev. **D73**, 023505 (2006).
54. L. J. Hall, T. Watari and T. T. Yanagida, Phys. Rev. **D73**, 103502 (2006).
55. M. Susperregi, Phys. Rev. **D55**, 560–572 (1997).
56. B. Feldstein, L. J. Hall and T. Watari, Phys. Rev. **D72**, 123506 (2005).
57. W. Lerche, D. Lust and A. N. Schellekens, Nucl. Phys. **B287**, 477 (1987).
58. R. Bousso and J. Polchinski, JHEP, **06**, 006 (2000).
59. S. Kachru, R. Kallosh, A. Linde and S. P. Trivedi, Phys. Rev. **D68**, 046005 (2003).
60. F. Denef and M. R. Douglas, JHEP **05**, 072 (2004).
61. D. Schwartz-Perlov and A. Vilenkin, JCAP **0606**, 010 (2006).
62. A. Aguirre, S. Gratton and M. C. Johnson, Hurdles for recent measures in eternal inflation, hep-th/0611221.
63. V. F. Mukhanov and G. V. Chibisov, JETP Lett. **33**, 532–535 (1981).
64. V. F. Mukhanov and G. V. Chibisov, Sov. Phys. JETP **56**, 258–265 (1982).
65. S. W. Hawking, Phys. Lett. **B115**, 295 (1982).
66. A. H. Guth and S. Y. Pi, Phys. Rev. Lett. **49**, 1110–1113 (1982).
67. V. F. Mukhanov, H. A. Feldman and R. H. Brandenberger, Phys. Rept. **215**, 203–333 (1992).
68. R. H. Brandenberger, Rev. Mod. Phys. **57**, 1 (1985).
69. D. Polarski and A. A. Starobinsky, Class. Quant. Grav. **13**, 377–392 (1996).
70. C. Kiefer and D. Polarski, Annalen Phys. **7**, 137–158 (1998).
71. C. Kiefer, J. Lesgourgues, D. Polarski and A. A. Starobinsky, Class. Quant. Grav. **15**, L67–L72 (1998).
72. C. Kiefer, D. Polarski and A. A. Starobinsky, Int. J. Mod. Phys. **D7**, 455–462 (1998).
73. C. Kiefer, D. Polarski and A. A. Starobinsky, Phys. Rev. **D62**, 043518 (2000).
74. C. Kiefer, I. Lohmar, D. Polarski and A. A. Starobinsky, Class. Quant. Grav. **24**, 1699 (2007).

75. L. Raul W. Abramo, R. H. Brandenberger and V. F. Mukhanov, Phys. Rev. **D56**, 3248–3257 (1997).
76. N. Afshordi and R. H. Brandenberger, Phys. Rev. **D63**, 123505 (2001).
77. L. R. Abramo and R. P. Woodard, Phys. Rev. **D65**, 043507 (2002).
78. L. R. Abramo and R. P. Woodard, Phys. Rev. **D65**, 063515 (2002).
79. L. R. Abramo and R. P. Woodard, Phys. Rev. **D65**, 063516 (2002).
80. G. Geshnizjani and R. Brandenberger, JCAP **0504**, 006 (2005).
81. G. Geshnizjani and N. Afshordi, JCAP **0501**, 011 (2005).
82. E. O. Kahya and V. K. Onemli, Quantum stability of a w < −1 phase of cosmic acceleration, gr-qc/0612026.
83. H. Kodama and M. Sasaki, Prog. Theor. Phys. Suppl. **78**, 1–166 (1984).
84. A. V. Frolov and L. Kofman, JCAP **0305**, 009 (2003).
85. A. Borde and A. Vilenkin, Phys. Rev. **D56**, 717–723 (1997).
86. S. Winitzki, Null energy condition violations in eternal inflation, gr-qc/0111109.
87. T. Vachaspati, Eternal inflation and energy conditions in de sitter spacetime, astro-ph/0305439.
88. S. Dutta and T. Vachaspati, Phys. Rev. **D71**, 083507 (2005).
89. Y.-S. Piao, Phys. Rev. **D72**, 103513 (2005).
90. S. Dutta, Phys. Rev. **D73**, 063524 (2006).
91. M. Morikawa, Prog. Theor. Phys. **77**, 1163–1177 (1987).
92. M. Morikawa, Phys. Rev. **D42**, 1027–1034 (1990).
93. B. L. Hu and A. Matacz, Phys. Rev. **D51**, 1577–1586 (1995).
94. E. Calzetta and B. L. Hu, Phys. Rev. **D52**, 6770–6788 (1995).
95. A. Matacz, Phys. Rev. **D55**, 1860–1874 (1997).
96. A. Matacz, Phys. Rev. **D56**, 1836–1840 (1997).
97. H. Kubotani, T. Uesugi, M. Morikawa and A. Sugamoto, Prog. Theor. Phys. **98**, 1063–1080 (1997).
98. R. P. Woodard, Nucl. Phys. Proc. Suppl. **148**, 108–119 (2005).
99. N. C. Tsamis and R. P. Woodard, Nucl. Phys. **B724**, 295–328 (2005).
100. R. P. Woodard, Generalizing starobinskii's formalism to yukawa theory and to scalar QED, gr-qc/0608037.
101. S.-P. Miao and R. P. Woodard, Phys. Rev. **D74**, 044019 (2006).
102. T. Prokopec, N. C. Tsamis and R. P. Woodard, Class. Quant. Grav. **24**, 201–230 (2007).
103. S.-J. Rey, Nucl. Phys. **B284**, 706 (1987).
104. K. Nakao, Y. Nambu and M. Sasaki, Prog. Theor. Phys. **80**, 1041 (1988).
105. H. E. Kandrup, Phys. Rev. **D39**, 2245 (1989).
106. Y. Nambu and M. Sasaki, Phys. Lett. **B219**, 240 (1989).
107. Y. Nambu, Prog. Theor. Phys. **81**, 1037 (1989).
108. M. Mijic, Phys. Rev. **D42**, 2469–2482 (1990).
109. D. S. Salopek and J. R. Bond, Phys. Rev. **D43**, 1005–1031 (1991).
110. A. D. Linde and A. Mezhlumian, Phys. Lett. **B307**, 25–33 (1993).
111. A. D. Linde, D. A. Linde and A. Mezhlumian, Phys. Rev. **D49**, 1783–1826 (1994).
112. D. S. Salopek and J. R. Bond, Phys. Rev. **D42**, 3936–3962 (1990).
113. A. R. Liddle, P. Parsons and J. D. Barrow, Phys. Rev. **D50**, 7222–7232 (1994).
114. M. Bellini, H. Casini, R. Montemayor and P. Sisterna, Phys. Rev. **D54**, 7172–7180 (1996).

115. H. Casini, R. Montemayor and P. Sisterna, Phys. Rev. **D59**, 063512 (1999).
116. S. Winitzki and A. Vilenkin, Phys. Rev. **D61**, 084008 (2000).
117. S. Matarrese, M. A. Musso and A. Riotto, JCAP **0405**, 008 (2004).
118. M. Liguori, S. Matarrese, M. Musso and A. Riotto, JCAP **0408**, 011 (2004).
119. N. G. van Kampen, *Stochastic Processes in Physics and Chemistry* (North-Holland, Amsterdam, 1981).
120. H. Risken, *The Fokker–Planck Equation* (Springer-Verlag, 1989).
121. J. Garriga and V. F. Mukhanov, Phys. Lett. **B458**, 219–225 (1999).
122. A. Vilenkin, Phys. Rev. **D59**, 123506 (1999).
123. S. Winitzki and A. Vilenkin, Phys. Rev. **D53**, 4298–4310 (1996).
124. S. Winitzki, Phys. Rev. **D71**, 123507 (2005).
125. S. Matarrese, A. Ortolan and F. Lucchin, Phys. Rev. **D40**, 290 (1989).
126. J. Garcia-Bellido and A. D. Linde, Phys. Rev. **D51**, 429–443 (1995).
127. A. D. Linde, Phys. Lett. **B108**, 389–393 (1982).
128. A. D. Linde, Phys. Lett. **B114**, 431 (1982).
129. A. D. Linde, Phys. Lett. **B116**, 340 (1982).
130. A. Albrecht and P. J. Steinhardt, Phys. Rev. Lett. **48**, 1220–1223 (1982).
131. H. M. Hodges, Phys. Rev. **D39**, 3568–3570 (1989).
132. I. Yi, E. T. Vishniac and S. Mineshige, Phys. Rev. **D43**, 362–368 (1991).
133. Y. V. Shtanov, Phys. Rev. **D52**, 4287–4294 (1995).
134. J. Martin and M. Musso, Phys. Rev. **D73**, 043516 (2006).
135. S. Gratton and N. Turok, Phys. Rev. **D72**, 043507 (2005).
136. J. Martin and M. Musso, Phys. Rev. **D73**, 043517 (2006).
137. M. Tegmark, JCAP **0504**, 001 (2005).
138. A. D. Linde, D. A. Linde and A. Mezhlumian, Phys. Lett. **B345**, 203–210 (1995).
139. S. Winitzki, Phys. Rev. **D65**, 083506 (2002).
140. M. Aryal and A. Vilenkin, Phys. Lett. **B199**, 351 (1987).
141. J. R. Gott, Nature **295**, 304–307 (1982).
142. J. R. Gott and T. S. Statler, Phys. Lett. **B136**, 157–161 (1984).
143. M. Bucher, A. S. Goldhaber and N. Turok, Phys. Rev. **D52**, 3314–3337 (1995).
144. M. Bucher, A. S. Goldhaber and N. Turok, Nucl. Phys. Proc. Suppl. **43**, 173–176 (1995).
145. K. Yamamoto, M. Sasaki and T. Tanaka, Astrophys. J. **455**, 412–418 (1995).
146. S. R. Coleman, Phys. Rev. **D15**, 2929–2936 (1977).
147. K.-M. Lee and E. J. Weinberg, Phys. Rev. **D36**, 1088 (1987).
148. J. Garriga, Phys. Rev. **D49**, 6327–6342 (1994).
149. E. Farhi and A. H. Guth, Phys. Lett. **B183**, 149 (1987).
150. E. Farhi, A. H. Guth and J. Guven, Nucl. Phys. **B339**, 417–490 (1990).
151. R. Bousso, Phys. Rev. **D71**, 064024 (2005).
152. L. F. Abbott and S. R. Coleman, Nucl. Phys. **B259**, 170 (1985).
153. J. Garriga, D. Schwartz-Perlov, A. Vilenkin and S. Winitzki, JCAP **0601**, 017 (2006).
154. S. Winitzki, Phys. Rev. **D71**, 123523 (2005).
155. V. Vanchurin and A. Vilenkin, Phys. Rev. **D74**, 043520 (2006).
156. R. Bousso, Phys. Rev. Lett. **97**, 191302 (2006).
157. P. Lancaster, *Theory of Matrices* (Academic Press, New York, 1969).
158. J. L. Doob, *Stochastic Processes* (Wiley, New York, 1953).
159. D. Kannan, *Introduction to Stochastic Processes* (North-Holland, New York, 1979).

160. A. D. Linde, JETP Lett. **38**, 176–179 (1983).
161. X. Chen, S. Sarangi, S. H. Henry Tye and J. Xu, JCAP **0611**, 015 (2006).
162. G. W. Gibbons and N. Turok,   The measure problem in cosmology, hep-th/0609095.
163. R. Easther, E. A. Lim and M. R. Martin, JCAP **0603**, 016 (2006).
164. J. Garriga and A. Vilenkin, Phys. Rev. **D64**, 023507 (2001).
165. R. Bousso, B. Freivogel and I-S. Yang, Phys. Rev. **D74**, 103516 (2006).
166. R. Bousso, Precision cosmology and the landscape, hep-th/0610211.
167. A. D. Linde and A. Mezhlumian, Phys. Rev. **D53**, 4267–4274 (1996).
168. A. Vilenkin and S. Winitzki, Phys. Rev. **D55**, 548–559 (1997).
169. A. Aguirre, S. Gratton and M. C. Johnson, Measures on transitions for cosmology from eternal inflation hep-th/0612195.
170. V. Vanchurin, Phys. Rev. **D75**, 023524 (2007).
171. A. D. Linde and D. A. Linde, Phys. Rev. **D50**, 2456–2468 (1994).

6

Inflationary Perturbations: The Cosmological Schwinger Effect

Jérôme Martin[1]

[1]Institut d'Astrophysique de Paris, UMR 7095 CNRS, Université Pierre et Marie Curie, 98bis boulevard Arago, 75014 Paris, France
jmartin@iap.fr

Abstract. This pedagogical review aims at presenting the fundamental aspects of the theory of inflationary cosmological perturbations of quantum-mechanical origin. The analogy with the well-known Schwinger effect is discussed in detail and a systematic comparison of the two physical phenomena is carried out. In particular, it is demonstrated that the two underlying formalisms differ only up to an irrelevant canonical transformation. Hence, the basic physical mechanisms at play are similar in both cases and can be reduced to the quantization of a parametric oscillator leading to particle creation due to the interaction with a classical source: pair production in vacuum is therefore equivalent to the appearance of a growing mode for the cosmological fluctuations. The only difference lies in the nature of the source: an electric field in the case of the Schwinger effect and the gravitational field in the case of inflationary perturbations. Although, in the laboratory, it is notoriously difficult to produce an electric field such that pairs extracted from the vacuum can be detected, the gravitational field in the early universe can be strong enough to lead to observable effects that ultimately reveal themselves as temperature fluctuations in the cosmic microwave background. Finally, the question of how quantum cosmological perturbations can be considered as classical is discussed at the end of this chapter.

6.1 Introduction

The scenario of inflation was invented in order to solve puzzling issues associated with the standard hot Big Bang theory [1, 2]. Soon after its advent, it was realized that this scenario also contains a remarkable extra bonus: it gives a well-motivated mechanism for structure formation that leads to a nearly scale-invariant power spectrum [3, 4], namely exactly what is needed in order to account for various astrophysical observations in a satisfactory way [5]. However, this is not the only aspect that deserves to be stressed. Indeed, even from a fundamental point of view, this mechanism appears quite remarkable in the sense that it combines general relativity with quantum mechanics. The

J. Martin: *Inflationary Perturbations: The Cosmological Schwinger Effect*, Lect. Notes Phys.
738, 193–241 (2008)
DOI 10.1007/978-3-540-74353-8_6   © Springer-Verlag Berlin Heidelberg 2008

main purpose of this chapter is to thoroughly discuss this aspect of the theory
of inflationary cosmological perturbations.

This theory is in fact remarkable at two levels. Firstly, because it relies on
the phenomenon of particle creation which is a non-trivial effect in quantum
field theory. In this sense, it is equivalent to the well-known Schwinger [6] effect
and this analogy will be made explicit in this paper. The basic ingredient is a
quantum scalar field Φ (in practice this is rather a fermionic field Ψ but, for
simplicity, we will restrict ourselves to the case of a scalar field) interacting
with a classical source, in the case of the Schwinger effect, an electric field
E. The Schwinger effect has not yet been observed in the laboratory as it is
difficult to produce an electric field with the required strength but there are
prospects to do so, in particular at DESY with a free electron laser (FEL)
in the X-ray band [7, 8, 9] but also at SLAC with the Linac Coherent Light
Source (LCLS) [10]. Even if there is absolutely no reason to doubt the reality of
the Schwinger effect, observing pair creation in the laboratory would clearly
be a breakthrough and, in some sense, a verification of the corresponding
inflationary mechanism.

Secondly, the theory of cosmological perturbations is also remarkable for
the following reason. In cosmology, what plays the role of the constant electric field E [originating from a time-dependent potential vector $A_\mu(t)$] is
the background gravitational field, i.e., the Friedmann–Lemaître–Robertson–
Walker (FLRW) scale factor $a(t)$, and what plays the role of the quantum
fermionic field $\Psi(t, \boldsymbol{x})$ is the quantum perturbed metric $\delta g_{\mu\nu}(t, \boldsymbol{x})$, that is to
say the small inhomogeneous fluctuations of the gravitational field itself [11].
In the early Universe, the gravitational field is quite strong, i.e., for instance
$H/m_{\mathrm{Pl}} \sim 10^{-5}$, where H is the Hubble parameter, and this is why the cosmological version of the Schwinger effect can be efficient. From the previous
considerations, it is also clear that, in some sense, the inflationary mechanism relies on quantum gravity which adds another interesting aspect to the
problem. Of course, we only deal with linearized quantum gravity and this
is why we do not have to face tricky questions associated with finiteness of
quantum gravity and/or renormalization. More precisely, in the case of scalar
perturbations, $\delta g_{\mu\nu}(t, \boldsymbol{x})$ is replaced by the Mukhanov–Sasaki variable $v(t, \boldsymbol{x})$
which is a combination of the Bardeen potential (the generalization of the
Newtonian potential in general relativity) and of the small fluctuations in the
inflaton field. For gravitational waves, the relevant quantity is $h_{ij}(t, \boldsymbol{x})$, the
transverse and traceless part of the perturbed metric.

Let us notice that the two above-mentioned aspects are features of the
theory of cosmological perturbations in general. The inflationary aspect is
in fact not necessary in order to have particles creation: only a dynamical
background is required. However, a quasi-exponential expansion is mandatory
if one wants to obtain a power spectrum which is close to scale invariance as
indicated by astrophysical observations.

The fact that the inflationary mechanism for structure formation relies on general relativity and quantum mechanics also raises fundamental

interpretational questions. In particular, the question of how classicality emerges is of special relevance in this context [12, 13]. Indeed, the perturbations are of quantum-mechanical origin but no astrophysical observations suggest any typical quantum-mechanical signature. Therefore, it is necessary to understand how the perturbations have become classical (and in which sense). This leads to very deep issues. For instance, if one invokes a mechanism based on the phenomenon of decoherence [14], then one has to discuss what plays the role of the environment. This question is clearly non-trivial in the cosmological context. At the end of this chapter, we will address these questions using the Wigner function [15] as a tool to understand when a system can be considered as classical.

This chapter can be viewed as the third of a series on the inflationary theory, the first two ones being [16, 17]. The topics developed in those last two references will be supposed to be known and we will often refer to them. This chapter is organized as follows. In Sect. 6.2, we briefly review the Schwinger effect. In particular, we derive the rate of pair production in the Schrödinger functional approach and stress the importance of the Wentzel–Kramers–Brillouin (WKB) approximation as a method to choose a well-defined initial state. In Sect. 6.3, we quantize a free scalar field in a FLRW Universe and show that the basic physical phenomenon at play is equivalent to that responsible for the Schwinger effect, namely particle creation under the influence of a classical source. In particular, we demonstrate that, up to a canonical transformation, the underlying formalisms are the same. Roughly speaking, in both cases, one has to deal with parametric oscillators. The only difference between the two systems lies in the time dependence of the corresponding effective frequencies. In Sect. 6.4, we argue that the equations obeyed by the cosmological perturbations (in particular during inflation) are equivalent to the equations of motion of a free scalar field. We emphasize that the relevant observable quantity is the two-point correlation function since it is directly linked to the cosmic microwave background (CMB) temperature fluctuations. Finally, as mentioned above, in Sect. 6.5, we address the question of the classicality of the cosmological perturbations.

6.2 The Schwinger Effect

6.2.1 General Formalism

The action of a complex (charged) scalar field interacting with an electromagnetic field is given by

$$S = -\int \mathrm{d}^4 x \left(\frac{1}{2} \eta^{\alpha\beta} \mathcal{D}_\alpha \Phi \mathcal{D}_\beta \Phi^* + \frac{1}{2} m^2 \Phi \Phi^* \right) , \qquad (6.1)$$

where $\eta^{\alpha\beta}$ is the flat (Minkowski) space–time metric with signature $(-, +, +, +)$ and where the covariant derivative can be expressed as

$$\mathcal{D}_\alpha \Phi \equiv \partial_\alpha \Phi + iqA_\alpha \Phi \, , \tag{6.2}$$

q being the charge of the field. The quantity m represents the mass of the scalar particle. Assuming the following configuration for the vector potential $A_\mu = (0,0,0,-Et)$, where E is the magnitude of the static electric field aligned along the z-direction (by convention), one obtains the following equation of motion

$$\partial_t^2 \Phi - \partial^i \partial_i \Phi + 2iqEt\partial_z \Phi + q^2 E^2 t^2 \Phi + m^2 \Phi = 0 \, . \tag{6.3}$$

It turns out to be more convenient to Fourier transform the field since this allows us to study the evolution of the system mode by mode. For this purpose, one decomposes the field according to

$$\Phi(t,\boldsymbol{x}) = \frac{1}{(2\pi)^{3/2}} \int \mathrm{d}^3 \boldsymbol{k} \, \Phi_{\boldsymbol{k}}(t) e^{i\boldsymbol{k}\cdot\boldsymbol{x}} \, . \tag{6.4}$$

In the above expression, $\Phi_{\boldsymbol{k}}(t)$ is the time-dependent Fourier amplitude of the mode characterized by the wave-vector $\boldsymbol{k}$. Inserting the Fourier transform (6.4) into (6.1), the action of the system takes the form

$$S = - \int \mathrm{d}t \int_{\mathbb{R}^3} \mathrm{d}\boldsymbol{k} \left[-\dot{\Phi}_{\boldsymbol{k}} \dot{\Phi}_{\boldsymbol{k}}^* + \left(k^2 - 2qEk_z t + q^2 E^2 t^2 + m^2 \right) \Phi_{\boldsymbol{k}} \Phi_{\boldsymbol{k}}^* \right] \, , \tag{6.5}$$

where a dot denotes a derivative with respect to time. The variation of this Lagrangian with respect to $\Phi_{\boldsymbol{k}}^*$ and $\dot{\Phi}_{\boldsymbol{k}}^*$ leads to

$$\frac{\delta\bar{\mathcal{L}}}{\delta\Phi_{\boldsymbol{k}}^*} = - \left(k^2 - 2qEk_z t + q^2 E^2 t^2 + m^2 \right) \Phi_{\boldsymbol{k}} \, , \qquad \frac{\delta\bar{\mathcal{L}}}{\delta\dot{\Phi}_{\boldsymbol{k}}^*} \equiv p_{\boldsymbol{k}} = \dot{\Phi}_{\boldsymbol{k}} \, , \tag{6.6}$$

where $p_{\boldsymbol{k}}$ is the conjugate momentum of the Fourier component of the field and $\bar{\mathcal{L}}$ denotes the Lagrangian density in Fourier space. Using the above two formula, the Euler–Lagrange equation of motion reads

$$\ddot{\Phi}_{\boldsymbol{k}} + \omega^2(k,t)\Phi_{\boldsymbol{k}} = 0 \, , \tag{6.7}$$

where the time-dependent frequency $\omega(k,t)$ can be expressed as

$$\omega^2(k,t) \equiv k^2 - 2qEk_z t + q^2 E^2 t^2 + m^2 \, . \tag{6.8}$$

Equation (6.7) is of course similar to the one obtained by directly substituting (6.4) into (6.3). It is the equation of motion of a parametric oscillator. Let us recall that a parametric oscillator is an harmonic oscillator whose frequency depends on time. A typical example is a pendulum with a varying length.

Let us now pass to the Hamiltonian formalism. The Hamiltonian is obtained from the Lagrangian by a standard Legendre transformation and can be expressed as

$$H = \int_{\mathbb{R}^3} \mathrm{d}\boldsymbol{k} \left(p_{\boldsymbol{k}} \dot{\Phi}_{\boldsymbol{k}}^* + p_{\boldsymbol{k}}^* \dot{\Phi}_{\boldsymbol{k}} - \bar{\mathcal{L}} \right) = \int_{\mathbb{R}^3} \mathrm{d}\boldsymbol{k} \left[p_{\boldsymbol{k}} p_{\boldsymbol{k}}^* + \omega^2(k,t) \Phi_{\boldsymbol{k}} \Phi_{\boldsymbol{k}}^* \right] \, . \tag{6.9}$$

For the following considerations, it turns out to be convenient to also work with real variables instead of the complex Φ_k. Therefore, we now introduce the definitions

$$\Phi_k \equiv \frac{1}{\sqrt{2}} \left(\Phi_k^R + i\Phi_k^I\right) , \quad p_k \equiv \frac{1}{\sqrt{2}} \left(p_k^R + ip_k^I\right) . \tag{6.10}$$

Then, the Hamiltonian can be written as

$$H = \int_{\mathbb{R}^3} d\mathbf{k} \left[\frac{1}{2} \left(p_k^R\right)^2 + \frac{1}{2} \omega^2 \left(k, t\right) \left(\Phi_k^R\right)^2 + \frac{1}{2} \left(p_k^I\right)^2 + \frac{1}{2} \omega^2 \left(k, t\right) \left(\Phi_k^I\right)^2\right] . \tag{6.11}$$

One recognizes the Hamiltonian of a collection of parametric oscillators with a time-dependent frequency given by (6.8). Again, one can check that the Hamilton equations deduced from (6.9) and (6.11) lead to an equation of motion similar to the one already derived before, namely (6.7).

6.2.2 Quantization

Our next step is to describe the quantization of the system. More precisely, the complex scalar field is quantized while the gauge field remains classical. Therefore, we have to deal with the interaction of a quantum field with a classical source. Quantization is achieved by requiring the following commutation relations (a hat symbol is put on letters denoting operators)

$$\left[\hat{\Phi}_k^R, \hat{p}_p^R\right] = i\delta^{(3)} \left(\mathbf{k} - \mathbf{p}\right) , \quad \left[\hat{\Phi}_k^I, \hat{p}_p^I\right] = i\delta^{(3)} \left(\mathbf{k} - \mathbf{p}\right) . \tag{6.12}$$

We choose to work in the Schrödinger picture where the states are time-dependent and the operators constant. The above commutation relations admit the following representation

$$\hat{\Phi}_k^R \Psi = \Phi_k^R \Psi , \quad \hat{p}_k^R \Psi = -i\frac{\partial \Psi}{\partial \Phi_k^R} . \tag{6.13}$$

The state of the system is described by a functional of the scalar field, $\Psi[\Phi(t, \mathbf{x})]$ (in the present context, Ψ is the field functional and has clearly nothing to do with the fermionic field mentioned before), which can also be viewed as a function of an infinite number of variables, namely the values of Φ at each point in space. Alternatively, one can also consider this functional as a function of the infinite number of Fourier components of the field and write

$$\Psi = \prod_k^n \Psi_k \left(\Phi_k^R, \Phi_k^I\right) = \prod_k^n \Psi_k^R \left(\Phi_k^R\right) \Psi_k^I \left(\Phi_k^I\right) . \tag{6.14}$$

In the above equation, n represents the number of modes, in the intermediate calculations, that is useful to keep finite (for instance, if we imagine the field lives in a finite box). However, at the end, we will always consider the continuous case and take the limit $n \to +\infty$.

In the framework described before, the Schrödinger equation is a functional differential equation. However, the Hamiltonian takes the form of an infinite sum over k, see for instance (6.9) and (6.11), without any interaction between different modes. As a consequence, each mode evolves independently and the corresponding Hamiltonian is represented by an ordinary differential operator. Explicitly, one has

$$
H_k \Psi = (H_k^{\mathrm{R}} + H_k^{\mathrm{I}}) \, \Psi = -\frac{1}{2} \frac{\partial^2 \Psi}{\partial (\Phi_k^{\mathrm{R}})^2} + \frac{1}{2} \omega^2(k,t) \, (\Phi_k^{\mathrm{R}})^2 \, \Psi - \frac{1}{2} \frac{\partial^2 \Psi}{\partial (\Phi_k^{\mathrm{I}})^2}
$$
$$
+ \frac{1}{2} \omega^2(k,t) \, (\Phi_k^{\mathrm{I}})^2 \, \Psi \, , \tag{6.15}
$$

where the frequency ω has been given by (6.8).

Let us now consider the ground state of the system described before. We will discuss the choice of the initial conditions and the meaning of the vacuum state in the following but, as is well-known, it is given by a Gaussian state

$$
\Psi_k^{\mathrm{R}} \left(t, \Phi_k^{\mathrm{R}} \right) = N_k \left(t \right) \mathrm{e}^{-\Omega_k(t)\left(\Phi_k^{\mathrm{R}}\right)^2} \, , \quad \Psi_k^{\mathrm{I}} \left(t, \Phi_k^{\mathrm{I}} \right) = N_k \left(t \right) \mathrm{e}^{-\Omega_k(t)\left(\Phi_k^{\mathrm{I}}\right)^2} \, , \tag{6.16}
$$

where $N_k(t)$ and $\Omega_k(t)$ are functions that can be determined using the Schrödinger equation $\mathrm{i}\partial_t \Psi = H_k \Psi$. This leads to

$$
\mathrm{i}\frac{\dot{N}_k}{N_k} = \Omega_k \, , \quad \dot{\Omega}_k = -2\mathrm{i}\Omega_k^2 + \frac{\mathrm{i}}{2}\omega^2 \left(k, t \right) \, . \tag{6.17}
$$

The equation for the complex quantity Ω_k is a non-linear Ricatti equation. When a particular solution is known, the general solution can be found by means of two successive quadratures [18]. But this non-linear first order differential equation can also be transformed into a linear second order differential equation. It turns out that this last one is exactly the equation for the Fourier mode function, (6.7). Therefore, the solutions to (6.17) read

$$
N_k = \left(\frac{2\Re\Omega_k}{\pi} \right)^{1/4} \, , \quad \Omega_k = -\frac{\mathrm{i}}{2}\frac{\dot{f}_k}{f_k} \, , \tag{6.18}
$$

where f_k obeys the equation $\ddot{f}_k + \omega^2(k,t)f_k = 0$, that is to say, as already mentioned, the same equation as the Fourier component of the field, namely (6.7). The quantity N_k is obtained by normalizing the wave-function. One can check that this leads to an equation consistent with the first formula in (6.17). Therefore, the ground quantum state of the field is given by

$$
\Psi = \prod_k^n \left(\frac{2\Re\Omega_k}{\pi} \right)^{1/2} \mathrm{e}^{-\Omega_k(t)\left[(\Phi_k^{\mathrm{R}})^2 + (\Phi_k^{\mathrm{I}})^2\right]} \, . \tag{6.19}
$$

Once a particular solution for the mode function has been singled out, the functions N_k and Ω_k, and hence the wave-function of the field, are completely specified.

One can now use the above-mentioned state in order to calculate the amplitude associated with the transition between two states Ψ_1 and Ψ_2. It is defined by

$$\langle \Psi_1 | \Psi_2 \rangle = \int \prod_k^n d\Phi_k^{\mathbb{R}} d\Phi_k^{\mathbb{I}} \left(\frac{2 \Re \Omega_{1,k}}{\pi} \right)^{1/2} \left(\frac{2 \Re \Omega_{2,k}}{\pi} \right)^{1/2} e^{-\sum_p^n (\Omega_{1,p}^* + \Omega_{2,p})(\Phi_p^{\mathbb{R}})^2}$$

$$\times e^{-\sum_p^n (\Omega_{1,p}^* + \Omega_{2,p})(\Phi_p^{\mathbb{I}})^2} , \tag{6.20}$$

from which, after having performed the Gaussian integration, one deduces that

$$|\langle \Psi_1 | \Psi_2 \rangle|^2 = \det \left[\frac{4 \Re \Omega_{1,k} \Re \Omega_{2,k}}{\left(\Omega_{1,k}^* + \Omega_{2,k} \right) \left(\Omega_{1,k} + \Omega_{2,k}^* \right)} \right] . \tag{6.21}$$

At this point, one has to use the specific form of Ω_k, in particular its expression given by (6.18) in terms of the function f_k. One obtains

$$|\langle \Psi_1 | \Psi_2 \rangle|^2 = \det \left[\frac{(\dot{f}_{1,k} f_{1,k}^* - \dot{f}_{1,k}^* f_{1,k})(\dot{f}_{2,k} f_{2,k}^* - \dot{f}_{2,k}^* f_{2,k})}{(\dot{f}_{1,k}^* f_{2,k} - \dot{f}_{2,k} f_{1,k}^*)(\dot{f}_{2,k}^* f_{1,k} - \dot{f}_{1,k} f_{2,k}^*)} \right] . \tag{6.22}$$

In this formula $f_{1,k}$ is the mode function for the initial state while $f_{2,k}$ is the same quantity but for the final state. As usual, one can always expand $f_{2,k}$ over the basis $(f_{1,k}, f_{1,k}^*)$ and write

$$f_{2,k} = \alpha_k f_{1,k} + \beta_k f_{1,k}^* . \tag{6.23}$$

Then, using the fact that the Wronskian $\dot{f}_{1,k} f_{1,k}^* - \dot{f}_{1,k}^* f_{1,k}$ is a conserved quantity (as can be easily checked by differentiating it and using the equation satisfied by f_k), one arrives at [19]

$$|\langle \Psi_1 | \Psi_2 \rangle|^2 = \det \left(\frac{1}{|\alpha_k|^2} \right) . \tag{6.24}$$

Therefore, the determination of the transition amplitude amounts to integrating the equation controlling the evolution of the mode function. Once this is done, the coefficient α_k is known and the quantity $|\langle \Psi_1 | \Psi_2 \rangle|^2$ can be determined. In the next section, we discuss an explicit example.

6.2.3 Particle Creation

We now use the formalism developed above in order to study the creation of quantum scalar particles due to the interaction with a classical source. This is the well-known Schwinger effect [6]. In the following, we will demonstrate that the inflationary mechanism for cosmological perturbations is exactly similar to the one discussed here, see also [19, 20].

In order to determine the coefficient α_k, one can proceed as follows. Let us use the dimensionless variable $\tau \equiv \sqrt{qE}t - k_z/\sqrt{qE}$. Then, the equation of motion for the Fourier component of the field, see (6.7), takes the form

$$\frac{d^2\Phi_k}{d\tau^2} + \left(\Upsilon + \tau^2\right)\Phi_k = 0 , \tag{6.25}$$

with $\Upsilon \equiv (k_\perp^2 + m^2)/(qE)$. The quantity $k_\perp$ is defined by $k_\perp^2 \equiv k^2 - k_z^2 = k_x^2 + k_y^2$ (let us recall that we have chosen an electrical field aligned along the z-direction). Equation (6.25) can be integrated exactly, see (9.255.2) of [21] and the solution can be expressed as

$$\Phi_k(\tau) = A_k D_{-(1+i\Upsilon)/2}\left[(1 + i)\tau\right] + B_k D_{-(1+i\Upsilon)/2}\left[-(1 + i)\tau\right] , \tag{6.26}$$

where A_k and B_k are two constants fixed by the initial conditions and $D_p(z)$ is a parabolic cylinder function of order p.

Despite the previous change of variable, (6.25) has retained the form of an equation for a parametric oscillator but the frequency is now given by

$$\omega(\tau) \equiv \sqrt{\Upsilon + \tau^2} . \tag{6.27}$$

This equation is well-suited to the WKB approximation. This approximation is not only useful to get an approximate form of the solution but can also be used in order to choose initial conditions that are well motivated. Here, since we do already know the exact solution, it is clearly this last application we shall be concerned with.

By definition, the WKB mode function $(2\omega)^{-1/2}e^{\pm i \int \omega d\tau}$ obeys the following equation of motion $\ddot{\Phi}_k + \left(\omega^2 - Q\right)\Phi_k = 0$, where the quantity Q is defined by

$$Q \equiv \frac{3}{4}\frac{1}{\omega^2}\left(\frac{d\omega}{d\tau}\right)^2 - \frac{1}{2}\frac{1}{\omega}\frac{d^2\omega}{d\tau^2} . \tag{6.28}$$

Therefore, one sees that the WKB mode function is a good approximation to the actual one as soon as $|Q/w^2| \ll 1$. This last condition defines the regime where the WKB approximation is valid. Let us compute this quantity in the case of (6.25). Straightforward calculations lead to the following expression

$$\left|\frac{Q}{\omega^2}\right| = \frac{1}{\Upsilon^2}\frac{1}{2\left(1 + \tau^2/\Upsilon\right)^2}\left|\frac{5\tau^2/\Upsilon}{2\left(1 + \tau^2/\Upsilon\right)} - 1\right| . \tag{6.29}$$

The quantity $|Q/\omega^2|$ is represented in Fig. 6.1. It is clear that, in the limits $\tau/\sqrt{\Upsilon} \to \pm\infty$, the WKB approximation is valid. This means that there exists a well-defined vacuum state (or adiabatic vacuum) in the "in" region, $|0^-\rangle$, and in the "out" region, $|0^+\rangle$.

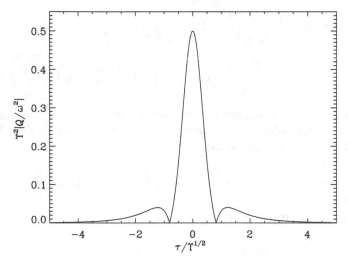

Fig. 6.1. Evolution of the quantity $|Q/\omega^2|$ with time τ in the case of the Schwinger effect. In the limit $\tau/\sqrt{\Upsilon} \to \pm\infty$, $|Q/\omega^2|$ vanishes and the notion of adiabatic vacuum is available

When the WKB approximation is satisfied, an approximate solution of the mode functions is available and, as already briefly mentioned above, is given by

$$\Phi_{\boldsymbol{k}}(\tau) \simeq \frac{\alpha_{\boldsymbol{k}}}{\sqrt{2\omega}} \exp\left[-i \int_{\tau_{\text{ini}}}^{\tau} \omega(\theta) d\theta\right] + \frac{\beta_{\boldsymbol{k}}}{\sqrt{2\omega}} \exp\left[+i \int_{\tau_{\text{ini}}}^{\tau} \omega(\theta) d\theta\right]$$
$$\equiv \alpha_{\boldsymbol{k}} \Phi_{\text{wkb},\boldsymbol{k}}(\tau) + \beta_{\boldsymbol{k}} \Phi_{\text{wkb},\boldsymbol{k}}^{*}(\tau) , \tag{6.30}$$

where $\tau_{\text{ini}} < 0$ is some arbitrary initial time.

One can now use the WKB approximation in order to choose an initial state in the following way. We require that the system is in the adiabatic vacuum in the "in" region, $|0^-\rangle$, that is to say when $\tau/\sqrt{\Upsilon} \to -\infty$. Technically, this means that one has $\lim_{\tau/\sqrt{\Upsilon}\to-\infty} \Phi_{\boldsymbol{k}} = \Phi_{\text{wkb},\boldsymbol{k}}$ or $\alpha_{\boldsymbol{k}} = 1$ and $\beta_{\boldsymbol{k}} = 0$ (hence satisfying $|\alpha_{\boldsymbol{k}}|^2 - |\beta_{\boldsymbol{k}}|^2 = 1$). This criterion completely specifies the coefficients $A_{\boldsymbol{k}}$ and $B_{\boldsymbol{k}}$ in (6.26) and, as a consequence, also completely determines the coefficients $\alpha_{\boldsymbol{k}}$ and $\beta_{\boldsymbol{k}}$ in the "out" region (when $\tau/\sqrt{\Upsilon} \to +\infty$) that are needed in order to compute the transition amplitude. Around $\tau \sim 0$, see Fig. 6.1, the WKB approximation is violated and we have particle creation. In the "out" region the vacuum is defined by $|0^+\rangle$ and, therefore, the number of particles present in this region is measured by evaluating the amplitude $\langle 0^-|0^+\rangle$.

We now briefly explain how this can be done at the technical level. The phase can be computed exactly and reads

$$\int_{\tau_{\text{ini}}}^{\tau} \omega(\theta)\, d\theta = \frac{\Upsilon}{2}\left[\frac{\tau}{\sqrt{\Upsilon}}\sqrt{1+\frac{\tau^2}{\Upsilon}} - \frac{\tau_{\text{ini}}}{\sqrt{\Upsilon}}\sqrt{1+\frac{\tau_{\text{ini}}^2}{\Upsilon}}\right.$$

$$\left. + \ln\left(\frac{\tau}{\sqrt{\Upsilon}} + \sqrt{1+\frac{\tau^2}{\Upsilon}}\right) - \ln\left(\frac{\tau_{\text{ini}}}{\sqrt{\Upsilon}} + \sqrt{1+\frac{\tau_{\text{ini}}^2}{\Upsilon}}\right)\right] \quad (6.31)$$

The arguments of the logarithms are always positive even if τ is negative (hence the corresponding expression with $\tau_{\text{ini}} < 0$ is also meaningful). In the limit $|\tau|\,\Upsilon^{-1/2} \gg 1$, the phase goes to

$$\int_{\tau_{\text{ini}}}^{\tau} \omega(\theta)\, d\theta \to \frac{1}{2}\left(\tau\,|\tau| + |\tau_{\text{ini}}|^2\right) + \frac{\Upsilon}{2}\left(\frac{|\tau|}{\tau}\ln|\tau| + \ln|\tau_{\text{ini}}|\right) \quad (6.32)$$

and, therefore, the WKB mode function takes the form

$$\Phi_{\text{wkb},k} = \frac{1}{\sqrt{2}} e^{-\mathrm{i}(\tau|\tau|+\tau_{\text{ini}}^2)/2}\, |\tau|^{-1/2-\mathrm{i}|\tau|\Upsilon/(2\tau)}\, |\tau_{\text{ini}}|^{-\mathrm{i}\Upsilon/2}. \quad (6.33)$$

Then, in the limit $\tau \to -\infty$, the exact solution given by (6.26) can be expressed as

$$\Phi_k \simeq A_k \frac{\sqrt{2\pi}}{\Gamma\left(\frac{1}{2}+\frac{\mathrm{i}\Upsilon}{2}\right)} e^{-\mathrm{i}\pi(1-\mathrm{i}\Upsilon)/4}\,(1-\mathrm{i})^{-(1-\mathrm{i}\Upsilon)/2}\sqrt{2}\, e^{\mathrm{i}|\tau_{\text{ini}}|^2/2}\, |\tau_{\text{ini}}|^{\mathrm{i}\Upsilon/2}\, \Phi_{\text{wkb},k}$$

$$+ \left[A_k e^{\mathrm{i}\pi(1+\mathrm{i}\Upsilon)/2} + B_k\right](1+\mathrm{i})^{-(1+\mathrm{i}\Upsilon)/2}\sqrt{2}\, e^{-\mathrm{i}|\tau_{\text{ini}}|^2/2}\, |\tau_{\text{ini}}|^{-\mathrm{i}\Upsilon/2}\, \Phi_{\text{wkb},k}^*.$$

$$(6.34)$$

Since, as explained above, we choose the initial state such that $\alpha_k = 1$ and $\beta_k = 0$, this amounts to requiring

$$A_k = \frac{\Gamma\left(\frac{1}{2}+\frac{\mathrm{i}\Upsilon}{2}\right)}{\sqrt{2\pi}} e^{\mathrm{i}\pi(1-\mathrm{i}\Upsilon)/4}\,(1-\mathrm{i})^{(1-\mathrm{i}\Upsilon)/2}\,\frac{1}{\sqrt{2}}e^{-\mathrm{i}|\tau_{\text{ini}}|^2/2}\,|\tau_{\text{ini}}|^{-\mathrm{i}\Upsilon/2} \quad (6.35)$$

$$B_k = -A_k e^{\mathrm{i}\pi(1+\mathrm{i}\Upsilon)/2}. \quad (6.36)$$

Finally, one considers the behavior of the mode function in the limit $\tau \to +\infty$ and, using again the WKB mode function in this regime, one can find the corresponding coefficients α_k and β_k. One obtains

$$|\alpha_k|^2 = 1 + e^{-\pi\Upsilon}, \quad |\beta_k|^2 = e^{-\pi\Upsilon}. \quad (6.37)$$

These expressions still satisfy $|\alpha_k|^2 - |\beta_k|^2 = 1$ as required.

We have now reached our final goal and can return to the calculation of the determinant in (6.24). Using the coefficient α_k obtained above, one has

$$|\langle 0^-|0^+\rangle|^2 = \det\left(\frac{1}{1+e^{-\pi\Upsilon}}\right) = \exp\left[-\text{Tr}\ln\left(1+e^{-\pi\Upsilon}\right)\right]. \quad (6.38)$$

The evaluation of the trace is standard and leads to the well-known result first obtained by Schwinger in the early 1950s [6]

$$|\langle 0^-|0^+\rangle|^2 = \exp\left[-\frac{VT}{(2\pi)^3}\sum_{n=1}^{\infty}\frac{(-1)^{n+1}}{n^2}(qE)^2\,e^{-n\pi m^2/(qE)}\right]. \quad (6.39)$$

The physical interpretation of this formula is clear. The argument of the exponential gives the number of pairs created in the space–time volume VT due to the interaction of the quantum scalar field with the classical electric field. One can define a critical electric field (we have restored the fundamental constants) by

$$E_{\text{cri}} = \frac{m^2c^3}{q\hbar}, \quad (6.40)$$

which is such that the number of particles created is significant only if $E \gg E_{\text{cri}}$. This condition can be understood by noting that this is just the requirement that the work performed by the force qE over the Compton length $\lambda = \hbar/(mc)$ is larger than the rest energy $2mc^2$. In the case of pairs e^+e^-, the critical electric field is given by $E_{\text{cri}} \sim 1.3 \times 10^{18}\,\text{V} \times \text{m}^{-1}$. It is also interesting to remark that the dependence of $|\langle 0^-|0^+\rangle|^2$ in E is non-perturbative. This is one of the few examples in quantum field theory where an exact result can be obtained (of course, this is not "full quantum theory" but rather "potential theory" since the radiative corrections to the Schwinger mechanism are not taken into account).

We will see that the inflationary mechanism of production of cosmological perturbations is similar to the Schwinger mechanism. Therefore, observing this latter effect in the laboratory could be seen as an indication that we are on the right track as far as the inflationary mechanism is concerned. For instance, at DESY, there are plans to construct a free electron laser (FEL) in the X-ray band which would effectively produce a very strong electric field and, hence, to observe the Schwinger mechanism [7, 8]. Unfortunately, even with a FEL, it is inconceivable to produce a static field with the required strength given present day technology. However, the situation is different for an oscillating electric field [22] (other configurations have been studied in [23, 24, 25]) and, in this case, it seems possible to extract pairs from the vacuum. This would also be a validation of the Schwinger mechanism since only the time dependence of $\omega(k,t)$ is changed but not the other basic ingredients. It is also interesting to notice that, in the context of the inflationary theory, this case is in fact very similar to the reheating [26, 27] stage where the effective frequency of the perturbations is alternating due to the oscillations of the inflaton field at the bottom of its potential.

To conclude this section, let us recall that the basic ingredient at play here is particle creation due to the interaction of a quantum field with a classical source. When the WKB approximation is valid, a well-defined notion of vacuum state exists, and when the WKB approximation is violated, particle creation occurs. We will see that, in the case of inflationary cosmological perturbations, exactly the same mechanism is available.

6.3 Quantization of a Free Scalar Field in Curved Space–Time

Before considering inflation itself, let us now discuss the case of a free real scalar field in curved space–time since this is the simplest example which allows us to capture all the essential features of the theory of inflationary cosmological perturbations.

6.3.1 General Formalism

We consider the question of quantizing a (massless) scalar field in curved space–time. The starting point is the following action

$$S = - \int \mathrm{d}^4 x \sqrt{-g} g^{\mu\nu} \frac{1}{2} \partial_\mu \Phi \, \partial_\nu \Phi \,, \qquad (6.41)$$

which, in a flat FLRW Universe whose metric is given by $\mathrm{d}s^2 = a^2(\eta)(-\mathrm{d}\eta^2 + \delta_{ij}\mathrm{d}x^i\mathrm{d}x^j)$, η being the conformal time, reads

$$S = \frac{1}{2} \int \mathrm{d}^4 x a^2(\eta) \left(\Phi'^2 - \delta^{ij} \partial_i \Phi \, \partial_j \Phi \right) \,. \qquad (6.42)$$

It follows immediately that the conjugate momentum to the scalar field can be expressed as

$$\Pi(\eta, \boldsymbol{x}) = a^2 \Phi'(\eta, \boldsymbol{x}) \,, \qquad (6.43)$$

where a prime denotes a derivative with respect to conformal time. As before, it is convenient to Fourier expand the field $\Phi(\eta, \boldsymbol{x})$ over the basis of plane waves (therefore, we make explicit use of the fact that the space-like hypersurfaces are flat). This gives

$$\Phi(\eta, \boldsymbol{x}) = \frac{1}{a(\eta)} \frac{1}{(2\pi)^{3/2}} \int \mathrm{d}\boldsymbol{k} \mu_{\boldsymbol{k}}(\eta) \mathrm{e}^{\mathrm{i}\boldsymbol{k}\cdot\boldsymbol{x}} \,. \qquad (6.44)$$

We have chosen to re-scale the Fourier component $\mu_{\boldsymbol{k}}$ with a factor $1/a(\eta)$ for future convenience. Since the scalar field is real, one has $\mu_{\boldsymbol{k}}^* = \mu_{-\boldsymbol{k}}$. The next step consists in inserting the expression of $\Phi(\eta, \boldsymbol{x})$ into the action (6.42). This leads to

$$S = \frac{1}{2} \int \mathrm{d}\eta \int_{\mathbb{R}^{3+}} \mathrm{d}^3 k \left[\mu_{\boldsymbol{k}}'^* \mu_{\boldsymbol{k}}' + \mu_{\boldsymbol{k}}' \mu_{\boldsymbol{k}}'^* - 2\frac{a'}{a} \left(\mu_{\boldsymbol{k}}' \mu_{\boldsymbol{k}}^* + \mu_{\boldsymbol{k}}'^* \mu_{\boldsymbol{k}} \right) \right. $$
$$\left. + \left(\frac{a'^2}{a^2} - k^2 \right) \left(\mu_{\boldsymbol{k}} \mu_{\boldsymbol{k}}^* + \mu_{\boldsymbol{k}}^* \mu_{\boldsymbol{k}} \right) \right] \,. \qquad (6.45)$$

Notice that the integral over the wave-numbers is calculated over half the space in order to sum over independent variables only [28]. This formula is similar to (6.5) for the case of the Schwinger effect (of course, in this last case, we do not have $\Phi_k = \Phi^*_{-k}$ since the field is charged and, hence, the integral is performed over all the momentum space).

Equipped with the Lagrangian in the momentum space (which, in the following, as it was the case in the previous section, we denote by $\bar{\mathcal{L}}$), one can check that it leads to the correct equation of motion. Since we have $\delta\bar{\mathcal{L}}/\delta\mu^*_k = 1/2[-2\mathcal{H}\mu'_k + 2(\mathcal{H}^2 - k^2)\mu_k]$, the Euler–Lagrange equation $d[\delta\bar{\mathcal{L}}/\delta\mu'_k{}^*]/d\eta - \delta\bar{\mathcal{L}}/\delta\mu^*_k = 0$ reproduces the correct equation of motion for the variable μ_k, namely

$$\frac{d^2\mu_k}{d\eta^2} + \omega^2(k,\eta)\mu_k = 0 \,, \tag{6.46}$$

that is to say, again, the equation of a parametric oscillator, as in (6.7), but with a frequency now given by

$$\omega^2(k,\eta) = k^2 - \frac{a''}{a} \,. \tag{6.47}$$

This last formula should be compared with (6.8). In the case of the Schwinger effect, the frequency was time-dependent because of the interaction of the scalar field with the time-dependent potential vector. Here, the frequency is time-dependent because the scalar field lives in a time-dependent background, or, in some sense, because the scalar field interacts with the classical gravitational background. Therefore, we already see at this stage that we can have particle creation due to the interaction with a classical gravitational field (instead of a classical electric field in the previous section). Of course, the two cases are not exactly similar in the sense that the time dependence of ω^2 is different. Indeed, in the Schwinger case, $\omega^2(k,t)$ typically contains terms proportional to t and t^2, see (6.8), while, in the inflationary case, the term a''/a is typically proportional to $1/\eta^2$. As a consequence, the solution to the mode equation and the particle creation rate will be different even if, again, the basic mechanism at play is exactly the same in both situations.

The mode amplitude μ_k is complex but one can also work with real variables μ^R_k and μ^I_k, as was done previously in (6.10), defined such that

$$\mu_k \equiv \frac{1}{\sqrt{2}}(\mu^R_k + i\mu^I_k) \,. \tag{6.48}$$

In terms of these variables, the relation $\mu^*_k = \mu_{-k}$ reads $\mu^R_k = \mu^R_{-k}$ and $\mu^I_k = -\mu^I_{-k}$. Then, the action (or Lagrangian) of the system takes the form

$$S = \frac{1}{2}\int d\eta \int_{\mathbb{R}^{3+}} d^3k \left\{ (\mu^R_k{}')^2 + (\mu^I_k{}')^2 - 2\frac{a'}{a}(\mu^R_k\mu^R_k{}' + \mu^I_k\mu^I_k{}') \right.$$
$$\left. + \left(\frac{a'^2}{a^2} - k^2\right)\left[(\mu^R_k)^2 + (\mu^I_k)^2\right] \right\} \,. \tag{6.49}$$

One can check that it also leads to the correct equations of motion for the two real variables μ_k^{R} and μ_k^{I}.

We can now pass to the Hamiltonian formalism. The conjugate momentum to μ_k is defined by the formula

$$p_k \equiv \frac{\delta \bar{\mathcal{L}}}{\delta \mu_k'^{\,*}} = \mu_k' - \frac{a'}{a}\mu_k \ . \tag{6.50}$$

One can check that the definitions of the conjugate momenta in the real and Fourier spaces are consistent in the sense that they are related by the (expected) expression

$$\Pi(\eta, \boldsymbol{x}) = \frac{a(\eta)}{(2\pi)^{3/2}} \int \mathrm{d}\boldsymbol{k}\, p_k \mathrm{e}^{\mathrm{i}\boldsymbol{k}\cdot\boldsymbol{x}} \ . \tag{6.51}$$

We see that the definition of the conjugate momentum p_k as the derivative of the Lagrangian in Fourier space with respect to $\mu_k'^{\,*}$ and not to μ_k' is consistent with the expression of the momentum in real space. Otherwise the momentum $\Pi(\eta, \boldsymbol{x})$ in real space would have been expressed in terms of p_k^* instead of p_k. Moreover, one can also check that

$$p_k^{\mathrm{R}} \equiv \frac{\delta \bar{\mathcal{L}}}{\delta \mu_k^{\mathrm{R}'}} = \mu_k^{\mathrm{R}'} - \frac{a'}{a}\mu_k^{\mathrm{R}} \ , \quad p_k^{\mathrm{I}} \equiv \frac{\delta \bar{\mathcal{L}}}{\delta \mu_k^{\mathrm{I}'}} = \mu_k^{\mathrm{I}'} - \frac{a'}{a}\mu_k^{\mathrm{I}} \ , \tag{6.52}$$

and, clearly, we have

$$p_k \equiv \frac{1}{\sqrt{2}}\left(p_k^{\mathrm{R}} + \mathrm{i}p_k^{\mathrm{I}}\right) \ , \tag{6.53}$$

as expected.

We are now in a position where we can compute explicitly the Hamiltonian in the momentum space. The Hamiltonian density, $\bar{\mathcal{H}}$, is defined in terms of the Hamiltonian H of the system through the relation

$$H = \int_{\mathbb{R}^{3+}} \mathrm{d}^3 k \bar{\mathcal{H}} = \int_{\mathbb{R}^{3+}} \mathrm{d}^3 k \left(p_k \mu_k'^{\,*} + p_k^* \mu_k' - \bar{\mathcal{L}}\right) \ , \tag{6.54}$$

and we obtain

$$\bar{\mathcal{H}} = p_k p_k^* + k^2 \mu_k \mu_k^* + \frac{a'}{a}\left(p_k \mu_k^* + p_k^* \mu_k\right) \ . \tag{6.55}$$

Let us make some comments on this expression. If the background gravitational field is not time-dependent, that is to say if the scalar field lives in Minkowski space–time where $a' = 0$, then the above Hamiltonian reduces to a free Hamiltonian: there is simply no classical "pump field." From the Schwinger effect point of view, this would be similar to a situation where there is no external classical electric field. In these two cases, no particle creation would occur. Moreover, one can also check that the Hamilton equations

$$\frac{d\mu_{\mathbf{k}}^*}{d\eta} = \frac{\partial \bar{\mathcal{H}}}{\partial p_{\mathbf{k}}} = c p_{\mathbf{k}}^* + \frac{a'}{a}\mu_{\mathbf{k}}^*, \quad \frac{dp_{\mathbf{k}}^*}{d\eta} = -\frac{\partial \bar{\mathcal{H}}}{\partial \mu_{\mathbf{k}}} = -\frac{a'}{a}p_{\mathbf{k}}^* - \frac{k^2}{c}\mu_{\mathbf{k}}^*, \quad (6.56)$$

lead to the correct equation of motion given by (6.46). Finally, in terms of the real variables, the Hamiltonian density reads

$$\bar{\mathcal{H}} = \frac{1}{2}\left[(p_{\mathbf{k}}^{\mathrm{R}})^2 + 2\frac{a'}{a}\mu_{\mathbf{k}}^{\mathrm{R}}p_{\mathbf{k}}^{\mathrm{R}} + k^2(\mu_{\mathbf{k}}^{\mathrm{R}})^2\right] + \frac{1}{2}\left[(p_{\mathbf{k}}^{\mathrm{I}})^2 + 2\frac{a'}{a}\mu_{\mathbf{k}}^{\mathrm{I}}p_{\mathbf{k}}^{\mathrm{I}} + k^2(\mu_{\mathbf{k}}^{\mathrm{I}})^2\right].$$
$$(6.57)$$

We notice that $\bar{\mathcal{H}}$ is simply the sum of two identical Hamiltonians for parametric oscillator, one for $\mu_{\mathbf{k}}^{\mathrm{R}}$ and the other for $\mu_{\mathbf{k}}^{\mathrm{I}}$.

The expressions (6.55) and (6.57) should be compared to (6.9) and (6.11). We see that, although similar, the formulae are not identical. However, as we are now going to show, this difference is only apparent. Indeed, let us now restart from the Lagrangian given by (6.49). One can always add a total derivative without modifying the underlying theory. If one adds the following term

$$\frac{1}{2}\frac{d}{d\eta}\left[\frac{a'}{a}(\mu_{\mathbf{k}}^{\mathrm{R}})^2 + \frac{a'}{a}(\mu_{\mathbf{k}}^{\mathrm{I}})^2\right], \quad (6.58)$$

then the Lagrangian takes the form

$$S = \frac{1}{2}\int d\eta \int_{\mathbb{R}^{3+}} d^3k \left\{(\mu_{\mathbf{k}}^{\mathrm{R}\prime})^2 + (\mu_{\mathbf{k}}^{\mathrm{I}\prime})^2 - \omega^2(k,\eta)\left[(\mu_{\mathbf{k}}^{\mathrm{R}})^2 + (\mu_{\mathbf{k}}^{\mathrm{I}})^2\right]\right\}, \quad (6.59)$$

where $\omega^2(k,\eta) = k^2 - a''/a$. In this case, the conjugate momenta are simply given by $p_{\mathbf{k}}^{\mathrm{R}} = \mu_{\mathbf{k}}^{\mathrm{R}\prime}$ and $p_{\mathbf{k}}^{\mathrm{I}} = \mu_{\mathbf{k}}^{\mathrm{I}\prime}$. As a consequence, the Hamiltonian now reads

$$H = \int_{\mathbb{R}^{3+}} d^3k \left\{\frac{1}{2}(\hat{p}_{\mathbf{k}}^{\mathrm{R}})^2 + \frac{1}{2}(\hat{p}_{\mathbf{k}}^{\mathrm{I}})^2 + \frac{1}{2}\omega^2(k,\eta)\left[(\hat{\mu}_{\mathbf{k}}^{\mathrm{R}})^2 + (\hat{\mu}_{\mathbf{k}}^{\mathrm{I}})^2\right]\right\}. \quad (6.60)$$

This time, the Hamiltonian is exactly similar to the Schwinger Hamiltonian given by (6.11). This is another manifestation of the fact that, except for the exact time dependence of the effective frequency, the physical phenomenon, namely particle creation under the influence of an external classical field, is the same in both cases.

Let us now investigate in more detail the relation between the Hamiltonian given by (6.57) and the one of (6.60). We have just seen that the two corresponding theories differ by a total derivative and, hence, are physically equivalent. Another way to discuss the same property is through a canonical transformation. For this purpose, let us consider the following Hamiltonian

$$H_1(p_1, q_1) = \frac{1}{2}p_1^2 + \frac{a'}{a}p_1 q_1 + \frac{1}{2}k^2 q_1^2, \quad (6.61)$$

where $a(\eta)$ is an arbitrary function of the time. Clearly, H_1 plays the role of the Hamiltonian in (6.57) and $a(\eta)$ is the scale factor. Then, let us consider a canonical transformation of type II [29] such that $(q_1, p_1) \rightarrow (q_2, p_2)$, the generating function of which is given by (a similar transformation has also been studied in [30, 31])

$$G_2(q_1, p_2, \eta) = q_1 p_2 - \frac{1}{2}\frac{a'}{a}q_1^2 . \tag{6.62}$$

From this function, it is easy to establish the relation between the "old" variables and the "new" ones. One obtains

$$p_1 = \frac{\partial G_2}{\partial q_1} = p_2 - \frac{a'}{a}q_1 , \quad q_2 = \frac{\partial G_2}{\partial p_2} = q_1 . \tag{6.63}$$

In particular, the first relation reproduces (6.50) with the correct sign. Finally, the "new" Hamiltonian is given by

$$H_2(p_2, q_2) = H_1 + \frac{\partial G_2}{\partial \eta} = \frac{1}{2}p_2^2 + \frac{1}{2}\left(k^2 - \frac{a''}{a}\right)q_2^2 . \tag{6.64}$$

Clearly this Hamiltonian is similar to the Hamiltonian of (6.60).

Therefore, the two versions of the theory, the one given by the Hamiltonian (6.57), which is what we naturally obtain in the case of cosmological perturbations (see Sect. 6.4), and the one which leads to the Hamiltonian (6.60) "à la Schwinger" are simply connected by a canonical transformation and, thus, are physically identical. In the following, we will see that this is also the case at the quantum level.

6.3.2 Quantization and the Squeezed States Formalism

So far, the discussion has been purely classical. We now study the quantization of the system starting with the Heisenberg picture. The quantization in the functional picture that we used for the Schwinger effect will be investigated in the next sub-section. At the quantum level, μ_k and p_k become operators satisfying the commutation relation

$$[\hat{\mu}_{\boldsymbol{k}}, \hat{p}_{\boldsymbol{p}}^\dagger] = i\delta^{(3)}(\boldsymbol{k} - \boldsymbol{p}) . \tag{6.65}$$

Clearly, factor ordering is now important. The quantum Hamiltonian is obtained from the classical one by properly symmetrizing the expression (6.55). This leads to

$$\hat{H} = \int_{\mathbb{R}^{3+}} \mathrm{d}^3\boldsymbol{k} \left[\hat{p}_{\boldsymbol{k}}\hat{p}_{\boldsymbol{k}}^\dagger + k^2\hat{\mu}_{\boldsymbol{k}}\hat{\mu}_{\boldsymbol{k}}^\dagger + \frac{a'}{2a}\left(\hat{p}_{\boldsymbol{k}}\hat{\mu}_{\boldsymbol{k}}^\dagger + \hat{\mu}_{\boldsymbol{k}}^\dagger\hat{p}_{\boldsymbol{k}} + \hat{p}_{\boldsymbol{k}}^\dagger\hat{\mu}_{\boldsymbol{k}} + \hat{\mu}_{\boldsymbol{k}}\hat{p}_{\boldsymbol{k}}^\dagger\right)\right] . \tag{6.66}$$

In addition, this guarantees the hermiticity of the Hamiltonian. The next step consists in introducing the normal variable $\hat{c}_{\boldsymbol{k}}$ [28] (which becomes the annihilation operator, $\hat{c}_{\boldsymbol{k}}^\dagger$ becoming the creation operator) defined by

$$\hat{c}_k(\eta) \equiv \sqrt{\frac{k}{2}}\hat{\mu}_k + \frac{i}{\sqrt{2k}}\hat{p}_k \ . \tag{6.67}$$

Equivalently, one can also express $\hat{\mu}_k$ and $\hat{p}_k$ in terms of the normal variable and its hermitic conjugate. This gives the following two relations

$$\hat{\mu}_k = \frac{1}{\sqrt{2k}}\left(\hat{c}_k + \hat{c}^\dagger_{-k}\right) \ , \quad \hat{p}_k = -i\sqrt{\frac{k}{2}}\left(\hat{c}_k - \hat{c}^\dagger_{-k}\right) \ . \tag{6.68}$$

Then, from the commutation relation (6.65), or equivalently from the relation in real space $[\hat{\Phi}(\eta, \boldsymbol{x}), \hat{\Pi}(\eta, \boldsymbol{y})] = i\delta^{(3)}(\boldsymbol{x} - \boldsymbol{y})$, it follows that $[c_k(\eta), c_p^\dagger(\eta)] = \delta^{(3)}(\boldsymbol{k} - \boldsymbol{p})$. In terms of the normal variables, the scalar field and its conjugate momentum can be expressed as

$$\hat{\Phi}(\eta, \boldsymbol{k}) = \frac{1}{a(\eta)}\frac{1}{(2\pi)^{3/2}}\int \frac{\mathrm{d}\boldsymbol{k}}{\sqrt{2k}}\left[\hat{c}_k(\eta)e^{i\boldsymbol{k}\cdot\boldsymbol{x}} + \hat{c}^\dagger_k(\eta)e^{-i\boldsymbol{k}\cdot\boldsymbol{x}}\right] \ , \tag{6.69}$$

$$\hat{\Pi}(\eta, \boldsymbol{x}) = -\frac{a(\eta)}{(2\pi)^{3/2}}\int \mathrm{d}\boldsymbol{k}\, i\sqrt{\frac{k}{2}}\left[\hat{c}_k(\eta)e^{i\boldsymbol{k}\cdot\boldsymbol{x}} - \hat{c}^\dagger_k(\eta)e^{-i\boldsymbol{k}\cdot\boldsymbol{x}}\right] \ . \tag{6.70}$$

Obviously, when $a' = 0$, we recover the flat space–time limit and so we expect the time dependence of the normal variables to be just $\hat{c}_k(\eta) \propto e^{ik\eta}$.

We can now calculate the Hamiltonian operator in terms of the creation and annihilation operators. Using (6.66) one obtains

$$\hat{H} = \frac{1}{2}\int_{\mathbb{R}^3} \mathrm{d}^3k \left[k\left(c_k c_k{}^\dagger + c_{-k}{}^\dagger c_{-k}\right) - i\frac{a'}{a}\left(c_k c_{-k} - c_{-k}{}^\dagger c_k{}^\dagger\right)\right] \ , \tag{6.71}$$

where it is important to notice that the integral is now calculated in $\mathbb{R}^3$ and not in $\mathbb{R}^{3+}$. Let us analyze this Hamiltonian. The first term is the standard one and represents a collection of harmonic oscillators. The most interesting part is the second term. This term is responsible for the quantum creation of particles in curved space–time. It can be viewed as an interacting term between the scalar field and the classical background. The coupling function ia'/a is proportional to the derivative of the scale factor and, therefore, vanishes in flat space–time. From the structure of the interacting term, i.e., in particular the product of two creation operators for the mode $\boldsymbol{k}$ and $-\boldsymbol{k}$, we can also see that we have creation of pairs of quanta with opposite momenta during the cosmological expansion (thus momentum is conserved as it should), exactly as we had particle creation due to the interaction of the scalar field with a classical electric field in the previous section.

We can now calculate the time evolution of the quantum operators (here, we are working in the Heisenberg picture). Everything is known if we can determine the temporal behavior of the creation and annihilation operators; this behavior is given by the Heisenberg equations which read

$$\frac{\mathrm{d}c_k}{\mathrm{d}\eta} = -i[c_k, \hat{H}] \ , \quad \frac{\mathrm{d}c_k{}^\dagger}{\mathrm{d}\eta} = -i[c_k{}^\dagger, \hat{H}] \ . \tag{6.72}$$

Inserting the expression for the Hamiltonian derived above, we arrive at the equations

$$i\frac{dc_{\boldsymbol{k}}}{d\eta} = kc_{\boldsymbol{k}} + i\frac{a'}{a}c_{-\boldsymbol{k}}{}^{\dagger} , \quad i\frac{dc_{\boldsymbol{k}}{}^{\dagger}}{d\eta} = -kc_{\boldsymbol{k}}{}^{\dagger} + i\frac{a'}{a}c_{-\boldsymbol{k}} . \tag{6.73}$$

This system of equations can be solved by means of a Bogoliubov transformation and the solution can be written as

$$c_{\boldsymbol{k}}(\eta) = u_k(\eta)c_{\boldsymbol{k}}(\eta_{\text{ini}}) + v_k(\eta)c_{-\boldsymbol{k}}{}^{\dagger}(\eta_{\text{ini}}) , \tag{6.74}$$

$$c_{\boldsymbol{k}}{}^{\dagger}(\eta) = u_k^*(\eta)c_{\boldsymbol{k}}{}^{\dagger}(\eta_{\text{ini}}) + v_k^*(\eta)c_{-\boldsymbol{k}}(\eta_{\text{ini}}) , \tag{6.75}$$

where η_{ini} is a given initial time and where the functions $u_k(\eta)$ and $v_k(\eta)$ satisfy the equations

$$i\frac{du_k(\eta)}{d\eta} = ku_k(\eta) + i\frac{a'}{a}v_k^*(\eta) , \quad i\frac{dv_k(\eta)}{d\eta} = kv_k(\eta) + i\frac{a'}{a}u_k^*(\eta) . \tag{6.76}$$

In addition, these two functions must satisfy $|u_k|^2 - |v_k|^2 = 1$ such that the commutation relation between the creation and annihilation operators is preserved in time. A very important fact is that the initial values of u_k and v_k are fixed and, from the Bogoliubov transformation, read

$$u_k(\eta_{\text{ini}}) = 1 , \quad v_k(\eta_{\text{ini}}) = 0 . \tag{6.77}$$

Therefore, we remark that, in some sense, the initial conditions are fixed by the procedure of quantization. In fact, (6.77) implies that the initial state has been chosen to be the vacuum $|0\rangle$ at time $\eta = \eta_{\text{ini}}$. A priori, it is not obvious that this choice is well-motivated but it turns out to be the case in an inflationary universe. This property constitutes one of the most important aspect of the inflationary scenario. Here, we do not discuss further this issue but we will come back to the problem of fixing the initial conditions at the beginning of inflation in the following.

At this point, the next move is to establish the link between the formalism exposed above and the classical picture. For this purpose, it is interesting to establish the equation of motion obeyed by the function $u_k + v_k^*$. Straightforward manipulations from (6.76) leads to

$$(u_k + v_k^*)'' + \left(k^2 - \frac{a''}{a}\right)(u_k + v_k^*) = 0 . \tag{6.78}$$

We see that the function $u_k + v_k^*$ obeys the same equation as the variable $\mu_{\boldsymbol{k}}$. This is to be expected since, using the Bogoliubov transformation, the scalar field operator can be re-written as

$$\hat{\Phi}(\eta, \mathbf{x}) = \frac{1}{a(\eta)}\frac{1}{(2\pi)^{3/2}}\int\frac{d\boldsymbol{k}}{\sqrt{2k}}\Big[(u_k + v_k^*)\,(\eta)c_{\boldsymbol{k}}(\eta_{\text{ini}})e^{i\boldsymbol{k}\cdot\boldsymbol{x}}$$
$$+ (u_k^* + v_k)\,(\eta)c_{\boldsymbol{k}}^{\dagger}(\eta_{\text{ini}})e^{-i\boldsymbol{k}\cdot\boldsymbol{x}}\Big] . \tag{6.79}$$

Therefore, if we are given a scale factor $a(\eta)$, we can now calculate the complete time evolution of the quantum scalar field by means of the formalism presented above.

In fact, the Bogoliubov transformation (6.74) and (6.75) can be expressed in a different manner which is useful in order to introduce the squeezed states formalism. For this purpose, let us come back to the functions u_k and v_k. We have seen that, in order for the commutator of the creation and annihilation operators to be preserved in time, these two functions must satisfy $|u_k|^2 - |v_k|^2 = 1$. This means that we can always write

$$u_k = e^{i\theta_k} \cosh r_k \;, \quad v_k = e^{-i(\theta_k - 2\phi_k)} \sinh r_k \;, \qquad (6.80)$$

where the quantities r_k, θ_k and ϕ_k are functions of time. They are called the squeezing parameter, rotation angle and squeezing angle, respectively. These functions obey the equations

$$\frac{\mathrm{d}r_k}{\mathrm{d}\eta} = \frac{a'}{a} \cos 2\phi_k \;, \quad \frac{\mathrm{d}\phi_k}{\mathrm{d}\eta} = -k - \frac{a'}{a} \sin 2\phi_k \coth 2r_k \;, \qquad (6.81)$$

$$\frac{\mathrm{d}\theta_k}{\mathrm{d}\eta} = -k - \frac{a'}{a} \sin 2\phi_k \tanh r_k \;. \qquad (6.82)$$

These expressions can be used for an explicit calculation of r_k, θ_k and ϕ_k when a specific scale factor $a(\eta)$ is given. Now, the crucial property is that the Bogoliubov transformation (6.74) and (6.75) which solves the perturbed Einstein equations can be cast into the following form [32, 33, 34, 35]

$$c_{\boldsymbol{k}}(\eta) = R(\theta)S(r,\varphi)c_{\boldsymbol{k}}(\eta_{\mathrm{i}})S^{\dagger}(r,\varphi)R^{\dagger}(\theta) \;, \qquad (6.83)$$

$$c_{\boldsymbol{k}}^{\dagger}(\eta) = R(\theta)S(r,\varphi)c_{\boldsymbol{k}}^{\dagger}(\eta_{\mathrm{i}})S^{\dagger}(r,\varphi)R^{\dagger}(\theta) \;, \qquad (6.84)$$

where the operators $R(\theta)$ and $S(r,\varphi)$ are given by

$$R(\theta) = \exp\left\{ -i\theta_k \left[c_{\boldsymbol{k}}^{\dagger}(\eta_{\mathrm{i}})c_{\boldsymbol{k}}(\eta_{\mathrm{i}}) + c_{-\boldsymbol{k}}^{\dagger}(\eta_{\mathrm{i}})c_{-\boldsymbol{k}}(\eta_{\mathrm{i}}) \right] \right\} \;, \qquad (6.85)$$

$$S(r,\varphi) = \exp\left\{ r_k \left[e^{-2i\phi_k} c_{\boldsymbol{k}}^{\dagger}(\eta_{\mathrm{i}})c_{\boldsymbol{k}}(\eta_{\mathrm{i}}) - e^{2i\phi_k} c_{-\boldsymbol{k}}^{\dagger}(\eta_{\mathrm{i}})c_{-\boldsymbol{k}}(\eta_{\mathrm{i}}) \right] \right\} \;. \qquad (6.86)$$

Equations (6.83) and (6.84) allow us to interpret the Bogoluibov transformation in a new manner: indeed we can also see the time evolution of the creation and annihilation operators as rotations in the Hilbert space.

The previous considerations are valid in the Heisenberg picture. What happens in the Schrödinger picture where the operators no longer evolve but the states become time-dependent? For the sake of simplicity, let us ignore θ_k and ϕ_k by setting $\theta_k = \phi_k = 0$. As mentioned above, let us also postulate that the system is originally placed in the vacuum state $|0\rangle$. Then, the previous results imply that, after the cosmological evolution, the mode characterized by the wave-vector $\boldsymbol{k}$ will evolve into the following state [32, 33, 34, 35]

$$\exp\{r_k[c_k^\dagger(\eta_i)c_k(\eta_i) - c_{-k}^\dagger(\eta_i)c_{-k}(\eta_i)]\}|0\rangle \,, \tag{6.87}$$

which is, by definition, a two-mode vacuum squeezed state. This state is a very peculiar state and is of particular relevance in other branches of physics as well, most notably in quantum optics [36].

We now discuss the properties of such a quantum state. For this purpose, it is interesting to recall that a state containing a fixed number of particles, $|n\rangle$, can be obtained by successive action of the creation operator on the vacuum. Explicitly, one has

$$|n\rangle = \frac{(c_k^\dagger)^n}{\sqrt{n!}}|0\rangle \,. \tag{6.88}$$

Let us also introduce the coherent (Glauber) quantum state [28]. It is defined by the following expression

$$|\alpha\rangle = e^{-|\alpha|^2/2} \sum_{n=0}^{\infty} \frac{\alpha^n}{\sqrt{n!}}|n\rangle \,, \tag{6.89}$$

where α is a complex number. The coherent state is especially important in quantum optics since they represent, in a sense to be specified, the most classical state. We will come back to this question in the last section. Finally we also define two new operators B and T by

$$B_k = \frac{r}{2}[(c_k)^2 - (c_k^\dagger)^2] \,, \quad T_k = e^{B_k} \,, \tag{6.90}$$

where r is a real number (in fact, our squeezing parameter). The operators B_k and T_k possess various interesting properties, in particular B_k is anti-unitary, $B_k^\dagger = -B_k$, and, as a consequence, T_k is unitary, $T_k T_k^\dagger = 1$. The general definition of a squeezed state $|s\rangle$ is given by

$$|s\rangle \equiv T_k^\dagger|\alpha\rangle = e^{-B_k}|\alpha\rangle \,. \tag{6.91}$$

Let us notice that this is the expression for a one-mode squeezed state while, in (6.87), we have to deal with a two-mode squeezed state (hence the presence of operators c_k and c_{-k} that arises from the fact that we have pair creation, while in the above definition we only have operators c_k^2 and $c_k^{\dagger 2}$). The properties of one and two-mode squeezed states are similar and, here, for simplicity, we focus on the one-mode state only. Moreover, in our case, $|\alpha\rangle = |0\rangle$ which means that, in the cosmological case, we have a two-mode vacuum state.

Why is this state called a squeezed state? To answer this question, we introduce two new operators that are linear combinations of the creation and annihilation operators, namely

$$(c_k)_P \equiv \frac{1}{2}(c_k + c_k^\dagger) \quad (c_k)_Q \equiv \frac{1}{2i}(c_k - c_k^\dagger) \,. \tag{6.92}$$

These new operators are annihilation and creation operators of standing waves since, in a Fourier expansion of the field, they would stand in front of

$\cos k\eta$ and $\sin k\eta$ rather than $e^{ik\eta}$ and $e^{-ik\eta}$ in the case of the standard creation and annihilation operators. Then, it is straightforward to demonstrate that

$$\langle s|(c_k)_P|s\rangle = \frac{\alpha + \alpha^*}{2}e^r , \quad \langle s|(c_k)_Q|s\rangle = \frac{\alpha - \alpha^*}{2i}e^{-r} . \qquad (6.93)$$

Let us now calculate the mean value of the squares of these operators. We have

$$\langle s|(c_k)_P^2|s\rangle = \frac{e^{2r}}{4}(\alpha^2 + \alpha^{*2} + 2\alpha\alpha^* + 1) , \qquad (6.94)$$

and a similar expression for $\langle s|(c_k)_Q^2|s\rangle$ (but with e^{-2r} instead of e^{2r}). We are now in a position where the dispersion in the squeezed state of the operators $(c_k)_P$ and $(c_k)_Q$ can be calculated. One finds

$$\Delta(c_k)_P = \sqrt{\langle s|(c_k)_P^2|s\rangle - \langle s|(c_k)_P|s\rangle^2} = \frac{e^r}{2} , \quad \Delta(c_k)_Q = \frac{e^{-r}}{2} , \qquad (6.95)$$

and, therefore, from these equations one deduces that

$$\Delta(c_k)_P\Delta(c_k)_Q = \frac{1}{4} . \qquad (6.96)$$

We see that the lower bound of the Heisenberg uncertainty relations is reached but, contrary to a coherent state, the dispersion is not equal for the two operators. On the contrary, the dispersion can be very small on one component and very large on the other hence the name "squeezed state." In the cosmological situation, this is actually the case. Indeed, authors of [32, 33, 34] have shown that, for modes whose wavelengths are of the order of the Hubble length today, that is to say the modes that contribute the most to the "large angle" CMB multipoles C_ℓ (corresponding to a frequency of $\omega \sim 10^{-17}$ Hz), one has $r \sim 120$. From (6.95), we see that this corresponds to a very strong squeezing, in fact much larger than what can be achieved in the laboratory [34].

It is also clear that a strongly squeezed state is not a classical state in the sense that it is very far from the coherent state for which $\Delta(c_k)_P = \Delta(c_k)_Q$. On the other hand, since the mean value of $N_k = c_k^\dagger c_k$ is given by

$$\langle s|N_k|s\rangle = \sinh^2 r , \qquad (6.97)$$

a strongly vacuum squeezed state contains a very large number of particles and this criterion is often taken as a criterion of classicality. Therefore, we see that the meaning of classicality for a strongly squeezed state is a subtle issue [12, 13, 32, 33, 34, 37, 38] since different criterions seem to give different answers. We will come back to this point in the last section of this chapter.

6.3.3 Quantization in the Functional Approach

Let us now discuss the quantization in the functional approach where each Fourier mode is described by a wave-function (see also [20]). For this purpose, we use the description in terms of real variables. This will allow us to emphasize again the complete analogy that exists between the Schwinger effect and the theory of inflationary cosmological perturbations of quantum-mechanical origin. We restart from (6.57) and, since we deal with quantum operators, we symmetrize the corresponding expressions. In this case, the quantum Hamiltonian reads

$$\hat{H} = \int_{\mathbb{R}^{3+}} d^3k \left[\frac{1}{2} (\hat{p}_{\boldsymbol{k}}^{\text{R}})^2 + \frac{a'}{2a} (\hat{\mu}_{\boldsymbol{k}}^{\text{R}} \hat{p}_{\boldsymbol{k}}^{\text{R}} + \hat{p}_{\boldsymbol{k}}^{\text{R}} \hat{\mu}_{\boldsymbol{k}}^{\text{R}}) + \frac{k^2}{2} (\hat{\mu}_{\boldsymbol{k}}^{\text{R}})^2 + \frac{1}{2} (\hat{p}_{\boldsymbol{k}}^{\text{I}})^2 \right.$$
$$\left. + \frac{a'}{2a} (\hat{\mu}_{\boldsymbol{k}}^{\text{I}} \hat{p}_{\boldsymbol{k}}^{\text{I}} + \hat{p}_{\boldsymbol{k}}^{\text{I}} \hat{\mu}_{\boldsymbol{k}}^{\text{I}}) + \frac{k^2}{2} (\hat{\mu}_{\boldsymbol{k}}^{\text{I}})^2 \right] \equiv H^{\text{R}} + H^{\text{I}} . \tag{6.98}$$

We also have the following commutation relations that are compatible with (6.65)

$$[\hat{\mu}_{\boldsymbol{k}}^{\text{R}}, \hat{p}_{\boldsymbol{p}}^{\text{R}}] = i\delta^{(3)} (\boldsymbol{k} - \boldsymbol{p}) , \quad [\hat{\mu}_{\boldsymbol{k}}^{\text{I}}, \hat{p}_{\boldsymbol{p}}^{\text{I}}] = i\delta^{(3)} (\boldsymbol{k} - \boldsymbol{p}) . \tag{6.99}$$

In the Schrödinger picture, similarly to (6.13), the above-mentioned operators admit the following representation

$$\hat{\mu}_{\boldsymbol{k}}^{\text{R}} \Psi = \mu_{\boldsymbol{k}}^{\text{R}} \Psi , \quad \hat{p}_{\boldsymbol{k}}^{\text{R}} \Psi = -i \frac{\partial \Psi}{\partial \mu_{\boldsymbol{k}}^{\text{R}}} . \tag{6.100}$$

Therefore, one deduces that the Hamiltonian (here, the Hamiltonian for the real part of $\mu_{\boldsymbol{k}}$, hence for a fixed Fourier mode) can be written as

$$H_{\boldsymbol{k}}^{\text{R}} \Psi = -\frac{1}{2} \frac{\partial^2 \Psi}{\partial (\mu_{\boldsymbol{k}}^{\text{R}})^2} - \frac{i}{2} \frac{a'}{a} \Psi - i \frac{a'}{a} \mu_{\boldsymbol{k}}^{\text{R}} \frac{\partial \Psi}{\partial \mu_{\boldsymbol{k}}^{\text{R}}} + \frac{k^2}{2} (\mu_{\boldsymbol{k}}^{\text{R}})^2 \Psi . \tag{6.101}$$

Again, if $a' = 0$, we recover the Hamiltonian of an harmonic oscillator (instead of the Hamiltonian of a parametric oscillator when $a' \neq 0$).

Let us now study the ground state of the theory. As done in (6.16), we have the following Gaussian state,

$$\Psi_{\boldsymbol{k}}^{\text{R}} (\eta, \mu_{\boldsymbol{k}}^{\text{R}}) = N_{\boldsymbol{k}} (\eta) e^{-\Omega_{\boldsymbol{k}}(\eta)(\mu_{\boldsymbol{k}}^{\text{R}})^2} , \tag{6.102}$$

where $N_{\boldsymbol{k}}$ and $\Omega_{\boldsymbol{k}}$ are two functions to be determined. They are found by means of the Schrödinger equation $i\partial_\eta \Psi_{\boldsymbol{k}}^{\text{R}} = H_{\boldsymbol{k}}^{\text{R}} \Psi_{\boldsymbol{k}}^{\text{R}}$ that leads to

$$i \frac{N_{\boldsymbol{k}}'}{N_{\boldsymbol{k}}} = \Omega_{\boldsymbol{k}} - \frac{i}{2} \frac{a'}{a} , \quad \Omega_{\boldsymbol{k}}' = -2i\Omega_{\boldsymbol{k}}^2 - 2\frac{a'}{a}\Omega_{\boldsymbol{k}} + i\frac{k^2}{2} . \tag{6.103}$$

The analogy with (6.17) is obvious. We notice, however, that the structure of the equations is not exactly similar. This is due to the presence of the terms proportional to a'/a in the Hamiltonian (6.101) that have no equivalent in the Hamiltonian (6.15). Below, we briefly come back to this point. These equations can be integrated and the solutions read

$$N_k = \left(\frac{2\Re\Omega_k}{\pi}\right)^{1/4} \, , \quad \Omega_k = -\frac{i}{2}\frac{(f_k/a)'}{(f_k/a)} \, , \qquad (6.104)$$

where f_k obeys the equation $f_k'' + \left(k^2 - a''/a\right) f_k = 0$. Therefore, the integration of the equation controlling the time evolution of the mode function leads to a complete determination of the quantum state of the system in full agreement with what was discussed before in the case of the Schwinger effect. Again, the fact that the solution for Ω_k is given in terms of the function f_k/a and not only in terms of f_k, as one could have guessed from (6.18), is due to the presence of the terms proportional to a'/a in (6.101).

Let us briefly come back to the derivation of the above solution. Equation (6.103) is a Ricatti equation and, therefore, can be solved in the usual way, namely by transforming this non-linear first order differential equation into a linear second order differential equation. In order to find N_k, one requires that the wave-function is normalized, that is to say

$$\int \Psi_k^R \Psi_k^{R\,*} d\mu_k^R = 1 \, , \qquad (6.105)$$

which leads to the previous expression of N_k. Moreover, there is also the following consistency check. The real part of the second part of (6.103) reads

$$(\Re\Omega_k)' = 4\Re\Omega_k \times \Im\Omega_k - 2\frac{a'}{a}\Re\Omega_k \, , \qquad (6.106)$$

and the imaginary part of the first part of (6.103) can be written as $N_k'/N_k = \Im\Omega_k - a'/(2a)$. It is straightforward to check that, inserting the above solution for N_k into the last equation, precisely leads to (6.106).

Let us now come back to the remark made before that the structure of (6.103) is not exactly similar to what we have in the Schwinger case due to the presence of the terms proportional to a'/a. The reason is clearly that we have used the Hamiltonian given by (6.57) which contains such terms. But, obviously, one can also use the Hamiltonian given by (6.60). Then, assuming again the Gaussian form (6.102) for the wave-function, the Schrödinger equation reduces to

$$i\frac{N_k'}{N_k} = \Omega_k \, , \quad \Omega_k' = -2i\Omega_k^2 + \frac{i}{2}\omega^2(k,\eta) \, . \qquad (6.107)$$

which are now exactly similar to (6.17). As a consequence, the solutions are also the same and read

$$N_k = \left(\frac{2\Re\Omega_k}{\pi}\right)^{1/4} \, , \quad \Omega_k = -\frac{i}{2}\frac{f_k'}{f_k} \, , \qquad (6.108)$$

where f_k obeys the mode function equation $f_k'' + \omega^2 f_k = 0$.

In Sect. 6.3.1, we have established, at the classical level, the equivalence between the two formulations discussed above, that is to say the one based on the Hamiltonian (6.57), which leads to a Gaussian wave-function with N_k and Ω_k given by (6.104), and the one based on the Hamiltonian (6.60), which also leads to a Gaussian wave-function but with N_k and Ω_k now given by (6.108). We now study this link at the quantum level and, for this purpose, we reconsider the simple model introduced after (6.61). In Sect. 6.3.1, we showed that the two formulations are connected by a canonical transformation and the question is now to implement this canonical transformation at the quantum level [30, 31, 39, 40, 41, 42]. For this purpose, one must find a unitary operator $\hat{\mathcal{U}}$ such that the relations

$$\hat{q}_2 = \hat{\mathcal{U}}\hat{q}_1\hat{\mathcal{U}}^\dagger , \quad \hat{p}_2 = \hat{\mathcal{U}}\hat{p}_1\hat{\mathcal{U}}^\dagger , \tag{6.109}$$

exactly reproduce the classical analogs (6.63). A natural candidate would be the following operator

$$\hat{\mathcal{U}} = e^{i\hat{G}_2} = \exp\left[-\frac{i}{2}\frac{a'}{a}\hat{q}_1^2 + \frac{i}{2}\left(\hat{q}_1\hat{p}_1 + \hat{p}_1\hat{q}_1\right)\right] , \tag{6.110}$$

where G_2 is the generating function introduced in (6.62). However, as already remarked in [30], this choice is too naive and does not work. In order to understand what is going on, let us introduce a generalized version of (6.61), following (2.21) of [30], which at the classical level reads

$$H_1\left(p_1, q_1\right) = \frac{1}{2}\beta_3 p_1^2 + \beta_2(\eta)p_1 q_1 + \frac{1}{2}\beta_1 k^2 q_1^2 , \tag{6.111}$$

where for simplicity we consider that β_3 and β_1 are constants while β_2 is a time-dependent function (in [30], all the β_i's are time-dependent functions). Clearly, our case corresponds to $\beta_1 = \beta_3 = 1$ and $\beta_2 = a'/a$. Then, as before, we consider a canonical transformation of type II such that $(q_1, p_1) \rightarrow (q_2, p_2)$ with the following generating function

$$G_2\left(q_1, p_2, \eta\right) = \beta_3^{-1/2}q_1 p_2 - \frac{\beta_2}{2\beta_3}q_1^2 . \tag{6.112}$$

Setting $\beta_1 = \beta_3 = 1$ and $\beta_2 = a'/a$ in the above expression reproduces (6.62) as expected. Performing standard calculations, one finds that the relation between the "old" variables and the "new" ones reads

$$p_1 = \frac{\partial G_2}{\partial q_1} = \beta_3^{-1/2}p_2 - \frac{\beta_2}{\beta_3}q_1 , \quad q_2 = \frac{\partial G_2}{\partial p_2} = \beta_3^{-1/2}q_1 , \tag{6.113}$$

and that the "new" Hamiltonian can now be expressed as

$$H_2\left(p_2, q_2\right) = H_1 + \frac{\partial G_2}{\partial \eta} = \frac{1}{2}p_2^2 + \frac{1}{2}\left(\beta_1\beta_3 k^2 - \beta_2' - \beta_2^2\right)q_2^2 . \tag{6.114}$$

Notice, in particular, that the coefficient β_3 is no longer present in the term $p_2^2/2$. Then, in agreement with (2.22) and (2.23) of [30], let us consider the following operator

$$\hat{\mathcal{U}}(\hat{q}_1, \hat{p}_1, \eta) = \exp\left(-\frac{i}{2}\beta_2 \hat{q}_1^2\right) \exp\left[-\frac{i}{4}(\ln\beta_3)(\hat{q}_1\hat{p}_1 + \hat{p}_1\hat{q}_1)\right] . \qquad (6.115)$$

Inserting this operator in (6.109) and using the Baker–Campbell–Hausdorff formula, $e^{\hat{A}}\hat{B}e^{-\hat{A}} = \hat{B} + [\hat{A}, \hat{B}] + \cdots$, leads to the transformation

$$\hat{p}_1 = \beta_3^{-1/2}\hat{p}_2 - \frac{\beta_2}{\beta_3}\hat{q}_1 , \qquad \hat{q}_2 = \beta_3^{-1/2}\hat{q}_1 , \qquad (6.116)$$

namely exactly (6.113), but now at the quantum level. Therefore, we conclude that $\hat{\mathcal{U}}$ in (6.115) is the operator generating the correct quantum canonical transformation. In addition, as one can check with the help of (6.112), this operator is different from $e^{i\hat{G}_2}$, in particular due to the presence of the factor $\ln\beta_3$. Let us also notice that a similar operator has been considered recently in [31, 39], which carries out an investigation very relevant for what is discussed here, and that a factor akin to $\ln\beta_3$ was also present in the operator $\hat{\mathcal{U}}$ of that paper [see (2.46) where this factor is written as "$\ln\sqrt{12}/a$"]. Moreover, and this is the main reason why we have considered a generalized version of (6.61), we notice that our case is in fact very special since it corresponds to $\beta_3 = 1$ or $\ln\beta_3 = 0$ (or "$\epsilon = 0$" in the language of Ref. [39]). This means that, in the operator (6.115), the second exponential totally "disappears" while, of course, the term proportional to $q_1 p_1$ remains present in the classical generating function. Therefore, in our case, the quantum generating operator is just given by

$$\hat{\mathcal{U}}(\hat{q}_1, \hat{p}_1, \eta) = \exp\left(-\frac{i}{2}\frac{a'}{a}\hat{q}_1^2\right) . \qquad (6.117)$$

Clearly, one can repeat the above calculations using this operator and show that this leads to (6.63) but at the quantum level.

Let us now turn to the transformation of the wave-function itself. It is given by

$$\Psi(q_2) = N_2 e^{-\Omega_2 q_2^2} = \hat{\mathcal{U}}^\dagger(\hat{q}_1, \hat{p}_1, \eta)\Psi(q_1) = \hat{\mathcal{U}}^\dagger(\hat{q}_1, \hat{p}_1, \eta)N_1 e^{-\Omega_1 q_1^2} , \qquad (6.118)$$

from which, using (6.117), one deduces that

$$N_2 = N_1 , \qquad \Omega_2 = \Omega_1 - i\frac{a'}{a} . \qquad (6.119)$$

The relation $N_2 = N_1$ comes from the fact that the quantity $N_{1,2}$ is given by the real part of the function $\Omega_{1,2}$ and that Ω_2 and Ω_1 differ by a complex factor only. The above relation exactly reproduces what was observed in (6.104) and (6.108). From these formulae, we see that $\Omega_1 = -i/2(f/a)'/(f/a)$, see (6.104), while $\Omega_2 = -if'/(2f)$, see (6.108), and they indeed satisfy (6.119). Of course, the wave-functions after the quantum canonical transformation is normalized because $\Psi_2^*\Psi_2 = \hat{\mathcal{U}}\hat{\mathcal{U}}^\dagger\Psi_1^*\Psi_1 = \Psi_1^*\Psi_1$, the operator $\hat{\mathcal{U}}$ being unitary.

6.3.4 The Power Spectrum

Let us now calculate the two-point correlation function in the quantum state where the scalar field is put by the cosmological evolution. As will be discussed in the following, this quantity is relevant in astrophysics because, in the case of cosmological perturbations, it is directly observable; in particular it is directly linked to CMB fluctuations. Its definition reads

$$\langle \hat{\Phi}(\eta, \boldsymbol{x}) \hat{\Phi}(\eta, \boldsymbol{x}+\boldsymbol{r}) \rangle = \int \prod_{\boldsymbol{k}}^{n} \mathrm{d}\mu_{\boldsymbol{k}}^{\mathrm{R}} \mathrm{d}\mu_{\boldsymbol{k}}^{\mathrm{I}} \Psi^{*}\left(\mu_{\boldsymbol{k}}^{\mathrm{R}}, \mu_{\boldsymbol{k}}^{\mathrm{I}}\right) \Phi(\eta, \boldsymbol{x})\Phi(\eta, \boldsymbol{x}+\boldsymbol{r})\Psi\left(\mu_{\boldsymbol{k}}^{\mathrm{R}}, \mu_{\boldsymbol{k}}^{\mathrm{I}}\right) . \tag{6.120}$$

In this formula, the brackets mean the quantum average according to the standard definition, i.e., $\langle \hat{A} \rangle \equiv \int \mathrm{d}x\, \Psi^{*} \hat{A}(x)\, \Psi$. Then, using the Fourier expansion of the scalar field and permuting the integrals, one obtains

$$\langle \hat{\Phi}(\eta, \boldsymbol{x}) \hat{\Phi}(\eta, \boldsymbol{x}+\boldsymbol{r}) \rangle = \frac{1}{a^2} \frac{1}{(2\pi)^3} \int \int \mathrm{d}\boldsymbol{p}\, \mathrm{d}\boldsymbol{q} \mathrm{e}^{\mathrm{i}\boldsymbol{p}\cdot\boldsymbol{x}} \mathrm{e}^{\mathrm{i}\boldsymbol{q}\cdot(\boldsymbol{x}+\boldsymbol{r})} \prod_{\boldsymbol{k}}^{n} \left(\frac{2\Re\Omega_{\boldsymbol{k}}}{\pi}\right)$$
$$\int \left(\prod_{\boldsymbol{k}}^{n} \mathrm{d}\mu_{\boldsymbol{k}}^{\mathrm{R}} \mathrm{d}\mu_{\boldsymbol{k}}^{\mathrm{I}}\right) \mathrm{e}^{-2\sum_{\boldsymbol{k}}^{n} \Re\Omega_{\boldsymbol{k}}\left[\left(\mu_{\boldsymbol{k}}^{\mathrm{R}}\right)^2 + \left(\mu_{\boldsymbol{k}}^{\mathrm{I}}\right)^2\right]} \mu_{\boldsymbol{p}} \mu_{\boldsymbol{q}} . \tag{6.121}$$

The above expression vanishes unless $\boldsymbol{p} = -\boldsymbol{q}$. Indeed, if $|\boldsymbol{p}| \neq |\boldsymbol{q}|$ then the quantity $\mu_{\boldsymbol{p}}\mu_{\boldsymbol{q}}$ is "linear" in $\mu_{\boldsymbol{p}}^{\mathrm{R,I}}$ and $\mu_{\boldsymbol{q}}^{\mathrm{R,I}}$ and, consequently, the Gaussian integral is zero. If $\boldsymbol{p} = \boldsymbol{q}$, then $\mu_{\boldsymbol{p}}\mu_{\boldsymbol{q}} = \left(\mu_{\boldsymbol{p}}^{\mathrm{R}}\right)^2 - \left(\mu_{\boldsymbol{p}}^{\mathrm{I}}\right)^2$ and each term is indeed non-vanishing but the sum is zero because of the minus sign. Therefore, one obtains

$$\langle \hat{\Phi}(\eta, \boldsymbol{x}) \hat{\Phi}(\eta, \boldsymbol{x}+\boldsymbol{r}) \rangle = \frac{2}{a^2} \frac{1}{(2\pi)^3} \int \mathrm{d}\boldsymbol{p}\, \mathrm{e}^{\mathrm{i}\boldsymbol{p}\cdot\boldsymbol{r}} \prod_{\boldsymbol{k}}^{n} \left(\frac{2\Re\Omega_{\boldsymbol{k}}}{\pi}\right)$$
$$\int \prod_{\boldsymbol{k}}^{n} \mathrm{d}\mu_{\boldsymbol{k}}^{\mathrm{R}} \mathrm{d}\mu_{\boldsymbol{k}}^{\mathrm{I}} \mathrm{e}^{-2\sum_{\boldsymbol{k}}^{n} \Re\Omega_{\boldsymbol{k}}\left[\left(\mu_{\boldsymbol{k}}^{\mathrm{R}}\right)^2 + \left(\mu_{\boldsymbol{k}}^{\mathrm{I}}\right)^2\right]} \left(\mu_{\boldsymbol{p}}^{\mathrm{R}}\right)^2 . \tag{6.122}$$

The overall factor of 2 originates from the fact that the integral over $\mu_{\boldsymbol{k}}^{\mathrm{R}}$ is equal to the integral over $\mu_{\boldsymbol{k}}^{\mathrm{I}}$. The next step is to perform the path integral. In the above infinite product of integrals, all of them are of the form "$\int \mathrm{d}x\, \mathrm{e}^{-\alpha x^2}$" except the one over $\mu_{\boldsymbol{p}}^{\mathrm{R}}$ which is of the form "$\int \mathrm{d}x\, x^2 \mathrm{e}^{-\alpha x^2}$." Using standard results for Gaussian integrals, one gets

$$\langle \hat{\Phi}(\eta, \boldsymbol{x}) \hat{\Phi}(\eta, \boldsymbol{x}+\boldsymbol{r}) \rangle = \frac{2}{a^2} \frac{1}{(2\pi)^3} \int \mathrm{d}\boldsymbol{p}\, \mathrm{e}^{\mathrm{i}\boldsymbol{p}\cdot\boldsymbol{r}} \prod_{\boldsymbol{k}}^{n} \left(\frac{2\Re\Omega_{\boldsymbol{k}}}{\pi}\right) \prod_{\boldsymbol{k}}^{n} \left(\frac{\sqrt{\pi}}{\sqrt{2\Re\Omega_{\boldsymbol{k}}}}\right)$$
$$\times \prod_{\boldsymbol{k}}^{n-1} \left(\frac{\sqrt{\pi}}{\sqrt{2\Re\Omega_{\boldsymbol{k}}}}\right) \frac{1}{2} \left[\frac{\sqrt{\pi}}{\sqrt{(2\Re\Omega_{\boldsymbol{p}})^3}}\right] . \tag{6.123}$$

The infinite product "$\prod_{k}^{n-1}$" means a product over all the wave-vectors but p. Clearly, one can complete this product by inserting an extra factor $\sqrt{\pi}/\sqrt{2\Re\Omega_p}$ coming from the last term in the integral. Then, the last two products exactly cancel the first one. Finally, one obtains the simple expression

$$\langle \hat{\Phi}(\eta, \boldsymbol{x}) \hat{\Phi}(\eta, \boldsymbol{x} + \boldsymbol{r}) \rangle = \frac{1}{a^2} \frac{1}{(2\pi)^3} \int d\boldsymbol{p}\, e^{i\boldsymbol{p}\cdot\boldsymbol{r}} \frac{1}{2\Re\Omega_p} . \qquad (6.124)$$

Using the form of Ω_p in the ground state wave-function, see (6.108), one obtains

$$2\Re\Omega_p = -\frac{i}{2} \frac{\mu_p' \mu_p^* - \mu_p \mu_p'^*}{\mu_p \mu_p^*} = \frac{1}{2\mu_p \mu_p^*} , \qquad (6.125)$$

where we have used the fact that, with the initial condition (we will return to this point in the following) $\mu_p(\eta) \to (2p)^{-1/2} e^{ip\eta}$ when $p\eta \to -\infty$, the Wronskian is equal to i.

At this point, one can also make the following remark about the canonical transformation discussed in the previous subsection. It is clear that the power spectrum must be the same before and after the canonical transformation. Above, we used the form of Ω_p given by (6.108). But one could have used the form given by (6.104) in the same manner and without affecting the final result. Technically, this can be seen in (6.124) where it is clear that the power spectrum only depends on $\Re\Omega_p$. Since we demonstrated before that the canonical transformation only modifies the imaginary part of Ω_p, the power spectrum remains indeed the same.

Therefore, the final expression reads

$$\langle \hat{\Phi}(\eta, \boldsymbol{x}) \hat{\Phi}(\eta, \boldsymbol{x} + \boldsymbol{r}) \rangle = \frac{2}{a^2} \frac{1}{(2\pi)^3} \int d\boldsymbol{p}\, e^{i\boldsymbol{p}\cdot\boldsymbol{r}} \mu_p \mu_p^* \qquad (6.126)$$

$$= \frac{1}{4\pi^2} \int_0^{+\infty} \frac{dp}{p} \frac{\sin pr}{pr} p^2 \left| \frac{\mu_p}{a} \right|^2 . \qquad (6.127)$$

This expression is the standard one, usually derived in the Heisenberg picture [16, 17]. Knowledge of the mode function (including the initial conditions) is sufficient to estimate the power spectrum. In the following, we consider the case of inflationary cosmological perturbations and investigate which quantity plays the role of μ_k in that framework. This will allow us to discuss the inflationary predictions.

6.4 Inflationary Cosmological Perturbations of Quantum-Mechanical Origin

6.4.1 General Formalism

In this section, we finally consider our main subject, namely the theory of inflationary cosmological perturbations of quantum-mechanical origin [11, 16,

17, 43]. Our goal is to go beyond the isotropic and homogeneous FLRW Universe, the metric of which can be written as

$$ds^2 = a^2(\eta) \left[-d\eta^2 + \delta_{ij}^{(3)} dx^i dx^j \right] , \qquad (6.128)$$

and to study how small quantum perturbations around the above-mentioned solution behave during inflation. As we will see, the basic physical phenomenon and, hence, the corresponding formalism are similar to what was discussed before. As already emphasized, we are mainly concerned with inflation, that is to say a phase of accelerated expansion that took place in the early universe. In general relativity, such a phase can be obtained if the matter content is dominated by a fluid whose pressure is negative. Since, at very high energies, quantum field theory is the natural candidate to describe matter, it is natural and simple to postulate that a scalar field (the "inflaton") was responsible for the evolution of the universe in this regime. Therefore, the total action of the system is given by

$$S = -\frac{m_{\rm Pl}^2}{16\pi} \int {\rm d}^4x\sqrt{-g}R - \int {\rm d}^4x\sqrt{-g}\left[\frac{1}{2}g^{\mu\nu}\partial_\mu\varphi\ \partial_\nu\varphi + V(\varphi)\right], \qquad (6.129)$$

where φ is the inflaton field. Our discussion will be (almost) independent of the detailed shape of $V(\varphi)$ but, clearly, deriving from high-energy physics (for instance string theory) what this shape could be (in particular explaining the required flatness of the) is a major issue [44, 45].

Beyond homogeneity and isotropy, the most general form of the perturbed line element can be expressed as [11]:

$$ds^2 = a^2(\eta)\{-(1 - 2\phi)d\eta^2 + 2(\partial_i B)dx^i d\eta + [(1 - 2\psi)\delta_{ij}^{(3)}$$
$$+2\partial_i\partial_j E + h_{ij}]dx^i dx^j\} . \qquad (6.130)$$

In the above expression, the functions ϕ, B, ψ and E represent the scalar sector whereas the tensor h_{ij}, satisfying $h_i{}^i = 0 = h_{ij}{}^{,j}$, represents the gravitational waves. These functions must be small in comparison to one in order for the perturbative treatment to be valid. There are no vector perturbations because a single scalar field cannot seed rotational perturbations. At the linear level, the two types of perturbations decouple and, therefore, can be treated separately.

In the case of scalar perturbations of the geometry evoked above, the four functions are in fact redundant (thanks to our freedom to choose the coordinate system) and, in fact, the scalar fluctuations of the geometry can be characterized by a single quantity, namely the gauge-invariant Bardeen potential $\Phi_{\rm B}$ [43] (not to be confused with the scalar field Φ considered before) defined by

$$\Phi_{\rm B}(\eta, \boldsymbol{x}) = \phi + \frac{1}{a}\left[a\left(B - E'\right)\right]' . \qquad (6.131)$$

On the other hand, the fluctuations in the inflaton scalar field are characterized by the following gauge-invariant quantity $\delta\varphi^{(\mathrm{gi})}$

$$\delta\varphi^{(\mathrm{gi})}(\eta, \boldsymbol{x}) = \delta\varphi + \varphi'(B - E') . \tag{6.132}$$

We have therefore two gauge-invariant quantities but only one degree of freedom since Φ_{B} and $\delta\varphi^{(\mathrm{gi})}$ are coupled through the perturbed Einstein equations. As a consequence, in the scalar sector of the theory, everything can be reduced to the study of a single gauge-invariant variable (the so-called Mukhanov–Sasaki variable) defined by [3]

$$v(\eta, \boldsymbol{x}) \equiv a\left[\delta\varphi^{(\mathrm{gi})} + \varphi'\frac{\Phi_{\mathrm{B}}}{\mathcal{H}}\right] . \tag{6.133}$$

Let us notice that we will also work with the rescaled variable μ_{s} defined by $\mu_{\mathrm{s}}(\eta, \boldsymbol{x}) \equiv -\sqrt{2\kappa}v$. Finally, density perturbations are also often characterized by the so-called conserved quantity $\zeta(\eta, \boldsymbol{x})$ [46, 47] defined by $\mu_{\mathrm{s}} = -2a\sqrt{\gamma}\zeta$, where $\gamma = 1 - \mathcal{H}'/\mathcal{H}^2$.

In the tensor sector (which is automatically gauge invariant), the quantity which plays the role of $\mu_{\mathrm{s}}(\eta, \boldsymbol{x})$ is $\mu_{\mathrm{T}}(\eta, \boldsymbol{x})$, defined according to $h_{ij} = (\mu_{\mathrm{T}}/a)Q_{ij}$, where Q_{ij} are the (transverse and traceless) eigentensors of the Laplace operator on the space-like sections [43].

As usual, it is more convenient to study the perturbations mode by mode and, for this purpose, we will follow the evolution of the perturbations in Fourier space. Therefore, the study of cosmological perturbations during inflation reduces to investigating the behaviors of only two variables: $\mu_{\mathrm{s}\boldsymbol{k}}(\eta)$ and $\mu_{\mathrm{T}\boldsymbol{k}}(\eta)$.

Let us now establish the equations of motion for our two basic quantities. Since we want the variation of the action (6.129) to give the first order equations of motion for $\mu_{\mathrm{s}\boldsymbol{k}}(\eta)$ and $\mu_{\mathrm{T}\boldsymbol{k}}(\eta)$, we have to expand the action pertubatively up to second order in the metric perturbations and in the scalar field fluctuations. After a lengthy and tedious calculation, one obtains [11]

$$^{(2)}\delta S = \frac{1}{2}\int \mathrm{d}^4x\left[(v')^2 - \delta^{ij}\partial_i v\partial_j v + \frac{(a\sqrt{\gamma})''}{a\sqrt{\gamma}}v^2\right]$$
$$+\frac{m_{\mathrm{Pl}}^2}{64\pi}\int \mathrm{d}^4xa^2(\eta)\left[(h^i{}_j)'(h^j{}_i)' - \partial_k(h^i{}_j)\partial^k(h^j{}_i)\right] , \tag{6.134}$$

Notice that the constant m_{Pl} does not appear explicitly in the scalar part of the action because it has been absorbed via the background Einstein equations (however, see also [31]). It is also important to stress again that the previous expression is valid for any $V(\varphi)$.

Variation of the action leads to the following equation of motion for the two quantities $\mu_{\mathrm{s}\boldsymbol{k}}(\eta)$ and $\mu_{\mathrm{T}\boldsymbol{k}}(\eta)$

$$\frac{\mathrm{d}^2\mu_{\mathrm{s},\mathrm{T}\boldsymbol{k}}}{\mathrm{d}\eta^2} + \omega_{\mathrm{s},\mathrm{T}}^2(k,\eta)\mu_{\mathrm{s},\mathrm{T}\boldsymbol{k}} = 0 , \tag{6.135}$$

with

$$\omega_{\mathrm{S}}^2(k,\eta) = k^2 - \frac{(a\sqrt{\gamma})''}{a\sqrt{\gamma}} \ , \quad \omega_{\mathrm{T}}^2(k,\eta) = k^2 - \frac{a''}{a} \ . \tag{6.136}$$

We have thus reached our goal and demonstrated that cosmological perturbations obey exactly the same type of equation as a scalar field interacting with a classical electric field (Schwinger effect), namely the equation of a parametric oscillator as can be checked by comparing (6.135) with (6.7). The only difference lies in the physical nature of the classical source. In the case of cosmological perturbations, the (background) gravitational field is the classical source. The time dependence of the frequencies ω_{S} and ω_{T} is also different (recall that, in the case of the Schwinger effect, ω^2 contains terms proportional to t and t^2). Here, the dependence is fixed as soon as the behavior of the scale factor $a(\eta)$ is known. It is also interesting to notice that, a priori, the time dependence of ω_{S} is not the same as the one of ω_{T}. Indeed, ω_{T} depends on a and its derivatives up to second order while ω_{S} depends on the scale factor and its derivatives up to the fourth order (since it contains a term γ'', the quantity γ itself containing a term a''). Finally, the quantization of the theory proceeds as before and, as a consequence of the interaction between the quantum cosmological perturbations and the classical background, this results in the phenomenon of particle creation, here graviton creation. Classically, this corresponds to the amplification ("growing mode") of the fluctuations.

In the next section, we describe this phenomenon for an inflationary scale factor.

6.4.2 The Inflationary Effective Frequencies

So far, we have never specified $a(\eta)$ and, a priori, the mechanism of graviton creation is valid for any scale factor provided it is time-dependent. However, clearly, the detailed properties of the transition amplitude $\langle\Psi_1|\Psi_2\rangle$ depend on the time behavior of the effective frequency $\omega^2(k,\eta)$ and, hence, on the form of $a(\eta)$. Obviously, in the case of the Schwinger effect, a frequency different from the one given by (6.8) would have led to a number of created pairs different from (6.39).

In order to evaluate $\omega_{\mathrm{S}}^2(k,\eta)$ and $\omega_{\mathrm{T}}^2(k,\eta)$ for a typical inflationary model, one can use the slow-roll approximation [48, 49, 50, 51]. Indeed, during inflation and by definition, the kinetic energy to energy ratio and the scalar field acceleration to the scalar field velocity ratio are small and this suggests to view these quantities as parameters in which a systematic expansion can be performed. Therefore, one introduces the two parameters ϵ_1 and ϵ_2 [51] according to

$$\epsilon_1 = 3\frac{\dot{\varphi}^2/2}{\dot{\varphi}^2/2 + V(\varphi)} \ , \quad \frac{\mathrm{d}}{\mathrm{d}t}\left(\frac{\dot{\varphi}^2}{2}\right) = H\dot{\varphi}^2\left(\frac{\epsilon_2}{2} - \epsilon_1\right) \ . \tag{6.137}$$

From the above expressions, one sees that $\epsilon_1/3$ measures the ratio of the kinetic energy to the total energy while $\epsilon_2 > 0$ (respectively $\epsilon_2 < 0$) represents a model where the kinetic energy itself increases (respectively decreases) with respect to the total energy. It is also interesting to notice that $\epsilon_2 = 2\epsilon_1$ marks the frontier between models where the kinetic energy increases ($\epsilon_2 > 2\epsilon_1$) and the models where it decreases ($\epsilon_2 < 2\epsilon_1$). Provided the slow-roll conditions are valid, that is to say $\epsilon_{1,2} \ll 1$, one can also invert the previous expressions and express the slow-roll parameters only in terms of the inflaton . This leads to

$$\epsilon_1 \simeq \frac{m_{\mathrm{Pl}}^2}{16\pi}\left(\frac{V'}{V}\right)^2 , \qquad \epsilon_2 \simeq \frac{m_{\mathrm{Pl}}^2}{4\pi}\left[\left(\frac{V'}{V}\right)^2 - \frac{V''}{V}\right] , \qquad (6.138)$$

where, in the present context, a prime denotes a derivative with respect to the scalar field φ. Concrete calculations of slow-roll parameters for specific models can be found in [16].

Then, one can show that the two effective frequencies, to first order in the slow-roll parameters, can be expressed as [11, 52]

$$\omega_{\mathrm{s}}^2(k,\eta) \simeq k^2 - \frac{2 + 3\epsilon_1 - 3\epsilon_2/2}{\eta^2} , \qquad \omega_{\mathrm{T}}^2(k,\eta) \simeq k^2 - \frac{2 + 3\epsilon_1}{\eta^2} . \qquad (6.139)$$

Several remarks are in order at this point. Firstly, and as already mentioned previously, the time dependence in the inflationary case is different from the Schwinger case: the effective frequency contains terms proportional to $1/\eta^2$. Therefore, although the basic physical phenomenon is the same, one can expect the detailed predictions to differ. Secondly, different inflationary models correspond to different inflatons (or to different time variations of the scale factor) and, hence, to different values for the slow-roll parameters. One notices that the effective frequencies are sensitive to the details of the inflationary models since $\omega_{\mathrm{s}}^2(k,\eta)$ and $\omega_{\mathrm{T}}^2(k,\eta)$ depend on ϵ_1 and ϵ_2.

6.4.3 The WKB Approximation

We have established the form of the effective frequencies in the case of inflation. One must now solve the mode equations (6.135). For this purpose, we now reiterate the analysis of Sect. 6.2.3 using the WKB approximation [53]. As was the case for the Schwinger effect, the mode function can be found exactly. It is given in terms of Bessel functions [instead of parabolic cylinder functions, see (6.26)]

$$\mu_k(\eta) = \sqrt{k\eta}\left[A_k J_\nu(k\eta) + B_k J_{-\nu}(k\eta)\right] , \qquad (6.140)$$

where the orders are now functions of the slow-roll parameters, $\nu_{\mathrm{s}} = -3/2 - \epsilon_1 - \epsilon_2/2$ and $\nu_{\mathrm{T}} = -3/2 - \epsilon_1$. Then, one must choose the initial conditions. As discussed in the case of the Schwinger effect, we use the WKB approximation to discuss this question. The first step is to calculate the quantity Q in order to identify the regime where an adiabatic vacuum is available. Straightforward calculations lead to

$$\left| \frac{Q_{\text{S,T}}}{\omega_{\text{S,T}}^2} \right| = \frac{1}{8} \left| \frac{1 - 3k^2\eta^2}{(1 - k^2\eta^2/2)^3} \right| + \mathcal{O}\left(\epsilon_1, \epsilon_2\right) . \tag{6.141}$$

This quantity is represented in Fig. 6.2.

Let us now discuss this plot in more detail. The problem is characterized by two scales: the wavelength of the corresponding Fourier mode given by

$$\lambda\left(\eta\right) = \frac{2\pi}{k} a(\eta) , \tag{6.142}$$

where k is the co-moving wave-number, and the Hubble radius which can be expressed as

$$\ell_{\text{H}}\left(\eta\right) = \frac{a^2}{a'} . \tag{6.143}$$

We notice that the quantity $|Q/\omega^2|$ vanishes in the limit $k\eta \to -\infty$. This limit corresponds to a regime where $\lambda \ll \ell_{\text{H}}$. In this case, the wavelength is so small in comparison with the scale ℓ_{H} characterizing the curvature of space–time that the Fourier mode does not feel the FLRW Universe but behaves as if it were in flat (Minkowski) space–time. Clearly, in this regime, an adiabatic vacuum state is available since we recover the standard quantum field theory description. In the limit $k\eta \to 0$, the quantity $|Q/\omega^2|$ goes to $1/8 = \mathcal{O}(1)$ as can be checked in Fig. 6.2. This regime corresponds to the case where $\lambda \gg \ell_{\text{H}}$, that is to say when the wavelength of the Fourier mode is outside the Hubble radius. In this case, the curvature of space–time is felt and, as a consequence,

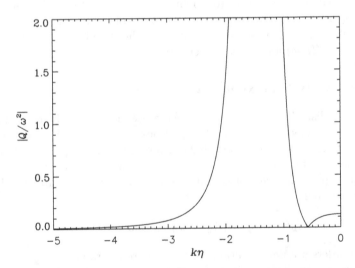

Fig. 6.2. Evolution of the quantity $|Q/\omega^2|$ with the quantity $k\eta$ for a typical model of inflation according to (6.141) (we have neglected the corrections proportional to the slow-roll parameters). In the limit $k\eta \to -\infty$, which corresponds to a wavelength much smaller than the Hubble radius, $|Q/\omega^2|$ vanishes and the notion of an adiabatic vacuum is available

the WKB approximation is violated and there is no unique vacuum state in this limit.

We have just seen that when a mode is sub-Hubble, that is to say $\lambda \ll \ell_{\mathrm{H}}$, the WKB approximation is valid. Let us notice that, without a phase of inflation, all the Fourier modes of astrophysical interest today would have been outside the Hubble radius in the early universe. It is only because, during inflation, the Hubble radius is constant that, initially, the Fourier modes are inside the Hubble radius. Therefore, although it was not designed for this purpose, a phase of inflation automatically implies that the WKB approximation is valid in the early universe and, as a consequence, ensures that we can choose a well-defined initial state. This is clearly an "extra bonus" of utmost importance. In the adiabatic regime, the solution for the mode function can be written as

$$\mu_{\boldsymbol{k}}(\eta) = \alpha_{\boldsymbol{k}} \mu_{\mathrm{wkb},\boldsymbol{k}}(\eta) + \beta_{\boldsymbol{k}} \mu_{\mathrm{wkb},\boldsymbol{k}}^{*}(\eta) , \tag{6.144}$$

where

$$\mu_{\mathrm{wkb},\boldsymbol{k}}(\eta) \equiv \frac{1}{\sqrt{2\omega(k,\eta)}} e^{-\mathrm{i} \int_{\eta_{\mathrm{ini}}}^{\eta} \omega(k,\tau)\mathrm{d}\tau} . \tag{6.145}$$

As done for the Schwinger effect, see (6.26), we now choose the initial conditions such that $\alpha_{\boldsymbol{k}} = 1$, $\beta_{\boldsymbol{k}} = 0$, corresponding to only one WKB branch in (6.144). This completely fixes the coefficients $A_{\boldsymbol{k}}$ and $B_{\boldsymbol{k}}$ in (6.140). One obtains [compare with (6.35) and (6.36)]

$$\frac{A_{\boldsymbol{k}}}{B_{\boldsymbol{k}}} = -e^{\mathrm{i}\pi\nu} , \quad B_{\boldsymbol{k}} = -\frac{2\mathrm{i}\pi}{m_{\mathrm{Pl}}} \frac{e^{-\mathrm{i}\nu(\pi/2)-\mathrm{i}(\pi/4)+\mathrm{i}k\eta_{\mathrm{ini}}}}{\sqrt{k}\sin(\pi\nu)} . \tag{6.146}$$

Equipped with the above exact solution for the mode function, the inflationary predictions can be determined.

Before turning to this calculation, let us quickly come back to the fact that the WKB approximation breaks down on super-Hubble scales. In fact, this problem bears a close resemblance with a situation discussed by atomic physicists at the time quantum mechanics was born. The subject debated was the application of the WKB approximation to the motion in a central field and, more specifically, how the Balmer formula for the energy levels of hydrogenic atoms can be recovered within the WKB approximation. The effective frequency for hydrogenic atoms is given by (obviously, in the atomic physics context, the wave equation is not a differential equation with respect to time but to the radial coordinate r)

$$\omega^{2}(E,r) = \frac{2m}{\hbar^{2}}\left(E + \frac{Ze^{2}}{r}\right) - \frac{\ell(\ell+1)}{r^{2}} , \tag{6.147}$$

where Ze is the (attractive) central charge and ℓ the quantum number of angular momentum. The symbol E denotes the energy of the particle and is negative in the case of a bound state. Apart from the term Ze^2/r and up to the identification $r \leftrightarrow \eta$, the effective frequency has exactly the same form as $\omega_{\mathrm{s,T}}(k,\eta)$ during inflation, see (6.136). Therefore, calculating the evolution of cosmological perturbations on super-Hubble scales, $|k\eta| \to 0$, is similar to determining the behavior of the hydrogen atom wave-function in the vicinity of the nucleus, namely $r \to 0$. The calculation of the energy levels by means of the WKB approximation was first addressed by Kramers [54] and by Young and Uhlenbeck [55]. They noticed that the Balmer formula was not properly recovered but did not realize that this was due to a misuse of the WKB approximation. In 1937 the problem was considered again by Langer [56]. In a remarkable article, he showed that the WKB approximation breaks down at small r, for an effective frequency given by (6.147) and, in addition, he suggested a method to circumvent this difficulty. Recently, this method has been applied to the calculation of the cosmological perturbations in [53, 57]. This gives rise to a new method of approximation, different from the more traditional slow-roll approximation.

6.4.4 The Inflationary Power Spectra

In this sub-section we turn to the calculation of the inflationary observables. The first step is to quantize the system. Obviously, this proceeds exactly as for the Schwinger effect or for a scalar field in curved space–time, the two cases that we have discussed before. We do not repeat the formalism here. As before, in the functional Schrödinger picture, the wave-function of the perturbations is given by

$$\Psi = \prod_{k}^{n} \Psi_{\boldsymbol{k}}\left(\mu_{\boldsymbol{k}}^{\mathrm{R}}, \mu_{\boldsymbol{k}}^{\mathrm{I}}\right) = \prod_{k}^{n} \Psi_{\boldsymbol{k}}^{\mathrm{R}}\left(\mu_{\boldsymbol{k}}^{\mathrm{R}}\right) \Psi_{\boldsymbol{k}}^{\mathrm{I}}\left(\mu_{\boldsymbol{k}}^{\mathrm{I}}\right) , \qquad (6.148)$$

with

$$\Psi_{\boldsymbol{k}}^{\mathrm{R}}\left(\eta, \mu_{\boldsymbol{k}}^{\mathrm{R}}\right) = N_{\boldsymbol{k}}\left(\eta\right) \mathrm{e}^{-\Omega_{\boldsymbol{k}}(\eta)\left(\mu_{\boldsymbol{k}}^{\mathrm{R}}\right)^2} , \quad \Psi_{\boldsymbol{k}}^{\mathrm{I}}\left(\eta, \mu_{\boldsymbol{k}}^{\mathrm{I}}\right) = N_{\boldsymbol{k}}\left(\eta\right) \mathrm{e}^{-\Omega_{\boldsymbol{k}}(\eta)\left(\mu_{\boldsymbol{k}}^{\mathrm{I}}\right)^2} ,$$

$$(6.149)$$

where the functions $N_{\boldsymbol{k}}(\eta)$ and $\Omega_{\boldsymbol{k}}(\eta)$ are functions that can be determined using the Schrödinger equation. This leads to expressions similar to (6.17) and (6.107) where now $\omega^2(k,t)$ should be replaced by $\omega_{\mathrm{s,T}}^2$ according to whether one considers the scalar perturbations or the gravitational waves. In particular, the function $\Omega_{\boldsymbol{k}}(\eta)$ is still given by $-i\mu_{\boldsymbol{k}}'/\mu_{\boldsymbol{k}}$, see (6.108), where, in the present context, $\mu_{\boldsymbol{k}}$ is given by the Bessel function of (6.140).

At this stage, one could compute the amplitude $\langle 0^-|0^+\rangle$ as one did in the case of the Schwinger effect. However, in the context of inflation, this is not the observable one is interested in. Indeed, we want to evaluate the amplitude of the fluctuations at the end of inflation and on super-Hubble scales. In this regime, as discussed before, there is no adiabatic state. So, in the context of inflation, there exists a "in" vacuum state $|0^-\rangle$ when $k\eta \to -\infty$ but there is no "out" region and, consequently, no $|0^+\rangle$ state. Of course, if one follows the evolution of the mode after inflation, then the unicity of the choice of the vacuum state is restored when the mode re-enters the Hubble radius either during the radiation or matter-dominated eras. But our goal is to compute the spectrum at the end of inflation. In other words, and contrary to the Schwinger effect, the quantity $\langle 0^-|0^+\rangle$ is not really relevant for the inflationary cosmological perturbations.

In fact, our goal is to calculate the anisotropies in the CMB (and/or to understand the distribution of galaxies). The key point is that the presence of cosmological perturbations causes anisotropies in the CMB: this is the Sachs–Wolfe effect [58, 59]. More precisely, on large scales, one has

$$\frac{\hat{\delta T}}{T}(e) \propto \hat{\zeta} = -\frac{\hat{\mu}_s}{2a\sqrt{\gamma}} , \tag{6.150}$$

where e represents a direction in the sky. The exact link is more complicated and has been discussed in details for instance in [17, 59]. In fact, it is convenient to expand this operator on the celestial sphere, i.e., on the basis of spherical harmonics

$$\frac{\hat{\delta T}}{T}(e) = \sum_{\ell=2}^{+\infty} \sum_{m=-\ell}^{m=\ell} \hat{a}_{\ell m} Y_{\ell m}(\theta, \varphi) . \tag{6.151}$$

This allows us to calculate the vacuum two-point correlation function of temperature fluctuations. One gets

$$\left\langle \frac{\hat{\delta T}}{T}(e_1)\frac{\hat{\delta T}}{T}(e_2) \right\rangle = \sum_{\ell=2}^{+\infty} \frac{(2\ell+1)}{4\pi} C_\ell P_\ell(\cos\gamma) , \tag{6.152}$$

where P_ℓ is a Legendre polynomial and γ is the angle between the two vectors e_1 and e_2. In the above expression, the brackets mean the standard quantum average. In practice, the observable two-point correlation function is rather defined by a spatial average over the celestial sphere. These two averages are of course not identical and the difference between them is at the origin of the concept of "cosmic variance"; see [60] for a detailed explanation. The C_ℓ 's are the multipole moments and have been measured with great accuracy by the WMAP experiment [5]. Clearly, as can be seen in (6.150), the above correlation function is related to the two-point correlation function of the cosmological fluctuations. Therefore, the relevant quantities to characterize the inflationary perturbations of quantum-mechanical origin are

$$\langle \hat{\zeta}(\eta, \boldsymbol{x})\hat{\zeta}(\eta, \boldsymbol{x}+\boldsymbol{r}) \rangle = \int_0^{+\infty} \frac{dk}{k} \frac{\sin kr}{kr} k^3 P_\zeta , \tag{6.153}$$

for scalar perturbations and, for tensor perturbations

$$\left\langle \hat{h}_{ij}(\eta, \boldsymbol{x})\hat{h}^{ij}(\eta, \boldsymbol{x}+\boldsymbol{r}) \right\rangle = \int_0^{+\infty} \frac{dk}{k} \frac{\sin kr}{kr} k^3 P_h . \tag{6.154}$$

One can then repeat the calculation done in Sect. 6.3.4 in order to evaluate the above quantities. Indeed, the calculation proceeds exactly in the same way since the wave-functional is still a Gaussian. This gives

$$k^3 P_\zeta(k) = \frac{k^3}{8\pi^2} \left| \frac{\mu_s}{a\sqrt{\gamma}} \right|^2 , \quad k^3 P_h(k) = \frac{2k^3}{\pi^2} \left| \frac{\mu_T}{a} \right|^2 . \tag{6.155}$$

These expressions should be compared with (6.127).

The two power spectra can be easily computed using the exact solution for the mode function, see (6.140). At first order in the slow-roll parameters, one arrives at [48, 49, 50, 51]

$$k^3 P_\zeta = \frac{H^2}{\pi \epsilon_1 m_{\rm Pl}^2} \left[1 - 2\left(C+1\right)\epsilon_1 - C\epsilon_2 - \left(2\epsilon_1 + \epsilon_2\right) \ln \frac{k}{k_*} \right] , \tag{6.156}$$

$$k^3 P_h = \frac{16 H^2}{\pi m_{\rm Pl}^2} \left[1 - 2\left(C+1\right)\epsilon_1 - 2\epsilon_1 \ln \frac{k}{k_*} \right] , \tag{6.157}$$

where C is a numerical constant, $C \simeq -0.73$, and k_* an arbitrary scale called the "pivot scale." We see that the amplitude of the scalar power spectrum is given by a scale-invariant piece (that is to say which does not depend on k), $H^2/(\pi \epsilon_1 m_{\rm Pl}^2)$, plus logarithmic corrections, the amplitude of which is controlled by the slow-roll parameters, namely by the micro-physics of inflation. The above remarks are also valid for tensor perturbations. The ratio of tensor over scalar is just given by $k^3 P_h / k^3 P_\zeta = 16\epsilon_1$. This means that the gravitational waves are always sub-dominant and that, when we measure the CMB anisotropies, we essentially see the scalar modes. This is rather unfortunate because this implies that one cannot measure the energy scale of inflation since the amplitude of the scalar power spectrum also depends on the slow-roll parameter ϵ_1. Only an independent measure of the gravitational waves contribution could allow us to break this degeneracy. On the other hand, the spectral indices are given by

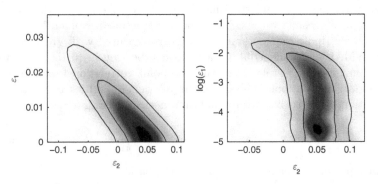

Fig. 6.3. The 68% and 95% confidence intervals of the two-dimensional marginalized posteriors in the slow-roll parameters plane, obtained at leading order in the slow-roll expansion [61]. The *shading* is the mean likelihood and the *left plot* is derived under an uniform prior on ϵ_1 while the *right panel* corresponds to an uniform prior on $\log \epsilon_1$

$$n_{\rm s} - 1 = \left.\frac{\ln k^3 P_\zeta}{{\rm d}\ln k}\right|_{k=k_*} = -2\epsilon_1 - \epsilon_2 , \quad n_{\rm T} = \left.\frac{\ln k^3 P_h}{{\rm d}\ln k}\right|_{k=k_*} = -2\epsilon_1 . \quad (6.158)$$

As expected, the power spectra are always close to scale invariance ($n_{\rm s} = 1$ and $n_{\rm T} = 0$) and the deviation from it is controlled by the magnitude of the two slow-roll parameters.

To conclude this section, let us signal that the slow-roll parameters ϵ_1 and ϵ_2 are already constrained by the astrophysical data, see Fig. 6.3 for the constraints coming from the WMAP data. A complete analysis can be found in [61, 62].

6.5 The Classical Limit of Quantum Perturbations

As discussed at length previously, the inflationary cosmological perturbations are of quantum-mechanical origin. However, from the observational point of view, it seems that we deal with a physical phenomenon where quantum mechanics does not play a crucial role (does not even a play a role at all). Therefore, from the conceptual point of view, it is important to understand how the system can become classical [12, 13, 46, 63] (see also [64, 65, 66, 67, 68]). We now turn to this question.

6.5.1 Coherent States

It seems natural to postulate that a quantum system behaves classically when it is placed in a state such that it follows (exactly or, at least, approximatively) the classical trajectory. For the sake of illustration, let us consider a simple one-dimensional system characterized by the Hamiltonian

$$H(p, q) = \frac{1}{2}p^2 + V(q) , \qquad (6.159)$$

where, for the moment, the potential $V(q)$ is arbitrary. Solving the classical Hamilton's equations (given some initial conditions)

$$\frac{{\rm d}p}{{\rm d}t} = -\frac{\partial V(q)}{\partial q} , \quad \frac{{\rm d}q}{{\rm d}t} = p , \qquad (6.160)$$

provides the classical solution $p_{\rm cl}$ and $q_{\rm cl}$. At the technical level, the above-mentioned criterion of classicality amounts to choosing a state $|\Psi\rangle$ such that

$$p_{\rm cl}(t) = \langle\Psi|\hat{p}(t)|\Psi\rangle , \quad q_{\rm cl}(t) = \langle\Psi|\hat{q}(t)|\Psi\rangle . \qquad (6.161)$$

This is clearly a non-trivial requirement as can be understood from the Ehrenfest theorem. Indeed, this theorem shows that, for any state $|\Psi\rangle$, one has

$$\frac{{\rm d}}{{\rm d}t}\langle\Psi|\hat{p}(t)|\Psi\rangle = -\left\langle\Psi\left|\frac{\partial\hat{V}(q)}{\partial q}\right|\Psi\right\rangle , \quad \frac{{\rm d}}{{\rm d}t}\langle\Psi|\hat{q}(t)|\Psi\rangle = \langle\Psi|\hat{p}|\Psi\rangle . \quad (6.162)$$

These equations resemble the Hamilton's equations (6.160) but are of course not identical. This implies that, placed in an arbitrary state, the quantum system does not behave classically (i.e., the means of the position and of the momentum do not obey the classical equations). It would be the case only for a state $|\Psi\rangle$ such that

$$\left\langle \Psi \left| \frac{\partial \hat{V}(q)}{\partial q} \right| \Psi \right\rangle = \frac{\partial}{\partial q} V \left(\langle \Psi | \hat{q} | \Psi \rangle \right) , \qquad (6.163)$$

which is not true in general since $\langle \Psi | \hat{q}^n | \Psi \rangle \neq \langle \Psi | \hat{q} | \Psi \rangle^n$, but obviously satisfied if the potential assumes the particular shape $V(q) \propto q^2$, i.e., for the harmonic oscillator. In this case, the means of the position and of the momentum do follow the classical trajectory whatever the state $|\Psi\rangle$ is. This means that (6.161) are in fact not sufficient to define classicality and that one needs to provide extra conditions. It seems natural to require that the wave packet is equally localized in coordinate and momentum $\Delta\hat{q} \equiv \sqrt{\langle\hat{q}^2\rangle - \langle\hat{q}\rangle^2} = \Delta\hat{p}$ to the minimum allowed by the Heisenberg bound, namely $\Delta\hat{q}\Delta\hat{p} = 1/2$. This is another way to define a coherent state, see (6.89) which, therefore, represents the "most classical" state of a quantum harmonic oscillator.

We now demonstrate that the state (6.89) indeed satisfies the above-mentioned properties. If the potential is given by $V(q) = k^2 q^2/2$, then the Hamilton's equations can be expressed as $\dot{p} = -k^2 q$ and $\dot{q} = p$ and the "normal variable" takes the form, see also (6.67),

$$\alpha \equiv \sqrt{\frac{k}{2}} \left(q + \frac{i}{k} p \right) , \qquad (6.164)$$

obeys the equation $\dot{\alpha} = -ik\alpha$ which allows us to write the classical trajectory in phase space as

$$p_{\rm cl}(t) = -i\sqrt{\frac{k}{2}} \left(\alpha_0 e^{-ikt} - \alpha_0^* e^{ikt} \right) , \qquad (6.165)$$

$$q_{\rm cl}(t) = \frac{1}{\sqrt{2k}} \left(\alpha_0 e^{-ikt} + \alpha_0^* e^{ikt} \right) , \qquad (6.166)$$

with

$$\alpha_0 \equiv \sqrt{\frac{k}{2}} \left[q(t = 0) + \frac{i}{k} p(t = 0) \right] , \qquad (6.167)$$

Let us consider that, at time $t = 0$, the system is placed in the state $|\alpha_0\rangle$. At time $t > 0$, the integration of the Schrödinger equation leads to

$$|\Psi(t)\rangle = e^{-ikt/2} |\alpha(t) = \alpha_0 e^{-ikt}\rangle . \qquad (6.168)$$

This result should be understood as follows. In the expression (6.89) which defines a coherent state, the factor α should be replaced with $\alpha_0 e^{-ikt}$ to get the formula expressing the above state $|\Psi\rangle$. As already mentioned, at the quantum level, the normal variable becomes the annihilation operator [obtained from (6.164) by simply replacing q and p with their quantum counterparts]. This implies

$$\hat{q} = \frac{1}{\sqrt{2k}}\left(\hat{a} + \hat{a}^\dagger\right) \quad \hat{p} = -i\sqrt{\frac{k}{2}}\left(\hat{a} - \hat{a}^\dagger\right) . \tag{6.169}$$

Then, the crucial step is that any coherent state $|\alpha\rangle$ is the eigenvector of $\hat{a}$ with the eigenvalue α, $\hat{a}|\alpha\rangle = \alpha|\alpha\rangle$. Using this property, it is easy to show that, for the state $|\Psi\rangle$ defined by (6.168), one has

$$\langle\Psi|\hat{p}|\Psi\rangle = -i\sqrt{\frac{k}{2}}\left(\alpha_0 e^{-ikt} - \alpha_0^* e^{ikt}\right), \tag{6.170}$$

$$\langle\Psi|\hat{q}|\Psi\rangle = \frac{1}{\sqrt{2k}}\left(\alpha_0 e^{-ikt} + \alpha_0^* e^{ikt}\right) . \tag{6.171}$$

In the same way, straightforward manipulations lead to

$$\langle\Psi|\hat{p}^2|\Psi\rangle = \frac{k}{2}\left\{1 + 4\Im^2\left[\alpha(t)\right]\right\} , \quad \langle\Psi|\hat{q}^2|\Psi\rangle = \frac{1}{2k}\left\{1 + 4\Re^2\left[\alpha(t)\right]\right\} \tag{6.172}$$

from which one deduces

$$\Delta\hat{q} = \sqrt{\frac{1}{2k}} \quad \Delta\hat{p} = \sqrt{\frac{k}{2}} . \tag{6.173}$$

We have thus reached our goal, i.e., we have shown that the state (6.168) follows the classical trajectory and that the quantum dispersion around this trajectory is the same in position and momentum and is minimal (that is to say the Heisenberg inequality is saturated). Therefore, as announced, the coherent state is indeed the "most classical" state. It is also interesting to give the explicit form of the wave-function. It reads

$$\Psi_\alpha(q,t) = e^{i\theta_\alpha}\left(\frac{k}{\pi}\right)^{1/4} e^{-ikt/2} e^{iqp_{cl}(t)} e^{-k[q-q_{cl}(t)]^2/2} , \tag{6.174}$$

where the phase factor is defined by $e^{i\theta_\alpha} \equiv e^{(\alpha^{*2}-\alpha^2)/4}$.

The above expression is defined in real space. However, if one wants to follow the evolution of the system in phase space, it is interesting to introduce the Wigner function [15, 69, 70, 71] defined by the expression (for a one-dimensional system)

$$W(q,p,t) \equiv \frac{1}{2\pi}\int du\, \Psi^*\left(q - \frac{u}{2}, t\right) e^{-ipu} \Psi\left(q + \frac{u}{2}, t\right) . \tag{6.175}$$

A system behaves classically if the Wigner function is positive-definite since, in this case, it can be interpreted as a classical distribution. In addition, if the Wigner function is localized in phase space over a small region corresponding to the classical position and momentum, then the corresponding quantum predictions become indistinguishable from their classical counterparts and we can indeed state that the system has "classicalized." For the wave-function (6.174), the Wigner function can be expressed as

$$W\left(q,p,t\right) = \frac{1}{\pi}e^{-k[q-q_{\mathrm{cl}}(t)]^2}e^{-\frac{1}{k}[p-p_{\mathrm{cl}}(t)]^2} . \tag{6.176}$$

It is represented in Fig. 6.4. We notice that the Wigner function is always positive and, therefore, according to the above considerations, the system can be considered as classical. Moreover, $W(p,q,t)$ is peaked over a small region in phase space and the wave packet follows exactly the classical trajectory

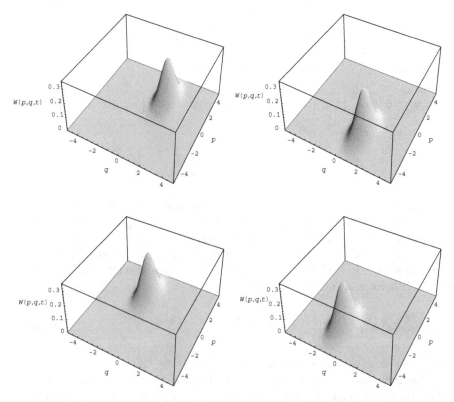

Fig. 6.4. Wigner function (6.176) for the coherent state $|\alpha\rangle$ at different times. The (arbitrary) values $q_0 = 1$, $p_0 = 1$ and $k = 2$ have been used for this figure. This implies $\alpha_0 = \sqrt{2}e^{i\pi/4}$ and, see (6.165), $p_{\mathrm{cl}} = 2\sqrt{2}\sin(\pi/4 - t)$ and $q_{\mathrm{cl}} = \sqrt{2}\cos(\pi/4 - t)$. The *upper left panel* represents the Wigner function (6.176) at time $t = 0$ while the *upper right, lower right* and *lower left panels* correspond to $W(p,q,t)$ at time $t = \pi/2, t = \pi$ and $t = 3\pi/2$, respectively. The wave packet follows the periodic (ellipsoidal) classical trajectory in phase space and its shape remains unchanged during the motion

(an ellipse), as is also clear from (6.176). This confirms our interpretation of the coherent state as the most classical state.

To conclude this sub-section, let us notice that the coherent state $|\alpha\rangle$ can be obtained by applying the following unitary operator on the vacuum state

$$|\alpha\rangle = e^{\alpha a^\dagger - \alpha^* a}|0\rangle . \tag{6.177}$$

This equation should be compared to (6.90) and (6.91). We see that the argument of the exponential is linear in the creation and annihilation operators while it was quadratic in the case of the squeezed state.

6.5.2 Wigner Function of the Cosmological Perturbations

In order to study whether the (super-Hubble) cosmological perturbations have "classicalized," we now use the technical tool of the Wigner function introduced before. The first application to cosmological perturbations was made in [12, 33]. For a two-dimensional system (here, we have in mind μ_k^R and μ_k^I for a fixed mode k), the generalization of (6.175) is straightforward and reads

$$W\left(\mu_k^R, \mu_k^I, p_k^R, p_k^I\right) \equiv \frac{1}{(2\pi)^2} \int \int du \, dv \, \Psi^*\left(\mu_k^R - \frac{u}{2}, \mu_k^I - \frac{v}{2}\right) e^{-ip_k^R u - ip_k^I v}$$

$$\times \Psi\left(\mu_k^R + \frac{u}{2}, \mu_k^I + \frac{v}{2}\right) , \tag{6.178}$$

where the wave-function is given by the expressions (6.149). Since we have to deal with Gaussian integrations only, the above Wigner function can be calculated exactly. One obtains

$$W\left(\mu_k^R, \mu_k^I, p_k^R, p_k^I\right) = \Psi\Psi^* \frac{1}{2\pi\Re\Omega_k} \exp\left[\frac{1}{2\Re\Omega_k}\left(p_k^R + 2\Im\Omega_k\mu_k^R\right)^2\right]$$

$$\times \exp\left[\frac{1}{2\Re\Omega_k}\left(p_k^I + 2\Im\Omega_k\mu_k^I\right)^2\right] . \tag{6.179}$$

It is represented in Fig. 6.5. The first remark is that the Wigner function is positive (as expected since we deal with a Gaussian state) and, therefore, can be interpreted as a classical distribution. However, as shown in Fig. 6.5, and contrary to the case of a coherent state, W is not peaked over a small region of phase space. We are interested in the behavior of the Wigner function for modes of astrophysical interest today. These modes have spent time outside the Hubble radius during inflation and, as a consequence, their squeezing parameter r is big. Therefore, it is convenient to express Ω_k in terms of the squeezing parameters

$$\Omega_k = \frac{k}{2} \frac{\cosh r - e^{-2i\phi} \sinh r}{\cosh r + e^{-2i\phi} \sinh r} , \tag{6.180}$$

and to take the strong squeezing limit, $r \to +\infty$. One has

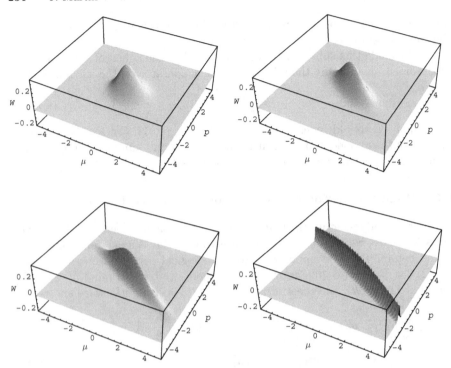

Fig. 6.5. Wigner function of cosmological perturbations obtained from (6.179) (for a one-dimensional system). The squeezing parameter r is chosen to be $r = 0.1$, 0.5, 1 and 2 for the *left upper*, *right upper*, *left lower* and *right lower panels*, respectively (it is not the same time ordering as in Fig. 6.4 because, in the present case, the motion is not periodic). The other squeezing parameters are taken to be $\phi = \pi/6$ and $\theta = 0$. As can be noticed in this figure, the Wigner function remains positive. Since the squeezing parameter increases with time, the different panels correspond in fact to the Wigner function at different times. At initial time, the quantum state is the vacuum and, therefore, the Wigner function is that of a coherent state, compare the *left upper panel* with Fig. 6.4. Then, the Wigner function develops the "Dirac function behavior" discussed in the text that clearly appears on this plot

$$\Re\Omega_{\boldsymbol{k}} \to 0 \ , \quad \Im\Omega_{\boldsymbol{k}} \to \frac{k}{2}\frac{\sin\phi}{\cos\phi} \ . \tag{6.181}$$

This implies

$$W\left(\mu_{\boldsymbol{k}}^{\mathrm{R}}, \mu_{\boldsymbol{k}}^{\mathrm{I}}, p_{\boldsymbol{k}}^{\mathrm{R}}, p_{\boldsymbol{k}}^{\mathrm{I}}\right) \to \Psi\Psi^*\delta\left(\frac{k}{2}\frac{\sin\phi}{\cos\phi}\mu_{\boldsymbol{k}}^{\mathrm{R}} + p_{\boldsymbol{k}}^{\mathrm{R}}\right)\delta\left(\frac{k}{2}\frac{\sin\phi}{\cos\phi}\mu_{\boldsymbol{k}}^{\mathrm{I}} + p_{\boldsymbol{k}}^{\mathrm{I}}\right) \ , \tag{6.182}$$

where δ denotes the Dirac function. The above limit is clearly visible in Fig. 6.5 for $r > 1$ (lower panels).

Therefore, in the large squeezing limit, the Wigner function is elongated along a very thin ellipse in phase space. At first sight, this means that the system is not classical since one cannot single out a small cell around some classical values (μ_{cl}, p_{cl}) that would follow a classical trajectory as it was discussed before. On the other hand, as already mentioned above, the Wigner function remains positive. This means that the interference term which makes the system quantum in the sense that the amplitudes rather than the probabilities should be summed up has become negligible. Therefore, in this sense, the system is classical or, more precisely, is in fact equivalent to a classical stochastic process with a Gaussian distribution (given by the term $\Psi\Psi^*$). We see that the nature of this classical limit is quite different to what happens in the case of a coherent state: we cannot predict a definite correlation between position and momentum but we can describe the system in terms of a classical random variable. In practice, this is what is done by astrophysicists: in particular, the quantity $a_{\ell m}$ in (6.151) is always treated as a Gaussian random variable and any detailed quantum-mechanical considerations avoided.

As argued in [13], the system has become classical (in the sense explained before) without any need to take into account its interaction with the environment. This is "decoherence without decoherence" as stressed in the above-referred article. More on this subtle issue can be found in [13, 38]. Of course, the question of whether the wave-function of the perturbation has collapsed or not (and the question of whether this question is meaningful in the present context and/or dependent on the interpretation of quantum mechanics that one chooses to consider) is even more delicate [63] and we will not touch upon this issue here.

6.5.3 Wigner Function of a Free Particle

We now consider the case of a free particle since it shares common points with the case of cosmological perturbations as first noticed in [38]. At the beginning of this section, we mentioned that, in the particular case of the harmonic oscillator $V(q) \propto q^2$, the quantum mean value of the position and momentum operators always follow the classical trajectory whatever the quantum state $|\Psi\rangle$ in which the system is placed. There is obviously another situation where this is also the case: the free massive particle where $V(q) = 0$. The wave-function is given by (we take $m = 1$)

$$\Psi(q,t) = \left(\frac{2a^2}{\pi}\right)^{1/4} \frac{1}{(a^4 + 4t^2)^{1/4}} e^{-i\tan^{-1}(2t/a^2)/2 + ik_0(q-q_0) - ik_0^2 t/2}$$
$$\times e^{-(q-q_0-k_0 t)^2/(a^2+2it)}, \tag{6.183}$$

where we have assumed that the wave packet is centered at $q = q_0$ at $t = t_0$. The parameter a represents the width of the wave packet while k_0 parameterizes its velocity. The means of $\hat{p}$ and $\hat{q}$ can be expressed as

$$\langle \hat{p} \rangle = k_0 = p_{cl}, \quad \langle \hat{q} \rangle = q_0 + k_0 t = q_{cl}. \tag{6.184}$$

Therefore, as announced, they follow exactly the classical trajectory. But, as argued in the sub-section 6.5.1, one must also compute the dispersions. Straightforward calculations lead to

$$\langle \hat{p}^2 \rangle = k_0^2 + \frac{1}{a^2} \,, \quad \langle \hat{q}^2 \rangle = \frac{1}{4a^2} \left(a^4 + 4t^2 \right) + \left(q_0 + k_0 t \right)^2 \,, \quad (6.185)$$

from which one deduces that

$$\Delta \hat{p} = \frac{1}{a} \,, \quad \Delta \hat{q} = \frac{a}{2} \sqrt{1 + 4\frac{t^2}{a^4}} \,. \quad (6.186)$$

At $t = 0$, one has $\Delta\hat{q}\Delta\hat{p} = 1/2$ and the Heisenberg bound is saturated: at initial time, the wave packet is minimal. But then, and contrary to the case of the harmonic oscillator, the dispersion on the position is increasing with time (while the dispersion on the momentum remains constant). The wave packet does not keep its shape unchanged while moving as it was the case for a potential $V(q) \propto q^2$. This is the well-known phenomenon dubbed "spreading of the wave packet." However, a quasi-classical interpretation of this situation exists. Indeed, when $t \gg a^2$, one has

$$\Delta\hat{q} \sim \frac{t}{a} = \Delta\hat{p}t = \Delta v_{\text{cl}}t \,, \quad (6.187)$$

which reproduces the classical motion. On the contrary, for small times, $\Delta\hat{q}$ must take values very different from the classical ones in order to satisfy the Heisenberg inequality. Therefore, in the regime $t \gg a^2$, the system is classical but in a sense slightly different from the one encountered in the harmonic oscillator case.

Let us now calculate the Wigner function of the free particle. One obtains

$$W(p, q, t) = \frac{1}{\pi} \exp\left[-\frac{2a^2}{a^4 + 4t^2} \left(q - q_{\text{cl}} \right)^2 \right]$$

$$\times \exp\left\{ -\frac{a^4 + 4t^2}{2a^2} \left[p - p_{\text{cl}} - \frac{4t}{a^4 + 4t^2} \left(q - q_{\text{cl}} \right) \right]^2 \right\} \quad (6.188)$$

This equation is similar to (6.179). The Wigner function is the product of one exponential factor whose argument is proportional to "$\left(q - q_{\text{cl}}^2 \right)$" [hidden in the term $\Psi^*\Psi$ in (6.179)] and of another exponential term whose argument has the form "$[p - p_{\text{cl}} - f(t) \left(q - q_{\text{cl}} \right)]^2$," where $f(t)$ is a function of time only. Therefore, the classical limit of cosmological perturbations can also be understood in terms of the (quasi-) classical limit of a free particle, as discussed in the previous paragraph.

The Wigner function (6.188) is represented in Fig. 6.6. This plot confirms the interpretation presented above. First of all, the Wigner function remains positive which indicates that a classical interpretation is meaningful. At initial time, the Wigner function is well-localized because the wave packet is minimal.

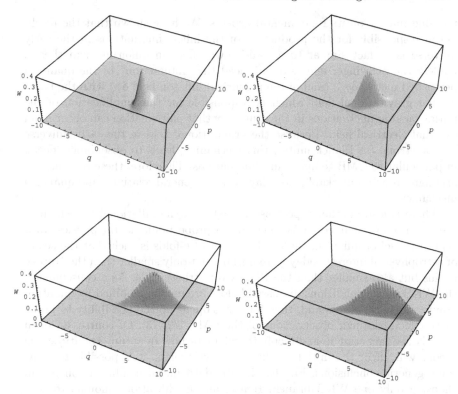

Fig. 6.6. Wigner function (6.188) of a free particle. The parameters chosen are $k_0 = 1$, $q_0 = 1$ and $a = 1$ which is in fact equivalent to considering the dimensionless quantities ap, q/a and t/a^2. The *left upper panel* corresponds to $t/a^2 = 0$ while the *right upper*, *left lower* and *right lower panels* represent the Wigner function at times $t/a^2 = 0.8, 2, 3$, respectively (here, the time ordering is similar to that in Fig. 6.5 and, therefore, different from that in Fig. 6.4). Initially, the wave packet is well-localized in phase space and, as time goes on, the spreading of the wave packet becomes apparent

Then, as time goes on, the spreading of the wave packet causes the spreading of the Wigner function in phase space. Clearly, this case bears some resemblance with that of cosmological perturbations, compare Figs. 6.5 and 6.6. Therefore, inflationary fluctuations on large scales become classical in the same sense that a free particle is classical far from the origin. A much more detailed description of this analogy can be found in [38].

6.6 Conclusions

In this review, we have presented a pedagogical introduction to the theory of inflationary cosmological perturbations of quantum-mechanical origin, focusing mainly on its fundamental aspects. We have shown that the mechanism responsible for the production of the initial fluctuations in the early universe is in fact similar to a well-known effect in quantum field theory, namely the Schwinger effect. It is indeed the "interaction" of the quantum perturbed metric $\delta g_{\mu\nu}$ (and of the perturbed inflaton field $\delta\varphi$) with the background gravitational field which is responsible for the amplification of the initial vacuum fluctuations in the same way that pair creation can occur in an external electrical field. Because the gravitational field in the early universe can be strong (in Planck units), this mechanism leads to observable effects, in particular to CMB temperature fluctuations. Therefore, these fluctuations originate from a remarkable interplay between general relativity and quantum mechanics.

There is also another aspect associated with the inflationary mechanism discussed above that could be relevant to probe fundamental physics. In a typical model of inflation, the total number of e-folds is such that the scales of astrophysical interest today were initially not only smaller than the Hubble radius but also smaller than the Planck (or string) scale. As a consequence, the WKB initial conditions discussed before are maybe modified by quantum gravity (stringy) effects and, therefore, this opens up the possibility to probe these effects through observations of the CMB [72, 73]. Of course, an open issue is the fact that it is difficult to calculate how quantum gravity/string theory will affect the initial conditions. Nevertheless, it is possible to draw some generic conclusions. Firstly, the standard initial condition consists in choosing only one WKB branch. Hence, any modification amounts to considering that the second branch is present. As a consequence, super-imposed oscillations in the power spectrum unavoidably appear, the amplitude and the frequency of these oscillations being unfortunately model dependent [72] (for the observational status of these oscillations, see Refs. [61, 73]). Secondly, the presence of the second WKB branch means, in some sense, the presence of particles in the initial state and, therefore, there is potentially a back-reaction problem. Generically, the larger the amplitude of the super-imposed oscillations, the more severe the back-reaction issue. On the other hand, predicting the effect of the energy density of the perturbations is difficult and it is not clear whether this will spoil inflation or, for instance, just renormalize the cosmological constant [74, 75]. These problems are still open questions but it is interesting to note that the inflationary scenario is rich enough to provide yet another means to learn about fundamental physics.

Acknowledgements

I would like to thank P. Brax, H. Fried, M. Lemoine, L. Lorenz, P. Peter and C. Ringeval for useful discussions and careful reading of the manuscript.

References

1. A. Guth, Phys. Rev. D **23**, 347 (1981).
2. A. Linde, Phys. Lett. B **108**, 389 (1982); A. Albrecht and P. J. Steinhardt, Phys. Rev. Lett. **48**, 1220 (1982); A. Linde, Phys. Lett. B **129**, 177 (1983).
3. V. Mukhanov and G. Chibisov, JETP Lett. **33**, 532 (1981); Sov. Phys. JETP **56**, 258 (1982).
4. S. Hawking, Phys. Lett. **115B**, 295 (1982); A. Starobinsky, Phys. Lett. **117B**, 175 (1982); A. Guth and S. Y. Pi, Phys. Rev. Lett. **49**, 1110 (1982); J. M. Bardeen, P. J. Steinhardt and M. S. Turner, Phys. Rev. D **28**, 679 (1983).
5. D. N. Spergel et al., astro-ph/0603449; C. L. Bennet et al., Astrophys. J. Suppl. **148**, 1 (2003), astro-ph/0302207; G. Hinshaw et al., Astrophys. J. Suppl. **148**, 135 (2003), astro-ph/0302217; L. Verde et al., Astrophys. J. Suppl. **148**, 195 (2003), astro-ph/0302218; H. V. Peiris et al., Astrophys. J. Suppl. **148**, 213 (2003), astro-ph/0302225.
6. J. Schwinger, Phys. Rev. **82**, 664 (1951).
7. A. Ringwald, hep-ph/0112254.
8. R. Alkofer, M. B. Hecht, C. D. Roberts, S. M. Schmidt and D. V. Vinnik, Phys. Rev. Lett. **87**, 193902 (2001), nucl-th/0108046.
9. Point 7 of the scientific case in the XFEL-STI Interim Report of January 11, 2005 available at http://xfel.desy.de/science/index_eng.html.
10. http://www-ssrl.slac.stanford.edu/lcls/.
11. V. F. Mukhanov, H. A. Feldman and R. H. Brandenberger, Phys. Rep. **215**, 203 (1992).
12. A. Guth and S. Pi, Phys. Rev. D **32**, 1899 (1985).
13. D. Polarski and A. Starobinsky, Class. Quant. Grav. **13**, 377 (1996), gr-qc/9504030.
14. W. Zurek, Phys. Rev. D **26**, 1862 (1982).
15. E. Wigner, Phys. Rev. **40**, 749 (1932).
16. J. Martin, Proceedings of the XXIV Brazilian National Meeting on Particles and Fields, Caxambu, Brazil (2004), astro-ph/0312492.
17. J. Martin, Lect. Notes Phys. **669**, 199 (2005), hep-th/0406011.
18. E. Ince, *Ordinary Differential Equations* (Dover, 1926).
19. C. Kiefer, Phys. Rev. D **45**, 2044 (1992).
20. K. Srinivasan and T. Padmanabhan, gr-qc/9807064.
21. I. S. Gradshteyn and I. M. Ryzhik, *Tables of Integrals, Series and Products* (Academic, New York, 1981).
22. E. Brezin and C. Itzykson, Phys. Rev. D **2**, 1191 (1970).
23. H. M. Fried, Y. Gabellini, B. H. McKellar and J. Avan, Phys. Rev. D **63**, 125001 (2001).
24. J. Avan, H. M. Fried and Y. Gabellini, Phys. Rev. D **67**, 016003 (2003), hep-th/0208053.
25. H. M. Fried and Y. Gabellini, Phys. Rev. D **73**, 011901 (2006), hep-th/0510233.
26. M. Turner, Phys. Rev. D **28**, 1243 (1983).
27. L. Kofman, A. Linde and A. Starobinsky, Phys. Rev. D **56**, 3258 (1997), hep-ph/9704452.
28. C. Cohen-Tannoudji, J. Dupont-Roc and G. Grynberg, *Photons et atomes, Introduction a l'électrodynamique quantique* (Editions du CNRS, 1987).
29. H. Goldstein, *Classical Mechanics* (Addisson-Wesley, 1990).

30. J. M. Cervero and A. Rodriguez, Int. J. Theor. Phys. **41**, 503 (2002), quant-ph/0106157.
31. P. Peter, E. Pinho and N. Pinto-Neto, JCAP **0507**, 014 (2005), hep-th/0509232.
32. L. P. Grishchuk and Y. V. Sidorov, Class. Quant. Grav. **6**, L155 (1989).
33. L. P. Grishchuk and Y. V. Sidorov, Phys. Rev. D **42**, 3413 (1990).
34. L. P. Grishchuk, H. A. Hauss and K. Bergman, Phys. Rev. D **46**, 1440 (1992).
35. A. Albrecht, P. Ferreira, M. Joyce and T. Prokopec, Phys. Rev. D **50**, 4807 (1994), astro-ph/9303001.
36. B. L. Schumacher, Phys. Rep. **135**, 317 (1986).
37. W. Schleich, D. F. Walls and J. A. Wheeler, Phys. Rev. A **38**, 1177 (1988).
38. C. Kiefer and D. Polarski, gr-qc/9805014.
39. A. Mostafazadeh, quant-ph/9612038.
40. Haewon Lee and W. S. l'Yi, Phys. Rev. A **51**, 982 (1995), hep-th/9406106.
41. Y. S. Kim and E. P. Wigner, Am. J. Phys. **58**, 439 (1990).
42. A. Ogura and M. Sekiguchi, quant-ph/0410173.
43. J. A. Bardeen, Phys. Rev. D **22**, 1882 (1980).
44. D. Lyth and A. Riotto, Phys. Rep. **314**, 1 (1999), hep-ph/9807278.
45. R. Kallosh, Chap., this volume.
46. D. H. Lyth, Phys. Rev. D **31**, 1792 (1985).
47. J. Martin and D. J. Schwarz, Phys. Rev. D **57**, 3302 (1998), gr-qc/9704049.
48. E. D. Stewart and D. H. Lyth, Phys. Lett. **B302**, 171 (1993).
49. J. Martin and D. J. Schwarz, Phys. Rev. D **62**, 103520 (2000), astro-ph/9911225; J. Martin, A. Riazuelo and D. J. Schwarz, Astrophys. J. **543**, L99 (2000), astro-ph/0006392.
50. S. Leach, A. R. Liddle, J. Martin and D. J. Schwarz, Phys. Rev. D **66**, 023515 (2002), astro-ph/0202094.
51. D. J. Schwarz, C. A. Terrero-Escalante and A. A. Garcia, Phys. Lett. **B517**, 243 (2001), astro-ph/0106020.
52. L. P. Grishchuk, Zh. Eksp. Teor. Fiz **67**, 825 (1974).
53. J. Martin and D. J. Schwarz, Phys. Rev. D **67**, 083512 (2003), astro-ph/0210533.
54. H. A. Kramers, Zeit. f. Phys. **39**, 836 (1926).
55. L. A. Young and G. E. Uhlenbeck, Phys. Rev. **36**, 1158 (1930).
56. R. E. Langer, Phys. Rev. **51**, 669 (1937).
57. R. Casadio, F. Finelli, M. Luzzi and G. Venturi, Phys. Rev. D **72**, 103516 (2005), gr-qc/0510103.
58. R. K. Sachs and A. M. Wolfe, Astrophys. J. **147**, 73 (1967).
59. M. Panek, Phys. Rev. D **49**, 648 (1986).
60. L. P. Grishchuk and J. Martin, Phys. Rev. D **56**, 1924 (1997), gr-qc/9702018.
61. J. Martin and C. Ringeval, JCAP **0608**, 009 (2006), astro-ph/0605367.
62. C. Ringeval, Chap., this volume.
63. A. Perez, H. Sahlmann and D. Sudarsky, Class. Quant. Grav. **23**, 2317 (2006), gr-qc/0508100.
64. A. Starobinsky, in *Field Theory, Quantum Gravity and Strings*, eds H. J. de Vega and N. Sanchez, **107** (1986).
65. D. Campo and R. Parentani, Phys. Rev. D **74**, 045015 (2005), astro-ph/0505376.
66. F. Lombardo, F. Mazzitelli and D. Monteoliva, Phys. Rev. D **62**, 045016 (2000), hep-ph/9912448.

67. F. Lombardo and D. Lopez Nacir, Phys. Rev. D **72**, 063506 (2005), gr-qc/0506051.
68. D. Lyth and D. Seery, astro-ph/0607647.
69. A. Anderson, Phys. Rev. D **42**, 585 (1990).
70. S. Habib, Phys. Rev. D **42**, 2566 (1990).
71. S. Habib and R. Laflamme, Phys. Rev. D **42**, 4056 (1990).
72. J. Martin and R. Brandenberger, Phys. Rev. D **63**, 123501 (2001), hep-th/0005209; R. Brandenberger and J. Martin, Mod. Phys. Lett. **A16**, 999 (2001), astro-ph/0005432; R. Brandenberger, S. Joras and J. Martin, Phys. Rev. D **66**, 083514 (2002), hep-th/0112122; J. Martin and R. Brandenberger, Phys. Rev. D **68**, 063513 (2003), hep-th/0305161.
73. J. Martin and C. Ringeval, Phys. Rev. D **69**, 083515 (2004), astro-ph/0310382; J. Martin and C. Ringeval, Phys. Rev. D **69**, 127303 (2004), astro-ph/0402609; J. Martin and C. Ringeval, JCAP **0501**, 007 (2005), hep-ph/0405249.
74. J. Martin and R. Brandenberger, Phys. Rev. D **71**, 023504 (2005), hep-th/0410223.
75. U. Danielsson, JCAP **0603**, 014 (2006), hep-th/0511273.

7

The Numerical Treatment of Inflationary Models

Christophe Ringeval

Theoretical and Mathematical Physics Group, Centre for Particle Physics and
Phenomenology, Louvain University, 2 Chemin du Cyclotron, 1348
Louvain-la-Neuve, Belgium

7.1 Motivations

The inflationary paradigm is currently passing all the tests raised by the so-called high-precision cosmology measurements [1]. Although this suggests that the existence of a quasi-exponential accelerated era in the early universe may be viewed as the standard lore, one has to keep in mind that almost all the inflationary field models lasting more than 60 efolds and leading to an almost scale-invariant power spectrum for adiabatic scalar perturbations may do the job. It is therefore of both theoretical and observational interest to look for inflationary properties that are, or will be in a foreseeable future, significant enough in the data to allow to discriminate the different models. Many studies are devoted to this task ranging from the details of the reheating era to the search for a theoretical embedding of inflation in supersymmetry or string theory [2, 3] (see in particular Chaps. 3 and 4 in this volume). In the following, we will be interested in the problem of discriminating models through the cosmological perturbations and the Cosmic Microwave Background (CMB) anisotropies.

Among the analytical tools available to study inflation in the cosmological context, the so-called slow-roll approximation provides analytical expressions for both the field evolution and the primordial scalar and tensor power spectra. It relies on an order by order expansion in terms of the so-called Hubble-flow functions $\epsilon_i(n)$, where $n = \ln(a/a_{\mathrm{ini}})$ is the number of efolds since the beginning of inflation and a the Friedman–Lemaître–Robertson–Walker (FLRW) scale factor. The Hubble flow functions are defined from the Hubble parameter $H(n)$ by

$$\epsilon_1 = -\frac{\mathrm{d}\ln H}{\mathrm{d}n}, \qquad \epsilon_{i+1} = \frac{\mathrm{d}\ln \epsilon_i}{\mathrm{d}n}. \qquad (7.1)$$

If the underlying field model is such that these functions remain small at the time where the length scales that are of cosmological interest today leave the Hubble radius, then the scalar and tensor power spectra can be Taylor

C. Ringeval: *The Numerical Treatment of Inflationary Models*, Lect. Notes Phys. **738**,
243–273 (2008)
DOI 10.1007/978-3-540-74353-8_7
© Springer-Verlag Berlin Heidelberg 2008

expanded around a given pivot wavenumber k_*. At first order, one gets [4, 5, 6, 7]

$$\mathcal{P}_\zeta(k) = \frac{\kappa^2 H^2}{8\pi^2 \epsilon_1} \left[1 - 2(C+1)\epsilon_1 - C\epsilon_2 - (2\epsilon_1 - \epsilon_2) \ln \frac{k}{k_*} \right], \qquad (7.2)$$

for the scalar modes and

$$\mathcal{P}_h(k) = \frac{2\kappa^2 H^2}{\pi^2} \left[1 - 2(C+1)\epsilon_1 - 2\epsilon_1 \ln \frac{k}{k_*} \right], \qquad (7.3)$$

for the tensor modes. In (7.2) and (7.3), $\kappa^2 = 8\pi/m_{\text{Pl}}^2$ is the gravitational coupling constant and C is a constant that results from the Taylor expansion ($C \simeq -0.73$). The Hubble parameter and the two first Hubble flow functions are evaluated at $N_* = n_{\text{end}} - n_*$: the number of efolds before the end of inflation at which the pivot length scale crosses the Hubble radius: $k_* = a(N_*)H(N_*)$. From these power spectra, assuming the conservation of the comoving curvature perturbation after Hubble exit ($k < aH$), the CMB anisotropies induced by the scalar and tensor perturbations can be derived and compared with the data. Using Markov–Chains–Monte–Carlo (MCMC) methods, one can extract constraints on the power spectra parameters, namely ϵ_1, ϵ_2 and $P_* = \kappa^2 H^2/(8\pi^2 \epsilon_1)$. Assuming a flat FLRW universe, the WMAP third year data lead to the posterior probability distributions plotted in Fig. 7.1.

A great advantage of the slow-roll approach is that one does not need to specify an explicit model of inflation [7, 9, 10, 11]. The constraints verified by ϵ_1 and ϵ_2 apply to all (single field) inflationary models verifying the slow-roll conditions $\epsilon_1 \ll 1$ and $\epsilon_2 \ll 1$. However, using these results for a given model of inflation requires the knowledge of N_* to determine the associated theoretical values of $\epsilon_i(N_*)$. As can be seen in Fig. 7.2, the value of N_* depends on the number of efolds N_{reh} during which the universe reheated before the radiation era. The reheating era depends on the microphysics associated with the decay of the inflaton field whose complexity renders the determination of N_{reh} difficult [12, 13, 14, 15, 16, 17, 18]. However, under reasonable assumptions, it has been shown in [19] that typically $40 \lesssim N_* \lesssim 60$, although these bounds may vary by a factor of two for extreme models. For the large field models represented in Fig. 7.1, the uncertainties on the reheating blur the theoretically predicted values of the slow-roll parameters (see the short segments in Fig. 7.1). Notice that the resulting errors in the ϵ_i remain small compared to the current CMB data accuracy, but this is not necessarily the case for other models, as for instance the small field models discussed in the following. The problem is expected to become even more significant with the next generation of more accurate CMB measurements.

Another difficulty that may show up in this slow-roll approach concerns the existence of features in the inflaton potential. Although current data support the almost scale invariance of the primordial power spectra, the presence of sharp localised deviations remains a possibility and might even be

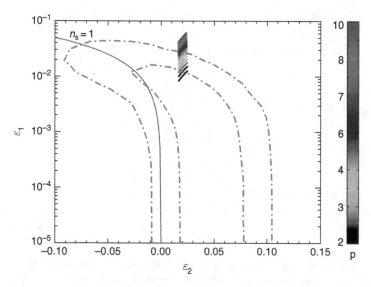

Fig. 7.1. WMAP third year data constraints on the first order Hubble flow parameters $\epsilon_1(N_*)$ and $\epsilon_2(N_*)$ [8]. The two *dashed contours* represent the 68 and 95% confidence intervals associated with the two-dimensional marginalised posterior probability distribution. The *solid curve* corresponds to a scale-invariant power spectrum whereas the short segments are the slow-roll predictions for the large field models $V(\varphi) \propto \varphi^p$. Note that the model predictions are not "dots" in the plane (ϵ_1, ϵ_2) due to their dependence with respect to N_*. Indeed, due to uncertainties on the reheating era, the efold N_* at which the observable pivot scale leaves the Hubble radius during inflation is not known (see Fig. 7.2). However, under reasonable assumptions, one may assume $40 \lesssim N_* \lesssim 60$ which leads to a "segment" in the plane (ϵ_1, ϵ_2)

favoured [8, 20, 21, 22, 23, 24, 25, 26]. In the framework of field inflation, features in the power spectra generically result from transient non-slow-rolling evolution associated with sharp features in the inflation potential. In these cases, deriving analytical approximations for the perturbations requires the use of more involved methods [27, 28, 29, 30]. Let us also stress that one of the key ingredients that renders analytical methods attractive is the conservation of the comoving curvature perturbation ζ on super-Hubble scales. As illustrated in Fig. 7.2, this allows one to identify the scalar and tensor power spectra deep in the radiation era, which seed the CMB anisotropies and structure formation, to those derived a few efolds after Hubble exit during inflation. However, if inflation is driven by more than one field, the existence of isocurvature modes that may source ζ after Hubble exit requires that the modes evolution should be traced till the end of reheating [31]. Although analytical methods can still be used, their use is restricted by their domain of validity [32, 33, 34].

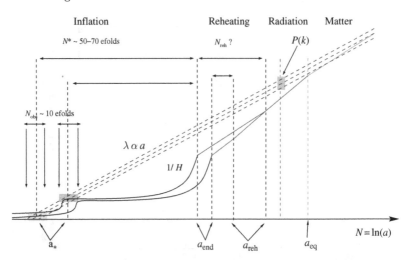

Inflation   Reheating   Radiation   Matter

Fig. 7.2. Sketch of length scale evolution in inflationary cosmology. The horizontal axis gives the number of efolds while the vertical axis represents a logarithmic measure of lengths. The cosmological stretching of the observable wavelengths is represented by the three *blue dashed lines*. The evolution of the Hubble radius is represented by a *solid line* from inflation to the matter era. In between, the reheating era connects the end of inflation to the radiation era. Although the redshift of equality is known, the uncertainties existing on the reheating lead to uncertainties on the redshift at which the observable wavelength today have left the Hubble radius during inflation. As a result, even for a given model of inflation, the resulting power spectra can be significantly different if the number of efolds during reheating is changed. This is illustrated on this plot by the observability of an inflaton potential feature that may accordingly be observable or not

The previous considerations suggest to use numerical methods to directly compute the inflationary perturbations and deduce the primordial power spectra. Numerical integrations in inflation are not new and have been used to test the validity of analytical approximations, or to derive the shape of the power spectra for some particular models [35, 36, 37, 38, 39]. However, with the advent of MCMC methods in cosmology, one may be interested in merging a full numerical integration of the perturbation during inflation with the CMB codes such as CAMB and COSMOMC [40, 41]. The advantage of such an approach is that the inflationary parameters become part of the cosmological model under scrutiny. When compared with the data, one may expect to get consistent marginalised constraints on both the parameters entering the inflaton potential and the usual cosmological parameters describing the radiation and matter content of the observed universe. From a Bayesian point of view, this is the method that should be used if one is interested in assessing the likelihood of one model to explain the data (statistical evidence). Moreover, since the underlying approximation is just linear perturbation theory, one may consider without additional complications the treatment of non-standard models as those involving several fields. Let us mention that our objective is to use numerical methods in a well-defined theoretical framework. For a given model, we use the CMB data to constrain the theoretical parameters and discuss the overall model ability to explain the data. Numerical (non-exact) methods have also been used in the context of the potential reconstruction problems where the goal is to constrain the shape of the inflaton potential along the observable window [42, 43, 44] (see Fig. 7.2).

However, as discussed before, to solve the cosmological perturbations from their creation as quantum fluctuations during inflation to now, it is necessary to model the reheating era. In fact, the importance of reheating for inflation is very similar to the importance of the reionisation for the CMB anisotropies. Although reionisation of the universe is a complex process, its basic effects on the CMB anisotropies can be modelled through the optical depth τ. Similarly, we will show that the basic effects induced by the reheating on the inflationary perturbations may be taken into account through a new parameter $\ln R$ which had not been considered so far.

The plan is as follows. In a first section, the theoretical setup is introduced. We use the sigma-model formalism which allows an easy implementation of any scalar field inflation models in General Relativity and multi-scalar tensor theories. The equations of motion for the background fields and their perturbations are presented in the first section. Their numerical integration is the subject of the second part. After having introduced our model of the reheating era, the last section illustrates the usefulness of the exact numerical method by an analysis of the third year WMAP data in the context of the small field models.

7.2 Multifield Inflation

It is out of the scope of this work to deal with all the inflationary models proposed so far. However, four-dimensional effective actions associated with many inflation models, and especially those being embedded in extra-dimensions, share the common feature that they involve several scalar fields that may be non-minimally coupled to gravity. This is for instance the case for the moduli associated with the position of the branes in various string-motivated inflation models (see Chap. 4 in this volume and [45]). It is therefore convenient to consider an action that may generically drive the dynamics of both minimally and non-minimally coupled scalar fields, as in the sigma-model [46, 47, 48, 49].

7.2.1 Sigma-Model Formalism

Denoting by $\mathcal{F}^a(x^\mu)$ the n_σ dimensionless scalar fields living on a sigma-model manifold with metric $\ell_{ab}(\mathcal{F}^c)$, we consider the action

$$S = \frac{1}{2\kappa^2} \int \left[R - \ell_{ab} g^{\mu\nu} \partial_\mu \mathcal{F}^a \partial_\nu \mathcal{F}^b - 2V(\mathcal{F}^c) \right] \sqrt{-g}\, \mathrm{d}^4 x \,, \tag{7.4}$$

where $g_{\mu\nu}$ is the usual four-dimensional metric tensor of determinant g, R the Ricci scalar and V the field potential.

For instance, if $\varphi = \mathcal{F}^{(1)}/\kappa$ and $\ell_{11} = 1$, this action describes a unique minimally coupled scalar field. In this case, the associated potential is $U(\varphi) = V(\varphi)/\kappa^2$ and we recover the standard form

$$S = \frac{1}{2\kappa^2} \int R\sqrt{-g}\, \mathrm{d}^4 x + \int \left[-\frac{1}{2} g^{\mu\nu} \partial_\mu \varphi \partial_\nu \varphi - U(\varphi) \right] \sqrt{-g}\, \mathrm{d}^4 x \,. \tag{7.5}$$

Another example is provided by the models of brane inflation where the inflaton field φ lives on the brane and a bulk field χ in the four-dimensional effective action couples to gravity in a non-minimal way [50, 51, 45]. With $n_\sigma = 2$, $\varphi = \mathcal{F}^{(1)}/\kappa$ and $\chi = \mathcal{F}^{(2)}$, (7.4) can be recast into

$$\begin{aligned} S &= \frac{1}{2\kappa^2} \int \left[R - g^{\mu\nu} \partial_\mu \chi \partial_\nu \chi - 2W(\chi) \right] \sqrt{-g}\, \mathrm{d}^4 x \\ &+ \int \left[-\frac{1}{2} A^2(\chi) \partial_\mu \varphi \partial_\nu \varphi - A^4(\chi) U(\varphi) \right] \sqrt{-g}\, \mathrm{d}^4 x \,, \end{aligned} \tag{7.6}$$

for

$$\ell_{ab} = \mathrm{diag}(A^2, 1), \qquad V(\varphi, \chi) = W(\chi) + \kappa^2 A^4(\chi) U(\varphi) \,. \tag{7.7}$$

This action describes the dynamics, in the Einstein frame, of the field φ evolving in a potential U in a scalar-tensor theory of gravity where χ is the scalar partner to the graviton [52]. The conformal function $A^2(\chi)$ and the self-interaction potential $W(\chi)$ depend on the brane setup considered [53, 54, 55, 56]. In the general case, (7.4) can be used to describe multifield inflation in a multi-scalar tensor theory of gravity.

Differentiating the action (7.4) with respect to the metric leads to the Einstein equations

$$G_{\mu\nu} = \mathcal{S}_{\mu\nu} \,, \tag{7.8}$$

with the source terms

$$\mathcal{S}_{\mu\nu} = \ell_{ab} \mathcal{S}_{\mu\nu}^{ab} - g_{\mu\nu} V \,, \tag{7.9}$$

where

$$\mathcal{S}_{\mu\nu}^{ab} = \partial_\mu \mathcal{F}^a \partial_\nu \mathcal{F}^b - \frac{1}{2} g_{\mu\nu} \partial_\rho \mathcal{F}^a \partial^\rho \mathcal{F}^b \,. \tag{7.10}$$

Similarly, the fields obey the Klein–Gordon-like equation

$$\Box \mathcal{F}^c + g^{\mu\nu} \Upsilon_{ab}^c \partial_\mu \mathcal{F}^a \partial_\nu \mathcal{F}^b = V^c \,, \tag{7.11}$$

where Υ denotes the Christoffel symbol on the field-manifold

$$\Upsilon_{ab}^c = \frac{1}{2} \ell^{cd} \left(\ell_{da,b} + \ell_{db,a} - \ell_{ab,d} \right) \,, \tag{7.12}$$

and V^c should be understood as the vector-like partial derivative of the potential

$$V^c = \ell^{cd} V_d = \ell^{cd} \frac{\partial V}{\partial \mathcal{F}^d} . \tag{7.13}$$

7.2.2 Background Evolution

In a flat (FLRW) universe with metric

$$ds^2 = g_{\mu\nu} dx^\mu dx^\nu = a^2(\eta) \left(-d\eta^2 + \delta_{ij} dx^i dx^j \right) , \tag{7.14}$$

η being the conformal time and i and j referring to the spatial coordinates, the equations of motion (7.8) and (7.11) simplify to

$$3\mathcal{H}^2 = \frac{1}{2} \ell_{ab} \mathcal{F}^{a\prime} \mathcal{F}^{b\prime} + a^2 V, \tag{7.15}$$

$$2\mathcal{H}' + \mathcal{H}^2 = -\frac{1}{2} \ell_{ab} \mathcal{F}^{a\prime} \mathcal{F}^{b\prime} + a^2 V, \tag{7.16}$$

$$\mathcal{F}^{c\prime\prime} + \Upsilon^c_{ab} \mathcal{F}^{a\prime} \mathcal{F}^{b\prime} + 2\mathcal{H} \mathcal{F}^{c\prime} = -a^2 V^c , \tag{7.17}$$

where a prime denotes differentiation with respect to the conformal time and $\mathcal{H} = aH$ is the conformal Hubble parameter. In terms of the efold time variable n, the field equations can be decoupled from the metric evolution and one gets

$$H^2 = \frac{V}{3 - \frac{1}{2} \dot{\sigma}^2}, \tag{7.18}$$

$$\frac{\dot{H}}{H} = -\frac{1}{2} \dot{\sigma}^2, \tag{7.19}$$

$$\frac{\ddot{\mathcal{F}}^c + \Upsilon^c_{ab} \dot{\mathcal{F}}^a \dot{\mathcal{F}}^b}{3 - \frac{1}{2} \dot{\sigma}^2} + \dot{\mathcal{F}}^c = -\frac{V^c}{V}, \tag{7.20}$$

a dot denoting differentiation with respect to n. We have introduced a velocity field

$$\dot{\sigma} = \sqrt{\ell_{ab} \dot{\mathcal{F}}^a \dot{\mathcal{F}}^b} . \tag{7.21}$$

In fact, σ is the so-called adiabatic field introduced in [31] which describes the collective evolution of all the fields along the classical trajectory. From (7.19) and (7.20) one may determine its equation of motion

$$\sigma'' + 2\mathcal{H}\sigma' + a^2 V_\sigma = 0 , \tag{7.22}$$

with $V_\sigma \equiv u^c V_c$ and where the u^a are unit vectors along the field trajectory:

$$u^a \equiv \frac{\mathcal{F}^{a\prime}}{\sigma'} = \frac{\dot{\mathcal{F}}^a}{\dot{\sigma}} . \tag{7.23}$$

In the Einstein frame, inflation occurs for $d^2a/dt^2 > 0$, or in terms of the first Hubble flow function, for[1]

$$\epsilon_1 \equiv -\frac{\dot{H}}{H} = \frac{1}{2}\dot{\sigma}^2 < 1 . \tag{7.24}$$

The multifield system induces an accelerated expansion of the universe if the resulting adiabatic field velocity $\dot{\sigma}$ remains less than $\sqrt{2}$. According to (7.20), the term in $1/(3-\dot{\sigma}^2/2)$ acts as a relativistic-like inertia for the fields evolution and thus $\dot{\sigma} < \sqrt{6}$ (for a positive potential). In this equation, the first term on the left hand side may be interpreted as a covariant acceleration on the curved field manifold, the second as a constant friction force and the right hand side as a driving force deriving from the potential $\ln V$. In fact, we recover the well-known attractor behaviour of the inflationary evolution: whatever the initial fields velocity, the friction term ensures that the terminal velocity of the fields will be, after a transient regime,

$$\dot{\mathcal{F}}^a \simeq -\frac{d\ln V}{d\mathcal{F}^a} . \tag{7.25}$$

Analytical integration of the previous expression lies at the roots of the slow-roll approximation when the effective potential $\ln V$ is flat enough. In the general case, the driving force always pushes the fields towards the minimum of $\ln V$. Let us note that this is why the monomial potentials $V \propto \varphi^p$ are actually "flat" for the large field values: $d\ln V/d\varphi = p/\varphi$ which goes to zero for $\varphi \to \infty$.

7.2.3 Linear Perturbations

Scalar Modes

In the longitudinal gauge, the scalar perturbations (with respect to the rotations of the three-dimensional space) of the FLRW metric can be expressed as

$$ds^2 = a^2 \left[-(1+2\Phi)\, d\eta^2 + (1-2\Psi)\,\gamma_{ij}dx^idx^j \right] , \tag{7.26}$$

where Φ and Ψ are the Bardeen potentials. With $\delta\mathcal{F}^a$ the field perturbations, the Einstein equations perturbed at first order read

$$3\mathcal{H}\Psi' + \left(\mathcal{H}' + 2\mathcal{H}^2\right)\Psi - \Delta\Psi = -\frac{1}{2}\ell_{ab}\mathcal{F}^{a\prime}\delta\mathcal{F}^{b\prime}$$

$$-\frac{1}{2}\left(\frac{1}{2}\ell_{ab,c}\mathcal{F}^{a\prime}\mathcal{F}^{b\prime} + a^2V_c\right)\delta\mathcal{F}^c, \tag{7.27}$$

---

[1] Notice that this does not imply that the universe is accelerating in the string frame and one has to verify that there are enough efolds of inflation to solve the homogeneity and flatness issues in that frame [49, 57, 58].

$$\Psi' + \mathcal{H}\Psi = \frac{1}{2}\ell_{ab}\mathcal{F}^{a\prime}\delta\mathcal{F}^b, \tag{7.28}$$

$$\Psi'' + 3\mathcal{H}\Psi' + \left(\mathcal{H}' + 2\mathcal{H}^2\right)\Psi = \frac{1}{2}\ell_{ab}\mathcal{F}^{a\prime}\delta\mathcal{F}^{b\prime}$$
$$+ \frac{1}{2}\left(\frac{1}{2}\ell_{ab,c}\mathcal{F}^{a\prime}\mathcal{F}^{b\prime} - a^2 V_c\right)\delta\mathcal{F}^c, \tag{7.29}$$

where use has been made of $\Phi = \Psi$ from the perturbed Einstein equation with $i \neq j$. For flat spacelike hypersurfaces,

$$\Delta \equiv \delta^{ij}\partial_i\partial_j. \tag{7.30}$$

Similarly, the perturbed Klein–Gordon equations read

$$\delta\mathcal{F}^{c\prime\prime} + 2\Upsilon^c_{ab}\mathcal{F}^{a\prime}\delta\mathcal{F}^{b\prime} + 2\mathcal{H}\delta\mathcal{F}^{c\prime}$$
$$+ \left(\Upsilon^c_{ab,d}\mathcal{F}^{a\prime}\mathcal{F}^{b\prime} + a^2 V^c_d - \ell^{ca}\ell_{ab,d}\,a^2 V^b\right)\delta\mathcal{F}^d$$
$$- \Delta\delta\mathcal{F}^c = 4\Psi'\mathcal{F}^{c\prime} - 2\Psi a^2 V^c. \tag{7.31}$$

As discussed in the introduction, if more than one scalar field is involved, the entropy perturbation modes can source the adiabatic mode even after Hubble exit. The equation that governs the evolution of the comoving curvature perturbation ζ can be obtained from (7.27) to (7.29), using the background equations. Firstly, the Bardeen potential verifies

$$\Psi'' + 6\mathcal{H}\Psi' + \left(2\mathcal{H}' + 4\mathcal{H}^2\right)\Psi - \Delta\Psi = -a^2 V_c\delta\mathcal{F}^c. \tag{7.32}$$

Using the geometrical definition for the comoving curvature perturbation [10]

$$\zeta \equiv \Psi - \frac{\mathcal{H}}{\mathcal{H}' - \mathcal{H}^2}\left(\Psi' + \mathcal{H}\Phi\right), \tag{7.33}$$

Equation (7.28) yields

$$\zeta = \Psi + \mathcal{H}\frac{\delta\sigma}{\sigma'}, \tag{7.34}$$

where the adiabatic perturbation $\delta\sigma$ is also the resulting perturbation of all fields projected onto the classical trajectory [see (7.21) and (7.23)]:

$$\delta\sigma = \frac{\ell_{ab}\mathcal{F}^{a\prime}\delta\mathcal{F}^b}{\sigma'} = u_a\delta\mathcal{F}^a. \tag{7.35}$$

The dynamical equation (7.32) now exhibits couplings between the adiabatic and entropy modes

$$\zeta' = \frac{2\mathcal{H}}{\sigma'^2}\Delta\Psi - \frac{2\mathcal{H}}{\sigma'^2}\left(a^2 V_a\delta\mathcal{F}^a - a^2\frac{V_c\mathcal{F}^{c\prime}}{\sigma'}\frac{\ell_{ab}\mathcal{F}^{a\prime}\delta\mathcal{F}^b}{\sigma'}\right), \tag{7.36}$$

which can be recast into

$$\zeta' = \frac{2\mathcal{H}}{\sigma'^2}\Delta\Psi - \frac{2\mathcal{H}}{\sigma'^2} \perp_d^c a^2 V_c \delta\mathcal{F}^d \ . \tag{7.37}$$

The orthogonal projector is defined by

$$\perp_{ab} = \ell_{ab} - \eta_{ab} \ , \tag{7.38}$$

where $\eta_{ab} \equiv u_a u_b$ is the first fundamental form of the one-dimensional manifold defined by the classical trajectory [59]. Clearly, the comoving curvature perturbation on super-Hubble scales for which $\Delta\Psi \simeq 0$ is only sourced by the entropy perturbations defined as the projections of all field perturbations on the field-manifold subspace orthogonal to the classical trajectory. If, on the contrary, there is a single field involved during inflation, then these terms vanish and we recover that ζ remains constant after Hubble exit.

Tensor Modes

In the Einstein frame, the scalar and tensor degrees of freedom are decoupled. Therefore, the equation of evolution for the tensor modes remains the same as in General Relativity. For a flat perturbed FLRW metric

$$ds^2 = -a^2 d\eta^2 + a^2 \left(\delta_{ij} + h_{ij}\right) dx^i dx^j \ , \tag{7.39}$$

where h_{ij} is a traceless and divergenceless tensor

$$\delta^{ij} h_{ij} = \delta^{ik}\partial_k h_{ij} = 0 \ , \tag{7.40}$$

one gets [5, 60]

$$h_{ij}'' + 2\mathcal{H}h_{ij}' - \Delta h_{ij} = 0 \ . \tag{7.41}$$

Primordial Power Spectra

The initial conditions for the cosmological perturbations require the knowledge of the two-point correlation functions for all of the observable scalar and tensor modes deep in the radiation era. In Fourier space, these are just the power spectra associated with the values taken by the adiabatic and entropy perturbations at the end of the reheating, i.e.

$$\mathcal{P}_{ab} = \frac{k^3}{2\pi^2} \left[\nu^a(k)\right]^* \left[\nu^b(k)\right] \ , \tag{7.42}$$

where ν^a stands for ζ or the entropy modes. Similarly, taking into account the polarisation degrees of freedom, the tensor power spectrum reads

$$\mathcal{P}_h(k) = \frac{2k^3}{\pi^2} |h(k)|^2 \ . \tag{7.43}$$

In the following, we summarise the numerical method used to solve the full set of Einstein and Klein–Gordon equations derived in this section. The power spectra can then be deduced from (7.42) and (7.43) by pushing the integration till the end of the reheating.

7.3 Numerical Method

7.3.1 Integrating the Background

As suggested by the form of (7.18)–(7.20), it is convenient to use the number of efolds n as the integration variable. In fact, the background evolution only requires the integration of the fields equation of motion (7.20). Cauchy's theorem guarantees that there is a unique solution provided all the $\mathcal{F}^a(0)$ and $\dot{\mathcal{F}}^a(0)$ are given at $n = 0$. Plugging the solutions for $\mathcal{F}^a(n)$ into (7.18) uniquely determines the Hubble parameter and thus the geometry during inflation.

Initial Conditions

However, as previously mentioned, the attractor behaviour induced by the friction term erases any effect associated with the initial field velocities after a few efolds. This is the very reason why initial conditions in inflation are essentially related to the initial field values only. On the numerical side, the attractor ensures the stability of almost all forward numerical integration schemes.[2] In the following, we have used a Runge–Kutta integration method of order five and the initial field velocities have been chosen on the attractor by setting

$$\dot{\mathcal{F}}^a(0) = - \left. \frac{\mathrm{d}\ln V}{\mathrm{d}\mathcal{F}^a} \right|_{\mathcal{F}^a(0)} . \tag{7.44}$$

The robustness of the attractor during inflation may be quantified by comparing the numerical solutions obtained from various arbitrary choices of the initial field velocities, at fixed value of $\mathcal{F}^a(0)$. As an illustration, we have plotted in Fig. 7.3 the efold evolution of $\dot{\mathcal{F}}^a(n)$ in a brane inflation model involving three scalar fields, two of them being non-minimally coupled to gravity and representing the positions of two branes in a five-dimensional bulk (see [49]). As can be seen on this plot, all the fields are on the attractor after a few efolds.

End of inflation

From the above initial conditions the fields evolve toward the minimum of the potential $\ln V$ while the expansion of the universe accelerates as long as $\epsilon_1 < 1$. It would therefore be natural to define the end of inflation by the efold n_{end} at which $\epsilon_1(n_{\mathrm{end}}) = 1$. However, this is usually not the end of the fields evolution since they have not yet reached the minimum of the potential. On the contrary, $\epsilon_1(n_{\mathrm{end}}) = 1$ just signals that the kinetic terms in (7.4) start to dominate over the potential. Since the expansion factor decelerates for $\epsilon_1(n_{\mathrm{end}}) > 1$, this late stage evolution takes place during a few efolds and the fields rapidly reach the minimum of the potential. In the standard

---

[2] as well as the instability of backward integrations.

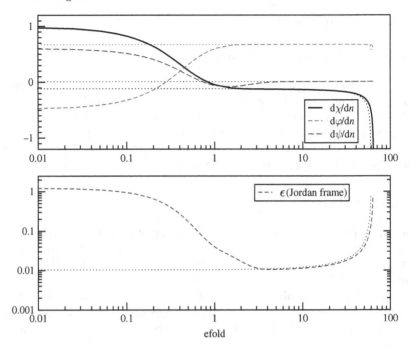

Fig. 7.3. Evolution of the field velocities $\dot{\mathcal{F}}^a = \{\dot{\chi}, \dot{\phi}, \dot{\psi}\}$ in the boundary inflation model of [49] which involves three fields, two of them being non-minimally coupled to gravity. The *dotted curves* are the solutions obtained by setting the initial field velocities on the attractor according to (7.44) whereas the solid and *dashed curves* are the solutions obtained from a random choice of the initial field velocities (ensuring however $H^2 > 0$). The first Hubble flow parameter in the string frame is plotted in the lower panel. Since this is also the adiabatic field velocity squared, the acceleration properties of the universe after a few efolds do not depend any longer on the initial field velocities [see (7.24)]

picture, the fields oscillate around the minimum of the potential and decay through parametric resonances into the relativistic fluids present during the radiation era [13, 61] (see Fig. 7.4) The details of the reheating process are very model dependent and require the knowledge of all the couplings between the inflaton and the standard model particles [2]. This implies that the number of efolds at which the universe reheated also depends on the model at hand. Further complications arise in multifield inflationary models in which the end of inflation and the reheating may be triggered by tachyonic instabilities. In these cases the condition $\epsilon_1(n_{\mathrm{end}}) = 1$ is not longer relevant and one should rather introduce a limiting field value $\mathcal{F}_{\mathrm{end}}$ to classically define the beginning of the reheating era (and the end of inflation).

Following the previous discussion, our phenomenological approach to the end of inflation is to assume an instantaneous transition to the reheating era. The efold n_{end} at which the transition occurs is either determined by the

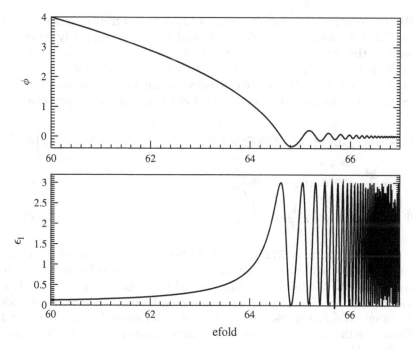

Fig. 7.4. End of inflation in the large field model $V \propto \varphi^2$. The field oscillations around the minimum of the potential trigger its decay and the reheating era

condition $\epsilon_1(n_{\text{end}}) = 1$ or when the relevant field crosses a limiting value $\mathcal{F}_{\text{end}}$; the choice being made according to the inflation model we are interested in. The value of $\mathcal{F}_{\text{end}}$ may be given by the underlying microphysics or considered as an additional parameter of the inflation model. For instance, as can be seen in Fig. 7.4, it is convenient to define the end of inflation for the large field models by $\epsilon_1 = 1$: the field evolution afterwards, i.e. its oscillations around the minimum of the potential and subsequent decay, is supposed to be part of the reheating stage.

Knowing the fields value $\mathcal{F}^a(n)$ from $n = 0$ to $n = n_{\text{end}}$, the background geometry is given by (7.18) and (7.19) and we can now numerically integrate the linear perturbations on the same efolding range. As discussed in the introduction, the link with the cosmological perturbations observed today still requires a reheating model that will be introduced in Sect. 7.4.1.

7.3.2 Integrating the Perturbations

As for the background, we have chosen to integrate the linear perturbations in efold time. Focusing on the scalar perturbations, their dynamics is driven by the Einstein and Klein–Gordon equations given in Sect. 7.2.3. These are however redundant due to the stress energy conservation already included in the Bianchi identities. As a result, it is necessary to integrate only a subset of (7.27)–(7.31). Although the Bardeen potential Ψ could be explicitly expressed in terms of the field perturbations $\delta\mathcal{F}^a$ only, such an expression is singular in the limit $k \to 0$ and $\epsilon_1 \to 0$, which is not appropriate for a numerical integration (see below). It is more convenient to simultaneously integrate the second order equations (7.31) and (7.32). Recast in efold time, they read

$$\delta\ddot{\mathcal{F}}^c + (3 - \epsilon_1)\delta\dot{\mathcal{F}}^c + 2\Upsilon^c_{ab}\dot{\mathcal{F}}^a\delta\dot{\mathcal{F}}^b + \left(\Upsilon^c_{ab,d}\dot{\mathcal{F}}^a\dot{\mathcal{F}}^b + \frac{V^c_d}{H^2} - \ell^{ca}\ell_{ab,d}\frac{V^b}{H^2}\right)\delta\mathcal{F}^d$$

$$+ \frac{k^2}{a^2H^2}\delta\mathcal{F}^c = 4\dot{\Psi}\dot{\mathcal{F}}^c - 2\Psi\frac{V^c}{H^2}, \tag{7.45}$$

$$\ddot{\Psi} + (7 - \epsilon_1)\dot{\Psi} + \left(2\frac{V}{H^2} + \frac{k^2}{a^2H^2}\right)\Psi = -\frac{V_c}{H^2}\delta\mathcal{F}^c. \tag{7.46}$$

The constraint equations (7.27) and (7.28) being first integrals of the above equations, there is still an integration constant that should be set to restore the equivalence to the full set of Einstein and Klein–Gordon equations. This one can be fixed by choosing the appropriate initial conditions at $n = n_{ic}$ for the Bardeen potential Ψ. Setting all the $\delta\mathcal{F}^a(n_{ic})$ and $\delta\dot{\mathcal{F}}^a(n_{ic})$, the initial conditions for the Bardeen potential are indeed uniquely given by (7.27) and (7.28). In efold time, one gets

$$\Psi = \frac{1}{2\left(\epsilon_1 - \dfrac{k^2}{a^2H^2}\right)}\left[\ell_{ab}\dot{\mathcal{F}}^a\delta\dot{\mathcal{F}}^b + \left(\frac{1}{2}\ell_{ab,c}\dot{\mathcal{F}}^a\dot{\mathcal{F}}^b + 3\ell_{ac}\dot{\mathcal{F}}^c + \frac{V_c}{H^2}\right)\delta\mathcal{F}^c\right],$$

$$\dot{\Psi} = \frac{1}{2}\ell_{ab}\dot{\mathcal{F}}^a\delta\mathcal{F}^b - \Psi,$$

$$\tag{7.47}$$

these expressions being evaluated at the initial efold time. As a result, the linear perturbations of both the fields and metric are uniquely determined by the initial conditions $\delta\mathcal{F}^a(n_{ic})$ and $\delta\dot{\mathcal{F}}^a(n_{ic})$. As discussed in the next section, the initial conditions are set on sub-Hubble scales for which $k \gg aH$ ensuring the regularity of (7.47).

Quantum Initial Conditions

In the context of single-field inflation, the initial conditions for the linear perturbations are given by the quantum fluctuations of the field-metric system on sub-Hubble scales $k \to \infty$. In this limit, the perturbations decouple from the expansion of the universe and a field quantisation can be performed along the lines described in [5, 11]. The canonically normalised quantum degrees of freedom are encoded in the Mukhanov–Sasaki variable

$$Q = \delta\sigma + \frac{\sigma'}{\mathcal{H}}\Psi = \delta\sigma + \sqrt{2\epsilon_1}\Psi, \tag{7.48}$$

with $\sigma = \mathcal{F}^{(1)} = \kappa\varphi$ for single field models. In terms of Q, the equation of motion (7.31) can be recast into

$$(aQ)'' + \left[k^2 - \frac{(a\sqrt{\epsilon_1})''}{a\sqrt{\epsilon_1}}\right] aQ = 0 , \qquad (7.49)$$

showing that in the small scale limit $k \to \infty$, the quantity aQ follows the dynamics of a free scalar field. Assuming a Bunch–Davies vacuum, aQ has a positive energy plane wave behaviour on small scales, and in Fourier space one gets

$$\lim_{k\to+\infty} aQ(\eta) = \kappa\frac{e^{-ik\eta}}{\sqrt{2k}} . \qquad (7.50)$$

This solution uniquely determines the subsequent evolution of the perturbations during inflation and will be our starting point for the numerical integration. According to (7.47), provided the initial conditions are set in the limit $k/\mathcal{H} \to \infty$, (7.50) also represents the small-scale behaviour of the rescaled adiabatic field perturbations $a\delta\sigma$.

For a multifield system the previous results can be generalised and the quantum modes identified with the adiabatic perturbations $\delta\sigma$ together with the canonically normalised entropy modes introduced in Sect. 7.2.3. In fact, if the original fields $\mathcal{F}^a$ are already canonically normalised, i.e. $\ell_{ab} = \delta_{ab}$, and independent dynamical variables in the small-scale limit, the adiabatic and entropy perturbations can be obtained from the original field perturbations by local rotations on the n_σ-dimensional field manifold [31, 33, 49]. In this case, denoting by $\delta\varsigma^a$ the adiabatic and entropy perturbations, with the convention $\delta\varsigma^{(1)} = \delta\sigma$, one has

$$\delta\varsigma^a = \mathcal{M}^a_b(\mathcal{F}^c)\delta\mathcal{F}^b , \qquad (7.51)$$

where $\mathcal{M}$ is an instantaneous rotation matrix, $\mathcal{M}^\dagger = \mathcal{M}^T = \mathcal{M}^{-1}$, depending on the background quantities only. Under these assumptions, the quantum modes are independent in the small scale limit and their two point correlators reduce to

$$\langle\delta\varsigma^{a*}(\boldsymbol{k})\delta\varsigma^b(\boldsymbol{k'})\rangle \underset{k\gg\mathcal{H}}{=} \delta^{ab}\mathcal{P}_\varsigma(k)\delta(\boldsymbol{k}-\boldsymbol{k'}) , \qquad (7.52)$$

$\mathcal{P}_\varsigma$ being the free field power spectrum given by the square modulus of (7.50). Consequently, all the correlators between the original field perturbations inherit these initial conditions:

$$\langle\delta\mathcal{F}^{a*}(\boldsymbol{k})\delta\mathcal{F}^b(\boldsymbol{k'})\rangle = \left(\mathcal{M}^{-1}\right)^{a*}_c \left(\mathcal{M}^{-1}\right)^b_d \langle\delta\varsigma^{c*}\delta\varsigma^d\rangle \underset{k\gg\mathcal{H}}{=} \delta^{ab}\mathcal{P}_\varsigma(k)\,\delta(\boldsymbol{k}-\boldsymbol{k'}) .$$
$$(7.53)$$

The previous results can be generalised for the sigma-models with a diagonal metric ℓ_{ab} different from the identity by the transformation $\delta\mathcal{F}^a \to \sqrt{\ell_{aa}}\delta\mathcal{F}^a$ (no summation). In the general case, the transformation matrix $\mathcal{M}$ would mix all the fields and the cross correlators. However, since it is always possible to diagonalise ℓ_{ab} through a field redefinition, we will now assume without loss of generality that ℓ_{ab} is diagonal.

Defining the normalised quantum modes by

$$\mu_{\rm s}^a = a\sqrt{2}\ell_{aa}^{1/2}k^{3/2}\delta\mathcal{F}^a \, , \tag{7.54}$$

with ℓ_{aa} evaluated along the background solution, in the small-scale limit (7.50) can be recast into the initial conditions

$$\mu_{\rm s}^a \underset{k\gg\mathcal{H}}{=} \kappa k, \qquad \frac{\mu_{\rm s}^{a\prime}}{k} \underset{k\gg\mathcal{H}}{=} -i\kappa k \, , \tag{7.55}$$

up to a phase factor. In terms of the field perturbations, the initial conditions in efold time therefore read

$$k^{3/2}\delta\mathcal{F}^a\Big|_{\rm ic} = \frac{\kappa k}{a_0}\frac{a_0}{a_{\rm end}}\frac{e^{n_{\rm end}-n_{\rm ic}}}{\sqrt{\ell_{aa}(n_{\rm ic})}} \, ,$$

$$k^{3/2}\delta\dot{\mathcal{F}}^a\Big|_{\rm ic} = -\frac{\kappa k}{a_0}\frac{a_0}{a_{\rm end}}\frac{e^{n_{\rm end}-n_{\rm ic}}}{\sqrt{\ell_{aa}(n_{\rm ic})}}\left[1+\frac{1}{2}\frac{\dot{\ell}_{aa}(n_{\rm ic})}{\ell_{aa}(n_{\rm ic})}+i\frac{k}{a_{\rm ic}H_{\rm ic}}\right] \, . \tag{7.56}$$

The efold $n_{\rm ic}$ at which these initial conditions should be set has not been specified yet. In fact, the limit $k/\mathcal{H}\to\infty$ would correspond to the infinite past and does not make sense for non-eternal field inflation models.[3] However, by definition of $n=\ln(a/a_{\rm ini})$, the condition $k/a\gg H$ is already satisfied a few efolds before Hubble exit. For all inflation models lasting more than N_* efolds, it would be natural to set the initial conditions for the perturbations at the beginning of inflation $n_{\rm ic}=0$ (see Fig. 7.2).

Choosing $n_{\rm ic}=0$ is however not appropriate for a numerical integration. Indeed, according to the initial values of the background fields, the total number of efolds $n_{\rm end}$ can be much greater than N_*. In such cases, most of the computing time for the perturbations would be spent into the deep sub-Hubble regime for which the modes behave as free plane waves. It is rather more convenient to set the initial conditions "closer" to the time n_k at which a given mode crosses the Hubble radius $k=\mathcal{H}(n_k)$. Following [35], a simple choice is to define $n_{\rm ic}$ for each mode according to

$$\frac{k}{\mathcal{H}(n_{\rm ic})} = C_{\rm q} \, , \tag{7.57}$$

$C_{\rm q}$ being a constant verifying $C_{\rm q}\gg 1$ and characterizing the decoupling limit. Strictly speaking, this choice introduces small trans-Planckian-like

---

[3] Eternal inflation may occur when the quantum fluctuations on Hubble length scales become dominant over the classical field evolution and the semi-classical approach used here would no longer be valid, at least in the self-reproducing regime (see Chap. 5 in this volume).

interferences between the modes[4] which remain however negligible provided C_q is big enough [62].

Mode Integration

For each perturbation mode of wavenumber k, (7.45) and (7.46) are numerically solved by setting the initial conditions (7.56) at the efold n_{ic}, solution of (7.57). In order to significantly speed-up the numerical integration, instead of using the already computed background solution it is more convenient to integrate both the background and the perturbations simultaneously.[5] This can be done along the following steps.

Firstly, (7.57) is solved to determine $n_{ic}(k_1)$ for the largest wavelength mode k_1 we are interested in. The background equations (7.20) are then integrated from $n = 0$ to $n = n_{ic}(k_1)$. At that efold, (7.20), (7.45) and (7.46) are simultaneously integrated till the end of inflation at $n = n_{end}$. This process is iterated for each of the $k_i > k_{i-1}$ mode wanted. However, to speed-up the integration, it is enough to re-integrate the background from $n_{ic}(k_{i-1})$ to $n_{ic}(k_i)$ rather than from $n = 0$ before switching on the perturbations (see Fig. 7.5). Such an integration gives the value of all the field and metric perturbations at the end of inflation $n = n_{end}$. In principle, the power spectra can then be deduced by using (7.42).

Primordial Power Spectra

For a multifield system, since the perturbation modes are supposed to be independent deep under the Hubble radius, they can be considered, from a classical point of view, as independent stochastic variables. As a result, the power spectra at the end of inflation are no longer given by (7.42) but should be computed as [38]

$$\mathcal{P}_{ab} = \frac{k^3}{2\pi^2} \sum_{m=1}^{n_\sigma} [\nu_m^a(k)]^* [\nu_m^b(k)] \ , \tag{7.58}$$

where, as before, ν^a stands for the observable perturbations one is interested in. The ν^a can be the field perturbations themselves but it is more customary for CMB analysis to use the comoving curvature perturbation ζ and the rescaled entropic perturbations $\delta\varsigma^a/\dot{\sigma}$ ($a > 1$). The index "m" in (7.58) refers

---

[4] In the free-field limit, a more accurate choice for n_{ic} is $k\eta(n_{ic}) = C_q$. This definition maintains the phase factor in (7.50) independent of k on the initial hypersurface.

[5] For direct numerical integrations, each step requires various forward and backward evaluations of the background functions. If these are not analytically known but precomputed, one has to use spline and interpolation methods which are heavily time consuming.

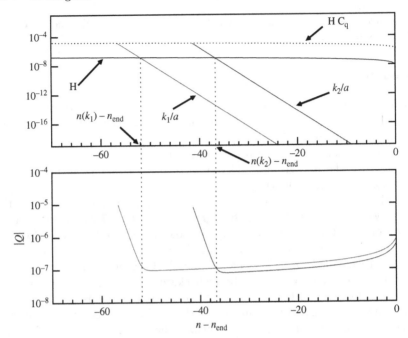

Fig. 7.5. Sketch of the numerical integration for the scalar perturbation in the $m^2\varphi^2$ single-field model. In the top frame, the initial conditions for each mode are set at the efold $n(k_i)$ solution of (7.57). The perturbations (and the background) are then integrated till the end of inflation at $n = n_{\mathrm{end}}$. The bottom frame represents the efold evolution of the Mukhavov–Sasaki variable Q for the two corresponding modes (7.48). Notice that only $\zeta = Q/\sqrt{2\epsilon_1}$ is conserved after Hubble exit

to the n_σ independent initial conditions obtained by setting only one perturbation mode μ_s^m in the Bunch–Davies vacuum at n_{ic}, the other $\mu_s^{q \neq m}$ vanishing. Notice that we have not explicitly written the entropy modes since various definitions are used in the literature. The definition of the entropy modes through the standard orthogonalisation procedure along the field trajectory can be found in [31, 33] and has the advantage to give canonically normalised perturbations. Another definitions introduce a reference field and define the entropy perturbations to be the relative perturbations of the other fields with respect to it. These differences come from the fact that one has to specify how the fields decay after inflation to know between which cosmological fluids entropy perturbations may exist. For instance, if all the cosmological fluids observed today are produced by the decay of one field only, then, although entropy perturbations exist during inflation, they are usually not observable afterwards [63, 64]. Their only effect would be to break the conservation of ζ on

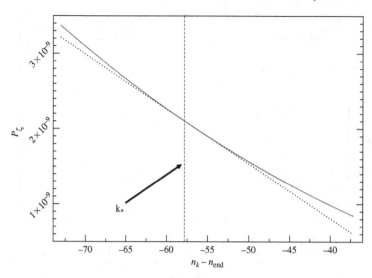

Fig. 7.6. Power spectra of the comoving curvature perturbation ζ at the end of inflation from the first order slow-roll approximation (*dotted line*) and an exact numerical integration ($V \propto \varphi^2$). The wavenumbers are expressed through n_k, the efold at which the mode k crosses the Hubble radius: $k = \mathcal{H}(n_k)$ (see Fig. 7.5)

super-Hubble scale thereby requiring the integration of all the perturbations till the end of inflation to determine the ζ power spectrum.

As an illustration, Fig. 7.6 shows the exact numerical power spectrum for the comoving curvature perturbation ζ for the single-field chaotic model $V \propto \varphi^2$. As can be seen on this plot, although the exact power spectrum differs from its first slow-roll approximated version given by (7.2), the differences remain small on a 10 efold observable range. Another example involving entropy perturbations is plotted in Fig. 7.7 for a two fields model of inflation. The presence of one entropy mode breaks the conservation of ζ on super-Hubble scale and the so-called consistency check of inflation $\mathcal{P}_h = 16\epsilon_1 \mathcal{P}_\zeta$ [see (7.2) and (7.3) at zero order].

Tensor Perturbations

Since in the Einstein frame the tensor and scalar degrees of freedom are decoupled, the numerical integration of the tensor modes does not present any difficulties. The equation of motion (7.41) can be recast into the equation of a parametric oscillator by defining the canonical mode function $\mu_{\mathrm{T}} = k^{3/2}ah$ satisfying the deep sub-Hubble initial conditions (7.50) for a Bunch–Davies vacuum [5]. The power spectrum $\mathcal{P}_h$ is readily obtained by evaluating (7.43) at the end of inflation.

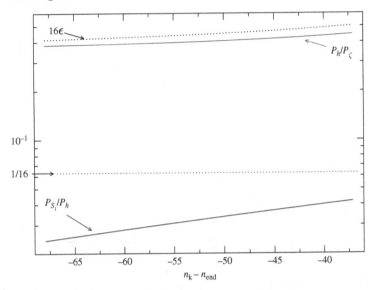

Fig. 7.7. Violation of the "consistency check" of inflation in a non-minimally coupled two-field model. The action is given by (7.6) with $W = 0$, $U \propto \varphi^2$ and $A^2 = \exp(-\alpha\chi)$. Even for a small value of $\alpha = 1/30$, the ratio between the scalar and tensor power spectra is no longer equal to $16\epsilon_1$. If the two fields were uncoupled on all scales, then the power spectrum of entropy perturbations would be the one of a free test scalar field $\mathcal{P}_{S_1} = \mathcal{P}_h/16$. As can be seen on the plot, this last condition is also violated

Physical Wavenumbers

Up to now, one may have noticed that all the power spectra have been plotted with respect to n_k and not with respect to the values of k. For astrophysical purposes, one needs to know the correspondence between the comoving k appearing in the above equations and the physical wavenumbers measured today k/a_0, whose typical unit is a Mpc^{-1}. As it appears in the initial conditions (7.56), rendering k/a_0 explicit requires the knowledge of a_0/a_{end}, i.e. the redshift z_{end} associated with the end of inflation. As discussed in Sect. 7.1, this can only be done if one knows the number of efolds during which the universe reheated. Let us also notice that the physics involved in the quantum generation of cosmological perturbations appears through these very numbers: $\kappa k/a_0$ are the wavenumbers measured today, usually of Mpc^{-1} size, expressed in unit of the Planck mass with $\kappa \equiv \sqrt{8\pi}/m_{\mathrm{Pl}}$ [see (7.56)].

7.4 Application to CMB Data Analysis

Figure 7.2 makes clear that from the integration of the perturbations described in the previous section, their power spectra are known at the end of inflation.

These primordial correlations then evolve through the reheating, radiation and matter eras to shape the universe into its current state. The theory of cosmological perturbations precisely predicts how such linear perturbations evolve in a FLRW universe from the tightly coupled regime deep inside the radiation era to today. As a result, we still have to know how the power spectra are modified through the reheating era. As already mentioned, reheating is very model dependent and a detailed analysis remains out of the scope of our current approach. Instead, remembering that the objective is to use the CMB anisotropies measurements as a probe to obtain information on the primordial correlations, we introduce a basic model of reheating described by some phenomenological parameters.

7.4.1 Reheating

Assuming that perturbations on super-Hubble scales are not significantly modified till the beginning of the radiation era,[6] the reheating may influence the observed power spectra through its effects on z_{end} (see Fig. 7.2). For instantaneous transitions between inflation, the reheating era and the radiation era [19], one has

$$\ln \frac{a_{\mathrm{end}}}{a_0} = \ln \frac{a_{\mathrm{end}}}{a_{\mathrm{reh}}} + \ln \frac{a_{\mathrm{reh}}}{a_{\mathrm{eq}}} + \ln \frac{a_{\mathrm{eq}}}{a_0} , \qquad (7.59)$$

where a_{end}, a_{reh} and a_{eq} are respectively the scale factor at the end of inflation, at the end of reheating and at equality between the energy density of radiation and the energy density of matter. The redshift of equality can be expressed in terms of the density parameter of radiation today Ω_{rad} and the Hubble parameter today H_0. Moreover, during the radiation era $\rho \propto a^{-4}$, and (7.59) can be recast into

$$\ln \frac{a_{\mathrm{end}}}{a_0} = \ln \frac{a_{\mathrm{end}}}{a_{\mathrm{reh}}} - \frac{1}{4} \ln \left(\kappa^4 \rho_{\mathrm{reh}} \right) + \frac{1}{2} \ln \left(\sqrt{3\Omega_{\mathrm{rad}}} \kappa H_0 \right) , \qquad (7.60)$$

where ρ_{reh} denotes the total energy density at the end of the reheating era. It is clear that the first two terms depend on the physics involved during the reheating. For instance, they exactly cancel for a radiation-like reheating era. This suggests to introduce a phenomenological parameter [8]

$$\ln R_{\mathrm{rad}} \equiv \ln \frac{a_{\mathrm{end}}}{a_{\mathrm{reh}}} - \frac{1}{4} \ln \left(\kappa^4 \rho_{\mathrm{reh}} \right) . \qquad (7.61)$$

From this parameter, the quantity $k/\mathcal{H}$ entering the equations of motion for the perturbations (7.45) and (7.46) can be evaluated in terms of the k/a_0 values measured today

---

[6] Although such an assumption is motivated by the fact that the physical processes involved during reheating are sub-Hubble, this assumption may not longer be true in the presence of entropy modes.

$$\frac{k}{aH} = \frac{\kappa k}{a_0} \frac{1}{\kappa H(n)} \frac{e^{n_{\mathrm{end}} - n}}{R_{\mathrm{rad}} \left(3\Omega_{\mathrm{rad}}\right)^{1/4} \sqrt{\kappa H_0}} , \tag{7.62}$$

and similarly for the initial conditions in (7.56).

7.4.2 CMB Anisotropies

For a given value of R_{rad}, the primordial power spectra deep in the radiation era are now uniquely determined by the numerical integration described in Sect. 7.3 and can be used as initial conditions for the subsequent evolution of the perturbations. The integration of the cosmological perturbations through the radiation and matter era, as well as the resulting CMB anisotropies, have been performed by using a modified version of the CAMB code [40]. The model parameters involved are both the inflation parameters and the usual cosmological parameters describing the FLRW model at late time. For instance, for a ΛCDM universe experiencing large field inflation in its earliest times, there are two parameters fixing the potential $V(\varphi) = M^4 \varphi^p$, one parameter describing the reheating era R_{rad}, plus the four cosmological base parameters: the number density of baryons Ω_{b}, of cold dark matter Ω_{c}, the Hubble parameter today H_0 and the redshift of reionisation of the universe z_{re} [41]. Let us recap that we have defined the end of large field inflation by $\epsilon_1(\varphi_{\mathrm{end}}) = 1$. Combined with the existence of the attractor during inflation, this ensures that φ_{end} is fixed by the potential parameters [see (7.25)]. The resulting angular power spectrum for the CMB temperature fluctuations is represented in Fig. 7.8 for a fiducial set of the parameters.

The next step is to use CMB measurements to constrain the models. For this purpose, the parameter space can be sampled by using Markov Chain Monte Carlo (MCMC) methods as implemented in the COSMOMC code [41] to extract the probability distributions satisfied by the model parameters. In the next section, we illustrate such a procedure for the ΛCDM model born under small field inflation by using the WMAP third year data [65, 66, 67].

7.4.3 WMAP3 Constraints on Small Field Models

small field inflation can be described by the action (7.5) when the potential reads

$$U(\varphi) = M^4 \left[1 - \left(\frac{\varphi}{\mu}\right)^p \right] . \tag{7.63}$$

The inflation model parameters are the energy scale M, the power p and the vacuum expectation value scale μ. As can be seen in Fig. 7.9, inflation proceeds for small initial field values and stops when $\epsilon_1(\varphi_{\mathrm{end}}) = 1$. Notice that the potential in (7.63) is negative for $\varphi > \mu$ which means that the above description is no longer correct. This is however not an issue since it occurs well after the end of inflation and the basic effects of the reheating are already encoded in our extra parameter R_{rad}.

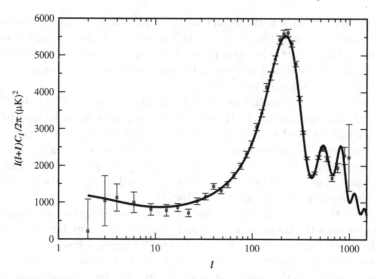

Fig. 7.8. Angular temperature power spectrum of the CMB anisotropies in a ΛCDM universe born under chaotic inflation. Fiducial values of the parameters have been used: $R_{\mathrm{rad}} = 1$, $p = 2$, $\kappa M = 2 \times 10^{-3}$, $\Omega_{\mathrm{b}} h^2 = 0.022$, $\Omega_{\mathrm{c}} = 0.12$, $h = 0.7$, $z_{\mathrm{re}} = 12$, where h is the reduced Hubble parameter and z_{re} the redshift of reionisation. The WMAP third year measurements are represented as blue squares

Observable Parameters

Since flatness is inherited from inflation, the ΛCDM cosmological model is described by the density parameters associated with the different cosmological fluids: Ω_{b}, Ω_{c} plus the Hubble parameter today H_0. The cosmological constant is fixed by $\Omega_\Lambda = 1 - \Omega_{\mathrm{b}} - \Omega_{\mathrm{c}}$. However, in order to minimise the

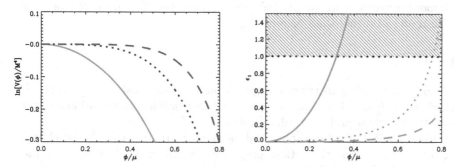

Fig. 7.9. The small field potential $\ln V$ on the left and the first Hubble flow function ϵ_1 on the right. Inflation occurs for $\epsilon_1 < 1$, for small values of the field. The three curves correspond respectively to $p = 2$, $p = 4$ and $p = 6$ from the left to the right (from [8])

parameter degeneracies with respect to the CMB angular power spectra, it is more convenient to perform the MCMC sampling on the equivalent set of parameters $\Omega_b h^2$, $\Omega_c h^2$, the optical depth τ and the quantity θ which measures the ratio of the sound horizon to the angular diameter distance (h is the reduced Hubble parameter today) [41].

Similarly for the potential parameters, as can be seen in (7.2) and (7.3), the overall amplitude of the CMB anisotropies is proportional to the Hubble parameter squared and thus to the potential V [see (7.18)]. Since the amplitude of the cosmological perturbations is a well measured quantity, the data may be more efficiently used by directly sampling the primordial amplitude of the scalar power spectrum $\mathcal{P}_* = \mathcal{P}_\zeta(k_*)$ instead of M (k_* being a fixed observable wavenumber: $k_*/a_0 = 0.05\,\mathrm{Mpc}^{-1}$).

However, the numerical method used to integrate the perturbations during inflation requires the input of a numerical value for M to predict the value of $\mathcal{P}_*$. In fact, one can use the trick described in [8]: under a rescaling $V \to sV$, the power spectrum scales as $\mathcal{P}_\zeta(k) \to s\mathcal{P}_\zeta(s^{1/2}k)$ at fixed R_{rad}. The idea is therefore to integrate the perturbations with an artificial normalisation of the potential, for instance $M = 1$, and then analytically rescale M from unity to its physical value that would be associated with $\mathcal{P}_*$. The required value of s is given by the ratio $\mathcal{P}_*/\mathcal{P}_\diamond^{(M=1)}$, where $\mathcal{P}_\diamond^{(M=1)}$ is the amplitude of the scalar power spectrum stemming from the numerical integration with $M = 1$ and evaluated at $k_\diamond = k_* s^{-1/2}$. Still, it is not really straightforward to determine s since both $\mathcal{P}_\zeta$ and k change simultaneously. The last subtlety is to remark that $k_\diamond = k_*$ if instead of considering R_{rad} fixed, one considers the rescaling of M at fixed $\ln R$, with $\ln R$ defined by

$$\ln R \equiv \ln R_{\mathrm{rad}} + \frac{1}{4}\ln\left(\kappa^4 \rho_{\mathrm{end}}\right) . \tag{7.64}$$

The parameters R and R_{rad} differ only by ρ_{end}, the energy density at the end of inflation which is uniquely determined from M, μ and p. It will therefore be more convenient to sample the model parameters $\mathcal{P}_*$ and $\ln R$ (together with μ and p) rather than M and R_{rad}.

Priors

The prior probability distributions for the base cosmological parameters $\Omega_b h^2$, $\Omega_c h^2$, τ and θ have been chosen as wide top hat uniform distribution centred over their current preferred value [1, 41].

Concerning the inflaton potential parameters, their priors can be chosen according to various theoretical prejudices [8]. Since $\mathcal{P}_*$ is related to the energy scale during inflation, a uniform prior has been considered around a value compatible with the amplitude of the cosmological perturbations: $\ln(10^{10}\mathcal{P}_*) \in [2.7, 4.0]$. For the scale μ associated with the vacuum expectation value of φ, we have considered two priors. The first includes smaller and larger values than the Planck mass: $\kappa\mu \in [1/10, 10]$. The second prior is also uniform but include values much larger than the Planck mass: $\kappa\mu \in [1/10, 100]$. Finally, a uniform prior is chosen for the power p on $[2.4, 10]$ ($p = 2$ is a particular case, see [8]).

It remains to express our prior knowledge on the reheating parameter $\ln R$. As previously mentioned, we are assuming that gravity can still be described classically which only makes sense if the energy densities involved remain smaller than the Planck energy scale, namely for $\kappa^4\rho_{\mathrm{end}} < 1$. On the other side of the energy spectrum, the success of big-bang nucleosynthesis (BBN) requires that the universe is radiation dominated at that time, thus $\rho_{\mathrm{reh}} > \rho_{\mathrm{nuc}}$ with $\rho_{\mathrm{nuc}} \simeq 1\,\mathrm{MeV}^4$ (and $\rho_{\mathrm{end}} > \rho_{\mathrm{reh}}$). Moreover, we will assume that during reheating the expansion of the universe can be described as dominated by a cosmological fluid of pressure P and energy density ρ. In this case, in order to satisfy the strong and dominant energy conditions in General Relativity, one has $-1/3 < P/\rho < 1$ (notice that $P/\rho \leq -1/3$ would be inflation). From (7.61) and (7.64), the resulting bounds read

$$\frac{1}{4}\ln\left(\kappa^4\rho_{\mathrm{nuc}}\right) < \ln R < -\frac{1}{12}\ln\left(\kappa^4\rho_{\mathrm{nuc}}\right) + \frac{1}{3}\ln\left(\kappa^4\rho_{\mathrm{end}}\right) , \qquad (7.65)$$

and an uniform prior on $\ln R$ have been chosen in between.

Results

The data sets used to constrain the small field ΛCDM model are the WMAP third year data together with the Hubble Space Telescope (HST) measurements ($H_0 = 72 \pm 8\,\mathrm{km/s/Mpc}$ [68]) and a top hat prior on the age of the universe between $10\,\mathrm{Gyrs}$ and $20\,\mathrm{Gyrs}$. The resulting marginalised posterior probability distributions for the base cosmological parameters are represented in Fig. 7.10. They are not significantly affected by the various prior choices on μ and their corresponding mean values and confidence intervals are compatible with the current state of the art [1].

The constraints obtained on the small field inflation parameters are shown in Fig. 7.11. As expected, the allowed range for the power spectra amplitude $\mathcal{P}_*$ is narrow. Concerning the above panels their interpretation require some precautions. Indeed, the probability that p takes small values depends on our theoretical prejudice on how big the field expectation value of the inflaton may be. If μ is allowed to be much greater than m_{Pl} then all p are equiprobable. On the other hand, if $\kappa\mu$ cannot take values bigger than 10 then small field inflation models with $p \simeq 2$ are disfavoured. The μ posterior shows on its own that, independently of the p values, $\kappa\mu > 10$ is slightly preferred by the data. Since these posteriors are marginalised over the other parameters, the previous statements are necessarily robust with respect to any reheating model, in the framework of our modelisation.

But more than being a nuisance parameter, Fig. 7.11 shows that $\ln R$ is also mildly constrained by the data: the probability distribution of $\ln R$ has

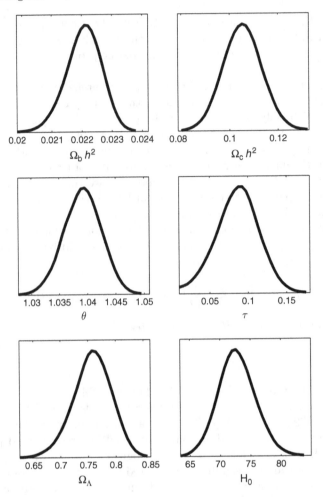

Fig. 7.10. Posterior probability distributions of the base cosmological parameters in the small field ΛCDM inflation model

a lower bound slightly above the prior $\ln R > -46$ given by (7.65). Although this is not obvious, the upper cut-off seen in the $\ln R$ posterior comes from the upper bound of (7.65) (see [8] for a more detailed discussion). For $\kappa\mu$ in $[0.1, 100]$, we finally obtain at 95% confidence level[7]

$$\ln R > -34 \,. \tag{7.66}$$

Plugging this inequality into (7.64) constrains some properties of the reheating era. This result can be understood by looking at Fig. 7.2. Since varying

---

[7] The dependence of the $\ln R$ posterior distribution with respect to the $\kappa\mu$ prior disappears as soon as $\kappa\mu$ is allowed to be greater than 10.

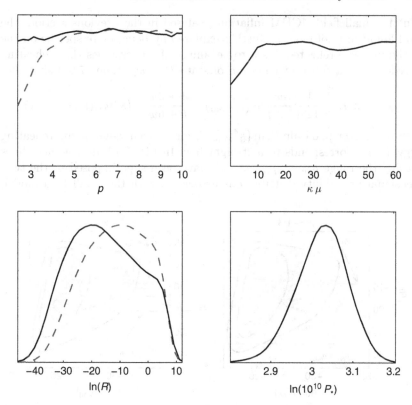

Fig. 7.11. Marginalised probability distributions for the small field inflation parameters. The results coming from the prior $\kappa\mu \in [0.1, 100]$ are represented by solid black lines whereas, when they differ, the posteriors associated with the $\kappa\mu$ prior in $[0.1, 10]$ are plotted as dashed red lines [8]

the reheating properties allow the observable window to move along the inflaton potential, it is not surprising that some part of the potential may be preferred from a data point of view. For the small field models, a more involved analysis would show that the spectral index of the scalar power spectrum departs from what is allowed by the data when $\ln R$ becomes too small [8]. This is precisely why the current WMAP data lead to the bound (7.66).

7.5 Conclusion

As a conclusion, we would like to discuss the future directions associated with the possibility of constraining some basic properties of the reheating era with the CMB data.

In the small field ΛCDM inflation analysed in the previous section, the bound found in (7.66) can be further explored by being more specific on the way the universe reheated. As a toy example, if one assumes that reheating proceeded with a constant equation of state $P = w\rho$, then (7.64) simplifies into

$$\ln R = \frac{1 - 3w}{12 + 12w} \ln\left(\kappa^4 \rho_{\mathrm{reh}}\right) + \frac{1 + 3w}{6 + 6w} \ln\left(\kappa^4 \rho_{\mathrm{end}}\right) . \qquad (7.67)$$

In the parameter plane $[\ln R, \ln\left(\kappa^4 \rho_{\mathrm{end}}\right)]$, for a given value of the reheating energy, (7.67) corresponds to a straight line. In Fig. 7.12, five of these lines exploring the range $\rho_{\mathrm{nuc}} < \rho_{\mathrm{reh}} < \rho_{\mathrm{end}}$ have been superimposed to the two-dimensional probability distributions associated with the small field model,

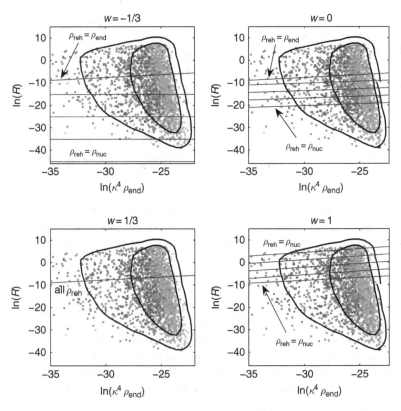

Fig. 7.12. One and two-sigma confidence intervals (solid contours) of the two-dimensional marginalised posteriors (point density) in the plane $[\ln R, \ln(\kappa^4 \rho_{\mathrm{end}})]$ for the small field models. The parameter $\kappa\mu$ varies in $[0.1, 100]$ from red (dark) to green (light). The four panels correspond to the situation in which the universe reheated with a constant equation of state $P = w\rho$. In each panel, the solid lines correspond to different values of the reheating temperature $(1/4) \ln(\kappa^4 \rho_{\mathrm{reh}})$ ranging from $-45, -35, -25, -15$ to $-(1/4) \ln(\kappa^4 \rho_{\mathrm{end}})$

and this for four different equations of state having $w \gtrsim -1/3, w = 0, w = 1/3$ and $w \lesssim 1$, respectively.

At can be seen on the top left frame, for $w \gtrsim -1/3$, low values of the reheating temperature lie out of the confidence contours. From a more robust analysis using importance sampling, one would find that in this case, at 95% confidence, $\rho_{\text{reh}} > 2\,\text{TeV}$ [8]. Of course this bound is not really impressive and close to the limits already set by BBN, moreover, it holds only for small field models and a quite extreme equation of state. However, is shows that it is already possible to get some information on the reheating era in a given model of inflation from the CMB data only. This is precisely on this point that particle physics models may be decisive. Indeed, once the inflaton couplings to the other particles are specified, the properties of the reheating era are fixed and a given particle physics model would appear as one curve parametrised by some coupling constants in the plane $[\ln R, \ln(\kappa^4 \rho_{\text{end}})]$ of Fig. 7.12. With the incoming flow of more accurate cosmological data, this may be an interesting way of constraining inflation as well as particle physics at very high energy.

References

1. D. N. Spergel et al. (2006), `astro-ph/0603449`.
2. B. A. Bassett, S. Tsujikawa and D. Wands Rev. Mod. Phys. **78**, 537–589 (2006), `astro-ph/0507632`.
3. S. H. H. Tye (2006), `hep-th/0610221`.
4. D. H. Lyth and A. Riotto, Phys. Rept. **314**, 1–146 (1999), `hep-ph/9807278`.
5. V. F. Mukhanov, H. A. Feldman and R. H. Brandenberger, Phys. Rept. **215**, 203–333 (1992).
6. E. D. Stewart and D. H. Lyth, Phys. Lett. **B302**, 171–175 (1993), `gr-qc/9302019`.
7. J. Martin and D. J. Schwarz, Phys. Rev. **D62**, 103520 (2000), `astro-ph/9911225`.
8. J. Martin and C. Ringeval, JCAP **0608**, 009 (2006), `astro-ph/0605367`.
9. A. R. Liddle, P. Parsons and J. D. Barrow, Phys. Rev. **D50**, 7222–7232 (1994), `astro-ph/9408015`.
10. J. Martin and D. J. Schwarz, Phys. Rev. **D57**, 3302–3316 (1998), `gr-qc/9704049`.
11. J. Martin, Lect. Notes Phys. **669**, 199–244 (2005), `hep-th/0406011`.
12. G. N. Felder et al., Phys. Rev. Lett. **87**, 011601 (2001), `hep-ph/0012142`.
13. L. Kofman, A. D. Linde and A. A. Starobinsky, Phys. Rev. **D56**, 3258–3295 (1997), `hep-ph/9704452`.
14. J. Garcia-Bellido and A. D. Linde, Phys. Rev. **D57**, 6075–6088 (1998), `hep-ph/9711360`.
15. V. N. Senoguz and Q. Shafi, Phys. Rev. **D71**, 043514 (2005), `hep-ph/0412102`.
16. D. I. Podolsky, G. N. Felder, L. Kofman and M. Peloso, Phys. Rev. **D73**, 023501 (2006), `hep-ph/0507096`.
17. M. Desroche, G. N. Felder, J. M. Kratochvil and A. Linde, Phys. Rev. **D71**, 103516 (2005), `hep-th/0501080`.

18. R. Allahverdi and A. Mazumdar (2006), hep-ph/0603244.
19. A. R. Liddle and S. M. Leach, Phys. Rev. **D68**, 103503 (2003), astro-ph/0305263.
20. J. Barriga, E. Gaztanaga, M. G. Santos and S. Sarkar, Mon. Not. Roy. Astron. Soc. **324**, 977 (2001), astro-ph/0011398.
21. J. Martin and C. Ringeval, Phys. Rev. **D69**, 083515 (2004), astro-ph/0310382.
22. J. Martin and C. Ringeval, Phys. Rev. **D69**, 127303 (2004), astro-ph/0402609.
23. J. Martin and C. Ringeval, JCAP **0501**, 007 (2005), hep-ph/0405249.
24. R. Easther, W. H. Kinney and H. Peiris, JCAP **0505**, 009 (2005), astro-ph/0412613.
25. P. Hunt and S. Sarkar, Phys. Rev. **D70**, 103518 (2004), astro-ph/0408138.
26. L. Covi, J. Hamann, A. Melchiorri, A. Slosar and I. Sorbera, Phys. Rev. **D74**, 083509 (2006), astro-ph/0606452.
27. J. Martin and D. J. Schwarz, Phys. Rev. **D67**, 083512 (2003), astro-ph/0210090.
28. R. Casadio, F. Finelli, M. Luzzi and G. Venturi, Phys. Rev. **D71**, 043517 (2005), gr-qc/0410092.
29. R. Casadio, F. Finelli, M. Luzzi and G. Venturi, Phys. Lett. **B625**, 1–6 (2005), gr-qc/0506043.
30. R. Casadio, F. Finelli, A. Kamenshchik, M. Luzzi and G. Venturi, JCAP **0604**, 011 (2006), gr-qc/0603026.
31. C. Gordon, D. Wands, B. A. Bassett and R. Maartens, Phys. Rev. **D63**, 023506 (2001), astro-ph/0009131.
32. H. Noh and J.-c. Hwang, Phys. Lett. **B515**, 231–237 (2001), astro-ph/0107069.
33. F. Di Marco, F. Finelli and R. Brandenberger, Phys. Rev. **D67**, 063512 (2003), astro-ph/0211276.
34. F. Di Marco and F. Finelli, Phys. Rev. **D71**, 123502 (2005), astro-ph/0505198.
35. D. S. Salopek, J. R. Bond and J. M. Bardeen, Phys. Rev. **D40**, 1753 (1989).
36. I. J. Grivell and A. R. Liddle, Phys. Rev. **D61** 081301 (2000), astro-ph/9906327.
37. J. A. Adams, B. Cresswell and R. Easther, Phys. Rev. **D64**, 123514 (2001), astro-ph/0102236.
38. S. Tsujikawa, D. Parkinson and B. A. Bassett, Phys. Rev. **D67**, 083516 (2003), astro-ph/0210322.
39. A. Makarov, Phys. Rev. **D72**, 083517 (2005), astro-ph/0506326.
40. A. Lewis, A. Challinor and A. Lasenby, Astrophys. J. **538**, 473–476 (2000), astro-ph/9911177.
41. A. Lewis and S. Bridle, Phys. Rev. **D66**, 103511 (2002), astro-ph/0205436.
42. H. Peiris and R. Easther, JCAP **0610**, 017 (2006), astro-ph/0609003.
43. H. Peiris and R. Easther (2006), astro-ph/0603587.
44. W. H. Kinney, E. W. Kolb, A. Melchiorri and A. Riotto, Phys. Rev. **D74**, 023502 (2006), astro-ph/0605338.
45. P. Brax, C. van de Bruck and A.-C. Davis, Rept. Prog. Phys. **67**, 2183–2232 (2004), hep-th/0404011.
46. T. Damour and G. Esposito-Farese, Class. Quant. Grav. **9**, 2093–2176 (1992).

47. T. Damour and K. Nordtvedt, Phys. Rev. **D48**, 3436–3450 (1993).
48. N. A. Koshelev, Grav. Cosmol. **10**, 289–294 (2004), astro-ph/0501600.
49. C. Ringeval, P. Brax, v. de Bruck, Carsten and A.-C. Davis, Phys. Rev. **D73**, 064035 (2006), astro-ph/0509727.
50. D. Langlois, Prog. Theor. Phys. Suppl. **148**, 181–212 (2003), hep-th/0209261.
51. R. Maartens, Living Rev. Rel. **7** 7 (2004), gr-qc/0312059.
52. C. Schimd, J.-P. Uzan and A. Riazuelo Phys. Rev. **D71** 083512 (2005), astro-ph/0412120.
53. A. Lukas, B. A. Ovrut, K. S. Stelle and D. Waldram, Nucl. Phys. **B552**, 246–290 (1999), hep-th/9806051.
54. A. Lukas, B. A. Ovrut and D. Waldram, Phys. Rev. **D61**, 023506 (2000), hep-th/9902071.
55. P. Brax and A. C. Davis, Phys. Lett. **B497**, 289–295 (2001), hep-th/0011045.
56. S. Kobayashi and K. Koyama, JHEP **12**, 056 (2002), hep-th/0210029.
57. G. Esposito-Farese and D. Polarski, Phys. Rev. **D63**, 063504 (2001), gr-qc/0009034.
58. J. Martin, C. Schimd and J.-P. Uzan, Phys. Rev. Lett. **96**, 061303 (2006), astro-ph/0510208.
59. B. Carter (1997), hep-th/9705172.
60. A. R. Liddle and D. H. Lyth, Phys. Rept. **231**, 1–105 (1993), astro-ph/9303019.
61. M. S. Turner, Phys. Rev. **D28**, 1243 (1983).
62. J. C. Niemeyer, R. Parentani and D. Campo, Phys. Rev. **D66**, 083510 (2002), hep-th/0206149.
63. S. Weinberg, Phys. Rev. **D70**, 083522 (2004), astro-ph/0405397.
64. M. Lemoine and J. Martin (2006), astro-ph/0611948.
65. L. Page et al. (2006), astro-ph/0603450.
66. G. Hinshaw et al. (2006), astro-ph/0603451.
67. N. Jarosik et al. (2006), astro-ph/0603452.
68. W. L. Freedman et al., Astrophys. J. **553**, 47–72 (2001), astro-ph/0012376.

8

Multiple Field Inflation

David Wands

Institute of Cosmology and Gravitation, Mercantile House, University
of Portsmouth, Portsmouth PO1 2EG, United Kingdom
david.wands@port.ac.uk

Abstract. Inflation offers a simple model for very early evolution of our universe
and the origin of primordial perturbations on large scales. Over the last 25 years
we have become familiar with the predictions of single-field models, but inflation
with more than one light scalar field can alter preconceptions about the inflationary
dynamics and our predictions for the primordial perturbations. I will discuss how
future observational data could distinguish between inflation driven by one field, or
many fields. As an example, I briefly review the curvation as an alternative to the
inflaton scenario for the origin of structure.

8.1 Introduction

Inflation provides an attractively simple model for the early evolution of our
universe, which can produce a large, spatially flat and largely homogeneous
observable universe. It also provides a source for small primordial perturbations which are the origin of the large-scale structure in our Universe today.
The vacuum fluctuations of any light scalar field present during inflation can
be swept up by the inflationary expansion to scales much larger than the
Hubble scale.

Inflation is most commonly discussed in terms of a potential energy which
is a function of a single, slowly rolling, scalar field. Single-field, slow-roll inflation produces an almost Gaussian distribution of adiabatic density perturbations on super-Hubble scales with an almost scale-invariant spectrum. But
supersymmetric field theories can contain many scalar fields that could play a
role during inflation and string theory, and other higher-dimensional theories,
yield four-dimensional effective actions with many moduli fields describing the
higher-dimensional degrees of freedom. One should be aware of the different
possibilities that open up in particle physics models containing more than one
light scalar field during inflation.

The presence of multiple fields during inflation can lead to quite different
inflationary dynamics that might appear unnatural in a single-field model and

D. Wands: *Multiple Field Inflation*, Lect. Notes Phys. **738**, 275–304 (2008)
DOI 10.1007/978-3-540-74353-8_8   © Springer-Verlag Berlin Heidelberg 2008

to spectra of primordial perturbations that would actually be impossible in single-field models. The presence of multiple light fields during inflation leads to the generation of non-adiabatic field perturbations during inflation. This can alter the evolution of the overall curvature perturbation, for instance leading to detectable non-Gaussianity, and may leave residual isocurvature fluctuations in the primordial density perturbation on large scales after inflation, which can be correlated with the curvature perturbation. Such alternative models are interesting not only as theoretical possibilities, but because they could be distinguished by increasingly precise observations in the near future.

In this chapter I will discuss some of the distinctive observational predictions of inflation in the presence of more than one scalar field. For a more comprehensive review of inflationary dynamics and reheating with multiple fields see [1].

8.2 Homogeneous Scalar Field Dynamics

The time-evolution of a single, spatially homogeneous scalar field is governed by the Klein–Gordon equation

$$\ddot{\phi} + 3H\dot{\phi} = -\frac{\mathrm{d}V}{\mathrm{d}\phi} , \tag{8.1}$$

where the Hubble expansion rate is given by the Friedmann constraint

$$3H^2 = 8\pi G \left[\frac{1}{2}\dot{\phi}^2 + V(\phi) \right] . \tag{8.2}$$

Multiple scalar fields obey the Klein–Gordon equation

$$\ddot{\varphi}_I + 3H\dot{\varphi}_I = -\frac{\partial}{\partial \varphi_I} \left(\sum_J U_J \right) , \tag{8.3}$$

where one must be allowed for the possibility that the potential energy is given by a sum over many terms

$$V = \sum_J U_J . \tag{8.4}$$

The wider range of interaction potentials possible in multiple field models leads to possibilities such as hybrid inflation.

In hybrid inflation models [2, 3] the *inflaton* field, φ_1, can roll towards the non-zero minimum of its potential, $U_1 = V_0 + m_1^2 \varphi_1^2/2$, which would lead to eternal inflation into the future in a single-field model. But in a hybrid model there is a second *waterfall* scalar field trapped during inflation in a local minimum, $\varphi_2 = 0$, with a potential, e.g. $U_2 = (g^2\varphi_1^2 - m_2^2)\varphi_2^2/2$ which becomes unstable below a critical value of the φ_1 field, triggering an instability of the vacuum energy driving inflation, rapidly bringing inflation to an end.

Another more subtle change enters through the Friedmann constraint:

$$3H^2 = 8\pi G \left(V + \sum_I \frac{1}{2}\dot{\varphi}_I^2 \right). \tag{8.5}$$

Even in the absence of explicit interactions in the scalar field Lagrangian, the fields will still be coupled gravitationally. In particular the Hubble expansion rate that enters the Klein–Gordon equation (8.3) is due to the sum over all fields in (8.5) and this can also alter the field dynamics even if the potential for each individual field is left unchanged. The additional Hubble damping present due to multiple fields can be used to drive slow-roll inflation in assisted inflation models [4] where the individual potentials would be too steep to drive inflation on their own.

The original assisted inflation model [4] considered n scalar fields with steep exponential potentials

$$V = \sum_I U_{I0} \exp\left(-\lambda_I \varphi_I / M_{\mathrm{Pl}}\right) \tag{8.6}$$

where I have used the reduced Planck mass $M_{\mathrm{Pl}}^2 = (8\pi G)^{-1}$. Each scalar field potential is too steep to drive inflation on its own if $\lambda_I^2 > 2$, but the additional damping effect due to the presence of the other scalar fields leads to a particular power-law inflation solution, $a \propto t^p$ where $p = 2/\lambda^2$ where the combined fields have an effective potential $V \propto \exp(-\lambda\sigma/M_{\mathrm{Pl}})$ with

$$\frac{1}{\lambda^2} = \sum_I \frac{1}{\lambda_I^2}. \tag{8.7}$$

Thus $\lambda \to 0$ for many fields as $n \to \infty$ and we can have slow-roll inflation even when each $\lambda_I^2 > 2$.

Even though the background dynamics can be reduced to an equivalent single field with a specified potential [5], there is an important qualitative difference between the inflationary dynamics in multiple field inflation with respect to the single-field case. The Hubble damping during inflation drives a single scalar field to a unique attractor solution during slow-roll inflation where the Hubble rate, field time-derivative and all local variables are a function of the local field value: $H(\phi)$, $\dot{\phi}(\phi)$, etc. This means that the evolution rapidly becomes independent of the initial conditions.

In multiple field models we may have a family of trajectories in phase space where, for example, the Hubble rate at a particular value of φ_1 is also dependent upon the value of φ_2. In this case the inflationary dynamics, and hence observational predictions, may be dependent upon the trajectory in phase space and thus the initial field values. It is this that allows non-adiabatic perturbations to survive on super-Hubble scales in multiple field inflation.

It is important to distinguish here between models, such as most hybrid models, with multiple fields but only one light direction in field space with small effective mass $\partial^2 V/\partial\varphi^2 \ll H^2$ during inflation, and models with many light fields, such as assisted inflation models. Only models with multiple *light* fields can have multiple slow-roll trajectories.

8.2.1 Inflation Field Direction During Inflation

It is convenient to identify the *inflation* field direction as the direction in field space corresponding to the evolution of the background (spatially homogeneous) fields during inflation [6] (see also [7, 8]). Thus for n scalar fields φ_I, where I runs from 1 to n, we have

$$\sigma = \int \sum_I \hat{\sigma}_I \dot{\varphi}_I \, dt , \qquad (8.8)$$

where the inflaton direction is defined by

$$\hat{\sigma}_I \equiv \frac{\dot{\varphi}_I}{\sqrt{\sum_J \dot{\varphi}_J^2}} . \qquad (8.9)$$

The n evolution equations for the homogeneous scalar fields (8.3) can then be written as the evolution for a single inflation field (8.1)

$$\ddot{\sigma} + 3H\dot{\sigma} + V_\sigma = 0 , \qquad (8.10)$$

where the potential gradient in the direction of the inflation is

$$V_\sigma \equiv \frac{\partial V}{\partial\sigma} = \sum_I \hat{\sigma}_I \frac{\partial V}{\partial\varphi_I} . \qquad (8.11)$$

The total energy density and pressure are then given by the usual single-field results for the inflaton.

8.2.2 An Example: Nflation

A topical example of multiple field inflation is Nflation. Dimopoulos et al. [9] proposed this model based on the very large number of axion fields predicted in low energy effective theories derived from string theory. Near the minimum of the effective potential the fields have a potential energy

$$V = \frac{1}{2} \sum_I m_I^2 \varphi_I^2 . \qquad (8.12)$$

This form of potential with a large number of massive fields was also previously studied by Kanti and Olive [10] and Kaloper and Liddle [11].

With a single scalar field the quadratic potential yields the familiar chaotic inflation model with a massive field $V = m^2\phi^2/2$. But to obtain inflation with a single massive field the initial value of the scalar field must be several times the Planck mass and there is a worry that we have no control over corrections to the potential at super-Planckian values in the effective field theory [12]. But with many scalar fields the collective dynamics can yield inflation even for sub-Planckian values if there are a sufficiently large number of fields. Kim and Liddle [13] have found that for random initial conditions, $-M_{\text{Pl}} < \varphi_I(0) < M_{\text{Pl}}$, the total number of e-folds is given by $n/12$, where n is the total number of fields. Thus we require $n > 600$ for sufficient inflation if none of the fields is allowed to exceed the Planck scale. This may seem to be a large number, but Dimopoulos et al. [9] cite string theory models of order 10^5 axion fields.

As remarked earlier, in the presence of more than one light field, the trajectory in field space at late times, and hence the observable predictions, may be dependent upon the initial conditions for the different fields. But Kim and Liddle [13] found evidence for what they called a "thermodynamic" regime where the predicted spectral index, $n_{\mathcal{R}}$, for the primordial curvature perturbations that arise from quantum fluctuations of the scalar fields, became independent of the precise initial conditions for a sufficiently large number of fields. In fact inflation with an arbitrary number of massive fields always yields a robust prediction for the tensor-to-scalar ratio r in terms of the number of e-foldings, N, from the end of inflation [12]

$$r = 8/N \, , \tag{8.13}$$

completely independent of the initial conditions. Thus Nflation seems to be an example of a multiple field model of inflation which makes observable predictions which need not depend upon the specific trajectory in field space.

In the limit where the masses become degenerate, $m_I^2 \to m$, the Nflation dynamics becomes particularly simple. The fields evolve radially towards the origin and the potential (8.12) reduces to that for a single field

$$V \to \frac{1}{2}m^2\sigma^2 \, , \tag{8.14}$$

where σ is the inflation field (8.8). Thus in this limit Nflation reproduces the single-field prediction for the tensor–scalar ratio $r = 0.16$ and the spectral index $n_{\mathcal{R}} = 0.96$. However, the presence of n light fields during Nflation also leads to $n - 1$ isocurvature modes during inflation and these have an exactly scale-invariant spectrum (up to first order in the slow-roll parameters) in the limit of degenerate masses [14]. In the following sections I will describe some of the distinctive predictions that can arise due to the existence of such non-adiabatic perturbations during inflation.

8.3 Primordial Perturbations from Inflation

I have so far only presented equations for the dynamics of homogeneous scalar fields driving inflation. But to test theoretical predictions against cosmological observations we need to consider inhomogeneous perturbations. It is the primordial perturbations produced during inflation that offer the possibility of determining the physical processes that drove the dynamical evolution of the very early universe. In the standard hot big bang model there seems to be no way to explain the existence of primordial perturbations, during the radiation-dominated era, on scales much larger than the causal horizon, or equivalently the Hubble scale. But inflation takes perturbations on small, sub-Hubble, scales and can stretch them up to arbitrarily large scales.

8.3.1 Scalar Field Perturbations, Without Interactions

Consider an inhomogeneous perturbation, $\varphi_I \rightarrow \varphi_I(t) + \delta\varphi_I(t, \mathbf{x})$, of the Klein–Gordon equation (8.3) for a non-interacting scalar field in an unperturbed FRW universe. (I will include perturbations of the metric and other fields later, but for simplicity I will neglect this complication for the moment.)

$$\ddot{\delta\varphi}_I + 3H\dot{\delta\varphi}_I + \left(m_I^2 - \nabla^2\right)\delta\varphi_I = 0\,, \tag{8.15}$$

where the effective mass-squared of the field is $m_I^2 = \partial^2 V/\partial\varphi_I^2$ and ∇^2 is the spatial Laplacian. Decomposing a field arbitrary perturbation into eigenmodes of the spatial Laplacian (Fourier modes in flat space) $\nabla^2\delta\varphi_I = -(k^2/a^2)\delta\varphi_I$, where k is the comoving wavenumber and a the FRW scale factor, we find that small-scale fluctuations in scalar fields on sub-Hubble scales (with comoving wavenumber $k > aH$) undergo under-damped oscillations, and on sufficiently small scales are essentially freely oscillating. Normalising the initial amplitude of these small-scale fluctuations to the zero-point fluctuations of a free field in flat spacetime we have [15]

$$\delta\varphi_I \simeq \frac{e^{-ikt/a}}{a\sqrt{2k}}\,. \tag{8.16}$$

During an accelerated expansion $\dot{a} = aH$ increases and modes that start on sub-Hubble scales ($k > aH$) are stretched up to super-Hubble scales ($k < aH$). Perturbations in light fields (with effective mass-squared $m^2 < 9H^2/4$) become over-damped (or "frozen-in") and (8.16) evaluated when $k \simeq aH$ gives the power spectrum for scalar field fluctuations at "Hubble-exit"

$$\mathcal{P}_{\delta\varphi_I} \equiv \frac{4\pi k^3}{(2\pi)^3}\left|\delta\varphi_I^2\right| \simeq \left(\frac{H}{2\pi}\right)^2\,. \tag{8.17}$$

Heavy fields with $m^2 > 9H^2/4$ remain over-damped and have essentially no perturbations on super-Hubble scales. But light fields become over-damped and can be treated as essentially classical perturbations with a Gaussian distribution on super-Hubble scales.

Thus inflation generates approximately scale-invariant perturbation spectra on super-Hubble scales in any light field (for $m^2 \ll H^2$ and $|\dot{H}| \ll H^2$).

8.3.2 Scalar Field and Metric Perturbations, with Interactions

The simplified discussion in the preceding subsection gives a good approximation to the scalar field perturbations generated around the time of Hubble-exit during slow-roll inflation, where field interactions and metric backreaction are small. However, to accurately track the evolution of perturbations through to the end of inflation and into the radiation-dominated era we need to include interactions between fields and, even in the absence of explicit interactions, we need to include gravitational backreaction.

For an inhomogeneous matter distribution the Einstein equations imply that we must also consider inhomogeneous metric perturbations about the spatially flat FRW metric. The perturbed FRW spacetime is described by the line element [1, 16]

$$
\begin{aligned}
ds^2 = &-(1 + 2A)dt^2 + 2a\partial_i B dx^i dt \\
&+ a^2 \left[(1 - 2\psi)\delta_{ij} + 2\partial_{ij} E + h_{ij} \right] dx^i dx^j \ ,
\end{aligned}
\tag{8.18}
$$

where ∂_i denotes the spatial partial derivative $\partial/\partial x^i$. We will use lower case Latin indices to run over the three spatial coordinates.

The metric perturbations have been split into scalar and tensor parts according to their transformation properties on the spatial hypersurfaces. The field equations for the scalar and tensor parts then decouple to linear order. Vector metric perturbations are automatically zero at first order if the matter content during inflation is described solely by scalar fields.

The tensor perturbations, h_{ij}, are transverse ($\partial^i h_{ij} = 0$) and trace-free ($\delta^{ij} h_{ij} = 0$). They are automatically independent of coordinate gauge transformations. These describe gravitational waves as they are the free part of the gravitational field and evolve independently of linear matter perturbations.

We can decompose arbitrary tensor perturbations into eigenmodes of the spatial Laplacian, $\nabla^2 e_{ij} = -(k^2/a^2)e_{ij}$, with comoving wavenumber k, and scalar amplitude $h(t)$:

$$
h_{ij} = h(t)e_{ij}^{(+,\times)}(x) \ ,
\tag{8.19}
$$

with two possible polarisation states, $+$ and $\times$. The Einstein equations yield a wave equation for the amplitude of the tensor metric perturbations

$$
\ddot{h} + 3H\dot{h} + \frac{k^2}{a^2}h = 0 \ ,
\tag{8.20}
$$

This is the same as the wave equation (8.15) for a massless scalar field in an unperturbed FRW metric.

The four scalar metric perturbations A, $\partial_i B$, $\psi \delta_{ij}$ and $\partial_{ij} E$ are constructed from 3-scalars, their derivatives and the background spatial metric. The intrinsic Ricci scalar curvature of constant time hypersurfaces is given by

$$^{(3)}R = \frac{4}{a^2} \nabla^2 \psi \;. \tag{8.21}$$

Hence we refer to ψ as the curvature perturbation.

First-order scalar field perturbations in a first-order perturbed FRW universe obey the wave equation [1]

$$\ddot{\delta\varphi}_I + 3H\dot{\delta\varphi}_I + \frac{k^2}{a^2}\delta\varphi_I + \sum_J V_{IJ}\delta\varphi_J$$
$$= -2V_I A + \dot{\varphi}_I \left[\dot{A} + 3\dot{\psi} + \frac{k^2}{a^2}(a^2\dot{E} - aB) \right] . \tag{8.22}$$

where the mass-matrix $V_{IJ} \equiv \partial^2 V / \partial\varphi_I \partial\varphi_J$. The Einstein equations relate the scalar metric perturbations to matter perturbations via the energy and momentum constraints [16]

$$3H\left(\dot{\psi} + HA\right) + \frac{k^2}{a^2}\left[\psi + H(a^2\dot{E} - aB)\right] = -4\pi G\,\delta\rho\,, \tag{8.23}$$

$$\dot{\psi} + HA = -4\pi G\,\delta q\,, \tag{8.24}$$

where the energy and pressure perturbations and momentum for n scalar fields are given by [1]

$$\delta\rho = \sum_I \left[\dot{\varphi}_I \left(\dot{\delta\varphi}_I - \dot{\varphi}_I A \right) + V_I \delta\varphi_I \right] , \tag{8.25}$$

$$\delta P = \sum_I \left[\dot{\varphi}_I \left(\dot{\delta\varphi}_I - \dot{\varphi}_I A \right) - V_I \delta\varphi_I \right] , \tag{8.26}$$

$$\delta q_{,i} = -\sum_I \dot{\varphi}_I \delta\varphi_{I,i} , \tag{8.27}$$

where $V_I \equiv \partial V / \partial\varphi_I$.

We can construct a variety of gauge-invariant combinations of the scalar metric perturbations. The longitudinal gauge corresponds to a specific gauge transformation to a (zero-shear) frame such that $E = B = 0$, leaving the gauge-invariant variables

$$\Phi \equiv A - \frac{d}{dt}\left[a^2(\dot{E} - B/a)\right] , \tag{8.28}$$

$$\Psi \equiv \psi + a^2 H(\dot{E} - B/a) \;. \tag{8.29}$$

Another variable commonly used to describe scalar perturbations during inflation is the field perturbation in the spatially flat gauge (where $\psi = 0$). This has the gauge-invariant definition [17, 18]:

$$\delta\varphi_{I\psi} \equiv \delta\varphi_I + \frac{\dot{\varphi}}{H}\psi \; . \tag{8.30}$$

It is possible to use the Einstein equations to eliminate the metric perturbations from the perturbed Klein–Gordon equation (8.22) and write a wave equation solely in terms of the field perturbations in the spatially flat gauge [19]

$$\ddot{\delta\varphi}_{I\psi} + 3H\dot{\delta\varphi}_{I\psi} + \frac{k^2}{a^2}\delta\varphi_{I\psi} + \sum_J \left[V_{IJ} - \frac{8\pi G}{a^3}\frac{\mathrm{d}}{\mathrm{d}t}\left(\frac{a^3}{H}\dot{\varphi}_I\dot{\varphi}_J\right) \right]\delta\varphi_{J\psi} = 0 \; . \tag{8.31}$$

Only at lowest order in the slow-roll expansion can the interaction terms be neglected and we recover the simplified wave equation (8.15) for a massless field in an unperturbed FRW universe.

8.3.3 Adiabatic and Entropy Perturbations

There are two more gauge-invariant scalars which are commonly used to describe the overall curvature perturbation. The curvature perturbation on uniform-density hypersurfaces is given by

$$-\zeta \equiv \psi + \frac{H}{\dot{\rho}}\delta\rho \; , \tag{8.32}$$

first introduced by Bardeen et al. [20] (see also [21, 22, 23]). The comoving curvature perturbation (strictly speaking the curvature perturbation on hypersurfaces orthogonal to comoving worldlines)

$$\mathcal{R} \equiv \psi - \frac{H}{\rho + P}\delta q \; , \tag{8.33}$$

where the scalar part of the 3-momentum is given by $\partial_i\delta q$. $\mathcal{R}$ has been used by Lukash [24], Lyth [25] and many others. For single-field inflation we have $\delta q = -\dot{\phi}\,\delta\phi$ and hence

$$\mathcal{R} = \psi + \frac{H}{\dot{\phi}}\delta\phi \; . \tag{8.34}$$

The difference between the two curvature perturbations $-\zeta$ and $\mathcal{R}$,

$$-\zeta - \mathcal{R} = \frac{H}{\dot{\rho}}\delta\rho_{\mathrm{m}} \; , \tag{8.35}$$

is proportional to the comoving density perturbation [26]

$$\delta\rho_{\mathrm{m}} \equiv \delta\rho - 3H\delta q \; . \tag{8.36}$$

The energy and momentum constraints (8.23) and (8.24) can be combined to give a generalisation of the Poisson equation

$$\frac{\delta\rho_{\rm m}}{\rho} = -\frac{2}{3}\left(\frac{k}{aH}\right)^2 \Psi \ , \tag{8.37}$$

relating the longitudinal gauge metric perturbation (8.29) to the comoving density perturbation (8.36). Thus the two curvature perturbations, $\mathcal{R}$ and $-\zeta$, coincide on large scales ($k/aH \ll 1$) so long as the longitudinal gauge metric perturbation, Ψ, remains finite – which is generally true during slow-roll inflation.

The energy conservation equation can be written in terms of the curvature perturbation on uniform-density hypersurfaces, defined in (8.32), to obtain the first-order evolution equation [1, 23]

$$\dot{\zeta} = -H\frac{\delta P_{\rm nad}}{\rho + P} - \Sigma \ , \tag{8.38}$$

where $\delta P_{\rm nad}$ is the non-adiabatic pressure perturbation,

$$\delta P_{\rm nad} = \delta P - \frac{\dot{P}}{\dot{\rho}}\delta\rho \ , \tag{8.39}$$

and Σ is the scalar shear along comoving worldlines [27], which can be given relative to the Hubble rate as

$$\begin{aligned}
\frac{\Sigma}{H} &\equiv -\frac{k^2}{3H}\left[\dot{E} - \frac{B}{a} + \frac{\delta q}{a^2(\rho + P)}\right] \\
&= -\frac{k^2}{3a^2H^2}\zeta - \frac{k^2\Psi}{3a^2H^2}\left[1 - \frac{2\rho}{9(\rho + P)}\frac{k^2}{a^2H^2}\right] \ .
\end{aligned} \tag{8.40}$$

Thus ζ and $\mathcal{R}$ are constant (and equal) for adiabatic perturbations on super-Hubble scales ($k/aH \ll 1$), so long as Ψ remains finite, in which case the shear of comoving worldlines can be neglected.

More generally we can define adiabatic perturbations to be perturbations which lie along the background trajectory in the phase space of spatially homogeneous fields [6, 23]. That is, we generalise (8.39) so that for adiabatic linear perturbations of any two variables x and y we require

$$\frac{\delta x}{\dot{x}} = \frac{\delta y}{\dot{y}} : \quad \text{adiabatic} \tag{8.41}$$

Thus adiabatic perturbations in a multiple field inflation can be characterised by a unique shift along the background trajectory $\delta N = -H\delta x/\dot{x} = -H\delta y/\dot{y}$. For example, for adiabatic perturbations of the primordial plasma we require that the baryon–photon ratio, $n_{\rm B}/n_\gamma$, remains unperturbed, and hence

$$\frac{\delta(n_{\rm B}/n_\gamma)}{n_{\rm B}/n_\gamma} = -3H\left(\frac{\delta n_{\rm B}}{\dot{n}_{\rm B}} - \frac{\delta n_\gamma}{\dot{n}_\gamma}\right) = 0 \ , \tag{8.42}$$

where we have used $\dot{n}_x = -3Hn_x$ for baryon number density, and photon number density below about 1 MeV.

For a single scalar field the non-adiabatic pressure (8.39) can be related to the comoving density perturbation (8.36) [6]

$$\delta P_{\mathrm{nad}} = -\frac{2V_{,\varphi}}{3H\dot{\varphi}}\delta\rho_{\mathrm{m}} \ . \tag{8.43}$$

From the Einstein constraint (8.37) this will vanish on large scales ($k/aH \to 0$) if Ψ remains finite, and hence single scalar field perturbations become adiabatic in this large-scale limit. In particular, we have from (8.38) that ζ becomes constant for adiabatic perturbations in this large-scale limit, and hence the curvature perturbation can be calculated shortly after Hubble-exit in single-field inflation and equated directly with the primordial curvature perturbation, independently of the details of reheating, etc. at the end of inflation. But in the presence of more than one light field the vacuum fluctuations stretched to super-Hubble scales will inevitably include non-adiabatic perturbations due to the presence of multiple trajectories in the phase space.

We define entropy perturbations to be fluctuations orthogonal to the background trajectory

$$S_{xy} \propto \frac{\delta x}{\dot{x}} - \frac{\delta y}{\dot{y}} \ : \quad \text{entropy} \tag{8.44}$$

For example, in the primordial era we can have entropy perturbations in the primordial plasma

$$S_{\mathrm{B}} = \frac{\delta(n_{\mathrm{B}}/n_\gamma)}{n_{\mathrm{B}}/n_\gamma} = -3H\left(\frac{\delta n_{\mathrm{B}}}{\dot{n}_{\mathrm{B}}} - \frac{\delta n_\gamma}{\dot{n}_\gamma}\right) \ . \tag{8.45}$$

which are also referred to as baryon isocurvature perturbations.

In the radiation-dominated era we can define a gauge-invariant primordial curvature perturbation associated with each of the component fluids [23, 28] in analogy with the total curvature perturbation (8.32)

$$\zeta_I \equiv -\psi - H\frac{\delta\rho_I}{\dot{\rho}_I} \ , \tag{8.46}$$

and these will be constant in the large-scale limit for non-interacting fluids with barotropic equation of state $P_i(\rho_i)$ (and hence vanishing non-adiabatic pressure perturbations for each fluid) [23]. We can identify isocurvature perturbations, such as the baryon isocurvature perturbation (8.45), with the difference between each ζ_I and, by convention, ζ_γ for the photons

$$S_I \equiv 3\left(\zeta_I - \zeta_\gamma\right) \ , \tag{8.47}$$

and the total curvature perturbation (8.32) is given by the weighted sum

$$\zeta = \sum_I \frac{\dot{\rho}_I}{\dot{\rho}}\zeta_I \ . \tag{8.48}$$

Arbitrary field perturbations in multiple field inflation can be decomposed into adiabatic perturbations along the inflaton trajectory and $n-1$ entropy perturbations orthogonal to the inflaton direction (8.9) in field space:

$$\delta\sigma = \sum_I \hat{\sigma}_I \delta\varphi_I , \qquad (8.49)$$

$$\delta s_I = \sum_J \hat{s}_{IJ} \delta\varphi_J , \qquad (8.50)$$

where $\sum_I \hat{s}_{JI}\hat{\sigma}_I = 0$. Without loss of generality I will assume that the entropy fields are also mutually orthogonal in field space. Note that I have assumed that the fields have canonical kinetic terms, that is, the field space metric is flat. See [7, 29, 30] for the generalisation to non-canonical kinetic terms.

The total momentum and pressure perturbation (8.27) and (8.26) for n scalar field perturbations can be written in the same form as for a single inflation field

$$\delta q = -\dot{\sigma}\delta\sigma , \qquad (8.51)$$

$$\delta P = \dot{\sigma}(\dot{\delta\sigma} - \dot{\sigma}A) - V_\sigma\delta\sigma . \qquad (8.52)$$

However, the density perturbation (8.25) is given by

$$\delta\rho = \dot{\sigma}(\dot{\delta\sigma} - \dot{\sigma}A) + V_\sigma\delta\sigma + 2\delta_s V , \qquad (8.53)$$

where the deviation from the single-field result arises due to the non-adiabatic perturbation of the potential orthogonal to the inflaton trajectory:

$$\delta_s V \equiv \sum_I V_I \delta\varphi_I - V_\sigma\delta\sigma . \qquad (8.54)$$

The non-adiabatic pressure perturbation (8.39) is written as [1, 6]

$$\delta P_{\mathrm{nad}} = -\frac{2V_\sigma}{3H\dot{\sigma}}\delta\rho_{\mathrm{m}} - 2\delta_s V , \qquad (8.55)$$

where the comoving density perturbation, $\delta\rho_{\mathrm{m}}$, is given by (8.36). Although the constraint (8.37) requires the comoving density perturbation to become small on large scales, as in the single-field case, there is now an additional contribution to the non-adiabatic pressure due to non-adiabatic perturbations of the potential which need not be small on large scales.

It is important to emphasise that the presence of entropy perturbations during inflation does not mean that the "primordial" density perturbation (at the epoch of primordial nucleosynthesis) will contain isocurvature modes. In particular, if the universe undergoes conventional reheating at the end of inflation and all particle species are driven towards thermal equilibrium with their abundances determined by a single temperature (with no non-zero chemical potentials) then the primordial perturbations must be adiabatic [31]. It is these primordial perturbations that set the initial conditions for the evolution of the radiation-matter fluid that determines the anisotropies in the cosmic microwave background and large-scale structure in our universe, and thus are directly constrained by observations. We will see that while the existence of non-adiabatic perturbations after inflation requires the existence of non-adiabatic perturbations during inflation [32], it is not true that non-adiabatic modes during inflation necessarily give primordial isocurvature modes [31].

8.4 Perturbations from Two-Field Inflation

In this section I will consider the specific example of the coupled evolution of two canonical scalar fields, ϕ and χ, during inflation and how this can give rise to correlated curvature and entropy perturbations on large scales after inflation [33]. I will use the local rotation in field space defined by (8.49) and (8.50) to describe the instantaneous adiabatic and entropy field perturbations.

The inflation field perturbation (8.49) is gauge-dependent, but we can choose to work with the inflaton perturbation in the spatially flat ($\psi = 0$) gauge:

$$\delta\sigma_\psi \equiv \delta\sigma + \frac{\dot\sigma}{H}\psi \ . \tag{8.56}$$

On the other hand, the orthogonal entropy perturbation (8.50) is automatically gauge-invariant.

The generalisation to two fields of the evolution equation for the inflation field perturbation in the spatially flat gauge, obtained from the perturbed Klein–Gordon equations (8.22), is [6]

$$\ddot{\delta\sigma}_\psi + 3H\dot{\delta\sigma}_\psi + \left[\frac{k^2}{a^2} + V_{\sigma\sigma} - \dot\theta^2 - \frac{8\pi G}{a^3}\frac{d}{dt}\left(\frac{a^3\dot\sigma^2}{H}\right)\right]\delta\sigma_\psi$$

$$= 2\frac{d}{dt}(\dot\theta\,\delta s) - 2\left(\frac{V_\sigma}{\dot\sigma} + \frac{\dot H}{H}\right)\dot\theta\,\delta s \ , \tag{8.57}$$

and the entropy perturbation obeys

$$\ddot{\delta s} + 3H\dot{\delta s} + \left(\frac{k^2}{a^2} + V_{ss} + 3\dot\theta^2\right)\delta s = \frac{\dot\theta}{\dot\sigma}\frac{k^2}{2\pi Ga^2}\Psi \ , \tag{8.58}$$

where $\tan\theta = \dot\chi/\dot\phi$ and

$$V_{\sigma\sigma} \equiv (\cos^2\theta)V_{\phi\phi} + (\sin 2\theta)V_{\phi\chi} + (\sin^2\theta)V_{\chi\chi}, \tag{8.59}$$

$$V_{ss} \equiv (\sin^2\theta)V_{\phi\phi} - (\sin 2\theta)V_{\phi\chi} + (\cos^2\theta)V_{\chi\chi}. \tag{8.60}$$

We can identify a purely adiabatic mode where $\delta s = 0$ on large scales. However, a non-zero entropy perturbation does appear as a source term in the perturbed inflaton equation whenever the inflaton trajectory is curved in field space, i.e. $\dot\theta \neq 0$. We note that $\dot\theta$ is given by [6]

$$\dot\theta = -\frac{V_{\rm s}}{\dot\sigma} \, , \tag{8.61}$$

where $V_{\rm s}$ is the potential gradient orthogonal to the inflaton trajectory in field space.

The entropy perturbation evolves independently of the curvature perturbation on large scales. It couples to the curvature perturbation only through the gradient of the longitudinal gauge metric potential, Ψ. Thus entropy perturbations are also described as "isocurvature" perturbations on large scales. Equation (8.57) shows that the entropy perturbation δs works as a source term for the adiabatic perturbation. This is in fact clearly seen if we take the time derivative of the curvature perturbation [6]:

$$\dot{\mathcal{R}} = \frac{H}{\dot H}\frac{k^2}{a^2}\Psi + \frac{2H}{\dot\sigma}\dot\theta\delta s \, . \tag{8.62}$$

Therefore $\mathcal{R}$ (or ζ) is not conserved even in the large-scale limit in the presence of the entropy perturbation δs with a non-straight trajectory in field space ($\dot\theta \neq 0$).

Analogous to the single-field case we can introduce slow-roll parameters for light, weakly coupled fields [34]. At first order in a slow-roll expansion, the inflaton rolls directly down the potential slope, that is $V_{\rm s} \simeq 0$. Thus we have only one slope parameter

$$\epsilon \equiv -\frac{\dot H}{H^2} \simeq \frac{1}{16\pi G}\left(\frac{V_\sigma}{V}\right)^2 \, , \tag{8.63}$$

but three parameters, $\eta_{\sigma\sigma}$, $\eta_{\sigma s}$ and η_{ss}, describing the curvature of the potential, where

$$\eta_{IJ} \equiv \frac{1}{8\pi G}\frac{V_{IJ}}{V} \, . \tag{8.64}$$

The background slow-roll solution is described in terms of the slow-roll parameters by

$$\dot\sigma^2 \simeq \frac{2}{3}\epsilon V \, , \quad H^{-1}\dot\theta \simeq -\eta_{\sigma s} \, , \tag{8.65}$$

while the perturbations obey

$$H^{-1}\dot{\delta\sigma}_\psi \simeq (2\epsilon - \eta_{\sigma\sigma})\,\delta\sigma_\psi - 2\eta_{\sigma s}\delta s \, ,$$
$$H^{-1}\dot{\delta s} \simeq -\eta_{ss}\delta s \, , \tag{8.66}$$

on large scales, where we neglect spatial gradients. Although $V_{\rm s} \simeq 0$ at the lowest order in slow-roll, this does not mean that the inflaton and entropy perturbations decouple. $\dot\theta$ given by (8.65) is in general non-zero at first order in slow-roll, and large-scale entropy perturbations do affect the evolution of the adiabatic perturbations when $\eta_{\sigma s} \neq 0$.

While the general solution to the two second-order perturbation equations (8.57) and (8.58) has four independent modes, the two first-order slow-roll equations (8.66) give the approximate form of the squeezed state on large scales. This has only two modes which we can describe in terms of dimensionless curvature and isocurvature perturbations:

$$\mathcal{R} \equiv \frac{H}{\dot{\sigma}}\delta\sigma_\psi , \quad \mathcal{S} \equiv \frac{H}{\dot{\sigma}}\delta s . \tag{8.67}$$

The normalisation of $\mathcal{R}$ coincides with the standard definition of the comoving curvature perturbation, (8.33). The normalisation of the dimensionless entropy during inflation, $\mathcal{S}$, is chosen here to coincide with [34]. It can be related to the non-adiabatic pressure perturbation (8.39) on large scales

$$\delta P_{\text{nad}} \simeq -\epsilon\eta_{\sigma s}\frac{H^2}{2\pi G}\mathcal{S} . \tag{8.68}$$

The slow-roll approximation can provide a useful approximation to the instantaneous evolution of the fields and their perturbations on large scales during slow-roll inflation, but is not expected to remain accurate when integrated over many Hubble times, where inaccuracies can accumulate. In single-field inflation the constancy of the comoving curvature perturbation after Hubble-exit, which does not rely on the slow-roll approximation, is crucial in order to make accurate predictions of the primordial perturbations using the slow-roll approximation only around Hubble crossing. In a two-field model we must describe the evolution after Hubble-exit in terms of a general transfer matrix:

$$\begin{pmatrix} \mathcal{R} \\ \mathcal{S} \end{pmatrix} = \begin{pmatrix} 1 & T_{\mathcal{R}\mathcal{S}} \\ 0 & T_{\mathcal{S}\mathcal{S}} \end{pmatrix} \begin{pmatrix} \mathcal{R} \\ \mathcal{S} \end{pmatrix}_* . \tag{8.69}$$

On large scales the comoving curvature perturbation still remains constant for the purely adiabatic mode, corresponding to $\mathcal{S} = 0$, and adiabatic perturbations remain adiabatic. These general results are enough to fix two of the coefficients in the transfer matrix, but $T_{\mathcal{R}\mathcal{S}}$ and $T_{\mathcal{S}\mathcal{S}}$ remain to be determined either within a given theoretical model, or from observations, or ideally by both. The scale-dependence of the transfer functions depends upon the inflaton–entropy coupling at Hubble-exit during inflation and can be given in terms of the slow-roll parameters as [34]

$$\frac{\partial}{\partial \ln k}T_{\mathcal{R}\mathcal{S}} = 2\eta_{\sigma s} + (2\epsilon - \eta_{\sigma\sigma} + \eta_{ss})T_{\mathcal{R}\mathcal{S}} ,$$

$$\frac{\partial}{\partial \ln k}T_{\mathcal{S}\mathcal{S}} = (2\epsilon - \eta_{\sigma\sigma} + \eta_{ss})T_{\mathcal{S}\mathcal{S}} . \tag{8.70}$$

8.4.1 Initial Power Spectra

For weakly coupled, light fields we can neglect interactions on wavelengths below the Hubble scale, so that vacuum fluctuations give rise to a spectrum of uncorrelated field fluctuations on the Hubble scale ($k = aH$) during inflation given by (8.17):

$$\mathcal{P}_{\delta\phi} \simeq \mathcal{P}_{\delta\chi} \simeq \left(\frac{H}{2\pi}\right)^2_* , \qquad (8.71)$$

where we use a $*$ to denote quantities evaluated at Hubble-exit. If a field has a mass comparable to the Hubble scale or larger then the vacuum fluctuations on wavelengths greater than the effective Compton wavelength are suppressed. In addition fluctuations in strongly interacting fields may develop correlations before Hubble-exit. But during slow-roll inflation the correlation between vacuum fluctuations in weakly coupled, light fields at Hubble-exit is suppressed by slow-roll parameters. This remains true under a local rotation in field-space to another orthogonal basis such as the instantaneous inflaton and entropy directions (8.49) and (8.50) in field space.

The curvature and isocurvature power spectra at Hubble-exit are given by

$$\mathcal{P}_\mathcal{R}|_* \simeq \mathcal{P}_\mathcal{S}|_* \simeq \left(\frac{H^2}{2\pi\dot\sigma}\right)^2_* \simeq \frac{8}{3}\left(\frac{V}{\epsilon M_{\mathrm{Pl}}^4}\right)_* , \qquad (8.72)$$

while the cross-correlation is first order in slow-roll [35, 36],

$$\mathcal{C}_{\mathcal{RS}}|_* \simeq -2C\eta_{\sigma s}\,\mathcal{P}_\mathcal{R}|_* , \qquad (8.73)$$

where $C = 2 - \ln 2 - \gamma \approx 0.73$ and γ is the Euler number. The normalisation chosen for the dimensionless entropy perturbation in (8.67) ensures that the curvature and isocurvature fluctuations have the same power at horizon-exit [34]. The spectral tilts at horizon-exit are also the same and are given by

$$n_\mathcal{R}|_* - 1 \simeq n_\mathcal{S}|_* \simeq -6\epsilon + 2\eta_{\sigma\sigma} . \qquad (8.74)$$

where $n_\mathcal{R} - 1 \equiv \mathrm{d}\ln\mathcal{P}_\mathcal{R}/\mathrm{d}\ln k$ and $n_\mathcal{S} \equiv \mathrm{d}\ln\mathcal{P}_\mathcal{S}/\mathrm{d}\ln k$.

The tensor perturbations (8.20) are decoupled from scalar metric perturbations at first order and hence the power spectrum has the same form as in single-field inflation. Thus the power spectrum of gravitational waves on super-Hubble scales during inflation is given by

$$\mathcal{P}_\mathrm{T}|_* \simeq \frac{16H^2}{\pi M_{\mathrm{Pl}}^2} \simeq \frac{128}{3}\frac{V_*}{M_{\mathrm{Pl}}^4} , \qquad (8.75)$$

and the spectral tilt is

$$n_\mathrm{T}|_* \simeq -2\epsilon . \qquad (8.76)$$

8.4.2 Primordial Power Spectra

The resulting primordial power spectra on large scales can be obtained simply by applying the general transfer matrix (8.69) to the initial scalar perturbations. The scalar power spectra probed by astronomical observations are thus given by [34]

$$\mathcal{P}_{\mathcal{R}} = (1 + T_{\mathcal{R}S}^2)\mathcal{P}_{\mathcal{R}}|_* \tag{8.77}$$

$$\mathcal{P}_S = T_{SS}^2 \mathcal{P}_{\mathcal{R}}|_* \tag{8.78}$$

$$\mathcal{C}_{\mathcal{R}S} = T_{\mathcal{R}S} T_{SS} \mathcal{P}_{\mathcal{R}}|_* \ . \tag{8.79}$$

The cross-correlation can be given in terms of a dimensionless correlation angle:

$$\cos \Delta \equiv \frac{\mathcal{C}_{\mathcal{R}S}}{\sqrt{\mathcal{P}_{\mathcal{R}}\mathcal{P}_S}} = \frac{T_{\mathcal{R}S}}{\sqrt{1 + T_{\mathcal{R}S}^2}} \ . \tag{8.80}$$

We see that if we can determine the dimensionless correlation angle, Δ, from observations, then this determines the off-diagonal term in the transfer matrix

$$T_{\mathcal{R}S} = \cot \Delta \ , \tag{8.81}$$

and we can in effect measure the contribution of the entropy perturbation during two-field inflation to the resultant curvature primordial perturbation. In particular this allows us in principle to deduce from observations the power spectrum of the curvature perturbation at Hubble-exit during two-field slow-roll inflation [34]:

$$\mathcal{P}_{\mathcal{R}}|_* = \mathcal{P}_{\mathcal{R}} \sin^2 \Delta \ . \tag{8.82}$$

The scale dependence of the resulting scalar power spectra depends both upon the scale dependence of the initial power spectra and of the transfer coefficients. The spectral tilts are given from (8.77)–(8.79) by

$$\begin{aligned} n_{\mathcal{R}} - 1 &= n_{\mathcal{R}}|_* - 1 + H_*^{-1}(\partial T_{\mathcal{R}S}/\partial t_*) \sin 2\Delta \,, \\ n_S &= n_{\mathcal{R}}|_* - 1 + 2H_*^{-1}(\partial \ln T_{SS}/\partial t_*) \,, \\ n_{\mathcal{C}} &= n_{\mathcal{R}}|_* - 1 + H_*^{-1}[(\partial T_{\mathcal{R}S}/\partial t_*)\tan \Delta + (\partial \ln T_{SS}/\partial t_*)] \,, \end{aligned} \tag{8.83}$$

where we have used (8.81) to eliminate $T_{\mathcal{R}S}$ in favour of the observable correlation angle Δ. Substituting (8.74) for the tilt at Hubble-exit, and (8.70) for the scale dependence of the transfer functions, we obtain [34]

$$\begin{aligned} n_{\mathcal{R}} &\simeq 1 - (6 - 4\cos^2 \Delta)\epsilon \\ &\quad + 2\left(\eta_{\sigma\sigma} \sin^2 \Delta + 2\eta_{\sigma s} \sin \Delta \cos \Delta + \eta_{ss} \cos^2 \Delta\right) \,, \\ n_S &\simeq -2\epsilon + 2\eta_{ss} \,, \\ n_{\mathcal{C}} &\simeq -2\epsilon + 2\eta_{ss} + 2\eta_{\sigma s} \tan \Delta \ . \end{aligned} \tag{8.84}$$

Although the overall amplitude of the transfer functions are dependent upon the evolution after Hubble-exit and through reheating into the radiation era, the spectral tilts can be expressed solely in terms of the slow-roll parameters at Hubble-exit during inflation and the correlation angle, Δ, which can in principle be observed.

If the primordial curvature perturbation results solely from the adiabatic inflation field fluctuations during inflation then we have $T_{\mathcal{R}S} = 0$ in (8.77) and hence $\cos \Delta = 0$ in (8.84), which yields the standard single-field result

$$n_{\mathcal{R}} \simeq 1 - 6\epsilon + 2\eta_{\sigma\sigma} \ . \qquad (8.85)$$

Any residual isocurvature perturbations must be uncorrelated with the adiabatic curvature perturbation (at first-order in slow-roll) with spectral index

$$n_{\mathcal{S}} \simeq -2\epsilon + 2\eta_{ss} \ . \qquad (8.86)$$

On the other hand, if the observed primordial curvature perturbation is produced due to some entropy field fluctuations during inflation, we have $T_{\mathcal{RS}} \gg 1$ and $\sin\Delta \simeq 0$. In a two-field inflation model any residual primordial isocurvature perturbations will then be completely correlated (or anti-correlated) with the primordial curvature perturbation and we have

$$n_{\mathcal{R}} - 1 \simeq n_{\mathcal{C}} \simeq n_{\mathcal{S}} \simeq -2\epsilon + 2\eta_{ss} \ . \qquad (8.87)$$

The gravitational wave power spectrum is frozen-in on large scales, independent of the scalar perturbations, and hence

$$\mathcal{P}_{\mathrm{T}} = \mathcal{P}_{\mathrm{T}}|_* \ . \qquad (8.88)$$

Thus we can derive a modified consistency relation [15] between observables applicable in the case of two-field slow-roll inflation:

$$r = \frac{\mathcal{P}_{\mathrm{T}}}{\mathcal{P}_{\mathcal{R}}} \simeq -8n_{\mathrm{T}}\sin^2\Delta \ . \qquad (8.89)$$

This relation was first obtained in [37] at the end of two-field inflation, and verified in [38] for slow-roll models. But it was realised in [34] that this relation also applies to the primordial perturbation spectra in the radiation era long after two-field slow-roll inflation has ended, and hence may be tested observationally.

More generally, if there is any additional source of the scalar curvature perturbation, such as additional scalar fields during inflation, then this could give an additional contribution to the primordial scalar curvature spectrum without affecting the gravitational waves, and hence the more general result is the inequality [39]:

$$r \leq -8n_{\mathrm{T}}\sin\Delta \ . \qquad (8.90)$$

This leads to a fundamental difference when interpreting the observational constraints on the amplitude of primordial tensor perturbations in multiple inflation models. In single-field inflation, observations directly constrain n_{T} and hence, from (8.76), the slow-roll parameter ϵ. However, in multiple field inflation, non-adiabatic perturbations can enhance the power of scalar perturbations after Hubble-exit and hence observational constraints on the amplitude of primordial tensor perturbations do not directly constrain the slow-roll parameter ϵ.

Current CMB data alone require $r < 0.55$ (assuming power-law primordial spectra) [40] which in single-field models is interpreted as requiring $\epsilon < 0.04$. But in multiple field models ϵ could be larger if the primordial density perturbation comes from non-adiabatic perturbations during inflation.

8.5 Non-Gaussianity

A powerful technique to calculate the primordial curvature perturbation resulting from many inflation models, including multi-field models, is to note that the curvature perturbation ζ defined in (8.32) can be interpreted as a perturbation in the local expansion [39, 41, 42]

$$\zeta = \delta N \; , \tag{8.91}$$

where δN is the perturbed expansion to uniform-density hypersurfaces with respect to spatially flat hypersurfaces, which is given to first order by

$$\zeta = -H \frac{\delta\rho_\psi}{\dot{\rho}} \; , \tag{8.92}$$

where $\delta\rho_\psi$ must be evaluated on spatially flat ($\psi = 0$) hypersurfaces.

An important simplification arises on large scales where anisotropy and spatial gradients can be neglected, and the local density, expansion, etc. obeys the same evolution equations as the homogeneous FRW universe [23, 27, 39, 42, 43, 44]. Thus we can use the homogeneous FRW solutions to describe the local evolution, which is known as the "separate universe" approach [23, 39, 43, 44]. In particular we can evaluate the perturbed expansion in different parts of the universe resulting from different initial values for the fields during inflation using the homogeneous background solutions [39]. The integrated expansion from some initial spatially flat hypersurface up to a late-time fixed density hypersurface, say at the epoch of primordial nucleosynthesis, is some function of the field values on the initial hypersurface, $N(\varphi_I|_\psi)$. The resulting primordial curvature perturbation on the uniform-density hypersurface is then

$$\zeta = \sum_I \delta N_I \, \delta\varphi_{I\psi} \; , \tag{8.93}$$

where $N_I \equiv \partial N/\partial\varphi_I$ and $\delta\varphi_{I\psi}$ is the field perturbation on some initial spatially flat hypersurfaces during inflation (8.30). In particular the power spectrum for the primordial density perturbation in a multi-field inflation can be written (at leading order) in terms of the field perturbations after Hubble-exit as

$$\mathcal{P}_\zeta = \sum_I (\delta N_I)^2 \mathcal{P}_{\delta\varphi_{I\psi}} \; . \tag{8.94}$$

This approach is readily extended to estimate the non-linear effect of field perturbations on the metric perturbations [27, 42, 43]. We can take (8.91) as our definition of the non-linear primordial curvature perturbation, ζ, so that in the radiation-dominated era the non-linear extension of (8.92) is given by [42]

$$\zeta = \frac{1}{4} \ln \left(\frac{\tilde{\rho}}{\rho} \right)_\psi \; , \tag{8.95}$$

where $\tilde{\rho}(t, \mathbf{x})$ is the perturbed (inhomogeneous) density evaluated on a spatially flat hypersurface and $\rho(t)$ is the background (homogeneous) density. This non-linear curvature perturbation as a function of the initial field fluctuations can simply be expanded as a Taylor expansion [45, 46, 47, 48]

$$\zeta \simeq \sum_I N_I \,\delta\varphi_{I\psi} + \frac{1}{2}\sum_{I,J} N_{IJ} \,\delta\varphi_{I\psi}\,\delta\varphi_{J\psi} + \frac{1}{6}\sum_{I,J,K} N_{IJK}\,\delta\varphi_{I\psi}\,\delta\varphi_{J\psi}\,\delta\varphi_{K\psi} + \cdots \ .$$

(8.96)

where we now identify (8.93) as the leading-order term.

We expect the field perturbations at Hubble-exit to be close to Gaussian for weakly coupled scalar fields during inflation [47, 49, 50, 51, 52]. In this case the bispectrum of the primordial curvature perturbation at leading (fourth) order, can be written using the δN-formalism, as [46, 53]

$$B_\zeta(\mathbf{k}_1, \mathbf{k}_2, \mathbf{k}_3) = \frac{6}{5} f_{\mathrm{NL}}\left[P_\zeta(k_1)P_\zeta(k_2) + P_\zeta(k_2)P_\zeta(k_3) + P_\zeta(k_3)P_\zeta(k_1)\right]\ . \tag{8.97}$$

where $P_\zeta(k) = 2\pi^2 \mathcal{P}_\zeta(k)/k^3$, and the dimensionless non-linearity parameter is given by [46]

$$f_{\mathrm{NL}} = \frac{5}{6}\frac{N_A N_B N^{AB}}{(N_C N^C)^2} \ . \tag{8.98}$$

Similarly, the connected part of the trispectrum in this case can be written as [45, 48]

$$T_\zeta(\mathbf{k}_1, \mathbf{k}_2, \mathbf{k}_3, \mathbf{k}_4) = \tau_{\mathrm{NL}}\left[P_\zeta(|\mathbf{k}_1 - \mathbf{k}_3|)P_\zeta(k_3)P_\zeta(k_4) + (11\ \mathrm{perms})\right]$$
$$+ \frac{54}{25}g_{\mathrm{NL}}\left[P_\zeta(k_2)P_\zeta(k_3)P_\zeta(k_4) + (3\ \mathrm{perms})\right] \ , \quad (8.99)$$

where

$$\tau_{\mathrm{NL}} = \frac{N_{AB}N^{AC}N^B N_C}{(N_D N^D)^3} \ , \tag{8.100}$$

$$g_{\mathrm{NL}} = \frac{25}{54}\frac{N_{ABC}N^A N^B N^C}{(N_D N^D)^3} \ . \tag{8.101}$$

The expression for τ_{NL} was first given in [54]. Note that we have factored out products in the trispectrum with different k dependence in order to define the two k-independent non-linearity parameters τ_{NL} and g_{NL}. This gives the possibility that observations may be able to distinguish between the two parameters [55].

In many cases there is single direction in field-space, χ, which is responsible for perturbing the local expansion, $N(\chi)$, and hence generating the primordial curvature perturbation (8.96). For example this would be the inflaton field in single-field models of inflation, or it could be the late-decaying scalar field in the curvaton scenario [56, 57, 58] as will be discussed in the next section. In this case the curvature perturbation (8.96) is given by

$$\zeta \simeq N'\delta\chi_\psi + \frac{1}{2}N''\delta\chi_\psi^2 + \frac{1}{6}N'''\delta\chi_\psi^3 + \cdots , \tag{8.102}$$

and the non-Gaussianity of the primordial perturbation has the simplest "local" form

$$\zeta = \zeta_1 + \frac{3}{5}f_{\rm NL}\zeta_1^2 + \frac{9}{25}\zeta_1^3 + \cdots \tag{8.103}$$

where $\zeta_1 = N'\delta\chi_\psi$ is the leading-order Gaussian curvature perturbation and the non-linearity parameters $f_{\rm NL}$ and $g_{\rm NL}$, are given by [46, 59]

$$f_{\rm NL} = \frac{5}{6}\frac{N''}{(N')^2} , \tag{8.104}$$

$$g_{\rm NL} = \frac{25}{54}\frac{N'''}{(N')^3} , \tag{8.105}$$

The primordial bispectrum and trispectrum are then given by (8.97) and (8.99), where the non-linearity parameters $f_{\rm NL}$ and $g_{\rm NL}$, given in (8.98) and (8.101), reduce to (8.104) and (8.105), respectively, and $\tau_{\rm NL}$ given in (8.100) reduces to

$$\tau_{\rm NL} = \frac{(N'')^2}{(N')^4} = \frac{36}{25}f_{\rm NL}^2 . \tag{8.106}$$

Thus $\tau_{\rm NL}$ is proportional to $f_{\rm NL}^2$ (first shown in [55] using the Bardeen potential, and in [46] using this notation). However, the trispectrum could be large even when the bispectrum is small because of the $g_{\rm NL}$ term [55, 59].

In the case of where the primordial curvature perturbation is generated solely by adiabatic fluctuations in the inflation field, σ, the large-scale perturbation is non-linearly conserved on large scales [42, 60, 61] and we can calculate N', N'', N''', etc. at Hubble-exit. In terms of the slow-roll parameters, we find

$$N' = \frac{H}{\dot\varphi} \simeq \frac{1}{\sqrt{2}}\frac{1}{M_{\rm Pl}}\frac{1}{\sqrt{\epsilon}} \sim \mathcal{O}\left(\epsilon^{-\frac{1}{2}}\right) , \tag{8.107}$$

$$N'' \simeq -\frac{1}{2}\frac{1}{M_{\rm Pl}^2}\frac{1}{\epsilon}(\eta_{\sigma\sigma} - 2\epsilon) \sim \mathcal{O}(1) , \tag{8.108}$$

$$N''' \simeq \frac{1}{\sqrt{2}}\frac{1}{M_{\rm Pl}^3}\frac{1}{\epsilon\sqrt{\epsilon}}\left(\epsilon\eta_{\sigma\sigma} - \eta_{\sigma\sigma}^2 + \frac{1}{2}\xi_\sigma^2\right) \sim \mathcal{O}(\epsilon^{\frac{1}{2}}), \tag{8.109}$$

where we have introduced the second-order slow-roll parameter $\xi_\sigma^2 = M_{\rm Pl}^4 V_\sigma V_{\sigma\sigma\sigma}/V^2$. Hence the non-linearity parameters for single-field inflation, (8.104) and (8.105), are given by

$$f_{\rm NL} = \frac{5}{6}(\eta_{\sigma\sigma} - 2\epsilon) , \tag{8.110}$$

$$g_{\rm NL} = \frac{25}{54}\left(2\epsilon\eta_{\sigma\sigma} - 2\eta_{\sigma\sigma}^2 + \xi_\sigma^2\right) . \tag{8.111}$$

with τ_{NL} given by (8.106). Although there are additional contributions to the primordial bispectrum and trispectrum coming from the intrinsic non-Gaussianity of the field perturbations at Hubble-exit, these are also suppressed by slow-roll parameters in slow-roll inflation. Thus the primordial non-Gaussianity is likely to be too small to ever be observed in the conventional inflaton scenario of single-field slow-roll inflation. Indeed any detection of primordial non-Gaussianity $f_{NL} > 1$ would appear to rule out this inflaton scenario.

However, significant nonGaussianity can be generated due to non-adiabatic field fluctuations. Thus far it has proved difficult to generate significant non-Gaussianity in the curvature perturbation during slow-roll inflation, even in multiple field models. But detectable non-Gaussianity can be produced when the curvature perturbation is generated from isocurvature field perturbations at the end of inflation [12, 62], during inhomogeneous reheating [14, 63, 64] or after inflation in the curvaton model, which I will discuss next.

8.6 Curvaton Scenario

Consider a light, weakly coupled scalar field, χ, that decays some time after inflation has ended. There are many such scalar degrees of freedom in supersymmetric theories and if they are too weakly coupled, and their lifetime is too long, this leads to the "Polonyi problem." Assuming the field is displaced from the minimum of its effective potential at the end of inflation, the field evolves little until the Hubble rate drops below its effective mass. Then it oscillates, with a time-averaged equation of state of a pressureless fluid, $P_\chi = 0$, (or, equivalently, a collection of non-relativistic particles) and will eventually come to dominate the energy density of the Universe. To avoid disrupting the standard "hot big bang" model and in particular to preserve the successful radiation-dominated model of primordial nucleosynthesis, we require that such fields decay into radiation before $t \sim 1$ s. For a weakly coupled field that decays with only gravitational strength, $\Gamma \sim m_\chi^3/M_P^2$, this requires $m_\chi > 100$ TeV.

But there is a further important feature of late-decaying scalar fields that has only recently received serious consideration. If the field is inhomogeneous then it could lead to an inhomogeneous radiation density after it decays [65, 66]. This is the basis of the curvaton scenario [56, 57, 58].

If this field is light ($m < H$) during inflation then small-scale quantum fluctuations will lead to a spectrum of large-scale perturbations, whose initial amplitude at Hubble-exit is given (at leading order in slow-roll) by (8.17). When the Hubble rate drops and the field begins oscillating after inflation, this leads to a first-order density perturbation in the χ-field:

$$\zeta_\chi = -\psi + \frac{\delta\rho_\chi}{3\rho_\chi} \ . \tag{8.112}$$

where $\rho_\chi = m_\chi^2 \chi^2/2$. ζ_χ remains constant for the oscillating curvaton field on large scales, so long as we can neglect its energy loss due to decay. Using (8.17) for the field fluctuations at Hubble-exit and neglecting any non-linear evolution of the χ-field after inflation (consistent with our assumption that the field is weakly coupled), we have

$$\mathcal{P}_{\zeta_\chi} \simeq \left(\frac{H}{6\pi\chi} \right)^2_{k=aH} . \tag{8.113}$$

The total density perturbation (8.32), considering radiation, γ, and the curvaton, χ, is given by [57]

$$\zeta = \frac{4\rho_\gamma\zeta_\gamma + 3\rho_\chi\zeta_\chi}{4\rho_\gamma + 3\rho_\chi} . \tag{8.114}$$

Thus if the radiation generated by the decay of the inflaton at the end of inflation is unperturbed ($\mathcal{P}_{\zeta_\gamma}^{1/2} \ll 10^{-5}$) the total curvature perturbation grows as the density of the χ-field grows relative to the radiation: $\zeta \sim \Omega_\chi\zeta_\chi$.

Ultimately the χ-field must decay (when $H \sim \Gamma$) and transfer its energy density and, crucially, its perturbation to the radiation and/or other matter fields. In the simplest case that the non-relativistic χ-field decays directly to radiation a full analysis [67, 68] of the coupled evolution equation gives the primordial radiation perturbation (after the decay)

$$\zeta = r_\chi(p)\zeta_\chi , \tag{8.115}$$

where $p \equiv [\Omega_\chi/(\Gamma/H)^{1/2}]_{\text{initial}}$ is a dimensionless parameter which determines the maximum value of Ω_χ before it decays, and empirically we find [68]

$$r_\chi(p) \simeq 1 - \left(1 + \frac{0.924}{1.24} p \right)^{-1.24} . \tag{8.116}$$

For $p \gg 1$ the χ-field dominates the total energy density before it decays and $r_\chi \sim 1$, while for $p \ll 1$ we have $r_\chi \sim 0.924 p \ll 1$.

Finally combining (8.113) and (8.115) we have

$$\mathcal{P}_\zeta \simeq r_\chi^2(p) \left(\frac{H}{6\pi\chi} \right)^2_{k=aH} . \tag{8.117}$$

Note that the primordial curvature perturbation tends to have less power on small scales due to the decreasing Hubble rate at Hubble-exit in (8.117), but can also have more power on small scales due to the decreasing χ, for a positive effective mass-squared, during inflation. In terms of slow-roll parameters the actual tilt is given by (8.84) when $\sin\Delta = 0$

$$n_\mathcal{R} - 1 \simeq -2\epsilon + 2\eta_{\chi\chi} . \tag{8.118}$$

In the extreme slow-roll limit the spectrum becomes scale invariant, as in the inflaton scenario.

In contrast to the inflaton scenario the final density perturbation in the curvaton scenario is a very much dependent upon the physics after the field perturbation exited the Hubble scale during inflation. For instance, if the curvaton lifetime is too short then it will decay before it can significantly perturb the total energy density and $\mathcal{P}_{\zeta_\gamma}^{1/2} \ll 10^{-5}$. The observational constraint on the amplitude of the primordial perturbations gives a single constraint upon both the initial fluctuations during inflation and the post-inflationary decay time. This is in contrast to the inflaton scenario where the primordial perturbation gives a direct window onto the dynamics of inflation, independently of the physics at lower energies. In the curvaton scenario there is the possibility of connecting the generation of primordial perturbations to other aspects of cosmological physics. For instance, it may be possible to identify the curvaton with fields whose late-decay is responsible for the origin of the baryon asymmetry in the universe, in particular with sneutrino models of leptogenesis (in which an initial lepton asymmetry is converted into a baryon asymmetry at the electroweak transition) [69].

The curvaton scenario has re-invigorated attempts to embed models of inflation in the very early universe within minimal supersymmetric models of particle physics constrained by experiment [70]. It may be possible that the inflation field driving inflation can be completely decoupled from visible matter if the dominant radiation in the universe today comes from the curvaton decay rather than reheating at the end of inflation. Indeed the universe needs not be radiation-dominated at all until the curvaton decays if instead the inflaton fast-rolls at the end of inflation.

The curvaton offers a new range of theoretical possibilities, but ultimately we will require observational and/or experimental predictions to decide whether the curvaton, inflaton or some other field generated the primordial perturbation. I will discuss observational predictions of the curvaton scenario in the following subsection.

8.6.1 Non-Gaussianity

The best way to distinguish between different scenarios for the origin of structure could be the statistical properties of the primordial density perturbation. Primordial density perturbations in the curvaton scenario originate from the small-scale vacuum fluctuations of the weakly interacting curvaton field during inflation, which can be described on super-Hubble scales by a Gaussian random field. Thus deviations from Gaussianity in the primordial bispectrum and connected trispectrum can be parameterised by the dimensionless parameters f_{NL} and g_{NL} defined in (8.104) and (8.105).

When the curvaton field begins oscillating about a quadratic minimum of its potential we have $\rho_\chi = m_\chi^2 \chi^2 / 2$, and the time-averaged equation of state becomes $P_\chi = 0$. The non-linear generalisation of the primordial curvature perturbation (8.46) on hypersurfaces of uniform-curvaton density is then

$$\zeta_\chi = \frac{1}{3} \ln \left(\frac{\tilde{\rho}_\chi}{\rho_\chi} \right)_\psi , \qquad (8.119)$$

where we distinguish here between the inhomogeneous density $\tilde{\rho}_\chi$ on spatially flat hypersurfaces and the average density ρ_χ. Given that $\rho_\chi \propto \chi^2$ we thus have

$$\zeta_\chi = \frac{1}{3} \ln \left(1 + \frac{2\chi\, \delta\chi + \delta\chi^2}{\chi^2} \right)_\psi . \qquad (8.120)$$

This gives the full probability distribution function for ζ_χ for Gaussian field perturbations $\delta\chi$. Expanding to first order we obtain $\zeta_{\chi 1} = 2\, \delta\chi_\psi / 3\chi$, and then to second and third orders we obtain by analogy with (8.103) the non-linearity parameters for the curvaton density perturbation

$$\zeta_\chi \simeq \zeta_{\chi 1} + \frac{3}{5} f_{\mathrm{NL}}^\chi \zeta_{\chi 1}^2 + \frac{9}{25} g_{\mathrm{NL}}^\chi \zeta_{\chi 1}^3 + \cdots , \qquad (8.121)$$

where [59]

$$f_{\mathrm{NL}}^\chi = -\frac{5}{4}, \qquad (8.122)$$

$$g_{\mathrm{NL}}^\chi = \frac{25}{12} . \qquad (8.123)$$

If the curvaton dominates the total energy density before it decays into radiation, then this is the curvature perturbation, and specifically the non-linearity parameters, inherited by the primordial radiation density. Although not suppressed by slow-roll parameters, this non-Gaussianity is still smaller than the upper limits best expected from the Planck satellite [53].

On the other hand if the curvaton decays before it dominates over the energy density of the existing radiation, so the transfer function $r_\chi(p) \ll 1$ in (8.115), then the curvaton may lead to a large and detectable non-Gaussianity in the radiation density after it decays. Assuming the sudden decay of the curvaton on the $H = \Gamma$ uniform-density hypersurface leads to a non-linear relation between the local curvaton density and the radiation density before and after the decay [59]

$$\rho_\gamma e^{-4\zeta} + \rho_\chi e^{3(\zeta_\chi - \zeta)} = \rho_\gamma + \rho_\chi . \qquad (8.124)$$

Expanding this term by term yields [46, 59, 71]

$$f_{\mathrm{NL}} = \frac{5}{4r_\chi} - \frac{5}{3} - \frac{5r_\chi}{6}, \qquad (8.125)$$

$$g_{\mathrm{NL}} = -\frac{25}{6r_\chi} + \frac{25}{108} + \frac{125r_\chi}{27} + \frac{25r_\chi^2}{18} . \qquad (8.126)$$

These reduce to (8.122) and (8.123) as $r_\chi \to 1$, but become large for $r_\chi \ll 1$. These analytic results rely on the sudden decay approximation but have been tested against numerical solutions [59, 72] and give an excellent approximation for both $r_\chi \ll 1$ and $r_\chi \simeq 1$.

More generally one can use (8.119) and (8.124), or use (8.91) and solve the non-linear, but homogeneous equations of motion to determine $N(\chi)$ to give the full probability distribution function for the primordial curvature perturbation ζ [59].

Current bounds from the WMAP satellite require $-54 < f_{\mathrm{NL}} < 114$ at the 95% confidence limit [40], and hence require $r > 0.011$. But future cosmic microwave background (CMB) experiments such as Planck could detect f_{NL} as small as around 5 [53], and it has even been suggested that it might one day be possible to constrain $f_{\mathrm{NL}} \sim 0.01$ [73].

8.6.2 Residual Isocurvature Perturbations

In the curvaton scenario the initial curvaton perturbation is a non-adiabatic perturbation and hence can in principle leave behind a residual non-adiabatic component. Perturbations in this one field would be responsible for both the total primordial density perturbation and any isocurvature mode and hence there is the clear prediction that the two should be completely correlated, corresponding to $\cos \Delta = \pm 1$ in (8.80) and $n_{\mathcal{R}} - 1 = n_{\mathcal{S}} = n_{\mathcal{C}}$ in (8.84).

Using ζ_I defined in (8.46) for different matter components it is easy to see how the curvaton could leave residual isocurvature perturbations after the curvaton decays. If any fluid has decoupled before the curvaton contributes significantly to the total energy density that fluid remains unperturbed with $\zeta_I \simeq 0$, whereas after the curvaton decays into radiation the photons perturbation is given by (8.115). Thus a residual isocurvature perturbation (8.47) is left

$$S_I = -3\zeta , \qquad (8.127)$$

which remains constant for decoupled perfect fluids on large scales.

The observational bound on isocurvature matter perturbations completely correlated with the photon perturbation is [74]

$$-0.42 < \frac{S_{\mathrm{B}} + (\rho_{\mathrm{cdm}}/\rho_{\mathrm{B}})S_{\mathrm{cdm}}}{\zeta_\gamma} < 0.25 . \qquad (8.128)$$

In particular if the baryon asymmetry is generated while the total density perturbation is still negligible then the residual baryon isocurvature perturbation, $S_{\mathrm{B}} = -3\zeta_\gamma$ would be much larger than the observational bound and such models are thus ruled out. The observational bound on CDM isocurvature perturbations are stronger by the factor $\rho_{\mathrm{cdm}}/\rho_{\mathrm{B}}$ [75] although CDM is usually assumed to decouple relatively late.

An interesting amplitude of residual isocurvature perturbations might be realised if the decay of the curvaton itself is the non-equilibrium event that generates the baryon asymmetry. In this case the net baryon number density directly inherits the perturbation $\zeta_{\mathrm{B}} = \zeta_\chi$ while the photon perturbation

$\zeta_\gamma \leq \zeta_\chi$ may be diluted by pre-existing radiation and is given by (8.115). Note that so long as the net baryon number is locally conserved it defines a conserved perturbation on large scales, even though it may still be interacting with other fluids and fields [60]. Hence the primordial baryon isocurvature perturbation (8.45) in this case is given by

$$S_{\rm B} = 3(1 - r_\chi)\zeta_\chi = \frac{3(1 - r_\chi)}{r_\chi}\zeta_\gamma . \qquad (8.129)$$

Thus the observational bound (8.128) requires $r_\chi > 0.92$ if the baryon asymmetry is generated by the curvaton decay.

There is no lower bound on the predicted amplitude of residual non-adiabatic modes and, although the detection of completely correlated isocurvature perturbations would give strong support to the curvaton scenario, the non-detection of primordial isocurvature density perturbations cannot be used to rule out the curvaton scenario. In particular, if the curvaton decays at sufficiently high temperature and all the particles produced relax to a thermal equilibrium abundance, characterised by a common temperature (and vanishing chemical potentials), then no residual isocurvature perturbations survive. In full thermal equilibrium there is a unique attractor trajectory in phase-space and only adiabatic perturbations (along this trajectory) survive on large scales.

8.7 Conclusions

Inflation offers a beautifully simple origin for structure in our universe. The zero-point fluctuations of the quantum vacuum state on sub-atomic scales are swept up by the accelerated expansion to astronomical scales, seeding an almost Gaussian distribution of primordial density perturbations. The large-scale structure of our universe can then form simply due to the gravitational instability of over-dense regions.

Astronomical observations over recent years have given strong support to this simple picture. But increasingly precise astronomical data will increasingly allow us to probe not only the parameters of what has become the standard cosmological model, but also to probe the nature of the primordial perturbations from which the structure formed. Any evidence of primordial gravitational waves, primordial isocurvature fluctuations and/or non-Gaussianity of the primordial perturbations could provide valuable information about the inflationary dynamics that preceded the hot Big Bang.

Single-field slow-roll inflation predicts adiabatic density perturbations with negligible non-Gaussianity, but could produce a gravitational wave background which could be detected by upcoming CMB experiments.

On the other hand multiple field inflation can lead to a wider range of possibilities which could be distinguished by observations. A spectrum of non-adiabatic field fluctuations on large scales during inflation could leave residual isocurvature perturbations after inflation, which can be correlated with the primordial curvature perturbation, and can give rise to a detectable level of non-Gaussianity. In simple models, such as the curvaton scenario, where the primordial curvature perturbations originate from almost Gaussian fluctuations in a single scalar field, any residual isocurvature perturbation is expected to be completely correlated with the curvature perturbation and the non-Gaussianity is of a specific "local" form.

After 25 years studying inflation, we may for the first time have evidence of a weak scale dependence of the power spectrum of the primordial curvature perturbation [40] which would begin to reveal the slow-roll dynamics during inflation. Primordial perturbations may have much more to tell us about the physics of inflation in the future.

References

1. B. A. Bassett, S. Tsujikawa and D. Wands, Rev. Mod. Phys. **78**, 537 (2006), arXiv:astro-ph/0507632.
2. A. D. Linde, Phys. Rev. D **49**, 748 (1994), arXiv:astro-ph/9307002.
3. E. J. Copeland, A. R. Liddle, D. H. Lyth, E. D. Stewart and D. Wands, Phys. Rev. D **49**, 6410 (1994), arXiv:astro-ph/9401011.
4. A. R. Liddle, A. Mazumdar and F. E. Schunck, Phys. Rev. D **58**, 061301 (1998), arXiv:astro-ph/9804177.
5. K. A. Malik and D. Wands, Phys. Rev. D **59**, 123501 (1999), arXiv:astro-ph/9812204.
6. C. Gordon, D. Wands, B. A. Bassett and R. Maartens, Phys. Rev. D **63**, 023506 (2001).
7. S. Groot Nibbelink and B. J. W. van Tent, Class. Quant. Grav. **19**, 613 (2002).
8. G. Rigopoulos, Class. Quant. Grav. **21**, 1737 (2004).
9. S. Dimopoulos, S. Kachru, J. McGreevy and J. G. Wacker, arXiv:hep-th/0507205.
10. P. Kanti and K. A. Olive, Phys. Rev. D **60**, 043502 (1999), arXiv:hep-ph/9903524; Phys. Lett. B **464**, 192 (1999), arXiv:hep-ph/9906331.
11. N. Kaloper and A. R. Liddle, Phys. Rev. D **61**, 123513 (2000), arXiv:hep-ph/9910499.
12. D. H. Lyth and A. Riotto, Phys. Rep. **314**, 1 (1999), arXiv:hep-ph/9807278.
13. S. A. Kim and A. R. Liddle, Phys. Rev. D **74**, 023513 (2006), arXiv:astro-ph/0605604.
14. C. T. Byrnes and D. Wands, Phys. Rev. D **73**, 063509 (2006), arXiv:astro-ph/0512195.
15. J. E. Lidsey, A. R. Liddle, E. W. Kolb, E. J. Copeland, Rev. Mod. Phys. **69**, 373 (1997).
16. V. F. Mukhanov, H. A. Feldman and R. H. Brandenberger, Phys. Rep. **215**, 203 (1992).
17. V. F. Mukhanov, JETP Lett. **41**, 493 (1985) [Pisma Zh. Eksp. Teor. Fiz. **41**, 402 (1985)].
18. M. Sasaki, Prog. Theor. Phys. **76**, 1036 (1986).

19. A. Taruya and Y. Nambu, Phys. Lett. B **428**, 37 (1998), arXiv:gr-qc/9709035.
20. J. M. Bardeen, P. J. Steinhardt and M. S. Turner, Phys. Rev. D **28**, 679 (1983).
21. J. M. Bardeen, *Lectures Given at 2nd Guo Shou-jing Summer School on Particle Physics and Cosmology, Nanjing, China, Jul 1988.*
22. J. Martin and D. J. Schwarz, Phys. Rev. D **57**, 3302 (1998).
23. D. Wands, K. A. Malik, D. H. Lyth and A. R. Liddle, Phys. Rev. D **62**, 043527 (2000).
24. V. N. Lukash, Sov. Phys. JETP **52**, 807 (1980).
25. D. H. Lyth, Phys. Rev. D **31**, 1792 (1985).
26. J. M. Bardeen, Phys. Rev. D **22**, 1882 (1980).
27. D. H. Lyth and D. Wands, Phys. Rev. D **68**, 103515 (2003).
28. K. A. Malik and D. Wands, JCAP **0502**, 007 (2005), arXiv:astro-ph/0411703.
29. F. Di Marco, F. Finelli and R. Brandenberger, Phys. Rev. D **67**, 063512 (2003).
30. K. Y. Choi, L. M. H. Hall and C. v. de Bruck, arXiv:astro-ph/0701247.
31. S. Weinberg, Phys. Rev. D **70**, 083522 (2004).
32. S. Weinberg, Phys. Rev. D **70**, 043541 (2004).
33. D. Langlois, Phys. Rev. D **59**, 123512 (1999).
34. D. Wands, N. Bartolo, S. Matarrese and A. Riotto, Phys. Rev. D **66**, 043520 (2002).
35. B. van Tent, Class. Quant. Grav. **21**, 349 (2004), arXiv:astro-ph/0307048.
36. C. T. Byrnes and D. Wands, Phys. Rev. D **74**, 043529 (2006), arXiv:astro-ph/0605679.
37. N. Bartolo, S. Matarrese and A. Riotto, Phys. Rev. D **64**, 123504 (2001).
38. S. Tsujikawa, D. Parkinson and B. A. Bassett, Phys. Rev. D **67**, 083516 (2003).
39. M. Sasaki and E. D. Stewart, Prog. Theor. Phys. **95**, 71 (1996).
40. D. N. Spergel et al., arXiv:astro-ph/0603449.
41. A. A. Starobinsky, JETP Lett. **42**, 152 (1985) [Pisma Zh. Eksp. Teor. Fiz. **42**, 124 (1985)].
42. D. H. Lyth, K. A. Malik and M. Sasaki, JCAP **0505**, 004 (2005).
43. M. Sasaki and T. Tanaka, Prog. Theor. Phys. **99**, 763 (1998).
44. G. I. Rigopoulos and E. P. S. Shellard, Phys. Rev. D **68**, 123518 (2003).
45. C. T. Byrnes, M. Sasaki and D. Wands, Phys. Rev. D **74**, 123519 (2006), arXiv:astro-ph/0611075.
46. D. H. Lyth and Y. Rodriguez, Phys. Rev. Lett. **95**, 121302 (2005), arXiv:astro-ph/0504045.
47. D. Seery and J. E. Lidsey, JCAP **0509**, 011 (2005), arXiv:astro-ph/0506056.
48. D. Seery and J. E. Lidsey, arXiv:astro-ph/0611034.
49. D. Seery, J. E. Lidsey and M. S. Sloth, arXiv:astro-ph/0610210.
50. J. Maldacena, JHEP **0305**, 013 (2003).
51. D. Seery and J. E. Lidsey, JCAP **0506**, 003 (2005), arXiv:astro-ph/0503692.
52. G. I. Rigopoulos and E. P. S. Shellard, J. Phys. Conf. Ser. **8** (2005) 145.
53. E. Komatsu and D. N. Spergel, Phys. Rev. D **63**, 063002 (2001).
54. L. Alabidi and D. H. Lyth, JCAP **0605**, 016 (2006), arXiv:astro-ph/0510441.
55. T. Okamoto and W. Hu, Phys. Rev. D **66**, 063008 (2002), arXiv:astro-ph/0206155.
56. K. Enqvist and M. S. Sloth, Nucl. Phys. B **626**, 395 (2002).
57. D. H. Lyth and D. Wands, Phys. Lett. B **524**, 5 (2002).
58. T. Moroi and T. Takahashi, Phys. Lett. B **522**, 215 (2001) [Erratum-ibid. B **539**, 303 (2002)].

59. M. Sasaki, J. Valiviita and D. Wands, Phys. Rev. D **74**, 103003 (2006), arXiv:astro-ph/0607627.
60. D. H. Lyth, C. Ungarelli and D. Wands, Phys. Rev. D **67**, 023503 (2003).
61. D. Langlois and F. Vernizzi, Phys. Rev. Lett. **95**, 091303 (2005), arXiv:astro-ph/ 0503416.
62. F. Bernardeau and J. P. Uzan, Phys. Rev. D **66**, 103506 (2002), arXiv:hep-ph/ 0207295; Phys. Rev. D **67**, 121301 (2003), arXiv:astro-ph/0209330.
63. M. Zaldarriaga, Phys. Rev. D **69**, 043508 (2004), arXiv:astro-ph/0306006.
64. E. W. Kolb, A. Riotto and A. Vallinotto, Phys. Rev. D **73**, 023522 (2006), arXiv:astro-ph/0511198.
65. S. Mollerach, Phys. Rev. D **42**, 313 (1990).
66. A. D. Linde and V. Mukhanov, Phys. Rev. D **56**, 535 (1997).
67. K. A. Malik, D. Wands and C. Ungarelli, Phys. Rev. D **67**, 063516 (2003).
68. S. Gupta, K. A. Malik and D. Wands, Phys. Rev. D **69**, 063513 (2004).
69. K. Hamaguchi, H. Murayama and T. Yanagida, Phys. Rev. D **65**, 043512 (2002).
70. K. Enqvist and A. Mazumdar, Phys. Rep. **380**, 99 (2003).
71. N. Bartolo, S. Matarrese and A. Riotto, Phys. Rev. D **69**, 043503 (2004).
72. K. A. Malik and D. H. Lyth, JCAP **0609**, 008 (2006), arXiv:astro-ph/0604387.
73. A. Cooray, arXiv:astro-ph/0610257.
74. A. Lewis, arXiv:astro-ph/0603753.
75. C. Gordon and A. Lewis, Phys. Rev. D **67**, 123513 (2003).

9

The Quest for Non-gaussianity

Antonio Riotto[1,2]

[1] CERN, Theory Division, Geneva 23, CH-1211, Switzerland
[2] INFN, Sezione di Padova, Via Marzolo 8, 35131 Padova Italy
   riotto.antonio@pd.infn.it

Abstract. Non-Gaussianity emerges as a key observable to discriminate among competing scenarios for the generation of cosmological perturbations and is one of the primary targets of present and future Cosmic Microwave Background satellite missions. We discuss the state-of-the-art of the subject of non-gaussianity, both from the theoretical and the observational point of view.

9.1 Introduction: Why Is it so Interesting to Measure Non-gaussianity in Cosmological Perturbations?

One of the relevant ideas in modern cosmology is represented by the inflationary paradigm. It is widely believed that there was an early epoch in the history of the Universe – before the epoch of primordial nucleosynthesis – when the Universe expansion was accelerated. Such a period of *cosmological inflation* can be attained if the energy density of the Universe is dominated by the vacuum energy density associated with the potential of a scalar field φ, called the inflation field. Through its kinematic properties, namely the acceleration of the Universe, the inflationary paradigm can elegantly solve the flatness, the horizon and the monopole problems of the standard Big–Bang cosmology, and in fact the first model of inflation by Guth in 1981 was introduced to address such problems. However, over the years, inflation has also become so popular because of another compelling feature. It can explain the production of the first density perturbations in the early Universe which are the seeds for the Large–Scale Structure (LSS) in the distribution of galaxies and the underlying dark matter and for the Cosmic Microwave Background (CMB) temperature anisotropies that we observe today. In fact inflation has become the dominant paradigm to understand the initial conditions for structure formation and CMB anisotropies. In the inflationary picture, primordial density and gravity–wave fluctuations are created from quantum fluctuations "redshifted" out of the horizon during an early period of superluminal expansion of the Universe, where they are "frozen". Perturbations at the surface of

A. Riotto: *The Quest for Non-gaussianity*, Lect. Notes Phys. **738**, 305–358 (2008)
DOI 10.1007/978-3-540-74353-8_9   © Springer-Verlag Berlin Heidelberg 2008

last scattering are observable as temperature anisotropies in the CMB, which was first detected by the Cosmic Background Explorer (COBE) satellite. The last and most impressive confirmation of the inflationary paradigm has been recently provided by the data of the Wilkinson Microwave Anisotropy Probe (WMAP) mission. The WMAP collaboration has produced a full–sky map of the angular variations of the CMB, with unprecedented accuracy. WMAP data confirm the inflationary mechanism as responsible for the generation of curvature (adiabatic) superhorizon fluctuations.

Since the primordial cosmological perturbations are tiny, the generation and evolution of fluctuations during inflation has been studied using linear perturbation theory. Within this approach, the primordial density perturbation is Gaussian; in other words, its Fourier components are uncorrelated and have random phases. Despite the simplicity of the inflationary paradigm, the mechanism by which cosmological adiabatic perturbations are generated is not yet established. In the standard slow–roll scenario associated to single field models of inflation, the observed density perturbations are due to fluctuations of the inflation field itself when it slowly rolls down its potential. When inflation ends, the inflaton φ oscillates about the minimum of its potential $V(\varphi)$ and decays, thereby reheating the Universe. As a result of the fluctuations each region of the Universe goes through the same history but at slightly different times. The final temperature anisotropies are caused by inflation lasting for different amounts of time in different regions of the Universe leading to adiabatic perturbations. Under this hypothesis, the WMAP dataset already allows to extract the parameters relevant for distinguishing among single–field inflation models.

An alternative to the standard scenario is represented by the curvaton mechanism where the final curvature perturbations are produced from an initial isocurvature perturbation associated with the quantum fluctuations of a light scalar field (other than the inflaton), the curvaton, whose energy density is negligible during inflation. The curvaton isocurvature perturbations are transformed into adiabatic ones when the curvaton decays into radiation much after the end of inflation.

Recently, other mechanisms for the generation of cosmological perturbations have been proposed, the inhomogeneous reheating scenario, the ghost inflationary scenario and the D–cceleration scenario, just to mention a few. For instance, the inhomogeneous reheating scenario acts during the reheating stage after inflation if superhorizon spatial fluctuations in the decay rate of the inflation field are induced during inflation, causing adiabatic perturbations in the final reheating temperature in different regions of the Universe.

The generation of gravity–wave fluctuations is a generic prediction of an accelerated de Sitter expansion of the Universe whatever mechanism for the generation of cosmological perturbations is operative, gravitational waves, whose possible observation might come from the detection of the B-mode of polarization in the CMB anisotropy, may be viewed as ripples of space–time around the background metric.

Since curvature fluctuations are (nearly) frozen on superhorizon scales, a way of characterizing them is to compute their spectrum on scales larger than the horizon. In the standard slow–roll inflationary models where the fluctuations of the inflation field φ are responsible for the curvature perturbations, the power–spectrum $\mathcal{P}_\zeta$ of the comoving curvature perturbation ζ (which is a measure of the spatial curvature as seen by comoving observers) is given by

$$\mathcal{P}_\zeta(k) = \frac{1}{2M_{\rm P}^2 \epsilon} \left(\frac{H_*}{2\pi}\right)^2 \left(\frac{k}{aH_*}\right)^{n_\zeta - 1} , \tag{9.1}$$

where $n_\zeta = 1 - 6\epsilon + 2\eta \simeq 1$ is the spectral index, $M_{\rm P} \equiv (8\pi G_{\rm N})^{-1/2} \simeq 2.4 \times 10^{18}$ GeV is the reduced Planck scale. Here

$$\epsilon = \frac{M_{\rm P}^2}{2} \left(\frac{V'}{V}\right)^2 ,$$

$$\eta = M_{\rm P}^2 \left(\frac{V''}{V}\right) \tag{9.2}$$

are the so–called slow–roll parameters ($\epsilon, \eta \ll 1$ during inflation), $H_* = \dot{a}/a$ indicates the Hubble rate during inflation and primes here denote derivatives with respect to φ. WMAP has determined the amplitude of the power–spectrum as $\mathcal{P}_\zeta(k) \simeq 2.95 \times 10^{-9} A$ where $A = 0.6 - 1$ depending on the model under consideration, which implies that

$$\frac{1}{2M_{\rm P}^2 \epsilon} \left(\frac{H_*}{2\pi}\right)^2 \simeq (2 - 3) \times 10^{-9} , \tag{9.3}$$

or

$$H_* \simeq (0.9 - 1.2) \times 10^{15} \, \epsilon^{1/2} \text{ GeV} . \tag{9.4}$$

The Friedmann equation in the slow–roll limit, $H^2 = V/(3M_{\rm P}^2)$, then gives "the energy scale of inflation",

$$V^{1/4} \simeq (6.3 - 7.1) \times 10^{16} \, \epsilon^{1/4} \text{ GeV} . \tag{9.5}$$

On the other hand, the power–spectrum of gravity–wave modes h_{ij} is given by

$$\mathcal{P}_{\rm T}(k) = \frac{k^3}{2\pi^2} \langle h_{ij}^* h^{ij} \rangle = \frac{8}{M_{\rm P}^2} \left(\frac{H_*}{2\pi}\right)^2 \left(\frac{k}{aH_*}\right)^{n_{\rm T}} ,$$

where $n_{\rm T} = -2\epsilon$ is the tensor spectral index. Since the fractional change of the power–spectra with scale is much smaller than unity, one can safely consider the power–spectra as being roughly constant on the scales relevant for the CMB anisotropy and define a tensor–to–scalar amplitude ratio

$$r = \frac{\mathcal{P}_{\rm T}}{\mathcal{P}_\zeta} = 16\epsilon . \tag{9.6}$$

The spectra $\mathcal{P}_\zeta(k)$ and $\mathcal{P}_T(k)$ provide the contact between theory and observation. The present WMAP dataset allows to extract an upper bound, $r < 1.28$ (95%), or $\epsilon < 0.08$. This limit together with (9.5) provides an upper bound on the energy scale of inflation,

$$V^{1/4} < 3.8 \times 10^{16} \, \text{GeV} \,. \tag{9.7}$$

The corresponding upper bound on the Hubble rate during inflation is $H_* < 3.4 \times 10^{14}$ GeV. A positive detection of the B–mode in CMB polarization, and therefore an indirect evidence of gravitational waves from inflation, once foregrounds due to gravitational lensing from local sources have been properly treated, requires $\epsilon > 10^{-5}$ corresponding to $V^{1/4} > 3.5 \times 10^{15}$ GeV and $H_* > 3 \times 10^{12}$ GeV. However, *what if* the curvature perturbation is generated through the quantum fluctuations of a scalar field other than the inflaton? Then, what is the expected amplitude of gravity–wave fluctuations in such scenarios? Consider, for instance, the curvaton scenario and the inhomogeneous reheating scenario. They free the inflaton from having to generate the cosmological curvature perturbation and thereby avoid the necessity of slow–roll conditions. The basic assumption is that the initial curvature perturbation due to the inflation field is negligible. The common lore to achieve such a condition is to assume that the energy scale of the inflaton potential is too small to match the observed amplitude of CMB anisotropy, that is $V^{1/4} \ll 10^{16}$ GeV. Therefore – while certainly useful to construct low-scale models of inflation – it is usually thought that these mechanisms predict an amplitude of gravitational waves which is far too small to be detectable by future satellite experiments aimed at observing B-modes in CMB polarization. This implies that a future detection of B-modes would favour the slow–roll models of inflation as generators of the cosmological perturbations. On the other hand, the lack of signal of gravity waves in the CMB anisotropies will not give us any information about the mechanism by which cosmological perturbations are created.

A precise measurement of the spectral index of comoving curvature perturbations will be a powerful tool to constrain inflationary models. Slow–roll inflation predicts $|n_\zeta - 1|$ significantly below 1. Deviations of n_ζ from unity are generically (but not always) proportional to $1/N$, where N is the number of e-folds till the end of inflation. The predictions of different models for the spectral index n_ζ, and for its scale dependence, are well summarized in the review [1] within slow–roll inflationary models. Remarkably, the eventual accuracy $\Delta n_\zeta \sim 0.01$ offered by the *Planck* satellite[3] is just what one might have specified in order to distinguish between various slow–roll models of inflation. Observation will discriminate strongly between slow–roll models of inflation in the next 10 or 15 years. If cosmological perturbations are due to the inflation field, then in 10 or 15 years there may be a consensus on the

---

[3] See, for instance, http://www.rssd.esa.int/index.php?project=PLANCK

form of the inflationary potential, and at a deeper level we may learn something valuable about the nature of the fundamental interactions beyond the Standard Model. However, *what if* Nature has chosen the other mechanisms for the creation of the cosmological perturbations, which generically predict a value of n_ζ very close to unity with a negligible scale dependence? Then, it implies that a precise measurement of the spectral index will not allow us to efficiently discriminate among different scenarios.

These *"what if"* options would be discouraging if they turn out to be true. They would imply that all future efforts for measuring tensor modes in the CMB anisotropy and the spectral index of adiabatic perturbations are of no use to disentangle the various scenarios for the creation of cosmological perturbations. There is, however, a third observable which will prove fundamental in providing information about the mechanism chosen by Nature to produce the structures we see today. It is the deviation from a pure Gaussian statistics, i.e., the presence of higher–order connected correlation functions of CMB anisotropies. The angular n–point correlation function

$$\langle f(\hat{\mathbf{n}}_1) f(\hat{\mathbf{n}}_2) \ldots f(\hat{\mathbf{n}}_n) \rangle \; , \tag{9.8}$$

is a simple statistic characterizing a clustering pattern of fluctuations on the sky, $f(\hat{\mathbf{n}})$. The bracket denotes the ensemble average. If the fluctuation is Gaussian, then the two–point correlation function specifies all the statistical properties of $f(\hat{\mathbf{n}})$, for the two–point correlation function is the only parameter in a Gaussian distribution. If it is not Gaussian, then we need higher–order correlation functions to determine the statistical properties.

For instance, a non–vanishing three–point function of scalar perturbations, or its Fourier transform, the bispectrum, is an indicator of non–gaussian features in the cosmological perturbations. The importance of the bispectrum comes from the fact that it represents the lowest order statistics able to distinguish non–gaussian from Gaussian perturbations. Therefore *an accurate calculation of the primordial bispectrum of cosmological perturbations has become an extremely important issue: its detection will allow to discriminate among the various theoretical mechanisms which have been proposed to generate cosmological perturbations.*

A note on references now: instead of quoting a massive list of references, we will simply cite most of the times the review [1] where the reader can find a detailed list of references.

9.2 What Is Non-gaussianity?

Despite the importance of non-gaussianity (NG) in the CMB anisotropies, little effort has been made so far to provide accurate theoretical predictions of it. On the contrary, the vast majority of the literature has been devoted to the computation of the bispectrum of either the comoving curvature perturbation or the gravitational potential on large scales within given inflationary models. These, however, are not the physical quantities which are observed. One should instead provide a full prediction for *the second-order radiation transfer function*. One should therefore keep in mind that

$$\text{NG in the CMB} = \text{Primordial NG} + \text{non-primordial NG}$$

where, by definition, we define primordial NG as the one generated either during inflation or after it when the comoving curvature perturbation becomes constant on superhorizon scales.

A first step towards determining the contributions which are not primordial has been taken from [2] where the full second-order radiation transfer function for the CMB anisotropies on large angular scales in a flat universe filled with matter and cosmological constant was computed, including the second-order generalization of the Sachs–Wolfe effect, both the early and late Integrated Sachs–Wolfe (ISW) effects and the contribution of the second-order tensor modes.

The most relevant sources are the so-called secondary anisotropies, which arise after the epoch of last scattering. These anisotropies can be divided into two categories: scattering secondaries, when the CMB photons scatter with electrons along the line of sight, and gravitational secondaries when effects are mediated by gravity. Among the scattering secondaries we may list the thermal Sunyaev–Zeldovich effect, where hot electrons in clusters transfer energy to CMB photons, the kinetic Sunyaev–Zeldovich effect produced by the bulk motion of electrons in clusters, the Ostriker–Vishniac effect, produced by bulk motions modulated by linear density perturbations, and effects due to reionization processes. The scattering secondaries are most significant on small angular scales as density inhomogeneities, bulk and thermal motions grow and become sizable on small length scales when structure formation proceeds.

Gravitational secondaries arise from the change in energy of photons when the gravitational potential is time-dependent, the ISW effect, and gravitational lensing. At late times, when the Universe becomes dominated by dark energy, the gravitational potential on linear scales starts to decay, causing the ISW effect mainly on large angular scales. Other secondaries that result from a time-dependent potential are the Rees–Sciama effect, produced during the matter-dominated epoch at second-order and by the time evolution of the potential on non-linear scales.

The fact that the potential never grows appreciably means that most second-order effects created by gravitational secondaries are generically small compared to those created by scattering ones. However, when a photon propagates from the last scattering to us, its path may be deflected because of the gravitational lensing. This effect does not create anisotropies, but only modifies existing ones. Since photons with large wavenumber k are lensed in many regions ($\sim k/H$, where H is the Hubble rate) along the line of sight, the corresponding second-order effect may be sizable. The three-point function arising from the correlation of the gravitational lensing effect and the

ISW effect generated by the matter distribution along the line of sight and the Sunyaev–Zeldovich effect are large and detectable by Planck.

Another relevant source of NG comes from the physics operating at recombination. A naive estimate leads to think that these non-linearities are tiny, being suppressed by an extra power of the gravitational potential. However, the dynamics at recombination is quite involved because all the non-linearities in the evolution of the baryon–photon fluid at recombination and the ones coming from general relativity should be accounted for. This complicated dynamics might lead to unexpected suppressions or enhancements of the NG at recombination. The computation of the full system of Boltzmann equations at second order describing the evolution of the photon, baryon and cold dark matter fluids have been recently computed in [2]. These equations allow one to follow the time evolution of the Cosmic Microwave Background (CMB) anisotropies at second order at all angular scales from the early epoch, when the cosmological perturbations were generated, to the present through the recombination era. This chapter set the stage for the computation of the full second-order radiation transfer function at all scales and for a generic set of initial conditions specifying the level of primordial non-gaussianity. In [3] an analytical approach to the second-order CMB anisotropies generated by the non-linear dynamics taking place at last scattering was presented. We study the acoustic oscillations of the photon–baryon fluid in the tight coupling limit and we extend to second order the Meszaros effect. These results are useful to provide the full second-order radiation transfer function at all scales necessary to establish the level of non-gaussianity in the CMB. We will come back to these results later.

9.2.1 Gravity Is Non-linear

To convince the reader that the contribution to the NG in the CMB anisotropies does not reduce to the primordial one, let us evaluate the contribution to NG from gravity on super-horizon scales, i.e. the so-called non-linear Sachs–Wolfe effect. Gravity itself is non-linear and therefore must contribute to NG.

Our starting point is the Arnowitt–Deser–Misner (ADM) formalism which is particularly useful to deal with the non-linear evolution of cosmological perturbations. The line element is

$$ds^2 = -N^2 \, dt^2 + N_i \, dt \, dx^i + \gamma_{ij} \, dx^i \, dx^j \;,$$

where the three-metric γ_{ij}, the lapse N and the shift N_i functions describe the evolution of timelike hypersurfaces. In the ADM formalism the equations simplify considerably if we set $N^i = 0$. Moreover we are interested only in scalar perturbations in a flat Universe and therefore we find it convenient to recast the metric as

$$ds^2 = -e^{2\Phi} \, dt^2 + a^2(t) e^{-2\Psi} \delta_{ij} \, dx^i \, dx^j \;, \tag{9.9}$$

where $a(t)$ is the scale factor describing the evolution of the homogeneous and isotropic Universe and we have introduced two gravitational potentials Φ and Ψ. The expression (9.9) holds at any order in perturbation theory. To make contact with the usual perturbative approach, one may expand the gravitational potentials at first and second order, e.g, $\Phi = \Phi_1 + \Phi_2/2$. From (9.9) one recovers at linear order the well-known longitudinal gauge, $N^2 = (1+ 2\Phi_1)$ and $\gamma_{ij} = a^2(1 - 2\Psi_1)\delta_{ij}$. At second order, one finds $\Phi_2 = \phi_2 - 2\phi_1^2$ and $\Psi_2 = \psi_2 + 2\psi_1^2$ where ϕ_1, ψ_1 and ϕ_2, ψ_2 (with $\phi_1 = \Phi_1$ and $\psi_1 = \Psi_1$) are the first- and second-order gravitational potentials in the longitudinal (Poisson) gauge adopted, $N^2 = (1 + 2\phi_1 + \phi_2)$ and $\gamma_{ij} = a^2(1 - 2\psi_1 - \psi_2)\delta_{ij}$ as far as scalar perturbations are concerned. In writing (9.9) we have neglected vector and tensor perturbation modes. For the vector perturbations the reason is that we are interested in long-wavelength perturbations, i.e. on scales larger than the horizon at last scattering, while vector modes will contain gradient terms being produced as non-linear combination of scalar modes and thus they will be more important on small scales (linear vector modes are not generated in standard mechanisms for cosmological perturbations, as inflation). The tensor contribution can be neglected for two reasons. First, the tensor perturbations produced from inflation on large scales give a negligible contribution to the higher-order statistics of the Sachs–Wolfe effect being of the order of (powers of) the slow-roll parameters during inflation (this holds for linear tensor modes as well as for tensor modes generated by the non-linear evolution of scalar perturbations during inflation). Moreover, while on large scales the tensor modes have been proven to remain constant in time, when they approach the horizon they have a wavelike contribution which oscillates with decreasing amplitude.

Since we are interested in cosmological perturbations on large scales, that is in perturbations whose wavelength is larger than the Hubble radius at last scattering, a local observer would see them in the form of a classical – possibly time-dependent – (nearly zero-momentum) homogeneous and isotropic background. Therefore, it should be possible to perform a change of coordinates in such a way as to absorb the super-Hubble modes and work with the metric for a homogeneous and isotropic Universe (plus, of course, cosmological perturbations on scales smaller than the horizon). We split the gravitational potential Φ as

$$\Phi = \Phi_\ell + \Phi_s , \tag{9.10}$$

where Φ_ℓ stands for the part of the gravitational potential receiving contributions only from the super-Hubble modes; Φ_s receives contributions only from the sub-horizon modes

$$\Phi_\ell = \int \frac{d^3k}{(2\pi)^3} \theta(aH - k) \Phi_k e^{ik\cdot x} ,$$

$$\Phi_s = \int \frac{d^3k}{(2\pi)^3} \theta(k - aH) \Phi_k e^{ik\cdot x} , \tag{9.11}$$

where H is the Hubble rate computed with respect to the cosmic time, $H = \dot{a}/a$, and $\theta(x)$ is the step function. Analogous definitions hold for the other gravitational potential Ψ.

By construction Φ_ℓ and Ψ_ℓ are a collection of Fourier modes whose wavelengths are larger than the horizon length, and we may safely neglect their spatial gradients. Therefore Φ_ℓ and Ψ_ℓ are only functions of time. This amounts to saying that we can absorb the large-scale perturbations in the metric (9.9) by the following redefinitions

$$d\bar{t} = e^{\Phi_\ell} dt \,, \tag{9.12}$$

$$\bar{a} = a\,e^{-\Psi_\ell} \,. \tag{9.13}$$

The new metric describes a homogeneous and isotropic Universe

$$ds^2 = -d\bar{t}^2 + \bar{a}^2 \delta_{ij} \, dx^i \, dx^j \,, \tag{9.14}$$

where for simplicity we have not included the sub-horizon modes. On super-horizon scales one can regard the Universe as a collection of regions of size of the Hubble radius evolving like unperturbed patches with metric (9.14).

Let us now go back to the quantity we are interested in, namely the anisotropies of the CMB as measured today by an observer $\mathcal{O}$. If one is interested in the CMB anisotropies at large scales, the effect of super-Hubble modes is encoded in the metric (9.14). During the travel from the last scattering surface – to be considered as the emitter point $\mathcal{E}$ – to the observer, CMB photons suffer a redshift determined by the ratio of the emitted frequency $\bar{\omega}_\mathcal{E}$ to the observed one $\bar{\omega}_\mathcal{O}$

$$\overline{T}_\mathcal{O} = \overline{T}_\mathcal{E} \frac{\bar{\omega}_\mathcal{O}}{\bar{\omega}_\mathcal{E}} \,, \tag{9.15}$$

where $\overline{T}_\mathcal{O}$ and $\overline{T}_\mathcal{E}$ are the temperatures at the observer point and at the last scattering surface, respectively.

What is then the temperature anisotropy measured by the observer? The expression (9.15) shows that the measured large-scale anisotropies are made of two contributions: the intrinsic inhomogeneities in the temperature at the last scattering surface and the inhomogeneities in the scaling factor provided by the ratio of the frequencies of the photons at the departure and arrival points. Let us first consider the second contribution. As the frequency of the photon is the inverse of a time period, we immediately get the fully non-linear relation

$$\frac{\bar{\omega}_\mathcal{E}}{\bar{\omega}_\mathcal{O}} = \frac{\omega_\mathcal{E}}{\omega_\mathcal{O}} e^{-\Phi_{\ell\mathcal{E}} + \Phi_{\ell\mathcal{O}}} \,. \tag{9.16}$$

As for temperature anisotropies coming from the intrinsic temperature fluctuation at the emission point, it is worth recalling how to obtain this quantity in the longitudinal gauge at first order. By expanding the photon energy density $\rho_\gamma \propto T_\gamma^4$, the intrinsic temperature anisotropies at last scattering are given by $\delta_1 T_\mathcal{E}/T_\mathcal{E} = (1/4)\delta_1\rho_\gamma/\rho_\gamma$. One relates the photon energy density fluctuation to the gravitational perturbation first by implementing the adiabaticity condition $\delta_1\rho_\gamma/\rho_\gamma = (4/3)\delta_1\rho_m/\rho_m$, where $\delta_1\rho_m/\rho_m$ is the relative fluctuation in the matter component, and then by using the energy constraint of Einstein equations $\Phi_1 = -(1/2)\delta_1\rho_m/\rho_m$. The result is $\delta_1 T_\mathcal{E}/T_\mathcal{E} = -2\Phi_{1\mathcal{E}}/3$. Adding this contribution to the anisotropies coming from the redshift factor (9.16) expanded at first order provides the standard (linear) Sachs–Wolfe effect $\delta_1 T_\mathcal{O}/T_\mathcal{O} = \Phi_{1\mathcal{E}}/3$. Following the same steps, we may easily obtain its full non-linear generalization.

Let us first relate the photon energy density $\bar{\rho}_\gamma$ to the energy density of the non-relativistic matter $\bar{\rho}_m$ by using the adiabaticity condition. Again here a bar indicates that we are considering quantities in the locally homogeneous Universe described by the metric (9.14). Using the energy continuity equation on large scales $\partial\bar{\rho}/\partial\bar{t} = -3\bar{H}(\bar{\rho}+\bar{P})$, where $\bar{H} = \mathrm{d}\ln\bar{a}/\mathrm{d}\bar{t}$ and $\bar{P}$ is the pressure of the fluid, one can easily show that there exists a conserved quantity in time at any order in perturbation theory.

$$F \equiv \ln\bar{a} + \frac{1}{3}\int^{\bar{\rho}} \frac{\mathrm{d}\bar{\rho}'}{(\bar{\rho}'+\bar{P}')} \,. \tag{9.17}$$

The perturbation δF is a gauge-invariant quantity representing the non-linear extension of the curvature perturbation ζ on uniform energy density hypersurfaces on superhorizon scales for adiabatic fluids. Indeed, expanding it at first and second order one gets the corresponding definition $\zeta_1 = -\psi_1 - \delta_1\rho/\dot{\rho}$ and the quantity ζ_2 introduced in [4]. At first order the adiabaticity condition corresponds to setting $\zeta_{1\gamma} = \zeta_{1m}$ for the curvature perturbations relative to each component. At the non-linear level the adiabaticity condition generalizes to

$$\frac{1}{3}\int \frac{\mathrm{d}\bar{\rho}_m}{\bar{\rho}_m} = \frac{1}{4}\int \frac{\mathrm{d}\bar{\rho}_\gamma}{\bar{\rho}_\gamma} \,, \tag{9.18}$$

or

$$\ln\bar{\rho}_m = \ln\bar{\rho}_\gamma^{3/4} \,. \tag{9.19}$$

To make contact with the standard second-order result, we may expand in (9.16) the photon energy density perturbations as $\delta\bar{\rho}_\gamma/\rho_\gamma = \delta_1\rho_\gamma/\rho_\gamma + \frac{1}{2}\delta_2\rho_\gamma/\rho_\gamma$, and similarly for the matter component. We immediately recover the adiabaticity condition

$$\frac{\delta_2\rho_\gamma}{\rho_\gamma} = \frac{4}{3}\frac{\delta_2\rho_m}{\rho_m} + \frac{4}{9}\left(\frac{\delta_1\rho_m}{\rho_m}\right)^2 \tag{9.20}$$

given in [1].

Next we need to relate the photon energy density to the gravitational potentials at the non-linear level. The energy constraint inferred from the $(0-0)$ component of Einstein equations in the matter-dominated era with the "barred" metric (9.14) is

$$\bar{H}^2 = \frac{8\pi G_N}{3}\bar{\rho}_m \,. \tag{9.21}$$

Using (9.12) and (9.13) the Hubble parameter $\overline{H}$ reads

$$\overline{H} = \frac{1}{\overline{a}}\frac{d\overline{a}}{d\overline{t}} = e^{-\Phi_\ell}(H - \dot{\Psi}_\ell) \,, \tag{9.22}$$

where $H = d\ln a/dt$ is the Hubble parameter in the "unbarred" metric. (9.21) thus yields an expression for the energy density of the non-relativistic matter which is fully non-linear, being expressed in terms of the gravitational potential Φ_ℓ

$$\overline{\rho}_m = \rho_m e^{-2\Phi_\ell} \,, \tag{9.23}$$

where we have dropped $\dot{\Psi}_\ell$ which is negligible on large scales. By perturbing the expression (9.23) we are able to recover in a straightforward way the solutions of the $(0-0)$ component of Einstein equations for a matter-dominated Universe in the large-scale limit obtained at second-order in perturbation theory. Indeed, recalling that Φ is perturbatively related to the quantity $\phi = \phi_1 + \phi_2/2$ used in [1] by $\Phi_1 = \phi_1$ and $\Phi_2 = \phi_2 - 2(\phi_1)^2$, one immediately obtains

$$\frac{\delta_1\rho_m}{\rho_m} = -2\phi_1 \,,$$

$$\frac{1}{2}\frac{\delta_2\rho_m}{\rho_m} = -\phi_2 + 4(\phi_1)^2 \,. \tag{9.24}$$

The expression for the intrinsic temperature of the photons at the last scattering surface $\overline{T}_\mathcal{E} \propto \overline{\rho}_\gamma^{1/4}$ follows from (9.19)

$$\overline{T}_\mathcal{E} = T_\mathcal{E}\, e^{-2\Phi_\ell/3} \,. \tag{9.25}$$

We are finally able to provide the expression for the CMB temperature which is fully non-linear and takes into account both the gravitational redshift of the photons due to the metric perturbations at last scattering and the intrinsic temperature anisotropies

$$\overline{T}_\mathcal{O} = \left(\frac{\omega_\mathcal{O}}{\omega_\mathcal{E}}\right) T_\mathcal{E}\, e^{\Phi_\ell/3} \,. \tag{9.26}$$

From (9.26) we read the *non-perturbative* anisotropy corresponding to the Sachs–Wolfe effect

$$\frac{\delta_{np}\overline{T}_\mathcal{O}}{T_\mathcal{O}} = e^{\Phi_\ell/3} - 1 \,. \tag{9.27}$$

Equation (9.27) is one of the main results of this paper and represents *at any order in perturbation theory* the extension of the linear Sachs–Wolfe effect. At first order one gets

$$\frac{\delta_1 T_\mathcal{O}}{T_\mathcal{O}} = \frac{1}{3}\Phi_1 \,, \tag{9.28}$$

and at second order

$$\frac{1}{2}\frac{\delta_2 T_\mathcal{O}}{T_\mathcal{O}} = \frac{1}{6}\Phi_2 + \frac{1}{18}(\Phi_1)^2 , \qquad (9.29)$$

which exactly reproduces the generalization of the Sachs–Wolfe effect at second order in the perturbations found in [1] (where $\Phi_1 = \phi_1$ and $\Phi_2 = \phi_2 - 2(\phi_1)^2$). This simple exercise tells us that *gravity introduces an order unity NG.*

9.2.2 A Basic Formula

From what we have learned so far, one phenomenological way of parametrizing the possible presence of non-gaussianity in the cosmological perturbations is to expand the Bardeen gravitational potential Φ as

$$\phi = \phi^{(1)} + f_{\mathrm{NL}}^\phi \star (\phi^{(1)})^2 , \qquad (9.30)$$

where $\phi^{(1)}$ represents the linear Gaussian contribution to the gravitational potential ϕ and f_{NL}^ϕ is the dimensionless *non-linearity* parameter. Here the $\star$-products reminds the fact that the non–linearity parameter might have a non–trivial scale dependence. As (9.30) shows, in order to compute and keep track of the non–gaussianity of the cosmological perturbations throughout the different stages of the evolution of the Universe, one has to perform a perturbation around the homogeneous background *up to second order.* In fact all the models that we are going to consider predict such distinctive quadratic non-linearities for cosmological perturbations. The bispectrum implied by such non-linearities is proportional to the non-linearity parameter $\langle\phi(\mathbf{k}_1)\phi(\mathbf{k}_2)\phi(\mathbf{k}_3)\rangle = (2\pi)^3\delta^{(3)}(\sum_i \mathbf{k}_i)[2f_{\mathrm{NL}} P_\Phi(k_1)P_\Phi(k_2) + \mathrm{cycl}]$ where $\phi(\mathbf{k})$ is the gravitational potential Fourier transform, and $P_\Phi(k)$ is the linear power spectrum defined as $\langle\phi^{(1)}(\mathbf{k}_1)\phi^{(1)}(\mathbf{k}_2)\rangle = (2\pi)^3\delta^{(3)}(\mathbf{k}_1 + \mathbf{k}_2)P_\phi(k_1)$. The non-linearity parameter has become the standard quantity to be observationally constrained by CMB experiments. The reader should remember that an alternative definition for the non-linear parameter frequently adopted in the literature is to use the comoving curvature perturbation

$$\zeta = \zeta^{(1)} + f_{\mathrm{NL}}^\zeta \star (\zeta^{(1)})^2 . \qquad (9.31)$$

The reader should also keep in mind that the relation between the gravitational potential ϕ and the comoving curvature perturbation ζ is not the linear one at first order $\zeta^{(1)} = -(5/3)\phi^{(1)}$.

9.3 Linear Perturbations on Large Scales

We now briefly recall how the density perturbations are generated and evolve on large scales from an early period of inflation/reheating through the standard radiation and matter dominated epochs for the standard single-field inflation and the curvaton and inhomogeneous reheating scenarios. The details can be found in the talks given by D. Wands and D. Lyth. Here we just want to note that all three scenarios can be understood by the evolution of the curvature perturbation $\zeta^{(1)} = -\psi^{(1)} - \mathcal{H}\delta^{(1)}\rho/\rho'$ which is a gauge-invariant definition of the curvature perturbation $\psi^{(1)}$ entering in the perturbed spatial components of the (flat) Friedman–Robertson–Walker metric $ds^2 = a^2(\tau)[-(1+2\phi)d\tau^2 + 2\omega_i dx^i d\tau + ((1+2\psi)\delta_{ij} + \chi_{ij})dx^i dx^j]$. Here each perturbation is expanded in a first and a second-order part as, for example, $\psi = \psi^{(1)} + \psi^{(2)}/2$, $a(\tau)$ is the scale factor as a function of the conformal time and $\mathcal{H} = a'/a$, with a prime denoting differentiation w.r.t conformal time. The evolution of ζ is obtained from the energy density continuity equation on large scales $\delta^{(1)}\rho' = -3\mathcal{H}(\delta^{(1)}\rho + \delta^{(1)}P) - 3\psi^{(1)'}(\rho + P)$ which brings

$$\zeta^{(1)'} = -\frac{\mathcal{H}}{\rho + P}\delta^{(1)}P_{\mathrm{nad}} , \qquad (9.32)$$

where $\delta\rho = \delta^{(1)}\rho + \delta^{(2)}\rho/2$ and δP are the total energy density and pressure perturbations, respectively, and $\delta^{(1)}P_{\mathrm{nad}} = \delta^{(1)}P - c_s^2\delta^{(1)}\rho$ is the so called non-adiabatic pressure perturbation, with $c_s^2 = p'/\rho'$. In all three scenarios the primordial density perturbations can be traced back to initial fluctuations of scalar fields. In the standard scenario during inflation $\zeta^{(1)} = \zeta_\varphi^{(1)}$ and on large scales it remains constant during inflation and also during the reheating phase. On the other hand the curvaton and the inhomogeneous reheating scenarios exploit the fact that the curvature perturbation can be sourced even on superhorizon scales by a non-adiabatic pressure perturbation via (9.32). The non-adiabatic pressure perturbation is essentially due to the relative fluctuations between the inflaton and the second scalar field. In the curvaton scenario after inflation the curvaton σ starts oscillating and decaying into radiation (γ) with $\rho_\sigma' + (3\mathcal{H} + \Gamma_\sigma)\rho_\sigma = 0$ and $\rho_\gamma' + 4\mathcal{H}\rho_\gamma = \Gamma_\sigma\rho_\sigma$, where Γ_σ is the curvaton decay rate. The total curvature perturbation is a weighted sum of the single curvature perturbations $\zeta^{(1)} = \rho_\sigma'\zeta_\sigma^{(1)}/\rho' + \rho_\gamma'\zeta_\gamma^{(1)}/\rho'$ and evolves according to

$$\zeta^{(1)'} = -\mathcal{H}\frac{\rho_\sigma'}{\rho'}(\zeta^{(1)} - \zeta_\sigma^{(1)}) , \qquad (9.33)$$

$$\zeta_\sigma^{(1)'} = \frac{a\Gamma_\sigma}{2}\frac{\rho_\sigma}{\rho_\sigma'}\frac{\rho'}{\rho}(\zeta^{(1)} - \zeta_\sigma^{(1)}) . \qquad (9.34)$$

Since initially $(\zeta^{(1)} - \zeta_\sigma^{(1)})_{\mathrm{in}} \simeq \zeta_{\sigma,\mathrm{in}} \neq 0$ because of the curvaton fluctuations generated during inflation, the total curvature perturbation grows during the curvaton oscillations until σ decays generating a final adiabatic perturbation $\zeta^{(1)} \simeq r\zeta_{\sigma,\mathrm{in}}$, where $r \simeq (\rho_\sigma/\rho)_{\mathrm{decay}}$ is the ratio of the curvaton to the total energy density at the epoch of decay. In the inhomogeneous reheating scenario spatial fluctuations of the inflaton decay rate $\delta\Gamma_\varphi$ are induced by a different light scalar field χ. In this case $\zeta^{(1)}$ evolves according to the same equation (9.33) (where $\sigma \to \varphi$) and $\zeta_\varphi^{(1)'} = a\Gamma_\varphi\rho_\varphi\rho'(\zeta^{(1)} - \zeta_\varphi^{(1)})/(2\rho_\varphi'\rho) + a\mathcal{H}\rho_\varphi\delta^{(1)}\Gamma_\varphi/\rho_\varphi'$. The total curvature perturbation is thus sourced by $\delta^{(1)}\Gamma_\varphi$

and one finds that after reheating $\zeta^{(1)} \simeq -\delta^{(1)}\Gamma_\varphi/6\Gamma_\varphi$. In each of the three scenarios we discuss here the perturbations produced are adiabatic and hence $\zeta^{(1)}$ will be conserved in the following radiation and matter dominated epochs.

9.4 Non-linear Perturbation on Large Scales: Generation of NG

Let us now consider how non-linearities in the cosmological perturbations are generated and how they evolve from a primordial epoch to the matter epoch. In order to do that one has to perform a full calculation at second order in the perturbations, by expanding both the metric and the energy momentum tensor in a first-order and a second-order contributions. In particular one can define a gauge-invariant curvature perturbation $\zeta = \zeta^{(1)} + \zeta^{(2)}/2$ where

$$-\zeta^{(2)} = \psi^{(2)} + \mathcal{H}\frac{\delta^{(2)}\rho}{\rho'} - 2\mathcal{H}\frac{\delta^{(1)}\rho'}{\rho'}\frac{\delta^{(1)}\rho}{\rho'} - 2\frac{\delta^{(1)}\rho}{\rho'}\psi^{(1)\prime}$$
$$-4\mathcal{H}\frac{\delta^{(1)}\rho}{\rho'}\psi^{(1)} + \left(\frac{\delta^{(1)}\rho}{\rho'}\right)^2\left(\mathcal{H}\frac{\rho''}{\rho'} - \mathcal{H}' - 2\mathcal{H}^2\right).$$

Similarly to its linear counterpart, $\zeta^{(2)}$ can change on superhorizon scales due to a non-adiabatic pressure perturbation (the expression of which is not given here for simplicity)

$$\zeta^{(2)\prime} = -\frac{\mathcal{H}}{\rho + P}\delta^{(2)}P_{nad} \tag{9.35}$$
$$- \frac{2}{\rho + P}[\delta^{(1)}P_{nad} - 2(\rho + P)\zeta^{(1)}]\zeta^{(1)\prime}.$$

The key point is that once $\zeta^{(2)}$ has been generated (either during inflation as in the standard scenario, or after inflation as in the curvaton scenario) it remains constant on large scales during the radiation/matter eras because the resulting perturbations are adiabatic. Therefore the recipe to evaluate the level of non-gaussianity is the following:

- Evaluate the non-linearities generated during the primordial epoch (i.e. the conserved value of $\zeta^{(2)}$)
- Evolve the gravitational potentials to second order in the radiation/matter-dominated epochs (second-order Einstein equations) by matching the conserved variable $\zeta^{(2)}$ to its initial value
- Evaluate the additional second-order corrections that arise when the non-gaussianities produced in the gravitational potential are transferred to the large-scale CMB anisotropies, where additional second-order corrections arise.

Let us first consider the standard single field models of slow-roll inflation, as it emerges as a reference point for all the other mechanisms.

9.4.1 Standard Scenario

As long as one considers linear theory the perturbations turn out to be Gaussian. However inflaton self-interactions as well as non-linear evolution of the metric perturbations can give rise to a certain degree of non-gaussianity. Originally, the computation of the bispectrum of the perturbations generated during inflation was addressed by just looking at the inflaton self-interactions (which necessarily produce non-linearities in the quantum fluctuations) in a fixed de-Sitter Background. As a result the non-linearity parameter $f_{NL}^{\phi} \simeq \mathcal{O}(\xi^2)$, where $\xi^2 = M_P^2 (V^{(1)}V^{(3)}/V^{(2)})$ which is second order in the slow–roll parameters. Then the stochastic approach to inflation has been adopted, in which back-reaction effects of the field fluctuations on the background metric are partially taken into account. An interesting result is that the main contribution comes in fact from the non-linear gravitational perturbations rather than by inflaton self-interactions with $f_{NL}^{\phi} \sim \mathcal{O}(\epsilon)$. Recently [5, 6] confirmed in a more rigorous way such a finding and represent indeed a step forward in the computation of the bispectrum generated from inflation. In [5, 6] a complete analysis at second order in the perturbations takes into account both the self-interactions of the inflation field as well as the metric perturbations yielding [5, 6]

$$f_{NL}^{\phi} = -\frac{5}{12}(n_\zeta - 1) + f(\mathbf{k_1}, \mathbf{k_2}) , \qquad (9.36)$$

where the spectral index is expressed in terms of the usual slow-roll parameters as $n_\zeta - 1 = 2\eta - 6\epsilon \ll 1$. Notice that $f(\mathbf{k_1}, \mathbf{k_2})$ is a momentum dependent part of f_{NL}^{ϕ} which is still of order $\mathcal{O}(\epsilon)$. Let us stress that the result in (9.36) refers only to the non-gaussianity generated *during* inflation, characterized by a tiny f_{NL}, typically $f_{NL}^{\phi} \sim \mathcal{O}(10^{-1} - 10^{-2})$.

But what about the post-inflationary evolution of non-gaussianity? During the reheating phase the second-order curvature perturbation $\zeta^{(2)}$ is indeed conserved and, since the perturbations are adiabatic, it remains constant on superhorizon scales also during the radiation/matter eras. Therefore we evolve the scalar second-order perturbations by matching $\zeta^{(2)}$ to its value at the end of inflation

$$\zeta^{(2)} = -\psi^{(2)} + \frac{1}{3}\frac{\delta^{(2)}\rho}{\rho} + \frac{20}{9}(\psi^{(1)})^2 = \mathcal{C} , \qquad (9.37)$$

where we have specified the expression of $\zeta^{(2)}$ for a matter dominated-epoch using the Poisson gauge, which generalizes the usual longitudinal gauge used at first order (see [1]). The constant $\mathcal{C}$ has been computed in [5], $\mathcal{C} \simeq (\eta - 3\epsilon) - 2\left(\zeta^{(1)}\right)^2$. During the matter epoch $\zeta^{(1)} = -5/3\psi^{(1)}$. In this way we fix the initial conditions for the non-linear evolution of the large-scale gravitational potentials in the matter-dominated epoch

$$\phi^{(2)} = -\frac{1}{2}\frac{\delta^{(2)}\rho}{\rho} + 4(\psi^{(1)})^2 , \qquad (9.38)$$

$$\psi^{(2)} - \phi^{(2)} = -\frac{2}{3}(\psi^{(1)})^2 + \text{non-local}, \tag{9.39}$$

obtained from the (0–0)- and (i–j)-component of the Einstein equations in the Poisson gauge, respectively. Note that (9.39) is a constraint showing how, unlike the linear case, the two gravitational potentials differ at second order because of quadratic terms in the first-order perturbations. In (9.39) the non-local terms are given by quadratic terms with a non-local dependence on space, non-local $\equiv 10\nabla^{-2}(\psi^{(1)}\nabla^2\psi^{(1)})/3 - 10\nabla^{-4}(\partial^i\partial_j(\psi^{(1)}\partial_i\partial^j\psi^{(1)}))$, which do not vanish even on large scales. From (9.37), (9.38) and (9.39) we obtain an expression for $\phi^{(2)}$ in terms of quadratic contributions [1]

$$\phi^{(2)} = 2(\psi^{(1)})^2 + 2\nabla^{-2}(\partial^i\psi^{(1)}\partial_i\psi^{(1)})$$
$$- 6\nabla^{-4}\partial_i\partial^j(\partial^i\psi^{(1)}\partial_j\psi^{(1)}). \tag{9.40}$$

where $\phi^{(1)} = \psi^{(1)}$ in the Poisson gauge. The gravitational potential $\phi = \phi^{(1)} + \phi^{(2)}/2$ will then have a non-gaussian $(-\chi^2)$ component. It can be expressed in momentum space as

$$\phi(\mathbf{k}) = \phi^{(1)}(\mathbf{k}) + \frac{1}{(2\pi)^3}\int d\mathbf{k}_1 d\mathbf{k}_2 f_{\text{NL}}^\phi(\mathbf{k}_1, \mathbf{k}_2)$$
$$\times \phi^{(1)}(\mathbf{k}_1)\phi^{(1)}(\mathbf{k}_2)\delta^{(3)}(\mathbf{k}_1 + \mathbf{k}_2 - \mathbf{k}) \tag{9.41}$$

where we have defined an effective "momentum–dependent" non–linearity parameter f_{NL}^ϕ. Here the linear lapse function $\phi^{(1)} = \psi^{(1)}$ is a Gaussian random field. Note that a momentum–dependent function must be added to the R.H.S. of (9.41) in order to satisfy the requirement that $\langle\phi\rangle = 0$. From (9.41) the gravitational potential bispectrum reads

$$\langle\phi(\mathbf{k}_1)\phi(\mathbf{k}_2)\phi(\mathbf{k}_3)\rangle = (2\pi)^3\delta^{(3)}\left(\sum_i \mathbf{k}_i\right)$$
$$\times [2f_{\text{NL}}^\phi(\mathbf{k}_1, \mathbf{k}_2)\, P_\Phi(k_1)P_\phi(k_2) + \text{cycl}]. \tag{9.42}$$

From (9.40) we read the non-linearity parameter for the gravitational potential

$$f_{\text{NL}}^\phi(\mathbf{k}_1, \mathbf{k}_2) = -\frac{1}{2} + g(\mathbf{k}_1, \mathbf{k}_2) \tag{9.43}$$

where

$$g(\mathbf{k}_1, \mathbf{k}_2) = 4\frac{\mathbf{k}_1 \cdot \mathbf{k}_2}{k^2} - 3\frac{(\mathbf{k}_1 \cdot \mathbf{k}_2)^2}{k^4} + \frac{3}{2}\frac{k_1^4 + k_2^4}{k^4}, \tag{9.44}$$

with $\mathbf{k} = \mathbf{k}_1 + \mathbf{k}_2$. We thus conclude that the tiny primordial signal is enhanced by the second-order corrections *after inflation* leading to $f_{\text{NL}}^\phi \sim \mathcal{O}(1)$.

9.4.2 Curvaton Scenario

For the curvaton (and the inhomogeneous reheating scenarios) the steps to follow the evolution of the non-linearities in the post-inflationary epoch are the same as we described for the standard scenario. The difference is in the initial conditions which are set by a different value of the curvature $\zeta^{(2)}$ produced after the curvaton decay, at the beginning of the radiation epoch $\zeta^{(2)} = r(3/2 - r^2)(\zeta_\sigma^{(1)})^2$, where $r \simeq (\rho_\sigma/\rho)_{\text{decay}}$. Using this as an initial condition to get the value for the constant $\mathcal{C}$ in (9.37) (with $\psi^{(1)} = -3r\zeta_\sigma^{(1)}/5$) together with (9.38) and (9.39), we obtain $f_{\text{NL}}^\phi(\mathbf{k}_1, \mathbf{k}_2) = 7/6 + 5r/6 - 5/(4r) + g(\mathbf{k}_1, \mathbf{k}_2)$.

9.4.3 Inhomogeneous Reheating Scenario

The initial conditions for the post-inflationary evolution are set by solving the evolution equation during reheating $\zeta^{(2)'} = -\delta^{(2)}\Gamma_\varphi/3 - \zeta^{(1)}\delta^{(1)}\Gamma_\varphi - 2\zeta^{(1)'}\zeta^{(1)} - 2(\zeta^{(1)}\delta^{(1)}\Gamma_\varphi/\mathcal{H})'/3$. Note that we have expanded at second order also the decay rate of the inflation field. In the case of a decay rate which is quadratic in a light scalar field χ, $\Gamma_\varphi = \Gamma_0 + \Gamma_1(\chi/\chi_*)^2$, we obtain $f_{\text{NL}}^\phi(\mathbf{k}_1, \mathbf{k}_2) = 3/4 + I + g(\mathbf{k}_1, \mathbf{k}_2)$. Here $I = -5/2 + 5\bar{\Gamma}/12\alpha\Gamma_1$ is a coefficient which depends on some details of the process, such as the ratio of the fluctuating channel to the total decay rate (see Table 9.1). The "minimal" picture corresponds to $I = 0$.

9.4.4 Large-Scale Temperature Fluctuations at Second-Order

In order to make contact with the observational quantities, the bispectrum of the CMB anisotropies, a further step is needed. We have to take into account that the non-linearities in the gravitational potentials are transferred to the CMB temperature fluctuations. In order to do that we take the large-scale limit of the expression for the CMB fluctuations derived up to second-order

$$\frac{\Delta T}{T} = \phi_\varepsilon + \tau_\varepsilon - \frac{1}{2}\left(\phi_\varepsilon^{(1)}\right)^2 + \phi_\varepsilon^{(1)}\tau_\varepsilon^{(1)} , \qquad (9.45)$$

where $\phi_\varepsilon = \phi_\varepsilon^{(1)} + \phi_\varepsilon^{(2)}/2$ is the lapse perturbations at emission on the last scattering surface and $\tau_\varepsilon = \tau_\varepsilon^{(1)} + \tau_\varepsilon^{(2)}/2$ is the intrinsic fractional temperature fluctuation at emission. In this way we define a non-gaussian contribution that can be taken as an initial condition for the non-linear CMB temperature fluctuations by selecting all those effects that survive in the large scale limit. This effect will contribute also when the modes reenter the horizon. In fact it will add to second-order integrated Sachs–Wolfe effect, and effects which are relevant on smaller scales (e.g., Rees–Sciama contributions, or Doppler effects) which have been dropped in (9.45). These effects should be distinguished from the large scale part singled out in (9.45) thanks to their different scale dependence. By imposing the adiabaticity condition between the matter and radiation fluids $\zeta_m = \zeta_r$, one finds the fundamental relation

$$\frac{\Delta T}{T} = \frac{1}{3}\left[\psi_\varepsilon^{(1)} + \frac{1}{2}\left(\phi_\varepsilon^{(2)} - \frac{5}{3}(\psi_\varepsilon^{(1)})^2\right)\right] . \qquad (9.46)$$

Table 9.1. Predictions of the non-linearity parameter f_{NL} from different scenarios for the generation of cosmological perturbations

	$f_{\mathrm{NL}}(\mathbf{k}_1, \mathbf{k}_2)$	Comments
Single-field inflation	$\frac{7}{3} - g(\mathbf{k}_1, \mathbf{k}_2)$	$g(\mathbf{k}_1, \mathbf{k}_2) = 4\frac{\mathbf{k}_1\cdot\mathbf{k}_2}{k^2} - 3\frac{(\mathbf{k}_1\cdot\mathbf{k}_2)^2}{k^4} + \frac{3}{2}\frac{k_1^4+k_2^4}{k^4}$
Curvaton scenario	$-\left[\frac{2}{3} + \frac{5}{6}r - \frac{5}{4r}\right] - g(\mathbf{k}_1, \mathbf{k}_2)$	$r \approx \left(\frac{\rho_\sigma}{\rho}\right)_{\mathrm{decay}}$
Inhomogeneous reheating	$\frac{13}{12} - I - g(\mathbf{k}_1, \mathbf{k}_2)$	$I = -\frac{5}{2} + \frac{5}{12}\frac{\bar{\Gamma}}{\alpha\Gamma_1}$ "minimal case" $I = 0$ ($\alpha = \frac{1}{6}, \Gamma_1 = \bar{\Gamma}$)
Multiple scalar fields	$\frac{\mathcal{P}_S}{\mathcal{P}_R}\cos^2\Delta \left(4\times10^3 \cdot \frac{V_{XX}}{3H^2}\right)\cdot 60\frac{H}{\chi}$	order of magnitude estimate of the absolute value
"Unconventional" inflation set-ups		
Warm inflation	$-\frac{5}{6}\left(\frac{\dot{\phi}_0}{H^2}\right)\left[\ln\left(\frac{\Gamma}{H}\right)\frac{V'''}{\Gamma}\right]$	second-order corrections not included
Ghost inflation	$-85\cdot\beta\cdot\alpha^{-8/5}$	post-inflationary corrections not included
D-cceleration	$-0.06\gamma^2$	post-inflationary corrections not included

In the inhomogeneous reheating scenario $\Gamma_1/\bar{\Gamma} \leq 1$ and $0 < \alpha \lesssim 1/6$. In the multiple-field case $-1 \leq \cos\Delta \leq 1$ measures the correlation between the adiabatic and entropy perturbations, while $\mathcal{P}_S/\mathcal{P}_R$ is the isocurvature fraction. In ghost inflation the coefficients α and β are typically $\sim O(1)$. In the D-cceleration mechanism of inflation the coefficient γ is expected to be $\gamma > 1$. For the multiple-field case and the unconventional inflation set-ups, the estimates can receive relevant corrections in the range $f_{\mathrm{NL}} \sim 1$ from the post-inflationary evolution of the perturbations.

It is such an expression that allow us to give the exact definition for the non-linearity parameter f_{NL} which is usually adopted to phenomenologically parametrize the non-gaussianity level of cosmological perturbations and has become the standard quantity to be observationally constrained by CMB experiments. Let us stress that the non-linearity parameter singles out the primordial large-scale part of the second-order CMB anisotropies. The definition of f_{NL} adopted in the analyses goes through the conventional Sachs–Wolfe formula $\Delta/T = -\Phi/3$ where Φ is the Bardeen potential, which is conventionally expanded as (up to a constant offset, which only affects the temperature monopole) $\Phi = \Phi_L + f_{NL} \star (\Phi_L)^2$, with $\Phi_L = -\phi^{(1)}$. Therefore if we expand the gravitational potential ϕ at second order as in (9.41), or equivalently as a general convolution $\phi = \phi^{(1)} + \phi^{(2)}/2 = \psi^{(1)} + f_{NL}^\phi \star (\psi^{(1)})^2$ we find the connection between theory and observations

$$f_{NL} = -f_{NL}^\phi + \frac{5}{6} + 1 \,, \tag{9.47}$$

where we have also accounted for the $+1$ shift from the angular averaging with the first-order metric determinant [1].

9.5 What Do We Learn from What We Have Done so Far?

There are lessons we can draw from all the efforts made so far:

- The level of NG in the CMB anisotropies is not simply the primordial one: one should compute how the NG evolves from the primordial epoch to today when we detect it.
- One should compute the radiation transfer function at second order on all scales. So far, we have done this exercise only for scales beyond the horizon. In other words, we have set the initial condition for the NG before the cosmological perturbations reenter the horizon
- The level of NG predicted within the one-single field models is **NOT** as small as $\sim (n_\zeta - 1) \sim 10^{-2}$, but it is of order unity. This comes about because of the non-linearities introduced by the evolution after inflation.
- A detection of NG at the level $\ll 1$ will tell us that not only all models of inflation based on one-single fields are ruled out, but also that something is wrong with the computation performed so far. In other words, a NG of level unity is expected.

It is clear that the next crucial question is: what does it mean NG of order unity? Is it 10^{-1} or 10? In the following we will try to get a (partial) answer to this question. Before though, we will discuss what the present (and future) observational status of detecting the NG in the CMB is.

9.6 Observational Constraints on NG in the CMB

9.6.1 Theoretical Predictions for the CMB Bispectrum from Inflation

In this section we derive analytical predictions for the angular bispectrum from inflation. A note of caution is in order here: all the results obtained in this Section assume a first-order transfer function for radiation. The second-order transfer function will be dealt with in the next section.

We expand the observed CMB temperature fluctuation field, $\Delta T(\hat{\mathbf{n}})/T$, into the spherical harmonics,

$$a_{lm} = \int d^2\hat{\mathbf{n}} \frac{\Delta T(\hat{\mathbf{n}})}{T} Y_{lm}^*(\hat{\mathbf{n}}) , \qquad (9.48)$$

where the hats denote unit vectors. The CMB angular bispectrum is given by

$$B_{l_1 l_2 l_3}^{m_1 m_2 m_3} \equiv \langle a_{l_1 m_1} a_{l_2 m_2} a_{l_3 m_3} \rangle , \qquad (9.49)$$

and the angular averaged bispectrum is

$$B_{l_1 l_2 l_3} = \sum_{\text{all } m} \begin{pmatrix} l_1 & l_2 & l_3 \\ m_1 & m_2 & m_3 \end{pmatrix} B_{l_1 l_2 l_3}^{m_1 m_2 m_3} , \qquad (9.50)$$

where the matrix is the Wigner-$3j$ symbol. The bispectrum, $B_{l_1 l_2 l_3}^{m_1 m_2 m_3}$, satisfies the triangle conditions and parity invariance: $m_1 + m_2 + m_3 = 0$, $l_1 + l_2 + l_3 =$ even, and $|l_i - l_j| \le l_k \le l_i + l_j$ for all permutations of indices. It implies that $B_{l_1 l_2 l_3}^{m_1 m_2 m_3}$ consists of the Gaunt integral, $\mathcal{G}_{l_1 l_2 l_3}^{m_1 m_2 m_3}$, defined by

$$\mathcal{G}_{l_1 l_2 l_3}^{m_1 m_2 m_3} \equiv \int d^2\hat{\mathbf{n}} Y_{l_1 m_1}(\hat{\mathbf{n}}) Y_{l_2 m_2}(\hat{\mathbf{n}}) Y_{l_3 m_3}(\hat{\mathbf{n}})$$
$$= \sqrt{\frac{(2l_1 + 1)(2l_2 + 1)(2l_3 + 1)}{4\pi}} \begin{pmatrix} l_1 & l_2 & l_3 \\ 0 & 0 & 0 \end{pmatrix} \begin{pmatrix} l_1 & l_2 & l_3 \\ m_1 & m_2 & m_3 \end{pmatrix} (9.51)$$

$\mathcal{G}_{l_1 l_2 l_3}^{m_1 m_2 m_3}$ is real, and satisfies all the conditions mentioned above.

Rotational invariance of the angular three–point correlation function implies that $B_{l_1 l_2 l_3}$ is written as

$$B_{l_1 l_2 l_3}^{m_1 m_2 m_3} = \mathcal{G}_{l_1 l_2 l_3}^{m_1 m_2 m_3} b_{l_1 l_2 l_3} , \qquad (9.52)$$

where $b_{l_1 l_2 l_3}$ is an arbitrary real symmetric function of l_1, l_2, and l_3. This form, (9.52), is necessary and sufficient to construct generic $B_{l_1 l_2 l_3}^{m_1 m_2 m_3}$ under rotational invariance; thus, we will use $b_{l_1 l_2 l_3}$ more frequently than $B_{l_1 l_2 l_3}^{m_1 m_2 m_3}$ in this section, and call this function the *reduced* bispectrum, as $b_{l_1 l_2 l_3}$ contains all physical information in $B_{l_1 l_2 l_3}^{m_1 m_2 m_3}$. Since the reduced bispectrum does not contain the Wigner-$3j$ symbol, which merely ensures the triangle conditions and parity invariance, it is easier to calculate physical properties of the bispectrum.

We calculate the angular averaged bispectrum, $B_{l_1 l_2 l_3}$, by substituting (9.52) into (9.50),

$$B_{l_1 l_2 l_3} = \sqrt{\frac{(2l_1 + 1)(2l_2 + 1)(2l_3 + 1)}{4\pi}} \begin{pmatrix} l_1 & l_2 & l_3 \\ 0 & 0 & 0 \end{pmatrix} b_{l_1 l_2 l_3} , \qquad (9.53)$$

where we have used the identity,

$$\sum_{\text{all } m} \begin{pmatrix} l_1 & l_2 & l_3 \\ m_1 & m_2 & m_3 \end{pmatrix} \mathcal{G}_{l_1 l_2 l_3}^{m_1 m_2 m_3} = \sqrt{\frac{(2l_1 + 1)(2l_2 + 1)(2l_3 + 1)}{4\pi}} \begin{pmatrix} l_1 & l_2 & l_3 \\ 0 & 0 & 0 \end{pmatrix} .$$

$$(9.54)$$

Alternatively, one can define the bispectrum in the flat–sky approximation,

$$\langle a(\mathbf{l_1}) a(\mathbf{l_1}) a(\mathbf{l_3}) \rangle = (2\pi)^2 \delta^{(2)} (\mathbf{l_1} + \mathbf{l_2} + \mathbf{l_3}) B(\mathbf{l_1}, \mathbf{l_2}, \mathbf{l_3}) , \qquad (9.55)$$

where $\mathbf{l}$ is a two–dimensional wave vector on the sky. This definition of $B(\mathbf{l_1}, \mathbf{l_2}, \mathbf{l_3})$ reduces to (9.52) with the correspondence

$$\mathcal{G}_{l_1 l_2 l_3}^{m_1 m_2 m_3} \rightarrow (2\pi)^2 \delta^{(2)} (\mathbf{l_1} + \mathbf{l_2} + \mathbf{l_3}) , \qquad (9.56)$$

in the flat–sky limit. Thus, we have

$$b_{l_1 l_2 l_3} \approx B(\mathbf{l_1}, \mathbf{l_2}, \mathbf{l_3}) \qquad \text{(flat–sky approximation)} . \qquad (9.57)$$

This fact motivates our use of the reduced bispectrum, $b_{l_1 l_2 l_3}$, rather than the angular averaged bispectrum, $B_{l_1 l_2 l_3}$.

If primordial fluctuations are adiabatic scalar fluctuations, then

$$a_{lm} = 4\pi(-i)^l \int \frac{d^3 \mathbf{k}}{(2\pi)^3} \Phi(\mathbf{k}) g_{Tl}(k) Y_{lm}^*(\hat{\mathbf{k}}) , \qquad (9.58)$$

where $\Phi(\mathbf{k})$ is the primordial curvature perturbation in Fourier space, and $g_{Tl}(k)$ is the radiation transfer function. a_{lm} takes over the non–gaussianity, if any, from $\Phi(\mathbf{k})$. Although (9.58) is valid only if the Universe is flat, it is straightforward to extend this to an arbitrary geometry. We can calculate the isocurvature fluctuations similarly by using the entropy perturbation and the proper transfer function.

The primordial non–gaussianity may be parameterized as a linear plus quadratic term in the gravitational potential in the general form of (9.41), where the non–linearity parameter f_{NL} appears as a kernel in Fourier space, rather than a constant. This gives rise to angular modulation of the quadratic non–linearity, which might be used to search for specific signatures of inflationary non–gaussianity in the CMB. In this section, however, we restrict ourselves to the simplest weak non–linear coupling case, assuming that f_{NL} is merely a multiplicative constant, as done in data analyses so far. Hence we write

$$\Phi(\mathbf{x}) = \Phi_L(\mathbf{x}) + f_{NL} \left[\Phi_L^2(\mathbf{x}) - \langle \Phi_L^2(\mathbf{x}) \rangle \right] , \qquad (9.59)$$

in real space, where $\Phi_L(\mathbf{x})$ denotes the linear Gaussian part of the perturbation, and $\langle \Phi(\mathbf{x}) \rangle = 0$ is guaranteed.

In Fourier space, we decompose $\Phi(\mathbf{k})$ into two parts,

$$\Phi(\mathbf{k}) = \Phi_L(\mathbf{k}) + \Phi_{NL}(\mathbf{k}) , \tag{9.60}$$

and accordingly we have

$$a_{lm} = a_{lm}^L + a_{lm}^{NL} , \tag{9.61}$$

where $\Phi_{NL}(\mathbf{k})$ is a non–linear curvature perturbation defined by

$$\Phi_{NL}(\mathbf{k}) \equiv f_{NL} \left[\int \frac{d^3\mathbf{p}}{(2\pi)^3} \Phi_L(\mathbf{k}+\mathbf{p}) \Phi_L^*(\mathbf{p}) - (2\pi)^3 \delta^{(3)}(\mathbf{k}) \langle \Phi_L^2(\mathbf{x}) \rangle \right] . \tag{9.62}$$

One can immediately check that $\langle \Phi(\mathbf{k}) \rangle = 0$ is satisfied. In this model, a non–vanishing component of the $\Phi(\mathbf{k})$–field bispectrum is

$$\langle \Phi_L(\mathbf{k}_1)\Phi_L(\mathbf{k}_2)\Phi_{NL}(\mathbf{k}_3) \rangle = (2\pi)^3 \delta^{(3)}(\mathbf{k}_1 + \mathbf{k}_2 + \mathbf{k}_3)\, 2 f_{NL} P_\Phi(k_1) P_\Phi(k_2) , \tag{9.63}$$

where $P_\Phi(k)$ is Bardeen's potential linear power–spectrum given by

$$\langle \Phi_L(\mathbf{k}_1)\Phi_L(\mathbf{k}_2) \rangle = (2\pi)^3 P_\Phi(k_1) \delta^{(3)}(\mathbf{k}_1 + \mathbf{k}_2) . \tag{9.64}$$

We have also used

$$\langle \Phi_L(\mathbf{k}+\mathbf{p})\Phi_L^*(\mathbf{p}) \rangle = (2\pi)^3 P_\Phi(p) \delta^{(3)}(\mathbf{k}) , \tag{9.65}$$

and

$$\langle \Phi_L^2(\mathbf{x}) \rangle = (2\pi)^{-3} \int d^3\mathbf{k} P_\Phi(k) . \tag{9.66}$$

Substituting (9.58) into (9.49), using (9.63) for the $\Phi(\mathbf{k})$–field bispectrum, and then integrating over angles $\hat{\mathbf{k}}_1$, $\hat{\mathbf{k}}_3$, and $\hat{\mathbf{k}}_3$, we obtain the primordial CMB angular bispectrum,

$$
\begin{aligned}
B_{l_1 l_2 l_3}^{m_1 m_2 m_3} &= \langle a_{l_1 m_1}^L a_{l_2 m_2}^L a_{l_3 m_3}^{NL} \rangle + \langle a_{l_1 m_1}^L a_{l_2 m_2}^{NL} a_{l_3 m_3}^L \rangle + \langle a_{l_1 m_1}^{NL} a_{l_2 m_2}^L a_{l_3 m_3}^L \rangle \\
&= 2 \mathcal{G}_{l_1 l_2 l_3}^{m_1 m_2 m_3} \int_0^\infty r^2 dr \left[b_{l_1}^L(r) b_{l_2}^L(r) b_{l_3}^{NL}(r) + b_{l_1}^L(r) b_{l_2}^{NL}(r) b_{l_3}^L(r) \right. \\
&\quad \left. + b_{l_1}^{NL}(r) b_{l_2}^L(r) b_{l_3}^L(r) \right] ,
\end{aligned}
\tag{9.67}
$$

where

$$b_l^L(r) \equiv \frac{2}{\pi} \int_0^\infty k^2 dk\, P_\Phi(k) g_{Tl}(k) j_l(kr), \tag{9.68}$$

$$b_l^{NL}(r) \equiv \frac{2}{\pi} \int_0^\infty k^2 dk\, f_{NL} g_{Tl}(k) j_l(kr) . \tag{9.69}$$

Note that $b_l^L(r)$ is dimensionless, while $b_l^{NL}(r)$ has a dimension of L^{-3}.

One can immediately check that (9.52) holds; thus, the reduced bispectrum, $b_{l_1 l_2 l_3}$ (9.52), for the primordial non–gaussianity reads

$$b_{l_1 l_2 l_3}^{\text{prim}} = 2 \int_0^\infty r^2 dr \left[b_{l_1}^{\text{L}}(r) b_{l_2}^{\text{L}}(r) b_{l_3}^{\text{NL}}(r) + b_{l_1}^{\text{L}}(r) b_{l_2}^{\text{NL}}(r) b_{l_3}^{\text{L}}(r) \right.$$
$$\left. + b_{l_1}^{\text{NL}}(r) b_{l_2}^{\text{L}}(r) b_{l_3}^{\text{L}}(r) \right] . \tag{9.70}$$

We can fully specify $b_{l_1 l_2 l_3}^{\text{prim}}$ by a single constant parameter, f_{NL}, as the CMB angular power–spectrum, C_l, will precisely measure the cosmological parameters. We stress again that this formula is valid only when the scale–dependence of f_{NL} is weak, which is a good approximation if the momentum–independent part of f_{NL} is larger than unity.

One can calculate the primordial CMB bispectrum ((9.67)–(9.70)) numerically as follows. One computes the full radiation transfer function, $g_{Tl}(k)$, with the CMBFAST code, assuming a single power–law spectrum, $P_\Phi(k) \propto k^{n-4}$, for the primordial curvature fluctuations. After doing the integration over k ((9.68) and (9.69)) with the same algorithm of CMBFAST, one performs the integration over r (9.70), $r = c(\tau_0 - \tau)$, where τ is the conformal time. τ_0 is the present–day value. In our model, $c\tau_0 = 11.8$ Gpc, and the decoupling occurs at $c\tau_* = 235$ Mpc at which the differential visibility has a maximum. Our $c\tau_0$ includes radiation effects on the expansion of the Universe; otherwise, $c\tau_0 = 12.0$ Gpc. Since most of the primordial signal is generated at τ_*, we choose the r integration boundary as $c(\tau_0 - 2\tau_*) \le r \le c(\tau_0 - 0.1\tau_*)$. We use a step-size of $0.1c\tau_*$, as we have found that a step size of $0.01c\tau_*$ gives very similar results. As cosmological model, let us assume a scale–invariant Standard Cold Dark Matter (SCDM) model with $\Omega_m = 1$, $\Omega_\Lambda = 0$, $\Omega_b = 0.05$, $h = 0.5$, and $n = 1$, and with power–spectrum $P_\Phi(k)$ normalized to *COBE*. Although this model is almost excluded by current observations, it is still useful to depict the basic effects of the transfer function on the bispectrum.

Figure 9.1 shows $b_l^{\text{L}}(r)$ (9.68) and $b_l^{\text{NL}}(r)$ (9.69) for several different values of r. We find that $b_l^{\text{L}}(r)$ and C_l look very similar to each other in shape and amplitude at $l \gtrsim 100$, although the amplitude in the Sachs–Wolfe regime is different by a factor of -3. This is because $C_l \propto P_\Phi(k) g_{Tl}^2(k)$, while $b_l^{\text{L}}(r) \propto P_\Phi(k) g_{Tl}(k)$, where $g_{Tl} = -1/3$. We also find that $b_l^{\text{L}}(r)$ has a good phase coherence over a wide range of r, while the phase of $b_l^{\text{NL}}(r)$ in the high–l regime oscillates rapidly as a function of r. This strongly damps the integrated result (9.67) in the high-l regime. The main difference between C_l and $b_l(r)$ is that $b_l(r)$ changes the sign, while C_l does not.

Looking at Fig. 9.1, we find $l^2 b_l^{\text{L}} \sim 2 \times 10^{-9}$ and $b_l^{\text{NL}} f_{\text{NL}}^{-1} \sim 10^{-10}$ Mpc^{-3}. As most of the signal is coming from the decoupling epoch, the volume element at τ_* is $r_*^2 \Delta r_* \sim (10^4)^2 \times 10^2$ Mpc^3; thus, we can give an order–of–magnitude estimate of the primordial reduced bispectrum (9.70) as

$$b_{lll}^{\text{prim}} \sim l^{-4} \left[2 r_*^2 \Delta r_* \left(l^2 b_l^{\text{L}} \right)^2 b_l^{\text{NL}} \times 3 \right] \sim l^{-4} \times 2 \times 10^{-17} f_{\text{NL}} . \tag{9.71}$$

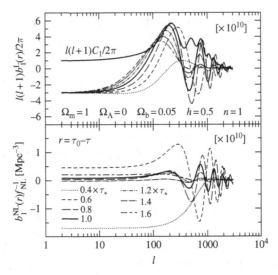

Fig. 9.1. Components of Primordial CMB Bispectrum. This figure shows $b_l^L(r)$ (9.68) and $b_l^{NL}(r)$ (9.69), the two terms in our calculation of the primordial CMB angular bispectrum, as a function of r. Various *lines* in the *top panel* show $\left[l(l+1)b_l^L(r)/2\pi\right] \times 10^{10}$, where $r = c(\tau_0 - \tau)$, at $\tau = 0.4, 0.6, 0.8, 1.0, 1.2, 1.4$, and $1.6 \times \tau_*$ (decoupling time); $\left[b_l^{NL}(r)f_{NL}^{-1}\right] \times 10^{10}$ are shown in the *bottom panel*. τ_0 is the present-day conformal time. Note that $c\tau_0 = 11.8$ Gpc, and $c\tau_* = 235$ Mpc in the cosmological model chosen here. The thickest *solid line* in the *top panel* is the CMB angular power–spectrum, $\left[l(l+1)C_l/2\pi\right] \times 10^{10}$. C_l is shown for comparison

Since $b_l^{NL}f_{NL}^{-1} \sim r_*^{-2}\delta(r - r_*)$ [see (9.74)], $r_*^2\Delta r_* b_l^{NL}f_{NL}^{-1} \sim 1$. This rough estimate agrees with the numerical result below (Fig. 9.2).

Figure 9.2 shows the integrated bispectrum (9.67) divided by the Gaunt integral, $\mathcal{G}_{l_1l_2l_3}^{m_1m_2m_3}$, which is the reduced bispectrum, $b_{l_1l_2l_3}^{prim}$. While the bispectrum is a 3D function, we show different 1D slices of the bispectrum in this figure. We plot

$$l_2(l_2 + 1)l_3(l_3 + 1) \left\langle a_{l_1m_1}^{NL} a_{l_2m_2}^L a_{l_3m_3}^L \right\rangle \left(\mathcal{G}_{l_1l_2l_3}^{m_1m_2m_3}\right)^{-1} / (2\pi)^2$$

as a function of l_3 in the top panel, while we plot

$$l_1(l_1 + 1)l_2(l_2 + 1) \left\langle a_{l_1m_1}^L a_{l_2m_2}^L a_{l_3m_3}^{NL} \right\rangle \left(\mathcal{G}_{l_1l_2l_3}^{m_1m_2m_3}\right)^{-1} / (2\pi)^2$$

in the bottom panel. We have multiplied each $b_l^L(r)$ which contains $P_\Phi(k)$ by $l(l+1)/(2\pi)$ so that the Sachs–Wolfe plateau at $l_3 \lesssim 10$ is easily seen. We have chosen l_1 and l_2 so as $(l_1, l_2) = (9, 11), (99, 101), (199, 201)$, and $(499, 501)$. We find that the $(l_1, l_2) = (199, 201)$ mode, the first acoustic peak mode, has the largest signal in this family of parameters. The top panel has a prominent first acoustic peak, and strongly damped oscillations in the high-l regime; the bottom panel also has a first peak, but damps more slowly. The typical

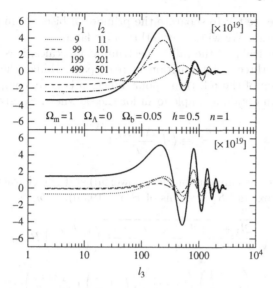

Fig. 9.2. Primordial CMB Bispectrum. The primordial angular bispectrum (9.67), divided by the Gaunt integral, $\mathcal{G}^{m_1 m_2 m_3}_{l_1 l_2 l_3}$ (9.51). The bispectrum is plotted as a function of l_3 for $(l_1, l_2) =$ (9,11), (99,101), (199,201), and (499,501). Each panel shows a different 1-dimensional slice of the bispectrum. The top panel shows $l_2(l_2+1)l_3(l_3+1)\langle a^{NL}_{l_1 m_1} a^{L}_{l_2 m_2} a^{L}_{l_3 m_3}\rangle f^{-1}_{NL} \left(\mathcal{G}^{m_1 m_2 m_3}_{l_1 l_2 l_3}\right)^{-1}/(2\pi)^2$, while the bottom panel shows $l_1(l_1+1)l_2(l_2+1)\langle a^{L}_{l_1 m_1} a^{L}_{l_2 m_2} a^{NL}_{l_3 m_3}\rangle f^{-1}_{NL} \left(\mathcal{G}^{m_1 m_2 m_3}_{l_1 l_2 l_3}\right)^{-1}/(2\pi)^2$. Note that we have multiplied the bispectrum in each panel by a factor of 10^{19}

amplitude of the reduced bispectrum is $l^4 b^{\text{prim}}_{lll} f^{-1}_{NL} \sim 10^{-17}$, which agrees with the order–of–magnitude estimate of (9.71).

$$b^{\text{prim}}_{l_1 l_2 l_3} \approx -6 f_{NL}\left(C^{SW}_{l_1} C^{SW}_{l_2} + C^{SW}_{l_1} C^{SW}_{l_3} + C^{SW}_{l_2} C^{SW}_{l_3}\right) \quad \text{(SW approximation)}. \tag{9.72}$$

Each term is of the same order as in (9.70). Here, C^{SW}_l is the CMB angular power–spectrum in the Sachs–Wolfe approximation,

$$C^{SW}_l \equiv \frac{2}{9\pi}\int_0^\infty k^2 dk\, P_\Phi(k) j_l^2(kr_*). \tag{9.73}$$

In deriving (9.72) from (9.70), we have approximated $b^{NL}_l(r)$ (9.69) with

$$b^{NL}_l(r) \approx \left(-\frac{f_{NL}}{3}\right)\frac{2}{\pi}\int_0^\infty k^2 dk\, j_l(kr_*) j_l(kr) = -\frac{f_{NL}}{3} r_*^{-2}\delta(r - r_*). \tag{9.74}$$

We stress again that the Sachs–Wolfe approximation gives a qualitatively different result from our full calculation (9.70) at $l_i \gtrsim 10$. The full bispectrum changes sign, while the approximation never changes sign because of the use of C^{SW}_l. The acoustic oscillation and the sign–change are actually great advantages when we try to separate the primordial bispectrum from various secondary bispectra. We will analyze this point later.

As we have calculated the full bispectrum at all scales, it is now possible to calculate the three–point function in real space. Unlike the bispectrum, however, the form of the full three–point function is fairly complicated; nevertheless, one can obtain a simple form for the skewness, S_3, given by

$$S_3 \equiv \left\langle \left(\frac{\Delta T(\hat{n})}{T} \right)^3 \right\rangle , \tag{9.75}$$

which is perhaps the simplest (but less powerful) statistic characterizing non–gaussianity. We expand S_3 in terms of $B_{l_1 l_2 l_3}$ (9.50), or $b_{l_1 l_2 l_3}$ (9.52), as

$$S_3 = \frac{1}{4\pi} \sum_{l_1 l_2 l_3} \sqrt{\frac{(2l_1+1)(2l_2+1)(2l_3+1)}{4\pi}} \begin{pmatrix} l_1 & l_2 & l_3 \\ 0 & 0 & 0 \end{pmatrix} B_{l_1 l_2 l_3} W_{l_1} W_{l_2} W_{l_3}$$

$$= \frac{1}{2\pi^2} \sum_{2 \leq l_1 l_2 l_3} \left(l_1 + \frac{1}{2} \right) \left(l_2 + \frac{1}{2} \right) \left(l_3 + \frac{1}{2} \right) \begin{pmatrix} l_1 & l_2 & l_3 \\ 0 & 0 & 0 \end{pmatrix}^2$$

$$\times b_{l_1 l_2 l_3} W_{l_1} W_{l_2} W_{l_3} , \tag{9.76}$$

where W_l is the experimental window function. We have used (9.53) to replace $B_{l_1 l_2 l_3}$ by the reduced bispectrum, $b_{l_1 l_2 l_3}$, in the last equality. Since $l = 0$ and 1 modes are not observable, we have excluded them from the summation. Throughout this section, we consider a single-beam window function, $W_l = \exp[-l(l+1)/(2\sigma_b^2)]$, where $\sigma_b = \text{FWHM}/\sqrt{8 \ln 2}$. Since $\begin{pmatrix} l_1 & l_2 & l_3 \\ 0 & 0 & 0 \end{pmatrix}^2 b_{l_1 l_2 l_3}$ is symmetric under permutation of indices, we change the way of summation as

$$\sum_{2 \leq l_1 l_2 l_3} \longrightarrow 6 \sum_{2 \leq l_1 \leq l_2 \leq l_3} . \tag{9.77}$$

This reduces the number of summations by a factor of $\simeq 6$. We will use this convention henceforth.

The top panel of Fig. 9.3 shows $S_3(< l_3)$, which is S_3 summed up to a certain l_3, for FWHM beam sizes of $7°$, $13'$ and $5'.5$. These values correspond to COBE, WMAP, and Planck beam sizes, respectively. Figure 9.3 also shows the infinitesimally thin beam case. We find that WMAP, Planck, and the ideal experiments, all measure very similar S_3 to one another, despite the fact that Planck and the ideal experiments can use many more modes than WMAP. The reason is the following. Looking at (9.76), one finds that S_3 is a linear integral of $b_{l_1 l_2 l_3}$ over l_i; thus, integrating oscillations in $b_{l_1 l_2 l_3}^{\text{prim}}$ around zero (see Fig. 9.2) damps the non–gaussian signal on small angular scales, $l \gtrsim 300$. Since the Sachs–Wolfe effect, which implies no oscillation, dominates the COBE–scale anisotropy, the cancellation on the COBE scale affects S_3 less significantly than on the WMAP and Planck scales. Planck suffers from severe

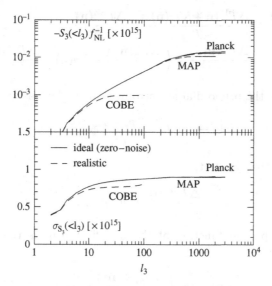

Fig. 9.3. Primordial Skewness. The *top panel* shows the primordial CMB skewness (9.76) summed up to a certain l_3, $-S_3(< l_3)f_{NL}^{-1} \times 10^{15}$. The *bottom panel* shows the error of S_3 (9.109) summed up to l_3, $\sigma_{S_3}(< l_3) \times 10^{15}$. The solid line represents the zero–noise ideal experiment, while the dotted lines show COBE, WMAP and Planck experiments

cancellation in small angular scales: Planck and the ideal experiments measure only the same amount of S_3 as WMAP does. As a result, the measured S_3 almost saturates at the WMAP resolution scale, $l \sim 500$.

We conclude this section by noting that when we can calculate the expected form of the bispectrum, then it becomes a "matched filter" for detecting non–gaussianity in the data, and thus much more powerful a tool than the skewness in which the information is lost through the coarse–graining.

9.6.2 Secondary Sources of CMB Bispectrum

Even if the CMB bispectrum were significantly detected in the CMB map, its origin would not necessarily be primordial, but rather there would be various secondary sources such as the Sunyaev–Zel'dovich (SZ) effect, the weak lensing effect, and so on, or foreground sources such as extragalactic radio sources. To isolate the primordial origin from the others, we have to know the accurate form of bispectra produced by secondary and foreground sources.

Coupling Between the Weak Lensing and the Sunyaev–Zel'dovich Effects

The coupling between the SZ effect and the weak lensing effect produces an observable effect in the bispectrum. We expand the CMB temperature field including the SZ and the lensing effect as

$$\frac{\Delta T(\hat{n})}{T} = \frac{\Delta T^{\mathrm{P}}\left(\hat{n}+\nabla\Theta(\hat{n})\right)}{T} + \frac{\Delta T^{\mathrm{SZ}}(\hat{n})}{T}$$

$$\approx \frac{\Delta T^{\mathrm{P}}(\hat{n})}{T} + \nabla\left(\frac{\Delta T^{\mathrm{P}}(\hat{n})}{T}\right)\cdot\nabla\Theta(\hat{n}) + \frac{\Delta T^{\mathrm{SZ}}(\hat{n})}{T}\ , \quad (9.78)$$

where P denotes the primordial anisotropy, $\Theta(\hat{n})$ is the lensing potential,

$$\Theta(\hat{n}) \equiv -2\int_0^{r_*} dr \frac{r_* - r}{rr_*}\Phi(r,\hat{n}r)\ , \quad (9.79)$$

and SZ denotes the SZ effect,

$$\frac{\Delta T^{\mathrm{SZ}}(\hat{n})}{T} = y(\hat{n})j_\nu\ , \quad (9.80)$$

where j_ν is a spectral function of the SZ effect $y(\hat{n})$ is the Compton y-parameter given by

$$y(\hat{n}) \equiv y_0 \int \frac{dr}{r_*}\frac{T_\rho(r,\hat{n}r)}{\overline{T}_{\rho 0}}a^{-2}(r)\ , \quad (9.81)$$

where

$$y_0 \equiv \frac{\sigma_T\bar{\rho}_{\mathrm{gas}0}k_{\mathrm{B}}\overline{T}_{\rho 0}r_*}{\mu_e m_{\mathrm{p}}m_e c^2} = 4.3\times 10^{-4}\mu_e^{-1}\left(\Omega_{\mathrm{b}}h^2\right)\left(\frac{k_{\mathrm{B}}\overline{T}_{\rho 0}}{1\ \mathrm{keV}}\right)\left(\frac{r_*}{10\ \mathrm{Gpc}}\right)\ . \quad (9.82)$$

$T_\rho \equiv \rho_{\mathrm{gas}}T_e/\bar{\rho}_{\mathrm{gas}}$ is the electron temperature weighted by the gas mass density, the overline denotes the volume average, and the subscript 0 means the present epoch. We adopt $\mu_e^{-1} = 0.88$, where $\mu_e^{-1} \equiv n_e/(\rho_{\mathrm{gas}}/m_{\mathrm{p}})$ is the number of electrons per proton mass in the fully ionized medium. Other quantities have their usual meaning.

Transforming (9.78) into harmonic space, we obtain

$$a_{lm} = a_{lm}^{\mathrm{P}} + \sum_{l'm'}\sum_{l''m''}(-1)^m \mathcal{G}_{ll'l''}^{-mm'm''}$$

$$\times\frac{l'(l'+1) - l(l+1) + l''(l''+1)}{2}a_{l'm'}^{\mathrm{P}}\Theta_{l''m''} + a_{lm}^{\mathrm{SZ}}$$

$$= a_{lm}^{\mathrm{P}} + \sum_{l'm'}\sum_{l''m''}(-1)^{m+m'+m''}\mathcal{G}_{ll'l''}^{-mm'm''}$$

$$\times\frac{l'(l'+1) - l(l+1) + l''(l''+1)}{2}a_{l'-m'}^{\mathrm{P}*}\Theta_{l''-m''}^* + a_{lm}^{\mathrm{SZ}}\ , \quad (9.83)$$

where $\mathcal{G}_{l_1l_2l_3}^{m_1m_2m_3}$ is the Gaunt integral (9.51). Substituting (9.83) into (9.49), and using the identity, $\mathcal{G}_{l_1l_2l_3}^{-m_1-m_2-m_3} = \mathcal{G}_{l_1l_2l_3}^{m_1m_2m_3}$, we obtain the bispectrum,

$$B_{l_1l_2l_3}^{m_1m_2m_3} = \mathcal{G}_{l_1l_2l_3}^{m_1m_2m_3}\left[\frac{l_1(l_1+1) - l_2(l_2+1) + l_3(l_3+1)}{2}C_{l_1}^{\mathrm{P}}\left\langle\Theta_{l_3m_3}^* a_{l_3m_3}^{\mathrm{SZ}}\right\rangle\right.$$

$$\left.+5\ \mathrm{permutations}\right]\ . \quad (9.84)$$

The form of (9.52) is confirmed; the reduced bispectrum $b_{l_1 l_2 l_3}^{\text{sz}-\text{lens}}$ includes the terms in square brackets.

While (9.84) is complicated, we can understand the physical effect producing the SZ–lensing bispectrum intuitively. Figure 9.4 shows how the SZ–lensing coupling produces the three–point correlation. Suppose that there are three CMB photons decoupled at the last scattering surface (LSS), and one of these photons penetrates through a SZ cluster between the LSS and us; the energy of the photon changes because of the SZ effect. When the other two photons pass near the SZ cluster, they are deflected by the gravitational lensing effect, changing their propagation directions, and coming toward us. What do we see after all? We see that the three CMB photons are correlated; we then measure a non–zero angular bispectrum. The cross–correlation strength between the SZ and lensing effects, $\langle \Theta_{l_3 m_3}^* a_{l_3 m_3}^{\text{SZ}} \rangle$, thus determines the bispectrum amplitude, as indicated by (9.84).

The quantity $\langle \Theta_{lm}^* a_{lm}^{\text{SZ}} \rangle$ was derived assuming the linear pressure bias model [1], $T_\rho = \overline{T}_\rho b_{\text{gas}} \delta$, and the mean temperature evolution, $\overline{T}_\rho \simeq \overline{T}_{\rho 0}(1 + z)^{-1}$, for $z < 2$, which is roughly suggested by recent hydrodynamic simulations

$$\langle \Theta_{lm}^* a_{lm}^{\text{SZ}} \rangle \simeq -j_\nu \frac{4 y_0 b_{\text{gas}} l^2}{3 \Omega_{\text{m}} H_0^2} \int_0^{z_*} \mathrm{d}z \frac{\mathrm{d}r}{\mathrm{d}z} D^2(z)(1+z)^2 \frac{r_* - r(z)}{r_*^2 r^5(z)} P_\Phi \left(k = \frac{l}{r(z)} \right),$$
(9.85)

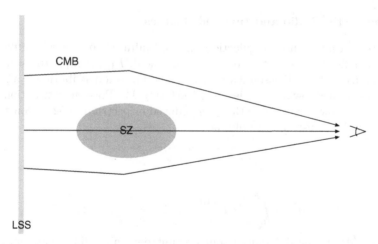

Fig. 9.4. SZ–lensing Coupling. A schematic view of the SZ–lensing coupling bispectrum. One of the three CMB photons, which are decoupled at the last scattering surface (LSS), penetrates through a SZ cluster, changing its temperature, and coming toward us. As the other two photons pass near the SZ cluster, they are deflected by the lensing effect, changing their propagation directions, and coming toward us. As a result, the three photons are correlated, generating a three–point correlation, the bispectrum

where $D(z)$ is the linear growth factor. Simulations without non–gravitational heating suggest that $\overline{T}_{\rho 0} \sim 0.2-0.4$ keV and $b_{\mathrm{gas}} \sim 5-10$; analytic estimations give similar numbers. In the pressure bias model, the free parameters (except cosmological ones) are $\overline{T}_{\rho 0}$ and b_{gas}; however, both actually depend upon the cosmological model. Since $l^3 \langle \Theta^*_{lm} a^{\mathrm{SZ}}_{lm} \rangle \sim 2 \times 10^{-10} j_\nu \overline{T}_{\rho 0} b_{\mathrm{gas}}$ and $l^2 C^{\mathrm{P}}_l \sim 6 \times 10^{-10}$, we have

$$b^{\mathrm{sz-lens}}_{lll} \sim l^{-3} \left[(l^2 C^{\mathrm{P}}_l) \left(l^3 \langle \Theta^*_{lm} a^{\mathrm{SZ}}_{lm} \rangle \right) \times 5/2 \right] \sim l^{-3} \times 3 \times 10^{-19} j_\nu \overline{T}_{\rho 0} b_{\mathrm{gas}} , \tag{9.86}$$

where $\overline{T}_{\rho 0}$ is in units of 1 keV, and $b_{l_1 l_2 l_3} = B^{m_1 m_2 m_3}_{l_1 l_2 l_3} \left(\mathcal{G}^{m_1 m_2 m_3}_{l_1 l_2 l_3} \right)^{-1}$ is the reduced bispectrum (9.52). Comparing this with (9.71), we obtain

$$\frac{b^{\mathrm{prim}}_{lll}}{b^{\mathrm{sz-lens}}_{lll}} \sim l^{-1} \times 10 \left(\frac{f_{\mathrm{NL}}}{j_\nu \overline{T}_{\rho 0} b_{\mathrm{gas}}} \right) . \tag{9.87}$$

This estimate suggests that the SZ–lensing bispectrum dominates the primordial bispectrum on small angular scales. This is why we have to separate the primordial from the SZ–lensing effect.

While the pressure bias model gives a rough estimate of the SZ power–spectrum, more accurate predictions exist. Several authors have predicted the SZ power–spectrum analytically using the Press–Schechter approach or hyper-extended perturbation theory. The predictions agree with hydrodynamic simulations.

Extragalactic Radio and Infrared Sources

The bispectrum from extragalactic radio and infrared sources whose fluxes, F, are smaller than a certain detection threshold, F_{d}, is simple to estimate, when we assume the Poisson distribution. The Poisson distribution is a good approximation at low frequencies ($\nu < 100$ GHz). The Poisson distribution has white–noise power–spectrum; thus, the reduced bispectrum (9.52) is constant, $b^{\mathrm{src}}_{l_1 l_2 l_3} = b^{\mathrm{src}} = \mathrm{const}$, and we obtain

$$B^{m_1 m_2 m_3}_{l_1 l_2 l_3} = \mathcal{G}^{m_1 m_2 m_3}_{l_1 l_2 l_3} b^{\mathrm{src}} , \tag{9.88}$$

where

$$b^{\mathrm{src}}(< F_{\mathrm{d}}) \equiv g^3(x) \int_0^{F_{\mathrm{d}}} dF F^3 \frac{dn}{dF} = g^3(x) \frac{\beta}{3 - \beta} n(> F_{\mathrm{d}}) F^3_{\mathrm{d}} . \tag{9.89}$$

Here, dn/dF is the differential source count per unit solid angle, and $n(> F_{\mathrm{d}}) \equiv \int_{F_{\mathrm{d}}}^\infty dF (dn/dF)$. We have assumed a power–law count, $dn/dF \propto F^{-\beta-1}$, for $\beta < 2$. The other symbols mean $x \equiv h\nu / k_{\mathrm{B}} T \simeq (\nu / 56.80 \text{ GHz}) (T/2.726 \text{ K})^{-1}$, and

$$g(x) \equiv 2 \frac{(hc)^2}{(k_{\mathrm{B}} T)^3} \left(\frac{\sinh x/2}{x^2} \right)^2 \simeq \frac{1}{67.55 \text{ MJy sr}^{-1}} \left(\frac{T}{2.726 \text{ K}} \right)^{-3} \left(\frac{\sinh x/2}{x^2} \right)^2 . \tag{9.90}$$

Using the Poisson angular power–spectrum, C^{ps}, given by

$$C^{\mathrm{ps}}(< F_{\mathrm{d}}) \equiv g^2(x) \int_0^{F_{\mathrm{d}}} \mathrm{d}F F^2 \frac{\mathrm{d}n}{\mathrm{d}F} = g^2(x) \frac{\beta}{2-\beta} n(> F_{\mathrm{d}}) F_{\mathrm{d}}^2 \,, \qquad (9.91)$$

we can rewrite b^{src} into a different form,

$$b^{\mathrm{src}}(< F_{\mathrm{d}}) = \frac{(2-\beta)^{3/2}}{\beta^{1/2}(3-\beta)} \left[n(> F_{\mathrm{d}})\right]^{-1/2} \left[C^{\mathrm{ps}}(< F_{\mathrm{d}})\right]^{3/2} \,. \qquad (9.92)$$

One can estimate that $n(> F_{\mathrm{d}}) \sim 300 \ \mathrm{sr}^{-1}$ for $F_{\mathrm{d}} \sim 0.2$ Jy at 217 GHz. This F_{d} corresponds to 5σ detection threshold for the Planck experiment at 217 GHz. Extrapolating to 94 GHz, one finds $n(> F_{\mathrm{d}}) \sim 7 \ \mathrm{sr}^{-1}$ for $F_{\mathrm{d}} \sim 2$ Jy, which corresponds to the WMAP 5σ threshold. These values yield

$$C^{\mathrm{ps}}(90 \ \mathrm{GHz}, < 2 \ \mathrm{Jy}) \sim 2 \times 10^{-16}, \qquad (9.93)$$
$$C^{\mathrm{ps}}(217 \ \mathrm{GHz}, < 0.2 \ \mathrm{Jy}) \sim 1 \times 10^{-17} \,. \qquad (9.94)$$

Thus, rough estimates for b^{src} are

$$b^{\mathrm{src}}(90 \ \mathrm{GHz}, < 2 \ \mathrm{Jy}) \sim 2 \times 10^{-25}, \qquad (9.95)$$
$$b^{\mathrm{src}}(217 \ \mathrm{GHz}, < 0.2 \ \mathrm{Jy}) \sim 5 \times 10^{-28} \,. \qquad (9.96)$$

While we have assumed the Euclidean source count ($\beta = 3/2$) for definiteness, this assumption does not affect order–of–magnitude estimates.

As the primordial reduced bispectrum is $\propto l^{-4}$ (9.71), and the SZ–lensing reduced bispectrum is $\propto l^{-3}$ (9.86), the point–source bispectrum rapidly comes to dominate the total bispectrum on small angular scales:

$$\frac{b_{lll}^{\mathrm{prim}}}{b^{\mathrm{src}}} \sim l^{-4} \times 10^7 \left(\frac{f_{\mathrm{NL}}}{b^{\mathrm{src}}/10^{-25}}\right), \qquad (9.97)$$

$$\frac{b_{lll}^{\mathrm{sz-lens}}}{b^{\mathrm{src}}} \sim l^{-3} \times 10^6 \left(\frac{j_\nu \overline{T} \rho_0 b_{\mathrm{gas}}}{b^{\mathrm{src}}/10^{-25}}\right) \,. \qquad (9.98)$$

For example, the point–sources dominates the SZ–lensing bispectrum measured by WMAP at $l \gtrsim 100$.

What do the SZ–lensing bispectrum and the point–source bispectrum look like? Figure 9.5 shows the primordial, the SZ–lensing, and the point–source reduced bispecta for the equilateral configurations, $l \equiv l_1 = l_2 = l_3$. We have plotted $l^2(l+1)^2 b_{lll}/(2\pi)^2$. We find that these bispecra are very different from each other in shape on small angular scales. It thus suggests that we can separate these three contributions on the basis of shape difference. We study this point in the next section.

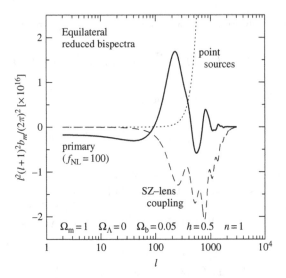

Fig. 9.5. Equilateral reduced bispectra. Comparison between the primordial (*solid line*), the SZ–lensing (*dashed line*), and the point–source (*dotted line*) reduced bispectra for the equilateral configurations, $l \equiv l_1 = l_2 = l_3$. We have plotted $[l^2(l+1)^2 b_{lll}/(2\pi)^2] \times 10^{16}$, which makes the Sachs–Wolfe plateau of the primordial reduced bispectrum on large angular scales, $l \lesssim 10$, easily seen

9.6.3 Measuring Bispectra: Signal–to–Noise Estimation

In this subsection we study how well we can measure the primordial bispectrum, and how well we can separate it from the secondary bispectra. Suppose that we fit the observed bispectrum, $B^{\mathrm{obs}}_{l_1 l_2 l_3}$, by theoretically calculated bispectra, which include both the primordial and secondary sources. We minimize χ^2 defined by

$$\chi^2 \equiv \sum_{2 \le l_1 \le l_2 \le l_3} \frac{\left(B^{\mathrm{obs}}_{l_1 l_2 l_3} - \sum_i A_i B^{(i)}_{l_1 l_2 l_3}\right)^2}{\sigma^2_{l_1 l_2 l_3}} , \qquad (9.99)$$

where i denotes a component such as the primordial, the SZ and lensing effects, extragalactic sources, and so on. We have removed unobservable modes, $l = 0$ and 1.

The variance of the bispectrum, $\sigma^2_{l_1 l_2 l_3}$, is the six–point function of a_{lm}. When non–gaussianity is weak, we calculate it as

$$\sigma^2_{l_1 l_2 l_3} \equiv \left\langle B^2_{l_1 l_2 l_3} \right\rangle - \left\langle B_{l_1 l_2 l_3} \right\rangle^2 \approx \mathcal{C}_{l_1} \mathcal{C}_{l_2} \mathcal{C}_{l_3} \Delta_{l_1 l_2 l_3} , \qquad (9.100)$$

where $\Delta_{l_1 l_2 l_3}$ takes values 1, 2 or 6 when all ls are different, two are the same, or all are the same, respectively. $\mathcal{C}_l \equiv C_l + C^N_l$ is the total CMB angular power–spectrum, which includes the power–spectrum of the detector noise, C_l^N. We calculate C_l^N analytically with the noise characteristics of relevant experiments. We do not include C_l from secondary sources, as they are subdominant compared with the primordial C_l and C_l^N for relevant experiments. Including C_l from extragalactic sources (9.93) or (9.94) changes our results by less than 10%.

Taking $\partial \chi^2 / \partial A_i = 0$, we obtain the equation

$$\sum_j \left[\sum_{2 \leq l_1 \leq l_2 \leq l_3} \frac{B_{l_1 l_2 l_3}^{(i)} B_{l_1 l_2 l_3}^{(j)}}{\sigma_{l_1 l_2 l_3}^2} \right] A_j = \sum_{2 \leq l_1 \leq l_2 \leq l_3} \frac{B_{l_1 l_2 l_3}^{\text{obs}} B_{l_1 l_2 l_3}^{(i)}}{\sigma_{l_1 l_2 l_3}^2} . \tag{9.101}$$

We then define the Fisher matrix, F_{ij}, as

$$F_{ij} \equiv \sum_{2 \leq l_1 \leq l_2 \leq l_3} \frac{B_{l_1 l_2 l_3}^{(i)} B_{l_1 l_2 l_3}^{(j)}}{\sigma_{l_1 l_2 l_3}^2}$$

$$= \frac{2}{\pi} \sum_{2 \leq l_1 \leq l_2 \leq l_3} \left(l_1 + \frac{1}{2} \right) \left(l_2 + \frac{1}{2} \right) \left(l_3 + \frac{1}{2} \right) \begin{pmatrix} l_1 & l_2 & l_3 \\ 0 & 0 & 0 \end{pmatrix}^2 \frac{b_{l_1 l_2 l_3}^{(i)} b_{l_1 l_2 l_3}^{(j)}}{\sigma_{l_1 l_2 l_3}^2} , \tag{9.102}$$

where we have used (9.53) to replace $B_{l_1 l_2 l_3}$ by the reduced bispectrum, $b_{l_1 l_2 l_3}$ [see (9.52) for definition]. Since the covariance matrix of A_i is F_{ij}^{-1}, we define the signal–to–noise ratio, $(S/N)_i$, for a component i, the correlation coefficient, r_{ij}, between different components i and j, and the degradation parameter, d_i, of $(S/N)_i$ due to r_{ij}, as

$$\left(\frac{S}{N} \right)_i \equiv \frac{1}{\sqrt{F_{ii}^{-1}}} , \tag{9.103}$$

$$r_{ij} \equiv \frac{F_{ij}^{-1}}{\sqrt{F_{ii}^{-1} F_{jj}^{-1}}} , \tag{9.104}$$

$$d_i \equiv F_{ii} F_{ii}^{-1} . \tag{9.105}$$

Note that r_{ij} does not depend upon the amplitude of the bispectra, but on their shape. We have defined d_i such that $d_i = 1$ for zero degradation, while $d_i > 1$ for degraded $(S/N)_i$. We study all the components to look at the separability between various bispectra.

We can give an order–of–magnitude estimate of S/N as a function of the angular resolution, l, as follows. Since the number of modes contributing to S/N increases as $l^{3/2}$, and $l^3 \begin{pmatrix} l & l & l \\ 0 & 0 & 0 \end{pmatrix}^2 \sim 0.36 \times l$, we estimate $(S/N)_i \sim (F_{ii})^{1/2}$ as

$$\left(\frac{S}{N}\right)_i \sim \frac{1}{3\pi} l^{3/2} \times l^{3/2} \left|\begin{pmatrix} l & l & l \\ 0 & 0 & 0 \end{pmatrix}\right| \times \frac{l^3 b_{lll}^{(i)}}{(l^2 C_l)^{3/2}} \sim l^5 b_{lll}^{(i)} \times 4 \times 10^{12}, \quad (9.106)$$

where we have used $l^2 C_l \sim 6 \times 10^{-10}$.

Table 9.2 tabulates F_{ij}, while Table 9.3 tabulates F_{ij}^{-1}; Table 9.4 tabulates $(S/N)_i$, while Table 9.5 tabulates d_i in the diagonal, and r_{ij} in the off-diagonal parts.

Measuring the Primordial Bispectrum

Figure 9.6 shows the signal–to–noise ratio, S/N. The top panel shows the differential S/N for the primordial bispectrum at $\ln l_3$ interval, $[d(S/N)^2/d\ln l_3]^{1/2} f_{NL}^{-1}$, and the bottom panel shows the cumulative S/N, $(S/N)(<l_3)f_{NL}^{-1}$, which is S/N summed up to a certain l_3. We have computed the detector noise power–spectrum, C_l^N, for COBE four–year map, WMAP 90 GHz channel, and Planck 217 GHz channel, and assumed full sky coverage. Figure 9.6 also shows the ideal experiment with no noise: $C_l^N = 0$. Both $[d(S/N)^2/d\ln l_3]^{1/2}$ and $(S/N)(<l_3)$ increase monotonically with l_3, roughly $\propto l_3$, up to $l_3 \sim 2000$ for the ideal experiment.

Beyond $l_3 \sim 2000$, an enhancement of the damping tail in C_l because of the weak lensing effect stops $[d(S/N)^2/d\ln l_3]^{1/2}$, and hence $(S/N)(<l_3)$, increasing. This leads to an important constraint on observations; even for the ideal noise–free, infinitesimally thin beam experiment, there is an upper limit on the value of $S/N \lesssim 0.3 f_{NL}$. For a given realistic experiment, $[d(S/N)^2/d\ln l_3]^{1/2}$ has a maximum at a scale near the beam size.

For COBE, WMAP and Planck experiments, the total $(S/N)f_{NL}^{-1}$ are 1.7×10^{-3}, 5.8×10^{-2}, and 0.19, respectively (see Table 9.4). To obtain $S/N > 1$, we need $f_{NL} > 600$, 20, and 5 respectively, while the ideal experiment requires $f_{NL} > 3$ (see Table 9.6). We can also roughly obtain these values by substituting (9.71) into (9.106),

$$\left(\frac{S}{N}\right)_{prim} \sim l \times 10^{-4} f_{NL}. \quad (9.107)$$

The degradation parameters, d_{prim}, are 1.46, 1.01, and 1.00 for the COBE, WMAP and Planck experiments, respectively (see Table 9.5), suggesting that WMAP and Planck experiments will separate the primordial bispectrum from the others with 1% or better accuracy. However, COBE cannot discriminate between them very well, as the primordial and the secondary sources change monotonically on the COBE angular scales. On WMAP and Planck scales, the primordial bispectrum starts oscillating around zero, being well separated in shape from the secondaries that do not oscillate. This is good news for the forthcoming high angular resolution CMB experiments.

Table 9.2. Fisher matrix. Fisher matrix, F_{ij} [see (9.102)]: i denotes a component in the first row; j denotes a component in the first column. $\overline{T}_{\rho 0}$ is in units of 1 keV, $b_{25}^{\mathrm{src}} \equiv b^{\mathrm{src}}/10^{-25}$, and $b_{27}^{\mathrm{src}} \equiv b^{\mathrm{src}}/10^{-27}$.

	Primordial	SZ-lensing	Point-sources
COBE primordial	$4.2 \times 10^{-6} \, f_{\mathrm{NL}}^2$	$-4.0 \times 10^{-7} \, f_{\mathrm{NL}} j_\nu \overline{T}_{\rho 0} b_{\mathrm{gas}}$	$-1.0 \times 10^{-9} \, f_{\mathrm{NL}} b_{25}^{\mathrm{src}}$
SZ-lensing		$1.3 \times 10^{-7} \, (j_\nu \overline{T}_{\rho 0} b_{\mathrm{gas}})^2$	$3.1 \times 10^{-10} \, j_\nu \overline{T}_{\rho 0} b_{\mathrm{gas}} b_{25}^{\mathrm{src}}$
point-sources			$1.1 \times 10^{-12} \, (b_{25}^{\mathrm{src}})^2$
WMAP			
primordial	$3.4 \times 10^{-3} \, f_{\mathrm{NL}}^2$	$2.6 \times 10^{-3} \, f_{\mathrm{NL}} j_\nu \overline{T}_{\rho 0} b_{\mathrm{gas}}$	$2.4 \times 10^{-3} \, f_{\mathrm{NL}} b_{25}^{\mathrm{src}}$
SZ-lensing		$0.14 \, (j_\nu \overline{T}_{\rho 0} b_{\mathrm{gas}})^2$	$0.31 \, j_\nu \overline{T}_{\rho 0} b_{\mathrm{gas}} b_{25}^{\mathrm{src}}$
point-sources			$5.6 \, (b_{25}^{\mathrm{src}})^2$
Planck			
primordial	$3.8 \times 10^{-2} \, f_{\mathrm{NL}}^2$	$7.2 \times 10^{-2} \, f_{\mathrm{NL}} j_\nu \overline{T}_{\rho 0} b_{\mathrm{gas}}$	$1.6 \times 10^{-2} \, f_{\mathrm{NL}} b_{27}^{\mathrm{src}}$
SZ-lensing		$39 \, (j_\nu \overline{T}_{\rho 0} b_{\mathrm{gas}})^2$	$5.7 \, j_\nu \overline{T}_{\rho 0} b_{\mathrm{gas}} b_{27}^{\mathrm{src}}$
point-sources			$2.7 \times 10^3 \, (b_{27}^{\mathrm{src}})^2$

Table 9.3. Inverted Fisher matrix, F_{ij}^{-1}. The meaning of the symbols is the same as in Table 9.2

	Primordial	SZ–lensing	Point–sources
COBE primordial	$3.5 \times 10^5\, f_{\mathrm{NL}}^{-2}$	$1.1 \times 10^6 \left(f_{\mathrm{NL}} j_\nu \overline{T} \rho_0 b_{\mathrm{gas}}\right)^{-1}$	$1.3 \times 10^7 \left(f_{\mathrm{NL}} b_{25}^{\mathrm{src}}\right)^{-1}$
SZ–lensing		$3.1 \times 10^7 \left(j_\nu \overline{T} \rho_0 b_{\mathrm{gas}}\right)^{-2}$	$-7.8 \times 10^9 \left(j_\nu \overline{T} \rho_0 b_{\mathrm{gas}} b_{25}^{\mathrm{src}}\right)^{-1}$
point sources			$3.1 \times 10^{12} \left(b_{25}^{\mathrm{src}}\right)^{-2}$
WMAP			
primordial	$3.0 \times 10^2\, f_{\mathrm{NL}}^{-2}$	$-6.1 \left(f_{\mathrm{NL}} j_\nu \overline{T} \rho_0 b_{\mathrm{gas}}\right)^{-1}$	$0.21 \left(f_{\mathrm{NL}} b_{25}^{\mathrm{src}}\right)^{-1}$
SZ–lensing		$8.4 \left(j_\nu \overline{T} \rho_0 b_{\mathrm{gas}}\right)^{-2}$	$-0.46 \left(j_\nu \overline{T} \rho_0 b_{\mathrm{gas}} b_{25}^{\mathrm{src}}\right)^{-1}$
point–sources			$0.21 \left(b_{25}^{\mathrm{src}}\right)^{-2}$
Planck			
primordial	$26\, f_{\mathrm{NL}}^{-2}$	$-4.9 \times 10^{-2} \left(f_{\mathrm{NL}} j_\nu \overline{T} \rho_0 b_{\mathrm{gas}}\right)^{-1}$	$-5.7 \times 10^{-5} \left(f_{\mathrm{NL}} b_{27}^{\mathrm{src}}\right)^{-1}$
SZ–lensing		$2.6 \times 10^{-2} \left(j_\nu \overline{T} \rho_0 b_{\mathrm{gas}}\right)^{-2}$	$-5.4 \times 10^{-5} \left(j_\nu \overline{T} \rho_0 b_{\mathrm{gas}} b_{27}^{\mathrm{src}}\right)^{-1}$
point–sources			$3.7 \times 10^{-4} \left(b_{27}^{\mathrm{src}}\right)^{-2}$

Table 9.4. Signal–to–noise ratio, $(S/N)_i$ [see (9.103)], of detecting the bispectrum. i denotes a component in the first row. The meaning of the symbols is the same as in Table 9.2

	Primordial	SZ–lensing	Point–sources
COBE	$1.7 \times 10^{-3} f_{\mathrm{NL}}$	$1.8 \times 10^{-4} \, |j_\nu| \overline{T}_\rho b_{\mathrm{gas}}$	$5.7 \times 10^{-7} \, b_{25}^{\mathrm{ps}}$
WMAP	$5.8 \times 10^{-2} f_{\mathrm{NL}}$	$0.34 \, |j_\nu| \overline{T}_\rho b_{\mathrm{gas}}$	$2.2 \, b_{25}^{\mathrm{ps}}$
Planck	$0.19 f_{\mathrm{NL}}$	$6.2 \, |j_\nu| \overline{T}_\rho b_{\mathrm{gas}}$	$52 \, b_{27}^{\mathrm{ps}}$

Measuring Primordial Skewness

For the skewness, we define S/N as

$$\left(\frac{S}{N}\right)^2 \equiv \frac{S_3^2}{\sigma_{S_3}^2} , \tag{9.108}$$

where the variance is

$$\sigma_{S_3}^2 \equiv \left\langle (S_3)^2 \right\rangle = 6 \int_{-1}^{1} \frac{d\cos\theta}{2} \, [\mathcal{C}(\theta)]^3$$

$$= 6 \sum_{l_1 l_2 l_3} \frac{(2l_1+1)(2l_2+1)(2l_3+1)}{(4\pi)^3} \begin{pmatrix} l_1 & l_2 & l_3 \\ 0 & 0 & 0 \end{pmatrix}^2 \mathcal{C}_{l_1} \mathcal{C}_{l_2} \mathcal{C}_{l_3} W_{l_1}^2 W_{l_2}^2 W_{l_3}^2$$

$$= \frac{9}{2\pi^3} \sum_{2 \le l_1 \le l_2 \le l_3} \left(l_1 + \frac{1}{2}\right)\left(l_2 + \frac{1}{2}\right)\left(l_3 + \frac{1}{2}\right) \begin{pmatrix} l_1 & l_2 & l_3 \\ 0 & 0 & 0 \end{pmatrix}^2$$
$$\times \mathcal{C}_{l_1} \mathcal{C}_{l_2} \mathcal{C}_{l_3} W_{l_1}^2 W_{l_2}^2 W_{l_3}^2 . \tag{9.109}$$

Table 9.5. Signal degradation parameter, d_i [see (9.105)], and correlation coefficient, r_{ij} [see (9.104)], matrix. i denotes a component in the first row; j denotes a component in the first column. d_i for $i = j$, while r_{ij} for $i \ne j$

	primordial	SZ–lensing	point–sources
COBE primordial	1.46	$0.33 \, \mathrm{sgn}(j_\nu)$	1.6×10^{-2}
SZ–lensing		3.89	$-0.79 \, \mathrm{sgn}(j_\nu)$
point–sources			3.45
WMAP			
primordial	1.01	$-0.12 \, \mathrm{sgn}(j_\nu)$	2.7×10^{-2}
SZ–lensing		1.16	$-0.35 \, \mathrm{sgn}(j_\nu)$
point–sources			1.14
Planck			
primordial	1.00	$-5.9 \times 10^{-2} \, \mathrm{sgn}(j_\nu)$	-5.8×10^{-4}
SZ–lensing		1.00	$-1.8 \times 10^{-2} \, \mathrm{sgn}(j_\nu)$
point–sources			1.00

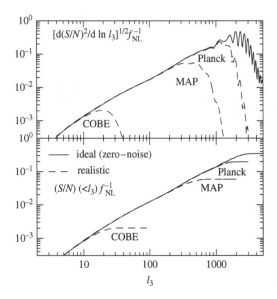

Fig. 9.6. The predictions of the signal–to–noise ratio, S/N, for the COBE, WMAP, and Planck experiments [see (9.103)]. The differential S/N at $\ln l_3$ interval is shown in the *upper panel*, while the cumulative S/N up to a certain l_3 is shown in the bottom panel. Both are in units of f_{NL}. *Solid line* represents the zero-noise ideal experiment, while *dotted lines* show the realistic experiments mentioned above. The total $(S/N)f_{NL}^{-1}$ are 1.7×10^{-3}, 5.8×10^{-2}, and 0.19 for COBE, WMAP and Planck experiments, respectively

In the last equality, we have used symmetry of the summed quantity with respect to indices (9.77), and removed unobservable modes, $l = 0$ and 1. Typically $\sigma_{S_3} \sim 10^{-15}$, as $\sigma_{S_3} \sim [\mathcal{C}(0)]^{3/2} \sim 10^{-15}$, where $\mathcal{C}(\theta)$ is the temperature auto–correlation function including the noise.

The bottom panel of Fig. 9.3 shows $\sigma_{S_3}(< l_3)$, which is σ_{S_3} summed up to a certain l_3, for COBE, WMAP and Planck experiments as well as for the ideal experiment. Since $C_l W_l^2 = C_l e^{-l(l+1)\sigma_b^2} + w^{-1}$, where w^{-1} is the white–noise power–spectrum of the detector noise, w^{-1} keeps $\sigma_{S_3}(< l_3)$ slightly increasing with l_3 beyond the experimental angular resolution scale, $l \sim \sigma_b^{-1}$. In contrast, $S_3(< l_3)$ becomes constant beyond $l \sim \sigma_b^{-1}$ (see the top panel of Fig. 9.3). As a result, S/N starts slightly decreasing beyond the resolution. We use the maximum S/N for calculating the minimum value of f_{NL} above which the primordial S_3 is detectable; we find that $f_{NL} > 800$, 80, 70 and 60 for COBE, WMAP, Planck and the ideal experiments, respectively, assuming full sky coverage.

These f_{NL} values are systematically larger than those for detecting $B_{l_1 l_2 l_3}$ by a factor of 1.3, 4, 14 and 20, respectively (see Table 9.6). The higher the angular resolution is, the less sensitive the primordial S_3 is to non–gaussianity

Table 9.6. The minimum non–linearity parameter, f_{NL}, needed for detecting the primordial non–gaussianity by the bispectrum or the skewness with signal–to–noise ratio greater than 1. These estimates include the effects of cosmic variance, detector noise, and foreground sources

Experiments	f_{NL} (Bispectrum)	f_{NL} (Skewness)
COBE	600	800
WMAP	20	80
Planck	5	70
Ideal	3	60

than $B_{l_1 l_2 l_3}$. This is because of the cancellation effect on smaller angular scales caused by the oscillation of $B_{l_1 l_2 l_3}$ damps S_3.

Figure 9.7 compares the expected signal–to–noise ratio of detecting the primordial non–gaussianity based on the bispectrum (9.103) with that based on the skewness (9.108). It shows that the bispectrum is almost an order of magnitude more sensitive to the non–gaussianity than the skewness. We conclude that when we can compute the predicted form of the bispectrum, it becomes a "matched filter" for detecting the non–gaussianity in data, and

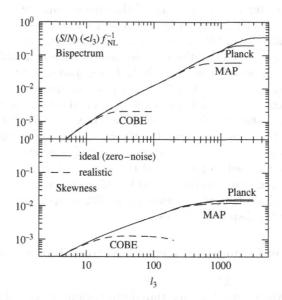

Fig. 9.7. Comparison of the signal–to–noise ratio summed up to a certain l_3, $S/N(< l_3)$, for the bispectrum (*top panel*; (9.103)) and the skewness (*bottom panel*; (9.108)). $S/N(< l_3)$ is in units of f_{NL}. The *dotted lines* show COBE, WMAP and Planck experiments (*dotted lines*), while the *solid line* shows the ideal experiment. See Table 9.6 for f_{NL} to obtain $S/N > 1$

thus much more a powerful tool than the skewness. Table 9.6 summarizes the minimum f_{NL} for detecting the primordial non–gaussianity using the bispectrum or the skewness for COBE, WMAP, Planck, and the ideal experiments. This shows that even the ideal experiment needs $f_{NL} > 3$ to detect the primordial bispectrum.

9.6.4 Measuring Primordial Non–gaussianity in the Cosmic Microwave Background

Measuring f_{NL} from nearly full–sky experiments is challenging. The bispectrum analysis requires $N^{5/2}$ operations ($N^{3/2}$ for computing three ls and N for averaging over the sky) where N is the number of pixels. The brute–force analysis is possible for the COBE data for which $N \sim 3000$, while it is quite challenging for mega–pixel experiments (e.g., $N \sim 3 \times 10^6$ for WMAP, 5×10^7 for Planck). In fact, just measuring all configurations of the bispectrum from the data is possible. What is challenging is to carry out many Monte Carlo simulations: in order to quantify the statistical significance of the measurements, one needs many simulations . It is the simulations that are computationally very expensive. Since the brute–force trispectrum analysis requires N^3, it is even more challenging.

Although we measure the individual triangle configurations of the bispectrum (or quadrilateral configurations of the trispectrum) at first, we eventually combine all of them to constrain model parameters such as f_{NL}, as the signal–to–noise per configuration is nearly zero. This may sound inefficient. Measuring all configurations is enormously time consuming. Is there any statistic which *already* combines all the configurations optimally, and fast to compute? Yes. A physical justification for our methodology is as follows. A model like (9.60) generates non–gaussianity in real space, and the Central–Limit Theorem makes the Fourier modes nearly Gaussian; thus, real-space statistics should be more sensitive. On the other hand, real-space statistics are weighted sum of Fourier-space statistics, which are often easier to predict. Therefore, we need to understand the shape of Fourier-space statistics to find sensitive real-space statistics, and for this purpose it is useful to have a specific, physically motivated non–gaussian model, compute Fourier statistics, and find optimal real-space statistics.

Reconstructing Primordial Fluctuations from Temperature Anisotropy

We begin with the primordial curvature perturbations $\Phi(\mathbf{x})$ and isocurvature perturbations $S(\mathbf{x})$. If we can reconstruct these primordial fluctuations from the observed CMB anisotropy, $\Delta T(\hat{\mathbf{n}})/T$, then we can improve the sensitivity to primordial non–gaussianity. We find that the harmonic coefficients of the CMB anisotropy, $a_{lm} = T^{-1} \int d^2\hat{\mathbf{n}} \Delta T(\hat{\mathbf{n}}) Y_{lm}^*(\hat{\mathbf{n}})$, are related to the primordial fluctuations as

$$a_{lm} = W_l \int r^2 dr \left[\Phi_{lm}(r) \alpha_l^{\mathrm{adi}}(r) + S_{lm}(r) \alpha_l^{\mathrm{iso}}(r) \right] + n_{lm} , \qquad (9.110)$$

where $\Phi_{lm}(r)$ and $S_{lm}(r)$ are the harmonic coefficients of the fluctuations at a given comoving distance, $r = |\mathbf{x}|$ from the observer. The beam function W_l and the harmonic coefficients of the noise n_{lm} represent instrumental effects. Since noise can be spatially inhomogeneous, the noise covariance matrix $\langle n_{lm} n_{l'm'}^* \rangle$ can be non–diagonal; however, we approximate it by $\simeq \sigma_0^2 \delta_{ll'} \delta_{mm'}$. We thus assume the "mildly inhomogeneous" noise for which this approximation holds. The function $\alpha_l(r)$ is defined by

$$\alpha_l(r) \equiv \frac{2}{\pi} \int k^2 dk \, g_{Tl}(k) j_l(kr) , \qquad (9.111)$$

where $g_{Tl}(k)$ is the radiation transfer function of either adiabatic (adi) or isocurvature (iso) perturbations. Note that this function is equal to $f_{\mathrm{NL}}^{-1} b_l^{\mathrm{NL}}(r)$ [see (9.68)].

Next, assuming that $\Phi(\mathbf{x})$ dominates, we try to reconstruct $\Phi(\mathbf{x})$ from the observed $\Delta T(\hat{\mathbf{n}})$. A linear filter, $\mathcal{O}_l(r)$, which reconstructs the underlying field, can be obtained by minimizing the variance of the difference between the filtered field $\mathcal{O}_l(r) a_{lm}$ and the underlying field $\Phi_{lm}(r)$. By evaluating

$$\frac{\partial}{\partial \mathcal{O}_l(r)} \langle |\mathcal{O}_l(r) a_{lm} - \Phi_{lm}(r)|^2 \rangle = 0 , \qquad (9.112)$$

one obtains a solution for the filter as

$$\mathcal{O}_l(r) = \frac{\beta_l(r) W_l}{\tilde{C}_l} , \qquad (9.113)$$

where the function $\beta_l(r)$ is given by

$$\beta_l(r) \equiv \frac{2}{\pi} \int k^2 dk \, P(k) g_{Tl}(k) j_l(kr) , \qquad (9.114)$$

and $P(k)$ is the power–spectrum of Φ. Of course, one can replace Φ with S when S dominates. This function is equal to $b_l^L(r)$ [see (9.69)]. Here, we put a tilde on a quantity that includes effects of W_l and noise such that $\tilde{C}_l \equiv C_l W_l^2 + \sigma_0^2$, where C_l is the theoretical power–spectrum that uses the same cosmological model as $g_{Tl}(k)$.

Finally, we transform the filtered field $\mathcal{O}_l(r) a_{lm}$ back to pixel space to obtain a Wiener–filtered, reconstructed map of $\Phi(r, \hat{\mathbf{n}})$ or $S(r, \hat{\mathbf{n}})$. We have assumed that there is no correlation between Φ and S.

Figure 6.4 shows $\mathcal{O}_l(r)$ as a function of l and r for (a) an adiabatic SCDM ($\Omega_m = 1$), (b) an adiabatic ΛCDM ($\Omega_m = 0.3$), (c) an isocurvature SCDM, and (d) an isocurvature ΛCDM. While we have used $P(k) \propto k^{-3}$ for both adiabatic and isocurvature modes, the specific choice of $P(k)$ does not affect $\mathcal{O}_l$ very much as $P(k)$ in β_l in the numerator approximately cancels out $P(k)$

in C_l in the denominator. On large angular scales (smaller l) the Sachs–Wolfe (SW) effect makes $\mathcal{O}_l$ equal to -3 for adiabatic modes and $-5/2$ for isocurvature modes in SCDM. For the ΛCDM models the late–time decay of the gravitational potential makes this limit different. Adiabatic and isocurvature modes are out of phase in l.

Figure 9.8 shows that $\mathcal{O}_l$ changes the sign of the fluctuations as a function of scales. This indicates that acoustic physics at the last scattering surface modulates fluctuations so that hot spots in the primordial fluctuations can be cold spots in the CMB, for example. Therefore, the shape of $\mathcal{O}_l$ "deconvolves" the sign change, recovering the phases of fluctuations. This is an intuitive reason why our cubic statistic derived below (9.117) works, and it proves more advantageous to measure primordial non–gaussianity on a filtered map than on a temperature map.

This property should be compared to that of real-space statistics measured on a temperature map. As we have shown in Sect. 9.6.3 the skewness of a temperature map is much less sensitive to the primordial non–gaussianity than

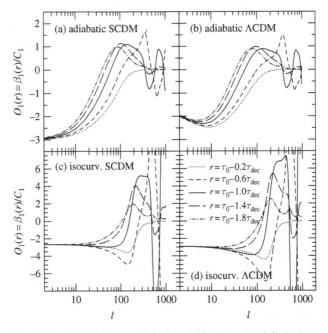

Fig. 9.8. Wiener filters, $\mathcal{O}_l(r) = \beta_l(r)/C_l$ (9.113). We plot (**a**) $\mathcal{O}_l$ for an adiabatic SCDM ($\Omega_m = 1$, $\Omega_\Lambda = 0$, $\Omega_b = 0.05$, $h = 0.5$), (**b**) for an adiabatic ΛCDM ($\Omega_m = 0.3$, $\Omega_\Lambda = 0.7$, $\Omega_b = 0.04$, $h = 0.7$), (**c**) for an isocurvature SCDM, and (**d**) for an isocurvature ΛCDM. The filters are plotted at five conformal distances $r = c(\tau_0 - \tau)$ as explained in the bottom-right panel. Here τ is the conformal time (τ_0 at the present). The SCDM models have $c\tau_0 = 11.84$ Gpc and $c\tau_{\rm dec} = 0.235$ Gpc, while the ΛCDM models $c\tau_0 = 13.89$ Gpc and $c\tau_{\rm dec} = 0.277$ Gpc, where $\tau_{\rm dec}$ is the photon decoupling epoch

the bispectrum, exactly because of the cancellation effect from the acoustic oscillations. The skewness of a filtered map, on the other hand, has a larger signal–to–noise ratio, and more optimal statistics like our cubic statistic derived below can be constructed. Other real–space statistics such as Minkowski functionals peak–peak correlations may also be more sensitive to the primordial non–gaussianity, when measured on the filtered maps.

Unfortunately, as g_{Tl} oscillates, our reconstruction of Φ or S from a temperature map alone is not perfect. While $\mathcal{O}_l$ reconstructs the primordial fluctuations very well on large scales via the Sachs–Wolfe effect, $\mathcal{O}_l \sim 0$ on intermediate scales ($l \sim 50$ for adiabatic and $l \sim 100$ for isocurvature), indicating loss of information on the phases of the underlying fluctuations. Then, toward smaller scales, we recover information, lose information, and so on. Exact scales at which $\mathcal{O}_l \sim 0$ depend on r and cosmology. A good news is that a high signal–to–noise map of the CMB polarization anisotropy will enable us to overcome the loss of information, as the polarization transfer function is out of phase in l compared to the temperature transfer function, filling up information at which $\mathcal{O}_l \sim 0$. In other words, the polarization anisotropy has finite information about the phases of the primordial perturbations, when the temperature anisotropy has zero information.

Measuring primordial non–gaussianity in adiabatic fluctuations

Using two functions introduced in the previous section, we construct a *cubic* statistic which is optimal for the primordial non–gaussianity. We apply filters to a_{lm}, and then transform the filtered a_{lm}'s to obtain two maps, A and B, given by

$$A(r, \hat{\mathbf{n}}) \equiv \sum_{lm} \frac{\alpha_l(r) W_l}{\tilde{C}_l} a_{lm} Y_{lm}(\hat{\mathbf{n}}), \qquad (9.115)$$

$$B(r, \hat{\mathbf{n}}) \equiv \sum_{lm} \frac{\beta_l(r) W_l}{\tilde{C}_l} a_{lm} Y_{lm}(\hat{\mathbf{n}}) . \qquad (9.116)$$

The latter map, $B(r, \hat{\mathbf{n}})$, is exactly the $\mathcal{O}_l$-filtered map, a Wiener–filtered map of the underlying primordial fluctuations. We then form a cubic statistic given by

$$\mathcal{S}_{\text{prim}} \equiv 4\pi \int r^2 dr \int \frac{d^2\hat{\mathbf{n}}}{4\pi} A(r, \hat{\mathbf{n}}) B^2(r, \hat{\mathbf{n}}) , \qquad (9.117)$$

where angular average is performed on the full sky, regardless of the sky cut. We find that $\mathcal{S}_{\text{prim}}$ reduces *exactly* to

$$\mathcal{S}_{\text{prim}} = \sum_{l_1 \leq l_2 \leq l_3} \frac{\tilde{B}^{obs}_{l_1 l_2 l_3} \tilde{B}^{prim}_{l_1 l_2 l_3}}{\tilde{C}_{l_1} \tilde{C}_{l_2} \tilde{C}_{l_3}} , \qquad (9.118)$$

where

$$\tilde{B}_{l_1 l_2 l_3} \equiv B_{l_1 l_2 l_3} W_{l_1} W_{l_2} W_{l_3} , \qquad (9.119)$$

and $B^{\text{obs}}_{l_1 l_2 l_3}$ is the observed bispectrum with the effect of W_l corrected while $B^{\text{prim}}_{l_1 l_2 l_3}$ is given by (9.70) and (9.53).

The denominator of (9.118) is the variance of $\tilde{B}^{\text{obs}}_{l_1 l_2 l_3}$ in the limit of weak non–gaussianity (say $|f_{\text{NL}}| < 10^3$) when all l's are different: $\left\langle \tilde{B}^2_{l_1 l_2 l_3} \right\rangle = \tilde{C}_{l_1} \tilde{C}_{l_2} \tilde{C}_{l_3} \Delta_{l_1 l_2 l_3}$, where $\Delta_{l_1 l_2 l_3}$ is 6 for $l_1 = l_2 = l_3$, 2 for $l_1 = l_2 \neq l_3$ etc., and 1 otherwise. The bispectrum configurations are thus summed up nearly optimally with the approximate inverse–variance weights, provided that $\Delta_{l_1 l_2 l_3}$ is approximated with $\simeq 1$. The least–square fit of $\tilde{B}^{\text{prim}}_{l_1 l_2 l_3}$ to $\tilde{B}^{\text{obs}}_{l_1 l_2 l_3}$ can be performed to yield

$$S_{\text{prim}} \simeq f_{\text{NL}} \sum_{l_1 \leq l_2 \leq l_3} \frac{(\tilde{B}^{\text{prim}}_{l_1 l_2 l_3})^2}{\tilde{C}_{l_1} \tilde{C}_{l_2} \tilde{C}_{l_3}} . \tag{9.120}$$

This equation gives an estimate of f_{NL} directly from S_{prim}.

The most time–consuming part is the back–and–forth harmonic transform necessary for pre–filtering [see (9.115) and (9.116)], taking $N^{3/2}$ operations times the number of sampling points of r, of order 100, for evaluating the integral (9.117). This is much faster than the full bispectrum analysis which takes $N^{5/2}$, enabling us to perform a more detailed analysis of the data in a reasonable amount of computational time. For example, measurements of all bispectrum configurations up to $l_{\text{max}} = 512$ take 8 h to compute on 16 processors of an SGI Origin 300; thus, even only 100 Monte Carlo simulations take 1 month to be carried out. On the other hand, S_{prim} takes only 30 s to compute, 1000 times faster. When we measure f_{NL} for $l_{\text{max}} = 1024$, we gain a factor of 4000 in computing time: 11 days for the bispectrum vs 4 min for S_{prim}. We can do 1000 simulations for $l_{\text{max}} = 1024$ in 3 days.

Point–Source Non–gaussianity

Next, we show that the filtering method is also useful for measuring foreground non–gaussianity arising from extragalactic point–sources. The residual point–sources left unsubtracted in a map can seriously contaminate both the power–spectrum and the bispectrum. We can, on the other hand, use multi–band observations as well as external template maps of dust, free–free, and synchrotron emission, to remove diffuse Galactic foreground. The radio sources with known positions can be safely masked.

The filtered map for the point–sources is

$$D(\hat{n}) \equiv \sum_{lm} \frac{W_l}{\tilde{C}_l} a_{lm} Y_{lm}(\hat{n}) . \tag{9.121}$$

This filtered map was actually used for detecting point–sources in the WMAP maps. Using $D(\hat{n})$, the cubic statistic is derived as

$$S_{\rm src} \equiv \int \frac{{\rm d}^2\hat{\mathbf{n}}}{4\pi} D^3(\hat{\mathbf{n}}) = \frac{3}{2\pi} \sum_{l_1 \le l_2 \le l_3} \frac{\tilde{B}^{\rm obs}_{l_1 l_2 l_3} \tilde{B}^{\rm src}_{l_1 l_2 l_3}}{\tilde{C}_{l_1} \tilde{C}_{l_2} \tilde{C}_{l_3}} . \tag{9.122}$$

Here, $B^{\rm src}_{l_1 l_2 l_3}$ is the point–source bispectrum for unit white–noise bispectrum (i.e., $b^{\rm src} = 1$ in (9.88)). When the covariance between $B^{\rm prim}_{l_1 l_2 l_3}$ and $B^{\rm src}_{l_1 l_2 l_3}$ is negligible as is the case for WMAP and Planck (see Table 9.5), we find

$$S_{\rm src} \simeq \frac{3 b^{\rm src}}{2\pi} \sum_{l_1 \le l_2 \le l_3} \frac{(\tilde{B}^{\rm src}_{l_1 l_2 l_3})^2}{\tilde{C}_{l_1} \tilde{C}_{l_2} \tilde{C}_{l_3}} . \tag{9.123}$$

Again, $S_{\rm src}$ measures $b^{\rm src}$ much faster than the full bispectrum analysis, constraining effects of residual point–sources on CMB sky maps. Since $S_{\rm src}$ does not contain the extra integral over r, it is even 100 times faster to compute than $S_{\rm prim}$. This statistic is particularly useful because it is sometimes difficult to tell how much of C_l is due to point–sources. In Sect. 9.6.5 we see how $S_{\rm src}$ (i.e., $b^{\rm src}$) is related to C_l due to the unsubtracted point–sources.

Incomplete Sky Coverage

Finally, we show how to incorporate incomplete sky coverage and pixel weights into our statistics. Suppose that we weight a sky map by $M(\hat{\mathbf{n}})$ to measure the harmonic coefficients,

$$a^{\rm obs}_{lm} = \frac{1}{T} \int {\rm d}^2\hat{\mathbf{n}} M(\hat{\mathbf{n}}) \Delta T(\hat{\mathbf{n}}) Y^*_{lm}(\hat{\mathbf{n}}) . \tag{9.124}$$

A full–sky a_{lm} is related to $a^{\rm obs}_{lm}$ through the coupling matrix $M_{ll'mm'} \equiv \int {\rm d}^2\hat{\mathbf{n}} M(\hat{\mathbf{n}}) Y^*_{lm}(\hat{\mathbf{n}}) Y_{l'm'}(\hat{\mathbf{n}})$ by $a^{\rm obs}_{lm} = \sum_{l'm'} a_{l'm'} M_{ll'mm'}$. In this case the observed bispectrum is biased by a factor of $\int {\rm d}^2\hat{\mathbf{n}} M^3(\hat{\mathbf{n}})/(4\pi)$; thus, we need to divide $S_{\rm prim}$ and S_{ps} by this factor. If only the sky cut is considered, then this factor is the fraction of the sky covered by observations.

Monte Carlo simulations of non–gaussian sky maps computed with (9.110) show that $S_{\rm prim}$ reproduces the input $f_{\rm NL}$ accurately both on full sky and incomplete sky with modest Galactic cut and inhomogeneous noise on the WMAP data, i.e., the statistic is unbiased. The error on $f_{\rm NL}$ from $S_{\rm prim}$ is as small as that from the full bispectrum analysis; however, one cannot make a sky cut very large, for example, more than 50% of the sky, as for it the covariance matrix of $\tilde{B}_{l_1 l_2 l_3}$ is no longer diagonal. The cubic statistic does not include the off–diagonal terms of the covariance matrix [see (9.118)]; however, it works fine for WMAP sky maps for which one can use more than 75% of the sky. Also, (9.123) correctly estimates $b^{\rm src}$ using simulated realizations of point–sources.

These fast methods allow to carry out extensive Monte Carlo simulations characterizing the effects of realistic noise properties of the experiments, sky cut, foreground sources, and so on. A reconstructed map of the primordial fluctuations, which plays a key role in the method, potentially gives other real–space statistics more sensitivity to primordial non–gaussianity. As it has been shown, the method can be applied to primordial non–gaussianity arising from inflation, gravity, or correlated isocurvature fluctuations, as well as to foreground non–gaussianity from radio point–sources, all of which can be important sources of non–gaussian fluctuations on the CMB sky maps.

9.6.5 Applications to Observational Data

There are two approaches to testing Gaussianity of the CMB.

- Blind tests (null tests) which make no assumption about the form of non–gaussianity. The simplest test would be measurements of deviation of one–point PDF. from a Gaussian distribution. (Measurements of the skewness, kurtosis, etc., e.g.) This approach is model–independent but its statistical power is weak. If we had no models to test, this approach would be the only choice.
- Testing specific models of non–gaussianity, constraining the model parameters. This approach is powerful in putting *quantitative* constraints on non–gaussianity, at the cost of being model–dependent. If we had a sensible (yet fairly generic) model to test, this approach would be more powerful than blind tests.

Both approaches have been applied to the CMB data on large angular scales ($\sim 7°$) on intermediate scales ($\sim 1°$), and on small scales ($\sim 10'$). So far, there is no compelling evidence for the cosmological non–gaussianity, and the pre-WMAP constraint on f_{NL} was weak, $f_{NL} < (2000\text{--}3000)$ at 95% confidence level.

In this section we briefly review results of Gaussianity tests on the WMAP data. The WMAP, Wilkinson Microwave Anisotropy Probe, has recently produced clean and precise sky maps of the CMB in five microwave bands, with the angular resolution 30 times better than that of the Differential Microwave Radiometer (DMR) aboard the COBE satellite. Detailed study of these sky maps offers a fundamental test of cosmology, as various cosmological effects change temperature and energy distribution of the CMB at all angular scales. The temperature and polarization power–spectra of the WMAP data have determined the best-fit cosmological model with errors in the parameter determinations being quite small ($< 10\%$). The systematic errors in the parameter determinations are minimized by both the careful instrumental design and data analysis techniques.

Apart from the CMB, there are a number of non–cosmological, "foreground" sources in the microwave sky. The emission from our Galaxy is the brightest component, which must be masked or subtracted out before any cosmological analysis of the CMB. Since the WMAP observes in five frequency bands, much of the Galactic emission can be reliably subtracted using the non–monochromatic nature of the Galaxy. The power–spectra measured in different bands coincide with each other after the foreground subtraction, which is reassuring. Actually, much more problematic a foreground component is the extragalactic radio sources. Although we can mask those positions of the sky which are known to have sources brighter than some threshold flux (which is determined by the sensitivity of observations), there always remain undetected sources. The undetected (unmasked) sources potentially contaminate the cosmological CMB signals. Since we cannot subtract them out individually, we must estimate the effect of the sources in a statistical manner.

The emission from the sources is highly non–gaussian and only important on small angular scales; thus, we can use the non–gaussian signals to directly estimate the source contribution. This example illustrates usefulness of the higher–order statistics in a real life.

Minkowski Functionals

For the first test, one can use (but is not limited to) the Minkowski functionals which measure morphological structures of the CMB, describing the properties of regions spatially bounded by a set of contours. The contours may be specified in terms of fixed temperature thresholds, $\nu = \Delta T/\sigma$, where σ is the standard deviation of the map, or in terms of the area. The three Minkowski functionals are (1) the total area above threshold, $A(\nu)$, (2) the total contour length, $C(\nu)$, and (3) the genus, $G(\nu)$, which is the number of hot spots minus the number of cold spots. Parameterization of contours by threshold is computationally simpler, while parameterization by area reduces the correlations between the Minkowski functionals; however, when a joint analysis of the three Minkowski functionals is performed, one has to explicitly include their covariance anyway. Therefore the simpler threshold parameterization will be used.

So far the Minkowski functionals at 5 different resolutions from the pixel size of 3.7° in diameter to 12 arcminutes have been measured. Fig. 9.9 shows one example at 28′ pixel resolution. The grey band shows the 68% confidence region derived from 1000 Gaussian Monte Carlo simulations. The *WMAP* data are in excellent agreement with the Gaussian simulations at all resolutions. But, *how Gaussian is it?*

Angular Bispectrum

For the second test, we use the fast cubic statistics derived in Sect. 9.6.4, which combine three–point (triangle) configurations of the angular bispectrum that are sensitive to the models under consideration.

Once again, we consider two components. The first one is the primordial non–gaussianity from inflation parametrized by $f_{\rm NL}$, which determines the amplitude of a quadratic term added to Bardeen's curvature perturbations:

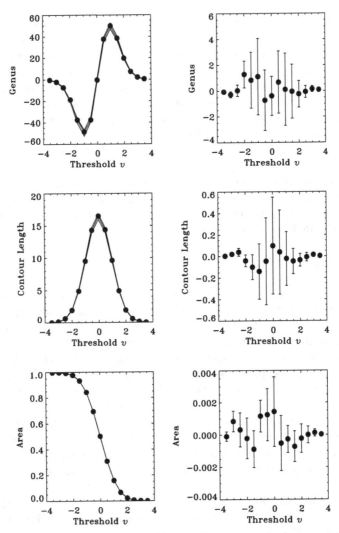

Fig. 9.9. The Minkowski functions at $28'$ pixel resolution (*filled circles*) and the residuals between the mean of the Gaussian simulations and the *WMAP* data. The *grey band shows* the 68% confidence interval for the Gaussian Monte Carlo simulations. The *WMAP* data are in excellent agreement with the Gaussian simulations

$\Phi(\mathbf{x}) = \Phi_L(\mathbf{x}) + f_{\mathrm{NL}} \left[\Phi_L^2(\mathbf{x}) - \left\langle \Phi_L^2(\mathbf{x}) \right\rangle \right]$, It is useful to estimate the *r.m.s.* amplitude of Φ to see how important the second–order term is. One obtains $\left\langle \Phi^2 \right\rangle^{1/2} \simeq \left\langle \Phi_L^2 \right\rangle^{1/2} \left(1 + f_{\mathrm{NL}}^2 \left\langle \Phi_L^2 \right\rangle \right)$, where $\left\langle \Phi^2 \right\rangle^{1/2} \simeq 3.3 \times 10^{-5}$; thus, a fractional contribution from the second term is

$$f_{\mathrm{NL}}^2 \left\langle \Phi_L^2 \right\rangle \simeq 10^{-5} (f_{\mathrm{NL}}/100)^2 \ . \tag{9.125}$$

We are talking about very small effects.

This parameterization is useful to find *quantitative* constraints on the amount of non–gaussianity allowed by the CMB data. Also, the form is general in that f_{NL} parameterizes the leading–order non–linear corrections to Φ.

Figure 9.10 shows f_{NL} measured from the foreground–cleaned Q+V+W coadded map using the cubic statistic, as a function of the maximum multipole l_{max} (for details of measurements. There is no significant detection of f_{NL} at any angular scale. There is no significant band–to–band variation, or significant detection in any band. The best constraint is $-58 < f_{NL} < 134$ (95%), which is equivalent to say that the fractional contribution to the *r.m.s.* value of Φ from the second–order term is smaller than 2×10^{-5}. These results support inflationary models, but still do not exclude the possibility of having a small contribution from non–linearities predicted by second–order perturbation theory.

Note that f_{NL} for $l_{max} = 265$ has a smaller error than that for $l_{max} = 512$, because the latter is dominated by the instrumental noise. Since all the pixels outside the cut region are uniformly weighted, the inhomogeneous noise in the map (pixels on the ecliptic equator are noisier than those on the north and south poles) is not accounted for. This leads to a noisier estimator than a minimum variance estimator. The constraint on f_{NL} for $l_{max} = 512$ will likely improve with more appropriate pixel-weighting schemes. Apparently,

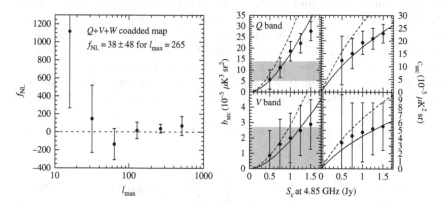

Fig. 9.10. (*left panel*) The non–linearity parameter, f_{NL}, as a function of the maximum multipole l_{max}, measured from the $Q + V + W$ coadded map using the bispectrum estimator. The error bars at each l_{max} are not independent. (*right panel*) The point–source angular bispectrum b^{src} and power–spectrum c^{src}. The left panels show b^{src} in Q band (*top panel*) and V band (*bottom panel*). The shaded areas show measurements from the *WMAP* sky maps with the standard source cut, while the filled circles show those with flux thresholds S_c defined at 4.85 GHz. The *dashed lines* show predictions from the source count model, while the *solid lines* are those multiplied by 0.65 to match the *WMAP* measurements. The right panels show c^{src}. The filled circles are computed from the measured b^{src} substituted into (9.126). The lines are predictions. The error bars are not independent

the fact that the constraint actually obtained from the data is worse than predicted (c.f., Table 9.6) should be due to sub-optimalness of the current estimator. The simple inverse noise (N^{-1}) weighting makes the constraints much worse than the uniform weighting, as it increases errors on large angular scales where the CMB signal dominates over the instrumental noise. (However, it works fine for the point–sources.) The uniform weighting is thus closer to optimal.

The Minkowski functionals shown in Fig. 9.9 also place constraints on f_{NL}, comparing the data to the predictions derived from Monte Carlo simulations of the non–gaussian CMB (for details of the simulations. It has been found that $f_{NL} < 139$ (95%), remarkably consistent with that from the bispectrum analysis.

Point–Source Non–gaussianity

The second component is the foreground non–gaussianity from radio point–sources, parameterized by the skewness, b^{src}. The filled circles in the right panels of Fig. 9.10 show b^{src} measured in Q (top panel) and V (bottom panel) band. We have used source masks for various flux cuts, S_c, defined at 4.85 GHz to make these measurements. (The masks are made from the GB6+PMN 5 GHz source catalogue.) We find that b^{src} increases as S_c: the brighter sources being unmasked, the more non–gaussianity is detected. On the other hand one can make predictions for b^{src} using the source count model. Comparing the measured values of b^{src} with the predicted counts (dashed lines) at 44 GHz, one finds that the measured values are smaller than the predicted values by a factor of 0.65. The solid lines show the predictions multiplied by 0.65. Our value for the correction factor matches well the one obtained from the WMAP source counts for 2–10 Jy in Q band.

The source bispectrum, b^{src}, is related to the source power–spectrum, c^{src}, by an integral relation,

$$c^{src}(S_c) = b^{src}(S_c)[g(\nu)S_c]^{-1} + \int_0^{S_c} \frac{dS}{S} b^{src}(S)[g(\nu)S]^{-1} , \qquad (9.126)$$

where $g(\nu)$ is a conversion factor from Jy sr^{-1} to μK which depends upon the observing frequency ν as $g(\nu) = (24.76 \text{ Jy } \mu\text{K}^{-1} \text{ sr}^{-1})^{-1}[(\sinh x/2)/x^2]^2$, $x \equiv h\nu/k_B T_0 \simeq \nu/(56.78 \text{ GHz})$. One can use this equation combined with the measured b^{src} as a function of the flux threshold S_c to directly determine c^{src} as a function of S_c, *without relying on any extrapolations*. The right panels of Fig. 9.10 also show the estimated c^{src} as filled circles. The measurements suggest that c^{src} for the standard source mask (indicated by the shaded area) is $c^{src} = (15 \pm 6) \times 10^{-3} \mu\text{K}^2$ sr in Q band. In V band, $c^{src} = (4.5 \pm 4) \times 10^{-3} \mu\text{K}^2$ sr.

In addition to the bispectrum, the WMAP team has carried out other methods to estimate the source contribution: (1) extrapolation from the number counts of detected sources in the WMAP data, and (2) the angular power–spectrum on small angular scales. These methods yield consistent results.

In summary the WMAP 1-year data has enormously improved the sensitivity for testing the Gaussianity of the CMB. Yet, we do not have any compelling evidence for primordial non–gaussianity. This result is consistent with what is predicted by inflation and the second–order perturbation theory. There may be some chance to find non–gaussian signals arising from second–order perturbations. Detection can be made possible by the *Planck* experiment combining the temperature and polarization anisotropies. While we can detect $f_{\rm NL} \sim 5$ by using the temperature alone (see Table 9.6), combining the polarization measurements increases our sensitivity: we have several observables for the bispectrum such as $\langle TTT \rangle$, $\langle TTE \rangle$, $\langle TEE \rangle$, and $\langle EEE \rangle$. The future polarization-dedicated satellite experiment (e.g., CMBPol) in combination with the *Planck* temperature map may enable us to detect $f_{\rm NL} \sim 3$.

9.7 Towards the Second-Order Transfer Function

From what we have said so far, it is clear that the most crucial theoretical step as far as the NG is concerned is to compute the full second-order radiation transfer function. Its determination will allow us to determine precisely the order unity NG coming from the post-inflationary evolution on all scales. The first step towards this determination has been recently taken in a couple of works [2, 3], where the computation of the full system of Boltzmann equations at second-order describing the evolution of the photon, baryon and cold dark matter fluids. These equations allow to follow the time evolution of the CMB anisotropies at second-order at all angular scales from the early epoch, when the cosmological perturbations were generated, to the present through the recombination era. In particular, in [3] an analytical approach to the second-order CMB anisotropies generated by the non-linear dynamics taking place at last scattering was provided. The acoustic oscillations of the photon–baryon fluid were studied in the tight coupling limit extending at second-order the Meszaros effect and with a generic set of initial conditions due to primordial non-gaussianity.

The starting point is the Boltzmann equation at first- and second-order [2]

$$\frac{\partial \Delta^{(1)}}{\partial \eta} + n^i \frac{\partial \Delta^{(1)}}{\partial x^i} + 4\frac{\partial \Phi^{(1)}}{\partial x^i} n^i - 4\frac{\partial \Psi^{(1)}}{\partial \eta} = -\tau' \left[\Delta_0^{(1)} + \frac{1}{2}\Delta_2^{(1)} P_2(\hat{\mathbf{v}} \cdot \mathbf{n}) \right.$$
$$\left. - \Delta^{(1)} + 4\mathbf{v} \cdot \mathbf{n} \right] , \qquad (9.127)$$

and at second-order

$$\frac{1}{2}\frac{d}{d\eta}\left[\Delta^{(2)}+4\Phi^{(2)}\right]+\frac{d}{d\eta}\left[\Delta^{(1)}+4\Phi^{(1)}\right]-4\Delta^{(1)}\left(\Psi^{(1)\prime}-\Phi^{(1)}_{,i}n^i\right)$$

$$-2\frac{\partial}{\partial\eta}\left(\Psi^{(2)}+\Phi^{(2)}\right)+4\frac{\partial\omega_i}{\partial\eta}n^i+2\frac{\partial\chi_{ij}}{\partial\eta}n^in^j$$

$$=-\frac{\tau'}{2}\left[\Delta^{(2)}_{00}-\Delta^{(2)}-\frac{1}{2}\sum_{m=-2}^{2}\frac{\sqrt{4\pi}}{5^{3/2}}\Delta^{(2)}_{2m}Y_{2m}(\mathbf{n})+2(\delta^{(1)}_e+\Phi^{(1)})\left(\Delta^{(1)}_0\right.\right.$$

$$+\frac{1}{2}\Delta^{(1)}_2P_2(\hat{\mathbf{v}}\cdot\mathbf{n})-\Delta^{(1)}+4\mathbf{v}\cdot\mathbf{n}\Bigg)$$

$$+4\mathbf{v}^{(2)}\cdot\mathbf{n}+2(\mathbf{v}\cdot\mathbf{n})\left[\Delta^{(1)}+3\Delta^{(1)}_0-\Delta^{(1)}_2\left(1-\frac{5}{2}P_2(\hat{\mathbf{v}}\cdot\mathbf{n})\right)\right]$$

$$-v\Delta^{(1)}_1\left(4+2P_2(\hat{\mathbf{v}}\cdot\mathbf{n})\right)+14(\mathbf{v}\cdot\mathbf{n})^2-2v^2\Bigg], \tag{9.128}$$

Let us recall some definitions of the quantities appearing in (9.127)–(9.128). $\Phi=\Phi^{(1)}+\Phi^{(2)}/2$ and $\Psi=\Psi^{(1)}+\Psi^{(2)}/2$ are the gravitational potentials in the Poisson gauge, while ω_i and χ_{ij} are the second-order vector and tensor perturbations of the metric. The photon temperature anisotropies are given by

$$\Delta^{(i)}(x^i,n^i,\tau)=\frac{\int dpp^3f^{(i)}}{\int dpp^3f^{(0)}}, \tag{9.129}$$

which represents the photon fractional energy perturbation (in a given direction) being the integral of the photon distribution function $f=f^{(1)}+f^{(2)}/2$ over the photon momentum magnitude p $(p^i=pn^i)$. The angular dependence of the photon anisotropies Δ can be expanded as

$$\Delta^{(i)}(\mathbf{x},\mathbf{n})=\sum_{\ell}\sum_{m=-\ell}^{\ell}\Delta^{(i)}_{\ell m}(\mathbf{x})(-i)^\ell\sqrt{\frac{4\pi}{2\ell+1}}Y_{\ell m}(\mathbf{n}), \tag{9.130}$$

with

$$\Delta^{(i)}_{\ell m}=(-i)^{-\ell}\sqrt{\frac{2\ell+1}{4\pi}}\int d\Omega\Delta^{(i)}Y^*_{\ell m}(\mathbf{n}), \tag{9.131}$$

where we warn the reader that the superscript stands by the order of the perturbation, while the subscripts indicate the order of the multipoles. At first order one can drop the dependence on m setting $m=0$ so that $\Delta^{(1)}_{\ell m}=(-i)^{-\ell}(2\ell+1)\delta_{m0}\Delta^{(1)}_\ell$. It is understood that on the left hand side of (9.128) one has to pick up for the total time derivatives only those terms which contribute to second-order. Thus we have to take

$$\frac{1}{2}\frac{d}{d\eta}\left[\Delta^{(2)}+4\Phi^{(2)}\right]+\frac{d}{d\eta}\left[\Delta^{(1)}+4\Phi^{(1)}\right]\Bigg|^{(2)}$$

$$=\frac{1}{2}\left(\frac{\partial}{\partial\eta}+n^i\frac{\partial}{\partial x^i}\right)\left(\Delta^{(2)}+4\Phi^{(2)}\right)$$

$$+n^i(\Phi^{(1)} + \Psi^{(1)})\partial_i(\Delta^{(1)} + 4\Phi^{(1)})$$

$$+ \left[(\Phi^{(1)}_{,j} + \Psi^{(1)}_{,j})n^i n^j - (\Phi^{,i} + \Psi^{,i}) \right] \frac{\partial \Delta^{(1)}}{\partial n^i} , \tag{9.132}$$

In (9.128) $\delta_e^{(1)}$ is the relative energy density perturbation of the electrons. These are in turn strongly coupled with protons (p) via Coulomb interactions, such that the density constrasts and the velocities are driven to a common value $\delta_e = \delta_p \equiv \delta_b$ and $\mathbf{v}_e = \mathbf{v}_b \equiv \mathbf{v}$ for what can then be called the baryon fluid. Finally

$$\tau' = -\bar{n}_e \sigma_T a . \tag{9.133}$$

is the differential optical depth for the Compton scatterings between photons and free electrons. The tightly coupled limit corresponds to the Compton interaction rate much bigger than the expansion of the universe, $\tau'/\mathcal{H} \gg 1$ (or $\tau \gg 1$). In this limit, one may proceed as it is done for the linear perturbations, i.e. expanding the equations in power of $1/\tau'$ and solving the equations for wavelengths $\lambda = 2\pi/k$ which are well above or beneath the horizon at the equality time. For instance and as an illustrative example, let us consider the photon perturbations which enter the horizon between the equality epoch and the recombination epoch, with wavelenghts $\eta_*^{-1} < k < \eta_{eq}^{-1}$. In fact, in order to find some analytical solutions, one assumes that by the time of recombination the universe is matter dominated $\eta_{eq} \ll \eta_*$. After a lengthy, but straightforward computation one finds that at second order

$$\Delta_{00}^{(2)} = \left[\frac{54}{5}(a_{NL} - 1) - \frac{2}{5}(9a_{NL} - 19)\cos(kc_s\eta) \right.$$

$$\left. - \frac{2}{7}\left(\frac{9}{10}\right)^2 G(\mathbf{k}_1, \mathbf{k}_2, \mathbf{k})\eta^2 \right] \Psi^{(1)}_{\mathbf{k}_1}(0)\Psi^{(1)}_{\mathbf{k}_2}(0) ,$$

$$G(\mathbf{k}_1, \mathbf{k}_2, \mathbf{k}) = \mathbf{k}_1 \cdot \mathbf{k}_2 - \frac{10}{3}\frac{(\mathbf{k} \cdot \mathbf{k}_1)(\mathbf{k} \cdot \mathbf{k}_2)}{k^2} . \tag{9.134}$$

In this expression c_s is the sound velocity at recombination, $\Psi^{(1)}_{\mathbf{k}}(0)$ is the primordial gravitational potential and the parameter a_{NL} depends on the physics of a given scenario. For example in the standard scenario $a_{NL} \simeq 1$, while in the curvaton case $a_{NL} = (3/4r) - r/2$, where $r \approx (\rho_\sigma/\rho)_D$ is the relative curvaton contribution to the total energy density at curvaton decay [1]. In the minimal picture for the inhomogeneous reheating scenario, $a_{NL} = 1/4$. For other scenarios we refer the reader to [1].

Expression (9.134) is illuminating and teaches us that, if the primordial NG is large, $a_{NL} \gg 1$, then the first-order radiation transfer function. However, if $a_{NL} = O(1)$, then computing the contribution to the NG from the post-inflationary era is crucial. Work along these lines is in progress.

References

1. N. Bartolo, E. Komatsu, S. Matarrese and A. Riotto, Phys. Rept. **402**, 103 (2004), arXiv:astro-ph/0406398.
2. N. Bartolo, S. Matarrese and A. Riotto, JCAP **0606**, 024 (2006), arXiv:astro-ph/0604416.
3. N. Bartolo, S. Matarrese and A. Riotto, arXiv:astro-ph/0610110.
4. K. A. Malik and D. Wands, Class. Quant. Grav. **21**, L65 (2004), arXiv:astro-ph/0307055.
5. V. Acquaviva, N. Bartolo, S. Matarrese and A. Riotto, Nucl. Phys. B **667**, 119 (2003), arXiv:astro-ph/0209156.
6. J. M. Maldacena, JHEP **0305**, 013 (2003), arXiv:astro-ph/0210603.

Production of Topological Defects at the End of Inflation

Mairi Sakellariadou

Department of Physics, King's College, University of London, Strand, London
WC2R 2LS, UK
Mairi.Sakellariadou@kcl.ac.uk

Abstract. Cosmological inflation and topological defects have been considered for a long time, either in disagreement or in competition. On the one hand an inflationary era is required to solve the shortcomings of the hot big bang model, while on the other hand cosmic strings and string-like objects are predicted to be formed in the early universe. Thus, one has to find ways so that both can coexist. I will discuss how to reconcile cosmological inflation with cosmic strings.

10.1 Introduction

For a number of years, inflation and cosmological defects have been considered either as two incompatible or as two competing aspects of modern cosmology. Let me explain why. Historically, one of the reasons for which inflation was proposed is to rescue the standard hot big bang model from the monopole problem. More precisely, setting an inflationary era after the formation of monopoles, these unwanted defects would have been diluted away. However, such a mechanism could also dilute cosmic strings unless they were produced at the end or after inflation. Later on, inflation and topological defects competed as the two alternative mechanisms to provide the generation of density perturbations leading to the observed large-scale structure and the anisotropies in the cosmic microwave background (CMB). However, the inconsistency between predictions from topological defect models and CMB data on the one hand, and the good agreement between adiabatic fluctuations generated by the amplification of the quantum fluctuations of the inflation field on the other hand, indicated a clear preference for inflation. Finally, the genericity of cosmic string formation in the framework of grand unified theories (GUTs) and the formation of defect-like objects in brane cosmologies, convinced us that cosmic strings have to play a role, which may be sub-dominant but it is definitely there. This conclusion led to the consideration of mixed models, where inflation and cosmic strings coexist. The study of such models,

M. Sakellariadou: *Production of Topological Defects at the End of Inflation*, Lect. Notes Phys.
738, 359–392 (2008)
DOI 10.1007/978-3-540-74353-8_10   © Springer-Verlag Berlin Heidelberg 2008

the comparison of their predictions against current data and the consequences for the theories within which we based our study are the aims of this study.

In Sect. 10.2, I briefly describe cosmological inflation, its success and its open questions. I then discuss hybrid inflation in general and then I focus on F-/D-term inflation in the framework of supersymmetry and supergravity theories. In Sect. 10.3, I discuss topological defects in general, and cosmic strings in particular. I then argue the genericity of string formation in the framework of GUTs. In Sect. 10.4, I briefly discuss braneworld cosmology, focusing on inflation within braneworld cosmologies and the generation of cosmic superstrings. In Sect. 10.5, I discuss observational consequences, and in particular the spectrum of CMB anisotropies and that of gravity waves. I compare the predictions of the models against current data, which allow me to constrain the parameter space of the models. I round up with the conclusions in Sect. 10.6.

10.2 Cosmological Inflation

Despite its success, the standard hot big bang cosmological model has a fairly severe drawback, namely the requirement, up to a high degree of accuracy, of an initially homogeneous and flat universe. An appealing solution to this problem is to introduce, during the very early stages of the evolution of the universe, a period of accelerated expansion, known as cosmological inflation [1]. The inflationary era took place when the universe was in an unstable vacuum-like state at a high energy density, leading to a quasi-exponential expansion. The combination of the hot big bang model and the inflationary scenario provides at present the most comprehensive picture of the universe at our disposal. Inflation ends when the Hubble parameter $H = \sqrt{8\pi\rho/(3m_{\mathrm{Pl}}^2)}$ (where ρ denotes the energy density and m_{Pl} stands for the Planck mass) starts decreasing rapidly. The energy stored in the vacuum-like state gets transformed into thermal energy, heating up the universe and leading to the beginning of the standard hot big bang radiation-dominated era.

Inflation is based on the basic principles of general relativity and field theory, and when the principles of quantum mechanics are also considered, it provides a successful explanation for the origin of the large-scale structure, associated with the measured temperature anisotropies in the CMB spectrum. Inflation is overall a very successful scenario and many different models have been proposed and studied over the last 25 years. Nevertheless, inflation still remains a paradigm in search of model. In principle, one should search for an inflationary model inspired from some fundamental theory and subsequently test its predictions against current data. Moreover, releasing the present universe from its acute dependence on the initial data, inflation is faced with the challenging task of proving itself generic, in the sense that inflation would take place without fine-tuning of the initial conditions. This issue, already addressed in the past [2], has been recently re-investigated [3].

10.2.1 Hybrid Inflation in SUSY GUTs

Chaotic inflation [4] is, to my opinion, the most elegant inflationary model. Nevertheless, in order for density inhomogeneities generated at the end of inflation to have the required amplitude $(\delta\rho/\rho) \sim 10^{-4} - 10^{-5}$, the model requires fine-tuning. In the simplest theory of a single scalar field minimally coupled to gravity, the coupling must be of the order of $\lambda \sim 10^{-13}$–10^{-14}; the same fine-tuning was required in the new inflationary model. This is a reason for which hybrid inflation [5] has been proposed.

Hybrid inflation is based on Einstein's gravity but is driven by false vacuum. The inflation field rolls down its potential while another scalar field is trapped in an unstable false vacuum. Once the inflaton field becomes much smaller than some critical value, a phase transition to the true vacuum takes place and inflation ends [for an illustration see Fig. 10.1]. Such a phase transition may leave behind topological defects as false vacuum remnants. In particular, the formation of topological defects may provide the mechanism to gracefully exit the inflationary era in a number of particle physics motivated inflationary models [6].

Theoretically motivated inflationary models can be built in the context of supersymmetry or supergravity. $N = 1$ supersymmetry models contain complex scalar fields which often have flat directions in their potential, thus offering natural candidates for inflationary models. In this framework, hybrid inflation driven by **F**-terms or **D**-terms is the standard inflationary model, leading [7] generically to cosmic string formation at the end of inflation. **F**-term inflation is potentially plagued with the η-problem, while **D**-term inflation avoids it. Let me briefly explain what this problem is. It is difficult to

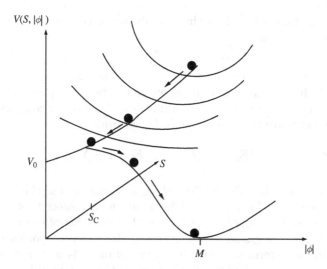

Fig. 10.1. A simplistic drawing of hybrid inflation

achieve slow-roll inflation within supergravity, however, inflation should last long enough to solve the shortcomings of the standard big bang model. The positive false vacuum of the inflation field breaks the global supersymmetry spontaneously, which gets restored once inflation has been completed. However, since in supergravity theories, supersymmetry breaking is transmitted by gravity, all scalar fields acquire an effective mass of the order of the expansion rate during inflation. Such a heavy mass for the scalar field playing the role of the inflaton spoils the slow-roll condition. It has been shown [8] that the *Hubble-induced* mass problem has its origin on the **F**-term interactions, while it disappears if the vacuum energy is instead dominated by the **D**-terms of the superfields.

F-Term Inflation

F-term inflation can be naturally accommodated in the framework of GUTs when a GUT gauge group, $\mathbf{G}_{\mathrm{GUT}}$, is broken down to the standard model (SM) gauge group, $\mathbf{G}_{\mathrm{SM}}$, at an energy scale M_{GUT} according to the scheme

$$\mathbf{G}_{\mathrm{GUT}} \xrightarrow{M_{\mathrm{GUT}}} \mathbf{H}_1 \xrightarrow[\Phi_+\Phi_-]{M_{\mathrm{infl}}} \mathbf{H}_2 \longrightarrow \mathbf{G}_{\mathrm{SM}} ; \tag{10.1}$$

Φ_+, Φ_- is a pair of GUT Higgs superfields in non-trivial complex conjugate representations, which lowers the rank of the group by one unit when acquiring non-zero vacuum expectation value. The inflationary phase takes place at the beginning of the symmetry breaking $\mathbf{H}_1 \xrightarrow{M_{\mathrm{infl}}} \mathbf{H}_2$. The gauge symmetry is spontaneously broken by adding **F**-terms to the superpotential. The Higgs mechanism leads generically [7] to Abrikosov–Nielsen–Olesen strings, called **F**-term strings.

F-term inflation is based on the globally supersymmetric renormalisable superpotential

$$W_{\mathrm{infl}}^F = \kappa S(\Phi_+\Phi_- - M^2) , \tag{10.2}$$

where S is a GUT gauge singlet left-handed superfield and κ, M are two constants (M has dimensions of mass) which can be taken positive with field redefinition. The scalar potential, as a function of the scalar complex component of the respective chiral superfields $\Phi_\pm, S$, reads

$$V(\phi_+, \phi_-, S) = |F_{\Phi_+}|^2 + |F_{\Phi_-}|^2 + |F_S|^2 + \frac{1}{2}\sum_a g_a^2 D_a^2 . \tag{10.3}$$

The **F**-term is such that $F_{\Phi_i} \equiv |\partial W/\partial \Phi_i|_{\theta=0}$, where we take the scalar component of the superfields once we differentiate with respect to $\Phi_i = \Phi_\pm, S$. The **D**-terms are $D_a = \bar\phi_i (T_a)^i{}_j \phi^j + \xi_a$, with a the label of the gauge group generators T_a, g_a the gauge coupling and ξ_a the Fayet–Iliopoulos term. By definition, in the **F**-term inflation the real constant ξ_a is zero; it can only be non-zero if T_a generates an extra U(1) group. In the context of **F**-term hybrid inflation the **F**-terms give rise to the inflationary potential energy density while the **D**-terms are flat along the inflationary trajectory, thus one may neglect them during inflation.

The potential, plotted in Fig. 10.2, has one valley of local minima, $V = \kappa^2 M^4$, for $S > M$ with $\phi_+ = \phi_- = 0$, and one global supersymmetric minimum, $V = 0$, at $S = 0$ and $\phi_+ = \phi_- = M$. Imposing initially $S \gg M$, the fields quickly settle down the valley of local minima. Since in the slow-roll inflationary valley the ground state of the scalar potential is non-zero, supersymmetry is broken. In the tree level, along the inflationary valley the potential is constant, therefore perfectly flat. A slope along the potential can be generated by including one-loop radiative corrections, which can be calculated using the Coleman–Weinberg expression [9]

$$\Delta V_{1\text{-loop}} = \frac{1}{64\pi^2} \sum_i (-1)^{F_i} m_i^4 \ln \frac{m_i^2}{\Lambda^2} , \qquad (10.4)$$

where the sum extends over all helicity states i, with fermion number F_i and mass squared m_i^2; Λ stands for a renormalisation scale. In this way, the scalar potential gets a little tilt which helps the inflation field S to slowly roll down the valley of minima. The one-loop radiative corrections to the scalar potential along the inflationary valley lead to the effective potential [10]

$$V_{\text{eff}}^F(|S|) = \kappa^2 M^4 \left\{ 1 + \frac{\kappa^2 \mathcal{N}}{32\pi^2} \left[2 \ln \frac{|S|^2 \kappa^2}{\Lambda^2} + (z+1)^2 \ln(1+z^{-1}) \right. \right.$$
$$\left. \left. + (z-1)^2 \ln(1-z^{-1}) \right] \right\} \quad \text{with} \quad z = \frac{|S|^2}{M^2} ; \qquad (10.5)$$

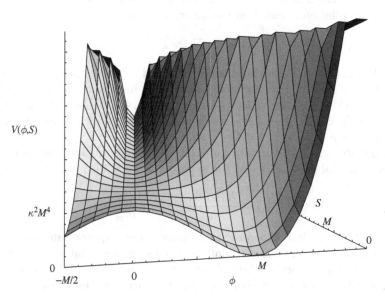

Fig. 10.2. A representation of the potential for F-term inflation in the context of supersymmetry

$\mathcal{N}$ stands for the dimensionality of the representation to which the complex scalar components ϕ_+, ϕ_- of the chiral superfields Φ_+, Φ_- belong. This implies that the effective potential, (10.5), depends on the particular symmetry breaking scheme considered [see (10.1)].

D-Term Inflation

D-term inflation is one of the most interesting models of inflation. It is possible to implement it naturally within high-energy physics, as for example supersymmetric GUTs (SUSY GUTs), supergravity (SUGRA) or string theories. Moreover, it avoids the Hubble-induced *mass* problem. In **D**-term inflation, the gauge symmetry is spontaneously broken by introducing Fayet–Iliopoulos (FI) **D**-terms. In standard **D**-term inflation, the constant FI term gets compensated by a single complex scalar field at the end of the inflationary era, which implies that standard **D**-term inflation ends with the formation of cosmic strings, called **D**-strings. More precisely, in its simplest form, the model requires a symmetry breaking scheme

$$G_{\text{GUT}} \times U(1) \xrightarrow{M_{\text{GUT}}} H \times U(1) \xrightarrow[\Phi_+\Phi_-]{M_{\text{infl}}} H \to G_{\text{SM}} \ . \tag{10.6}$$

A supersymmetric description of the standard **D**-term inflation is insufficient; the inflation field reaches values of the order of the Planck mass, or above it, even if one concentrates only around the last 60 e-folds of inflation; the correct analysis is therefore in the context of supergravity.

D-term inflation is based on the superpotential

$$W = \lambda S \Phi_+ \Phi_- \ , \tag{10.7}$$

where S, Φ_+, Φ_- are three chiral superfields and λ is the superpotential coupling. In its standard form, the model assumes an invariance under an Abelian gauge group $U(1)_\xi$, under which the superfields S, Φ_+, Φ_- have charges $0, +1$ and -1, respectively. It is also assumed the existence of a constant Fayet–Iliopoulos term ξ.

In the *standard* supergravity formulation the Lagrangian depends on the Kähler potential $K(\Phi_i, \bar{\Phi}_i)$ and the superpotential $W(\Phi_i)$ only through the combination

$$G(\Phi_i, \bar{\Phi}_i) = \frac{K(\Phi_i, \bar{\Phi}_i)}{m_{\text{Pl}}^2} + \ln \frac{|W(\Phi_i)|^2}{m_{\text{Pl}}^6} \ . \tag{10.8}$$

However, this standard supergravity formulation is inappropriate to describe **D**-term inflation [11]. In **D**-term inflation the superpotential vanishes at the unstable de Sitter vacuum (anywhere else the superpotential is non-zero). Thus, standard supergravity is inappropriate, since it is ill-defined at $W = 0$. In conclusion, **D**-term inflation must be described with a non-singular formulation of supergravity when the superpotential vanishes.

Various formulations of effective supergravity can be constructed from the superconformal field theory. One must first build a Lagrangian with full superconformal theory, and then the gauge symmetries that are absent in Poincaré supergravity must be gauge fixed. In this way, one can construct a non-singular theory at $W = 0$, where the action depends on all three functions: the Kähler potential $K(\Phi_i, \bar{\Phi}_i)$, the superpotential $W(\Phi_i)$ and the kinetic function $f_{ab}(\Phi_i)$ for the vector multiplets. To construct a formulation of supergravity with constant Fayet–Iliopoulos terms from superconformal theory, one finds [11] that under U(1) gauge transformations in the directions in which there are constant Fayet–Iliopoulos terms ξ_α, the superpotential W must transform as [11]

$$\delta_\alpha W = \eta_{\alpha i}\partial^i W = -i\frac{g\xi_\alpha}{m_{\mathrm{Pl}}^2}W \; ; \tag{10.9}$$

it is incorrect to keep the same charge assignments as in standard supergravity.

D-term inflationary models can be built with different choices of Kähler geometry. Let us first consider **D**-term inflation within minimal supergravity. It is based on

$$K_{\min} = \sum_i |\Phi_i|^2 = |\Phi_-|^2 + |\Phi_+|^2 + |S|^2 \; , \tag{10.10}$$

with $f_{ab}(\Phi_i) = \delta_{ab}$. The tree level scalar potential is [11]

$$\begin{aligned}
V_{\min} = \lambda^2 &\exp\left(\frac{|\phi_-|^2 + |\phi_+|^2 + |S|^2}{m_{\mathrm{Pl}}^2}\right)\left[|\phi_+\phi_-|^2\left(1 + \frac{|S|^4}{m_{\mathrm{Pl}}^4}\right)\right.\\
&+ |\phi_+ S|^2\left(1 + \frac{|\phi_-|^4}{m_{\mathrm{Pl}}^4}\right) + |\phi_- S|^2\left(1 + \frac{|\phi_+|^4}{m_{\mathrm{Pl}}^4}\right) + 3\frac{|\phi_-\phi_+ S|^2}{m_{\mathrm{Pl}}^2}\right]\\
&+ \frac{g^2}{2}\left(q_+|\phi_+|^2 + q_-|\phi_-|^2 + \xi\right)^2 \; ,
\end{aligned} \tag{10.11}$$

with

$$q_\pm = \pm 1 - \xi/(2m_{\mathrm{Pl}}^2) \; . \tag{10.12}$$

The potential has two minima: one global minimum at zero and one local minimum equal to $V_0 = (g^2/2)\xi^2$. For arbitrary large S the tree level value of the potential remains constant and equal to V_0; the S plays the role of the inflation field. Assuming chaotic initial conditions $|S| \gg S_{\mathrm{s}}$, inflation begins. Along the inflationary trajectory the **D**-term, which is the dominant one, splits the masses in the $\Phi_\pm$ superfields, leading to the one-loop effective potential for the inflation field. Considering the one-loop radiative corrections [10, 12]

$$V_{\min}^{\mathrm{eff}}(|S|) = \frac{g^2\xi^2}{2}\left\{1 + \frac{g^2}{16\pi^2}\left[2\ln\left(z\frac{g^2\xi}{\Lambda^2}\right) + f_V(z)\right]\right\} \; , \tag{10.13}$$

where

$$f_V(z) = (z+1)^2\ln\left(1 + \frac{1}{z}\right) + (z-1)^2\ln\left(1 - \frac{1}{z}\right) \; , \tag{10.14}$$

with

$$z \equiv \frac{\lambda^2}{g^2 \xi} |S|^2 \exp\left(\frac{|S|^2}{m_{\mathrm{Pl}}^2}\right) \ . \tag{10.15}$$

As a second example, consider **D**-term inflation based on Kähler geometry with a shift symmetry, $\phi \to \phi + c$ (where c is a real constant). Such models can lead [13] to flat enough potentials with stabilisation of the volume of the compactified space. They can therefore be used to build successful inflationary models in the framework of string theories. The Kähler potential is

$$K_{\mathrm{shift}} = \frac{1}{2}(S + \bar{S})^2 + |\phi_+|^2 + |\phi_-|^2 \ ; \tag{10.16}$$

the kinetic function has the minimal structure. The scalar potential reads [14]

$$
\begin{aligned}
V_{\mathrm{shift}} &\simeq \frac{g^2}{2}\left(|\phi_+|^2 - |\phi_-|^2 + \xi\right)^2 \\
&+ \lambda^2 \exp\left(\frac{|\phi_-|^2 + |\phi_+|^2}{m_{\mathrm{Pl}}^2}\right) \exp\left[\frac{(S + \bar{S})^2}{2m_{\mathrm{Pl}}^2}\right] \\
&\times \left[|\phi_+\phi_-|^2 \left(1 + \frac{S^2 + \bar{S}^2}{m_{\mathrm{Pl}}^2} + \frac{|S|^2|S + \bar{S}|^2}{m_{\mathrm{Pl}}^4}\right) + |\phi_+ S|^2 \left(1 + \frac{|\phi_-|^4}{m_{\mathrm{Pl}}^4}\right) \right. \\
&+ \left. |\phi_- S|^2 \left(1 + \frac{|\phi_+|^4}{m_{\mathrm{Pl}}^4}\right) + 3\frac{|\phi_-\phi_+ S|^2}{m_{\mathrm{Pl}}^2} \right] \ . \tag{10.17}
\end{aligned}
$$

As in **D**-term inflation within minimal supergravity, the potential has a global minimum at zero for $\langle \Phi_+ \rangle = 0$ and $\langle \Phi_- \rangle = \sqrt{\xi}$ and a local minimum equal to $V_0 = (g^2/2)\xi^2$ for $\langle S \rangle \gg S_c$ and $\langle \Phi_\pm \rangle = 0$.

The exponential factor $\mathrm{e}^{|S|^2}$, which we got in the case of minimal supergravity, has been replaced by $\mathrm{e}^{(S+\bar{S})^2/2}$. Writing $S = \eta + i\phi_0$ one gets $\mathrm{e}^{(S+\bar{S})^2/2} = \mathrm{e}^{\eta^2}$. If η plays the role of the inflation field, we obtain the same potential as for minimal **D**-term inflation. If instead ϕ_0 is the inflation field, the inflationary potential is identical to that of the usual **D**-term inflation within global supersymmetry [10]. The latter case is better adapted with the choice of K_{shift}, since then the exponential term is constant during inflation and thus it cannot spoil the slow-roll conditions.

As a last example, consider a Kähler potential with non-renormalisable terms:

$$
\begin{aligned}
K_{\mathrm{non\text{-}renorm}} &= |S|^2 + |\Phi_+|^2 + |\Phi_-|^2 \\
&+ f_+\left(\frac{|S|^2}{m_{\mathrm{Pl}}^2}\right)|\Phi_+|^2 + f_-\left(\frac{|S|^2}{m_{\mathrm{Pl}}^2}\right)|\Phi_-|^2 + b\frac{|S|^4}{m_{\mathrm{Pl}}^2} \ , \tag{10.18}
\end{aligned}
$$

where $f_\pm$ are arbitrary functions of $(|S|^2/m_{\mathrm{Pl}}^2)$ and the superpotential is given in (10.7). The effective potential reads [14]

$$V_{\mathrm{non\text{-}renorm}}^{\mathrm{eff}}(|S|) = \frac{g^2 \xi^2}{2}\left\{1 + \frac{g^2}{16\pi^2}\left[2\ln\left(z\frac{g^2\xi}{\Lambda^2}\right) + f_V(z)\right]\right\} \ , \tag{10.19}$$

where

$$f_V(z) = (z+1)^2 \ln\left(1+\frac{1}{z}\right) + (z-1)^2 \ln\left(1-\frac{1}{z}\right) \qquad (10.20)$$

with $z \equiv \dfrac{\lambda^2 |S|^2}{g^2 \xi} \exp\left(\dfrac{|S|^2}{m_{\rm Pl}^2} + b\dfrac{|S|^4}{m_{\rm Pl}^4}\right) \dfrac{1}{(1+f_+)(1+f_-)}$.   (10.21)

The cosmological consequences of these inflationary models will be presented in Sect. 10.5.

10.3 Topological Defects in GUTs

Following the standard version of the hot big bang model, the universe could have expanded from a very hot (with a temperature $T \gtrsim 10^{19}\,{\rm GeV}$) and dense state, cooling towards its present state. As the universe expands and cools down, it undergoes a number of phase transitions, breaking the symmetry between the different interactions. Such phase transitions may leave behind topological defects [15] as false vacuum remnants, via the Kibble mechanism [16]. Whether or not topological defects are formed during phase transitions followed by spontaneously broken symmetries (SSB) depend on the topology of the vacuum manifold $\mathcal{M}_n$, which also determines the type of the produced defects. The properties of $\mathcal{M}_n$ are usually described by the kth homotopy group $\pi_k(\mathcal{M}_n)$, which classifies distinct mappings from the k-dimensional sphere S^k into the manifold $\mathcal{M}_n$.

Let me consider the symmetry breaking of a group G down to a subgroup H of G . If $\mathcal{M}_n$ = G/H has disconnected components, or equivalently if the order k of the non-trivial homotopy group is $k = 0$, two-dimensional defects, *domain walls*, get formed. The spacetime dimension, d, of the defects is determined by the order of the non-trivial homotopy group by $d = 4 - 1 - k$. If $\mathcal{M}_n$ is not simply connected, meaning that $\mathcal{M}_n$ contains loops which cannot be continuously shrunk into a point, *cosmic strings* get produced. A necessary but not sufficient condition for the formation of stable strings is that the first (fundamental) homotopy group $\pi_1(\mathcal{M}_n)$ of $\mathcal{M}_n$, is non-trivial, or multiply connected. Cosmic strings are line-like ($d = 2$) defects. If $\mathcal{M}_n$ contains unshrinkable surfaces, then *monopoles* ($k = 1$, $d = 1$) get formed. Finally, if $\mathcal{M}_n$ contains non-contractible three spheres, then event-like defects, called *textures*, ($k = 3$, $d = 0$), arise.

Depending on whether the original symmetry is local (gauged) or global (rigid), topological defects are called local or global. The energy of local defects is strongly confined, while the gradient energy of global defects is spread out over the causal horizon at defect formation. Patterns of symmetry breaking which lead to the formation of local monopoles or local domain walls are ruled out, since they should soon dominate the energy density of the universe and close it, unless an inflationary era takes place after their formation. Local textures are insignificant in cosmology since their relative contribution to the energy density of the universe decreases rapidly with time [17].

Even if the non-trivial topology required for the existence of a defect is absent in a field theory, it may still be possible to have defect-like solutions. Defects may be *embedded* in such topologically trivial field theories [18]. While stability of topological defects is guaranteed by topology, embedded defects are in general unstable under small perturbations.

10.3.1 Cosmic Strings

Cosmic strings [19] are analogous to flux tubes in type II superconductors, or to vortex filaments in superfluid helium. Topologically stable strings do not have ends; they either form closed loops or they extend to infinity. The linear mass density ofstrings, μ, which in the simplest models also determines the string tension, specifies the energy scale, η, of the symmetry breaking, $\mu \sim \eta^2$. The strength of gravitational interactions of strings is expressed in terms of the dimensionless parameter $G\mu \sim \eta^2/m_{\rm Pl}^2$ (with G the gravitational Newton's constant). For grand unification strings, the energy per unit length is $\mu \sim 10^{22}\,{\rm kg/m}$, or equivalently, $G\mu \sim \mathcal{O}(10^{-6})$.

At formation, cosmic strings form a tangled network, made of Brownian infinitely long strings and a distribution of closed loops. Curved segments of strings moving under their tension reach almost relativistic speeds. When two string segments intersect, they exchange partners (*intercommute*) with a probability equal to 1. String–string and self-string intersections lead to daughter infinitely long strings and closed loops, as they can be seen in Fig. 10.3. Clearly, string intercommutations produce discontinuities on the new string segments at the intersection point. These discontinuities (*kinks*) are composed of right- and left-moving pieces travelling along the string at the speed of light.

Early analytic work [20] identified the key property of *scaling*, where at least the basic properties of the string network can be characterised by a single length scale, roughly the persistence length (defined as the distance beyond which the directions along the string are uncorrelated), $\xi(t)$, and the typical separation between string segments, $d(t)$, both grow with the cosmic horizon. This result was supported by subsequent numerical work [21]. However, further investigation revealed dynamical processes, including loop production, at scales much smaller than ξ [22].

Recent numerical simulations of cosmic string evolution in a expanding universe found evidence [23] of a scaling regime for the cosmic string loops in the radiation and matter-dominated eras down to the hundredth of the horizon time. It is important to note that the scaling was found without considering any gravitational back reaction effect; it was just the result of string intercommuting mechanism. As it was reported in [23], the scaling regime of string loops appears after a transient relaxation era, driven by a transient over-production of string loops with lengths close to the initial correlation length of the string network. Calculating the amount of energy-momentum tensor

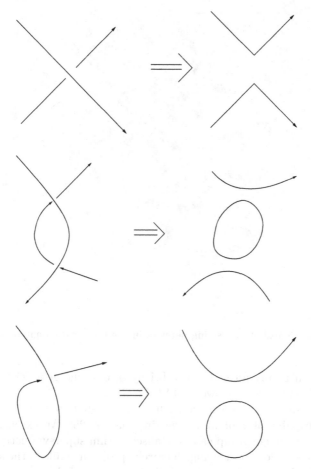

Fig. 10.3. At the *top*, string–string interactions at one point leading to the formation of two new long strings via exchange of partners. In the *middle*, string–string interactions at two points, leading to two new long strings and a loop. At the *bottom*, self–self interactions leading to the formation of a new long string and a loop [19]

lost from the string network, it was found [23] that a few percent of the total string energy density disappear in the very brief process of formation of numerically unresolved string loops during the very first timesteps of the string evolution. Subsequently, other studies supported these findings [24]. A snapshot of the evolution of a cosmic string network during the matter-dominated era is shown in Fig. 10.4.

10.3.2 Genericity of Cosmic String Formation Within SUSY GUTs

To investigate the cosmological consequences of cosmic strings formed at the end of hybrid inflation, one should first address the question of whether such

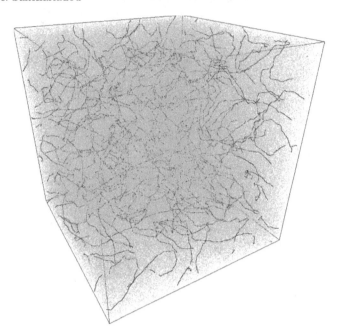

Fig. 10.4. Snapshot of a string network in the matter-dominated era [23]

objects are generically formed. I will briefly discuss the genericity of cosmic string formation in the framework of SUSY GUTs.

Even though the standard model has been tested to a very high precision, it is incapable of explaining neutrino masses [25]. An extension of the Standard Model gauge group can be realised within supersymmetry (SUSY). SUSY offers a solution to the gauge hierarchy problem, while in the supersymmetric standard model the gauge coupling constants of the strong, weak and electromagnetic interactions meet at a single point $M_{\mathrm{GUT}} \simeq (2\text{–}3) \times 10^{16}$ GeV. In addition, SUSY GUTs can provide the scalar field which could drive inflation, explain the matter–anti-matter asymmetry of the universe and propose a candidate, the lightest superparticle, for cold dark matter.

Within SUSY GUTs there is a large number of SSB patterns leading from a large gauge group G to the SM gauge group $G_{\mathrm{SM}} \equiv \mathrm{SU}(3)_{\mathrm{C}} \times \mathrm{SU}(2)_{\mathrm{L}} \times \mathrm{U}(1)_{\mathrm{Y}}$. The study of the homotopy group of the false vacuum for each SSB scheme will determine whether there is defect formation and it will identify the type of the defect formed. Clearly, if there is formation of domain walls or monopoles, one will have to place an era of supersymmetric hybrid inflation to dilute them. To consider a SSB scheme as a successful one, it should be able to explain the matter/anti-matter asymmetry of the universe and to account for the proton lifetime measurements [25]. In what follows, I consider a mechanism of baryogenesis via leptogenesis, which can be thermal or non-thermal one. In the case of non-thermal leptogenesis, $\mathrm{U}(1)_{\mathrm{B-L}}$ (B and L, are the baryon and lepton numbers, respectively) is a subgroup of the GUT gauge group, G_{GUT}, and B–L is broken at the end or after inflation. In the case of thermal leptogenesis, B–L is broken independently of inflation. If leptogenesis is thermal and B–L is broken before the inflationary era, then one should check whether the temperature at which B–L is broken, which will define the mass of the right-handed neutrinos, is smaller than the reheating temperature which should be lower than the limit imposed by the gravitino. To ensure the stability of proton, the discrete symmetry Z_2, which is contained in $U(1)_{B-L}$, must be kept unbroken down to low energies. This implies that the successful SSB schemes should end at $G_{SM} \times Z_2$. I will then examine how often cosmic strings have survived after the inflationary era, within all acceptable SSB patterns.

To accomplish this task one has to choose the large gauge group G_{GUT}. In [7] this study has been done explicitly for a large number of simple Lie groups. Since I consider GUTs based on simple gauge groups, the type of supersymmetric hybrid inflation will be of the F-type. The minimum rank of G_{GUT} has to be at least equal to 4, to contain the G_{SM} as a subgroup. Then one has to study the possible embeddings of G_{SM} in G_{GUT} to be in agreement with the standard model phenomenology and especially with the hypercharges of the known particles. Moreover, the group must include a complex representation, needed to describe the Standard Model fermions, and it must be anomaly free. Since, in principle, $SU(n)$ may not be anomaly free, I assume that the $SU(n)$ groups which I use have indeed a fermionic representation that certifies that the model is anomaly free. I set as the upper bound on the rank r of the group, $r \leq 8$. Clearly, the choice of the maximum rank is in principle arbitrary. This choice could, in a sense, be motivated by the Horava–Witten [26] model, based on $E_8 \times E_8$. Thus, the large gauge group G_{GUT} could be one of the following: $SO(10)$, E_6, $SO(14)$, $SU(8)$, $SU(9)$; flipped $SU(5)$ and $[SU(3)]^3$ are included within this list as subgroups of $SO(10)$ and E_6, respectively.

A detailed study of all the SSB schemes which bring us from G_{GUT} down to the standard model gauge group G_{SM}, by one or more intermediate steps, shows that cosmic strings are generically formed at the end of hybrid inflation. If the large gauge group G_{GUT} is $SO(10)$ then cosmic strings formation is unavoidable [7, 27]. For E_6 it depends whether one considers thermal or non-thermal leptogenesis. More precisely, under the assumption of non-thermal leptogenesis cosmic strings formation is unavoidable. If I consider thermal leptogenesis then cosmic strings formation at the end of hybrid inflation arises in 98% of the acceptable SSB schemes [28]. If the requirement of having Z_2 unbroken down to low energies is relaxed and thermal leptogenesis is considered as being the mechanism for baryogenesis, then cosmic strings formation accompanies hybrid inflation in 80% of the SSB schemes [28].

For an illustration I give below the list of the SSB schemes of E_6 down to the $G_{SM} \times Z_2$ via $SO(10) \times U(1)$ (the reader is referred to [7] for a full analysis). Every $\xrightarrow{n}$ represent an SSB during which there is formation of topological defects, whose type is denoted by n: 1 for monopoles, 2 for topological cosmic strings, $2'$ for embedded strings, 3 for domain walls. Note that for, e.g. $3_C\ 2_L\ 2_R\ 1_{B-L}$ stands for $SU(3)_C \times SU(2)_L \times SU(2)_R \times U(1)_{B-L}$.

$$
E_6 \xrightarrow{1} SO(10)\ 1_{V'}
\begin{cases}
\xrightarrow{2} SO(10) & \longrightarrow & (10.23) \\
\xrightarrow{1} 5\ 1_V\ 1_{V'} & \longrightarrow & (10.24) \\
\xrightarrow{1} 5_F\ 1_V\ 1_{V'} & \longrightarrow & (10.25) \\
\xrightarrow{1} 5_E\ 1_V\ 1_{V'} & \xrightarrow{2',2} G_{SM}\ Z_2 \\
\xrightarrow{2} 5\ 1_{V'}\ Z_2 & \longrightarrow & (10.24a) \\
\xrightarrow{1,2} 5\ 1_V & \longrightarrow & (10.23a) \\
\xrightarrow{1} 5_F\ 1_V & \xrightarrow{2',2} G_{SM}\ Z_2 \\
\xrightarrow{1} G_{SM}\ 1_V & \xrightarrow{2} G_{SM}\ Z_2 \\
\xrightarrow{1,2} G_{SM}\ 1_{V'}\ Z_2 & \xrightarrow{2} G_{SM}\ Z_2 \\
\xrightarrow{1,2} 4_C\ 2_L\ 2_R\ 1_{V'} & \longrightarrow & (10.26) \\
\xrightarrow{1} 4_C\ 2_L\ 2_R & \longrightarrow & (10.27) \\
\xrightarrow{1} 3_C\ 2_L\ 2_R\ 1_{B-L}\ 1_{V'} & \longrightarrow & (10.26c) \\
\xrightarrow{1} 3_C\ 2_L\ 1_R\ 1_{B-L}\ 1_{V'} & \longrightarrow & (10.26b)
\end{cases}
\tag{10.22}
$$

where

$$
SO(10)
\begin{cases}
\xrightarrow{1} & 5\ 1_V & \xrightarrow{1} 3_C\ 2_L\ 1_Z\ 1_V \xrightarrow{2} G_{SM}\ Z_2 \\
\xrightarrow{1} & 4_C\ 2_L\ 2_R & \longrightarrow & (10.27) \\
\xrightarrow{1,2} & 4_C\ 2_L\ 2_R\ Z_2^C & \longrightarrow & (10.28) \\
\xrightarrow{1,2} & 4_C\ 2_L\ 1_R\ Z_2^C & \longrightarrow & (10.28b) \\
\xrightarrow{1} & 4_C\ 2_L\ 1_R & \longrightarrow & (10.27b) \\
\xrightarrow{1,2} & 3_C\ 2_L\ 2_R\ 1_{B-L}\ Z_2^C & \longrightarrow & (10.28a) \\
\xrightarrow{1} & 3_C\ 2_L\ 2_R\ 1_{B-L} & \longrightarrow & (10.27a) \\
\xrightarrow{1} & 3_C\ 2_L\ 1_R\ 1_{B-L} & \xrightarrow{2} & G_{SM}\ Z_2
\end{cases}
\tag{10.23}
$$

$$
5\ 1_V\ 1_{V'}
\begin{cases}
\xrightarrow{2} 5\ 1_{V'}\ Z_2 &
\begin{cases}
\xrightarrow{1} G_{SM}\ 1_{V'}\ Z_2 \xrightarrow{2} G_{SM}\ Z_2 \\
\xrightarrow{2} G_{SM}\ 1_V \xrightarrow{2} G_{SM}\ Z_2 \\
\xrightarrow{2} G_{SM}\ 1_{V'}\ Z_2 \xrightarrow{2} G_{SM}\ Z_2
\end{cases} \\
\xrightarrow{1} G_{SM}\ 1_V\ 1_{V'} & \\
\xrightarrow{2} 5\ 1_V & \longrightarrow \quad (10.23a)
\end{cases}
\tag{10.24}
$$

$$
5_F\ 1_V\ 1_{V'}
\begin{cases}
\xrightarrow{2} 5_F\ 1_V \xrightarrow{2',2} G_{SM}\ Z_2 \\
\xrightarrow{2',2} G_{SM}\ Z_2
\end{cases}
\tag{10.25}
$$

$$4_C\,2_L\,2_R\,1_{V'}\left\{
\begin{array}{l}
\xrightarrow{2} 4_C\,2_L\,2_R \qquad\qquad \longrightarrow (10.27)\\[4pt]
\xrightarrow{1} 3_C\,2_L\,1_R\,1_{B-L}\,1_{V'}\left\{
\begin{array}{ll}
\xrightarrow{2} 3_C\,2_L\,1_R\,1_{B-L} \xrightarrow{2} G_{SM}\,Z_2\\
\xrightarrow{2',2} G_{SM}\,1_{V'}\,Z_2 & \xrightarrow{2} G_{SM}\,Z_2\\
\xrightarrow{2',2} G_{SM}\,Z_2
\end{array}\right.\\[14pt]
\xrightarrow{1} 3_C\,2_L\,2_R\,1_{B-L}\,1_{V'}\left\{
\begin{array}{ll}
\xrightarrow{1} 3_C\,2_L\,1_R\,1_{B-L}\,1_{V'} \longrightarrow & (10.26b)\\
\xrightarrow{2} 3_C\,2_L\,2_R\,1_{B-L} \longrightarrow & (10.27a)\\
\xrightarrow{2',2} G_{SM}\,1_{V'}\,Z_2 \quad \xrightarrow{2} & G_{SM}\,Z_2\\
\xrightarrow{1,2} 3_C\,2_L\,1_R\,1_{B-L} \quad \xrightarrow{2} & G_{SM}\,Z_2\\
\xrightarrow{2',2} G_{SM}\,Z_2
\end{array}\right.\\[20pt]
\xrightarrow{1} 4_C\,2_L\,1_R\,1_{V'}\left\{
\begin{array}{ll}
\xrightarrow{2} 4_C\,2_L\,1_R \longrightarrow & (10.27b)\\
\xrightarrow{1} 3_C\,2_L\,1_R\,1_{B-L}\,1_{V'} \longrightarrow & (10.26b)\\
\xrightarrow{2',2} G_{SM}\,1_{V'}\,Z_2 \quad \xrightarrow{2} & G_{SM}\,Z_2\\
\xrightarrow{1,2} 3_C\,2_L\,1_R\,1_{B-L} \quad \xrightarrow{2} & G_{SM}\,Z_2\\
\xrightarrow{2} G_{SM}\,Z_2
\end{array}\right.\\[20pt]
\xrightarrow{1,2} G_{SM}\,1_{V'}\,Z_2 \qquad \xrightarrow{2} \quad G_{SM}\,Z_2\\
\xrightarrow{1,2} 3_C\,2_L\,2_R\,1_{B-L} \qquad \longrightarrow (10.27a)\\
\xrightarrow{1} 3_C\,2_L\,1_R\,1_{B-L} \qquad \xrightarrow{2} \quad G_{SM}\,Z_2
\end{array}\right.$$

$$(10.26)$$

with

$$4_C\,2_L\,2_R\left\{
\begin{array}{l}
\xrightarrow{1} 3_C\,2_L\,2_R\,1_{B-L}\left\{
\begin{array}{l}
\xrightarrow{1} 3_C\,2_L\,1_R\,1_{B-L} \xrightarrow{2} G_{SM}\,Z_2\\
\xrightarrow{2',2} G_{SM}\,Z_2
\end{array}\right.\\[14pt]
\xrightarrow{1} 4_C\,2_L\,1_R\left\{
\begin{array}{l}
\xrightarrow{1} 3_C\,2_L\,1_R\,1_{B-L} \xrightarrow{2} G_{SM}\,Z_2\\
\xrightarrow{2',2} G_{SM}\,Z_2
\end{array}\right.\\[14pt]
\xrightarrow{1} 3_C\,2_L\,1_R\,1_{B-L} \qquad \xrightarrow{2} \quad G_{SM}\,Z_2
\end{array}\right.$$

$$(10.27)$$

$$4_C\,2_L\,2_R\,Z_2^C\left\{
\begin{array}{l}
\xrightarrow{1} 3_C\,2_L\,2_R\,1_{B-L}\,Z_2^C\left\{
\begin{array}{ll}
\xrightarrow{3} 3_C\,2_L\,2_R\,1_{B-L} \longrightarrow & (10.27a)\\
\xrightarrow{1,3} 3_C\,2_L\,1_R\,1_{B-L} \xrightarrow{2} G_{SM}\,Z_2
\end{array}\right.\\[14pt]
\xrightarrow{1} 4_C\,2_L\,1_R\,Z_2^C\left\{
\begin{array}{ll}
\xrightarrow{3} 4_C\,2_L\,1_R \longrightarrow & (10.27b)\\
\xrightarrow{1,3} 3_C\,2_L\,1_R\,1_{B-L} \xrightarrow{2} G_{SM}\,Z_2
\end{array}\right.\\[14pt]
\xrightarrow{3} 4_C\,2_L\,2_R \qquad\qquad \longrightarrow \quad (10.27)\\
\xrightarrow{1} 4_C\,2_L\,1_R \qquad\qquad \longrightarrow \quad (10.27b)\\
\xrightarrow{1,3} 3_C\,2_L\,2_R\,1_{B-L} \qquad \longrightarrow \quad (10.27a)\\
\xrightarrow{1,3} 3_C\,2_L\,1_R\,1_{B-L} \qquad \xrightarrow{2} \quad G_{SM}\,Z_2
\end{array}\right.$$

$$(10.28)$$

In addition, there are more direct schemes; they are listed below:

$$
E_6 \left\{
\begin{array}{lll}
\xrightarrow{1} 5\ 1_V\ 1_{V'} & \longrightarrow (10.24) \\
\xrightarrow{1} 5_F\ 1_V\ 1_{V'} & \longrightarrow (10.25) \\
\xrightarrow{1} 5_E\ 1_V\ 1_{V'} & \xrightarrow{2',2} G_{SM}\ Z_2 \\
\xrightarrow{1} 5\ 1_V & \longrightarrow (10.23a) \\
\xrightarrow{1} 5\ 1_{V'} & \longrightarrow (10.24a) \\
\xrightarrow{1} 5_F\ 1_V & \xrightarrow{2',2} G_{SM}\ Z_2 \\
\xrightarrow{1} 4_C\ 2_L\ 2_R\ 1_{V'} & \longrightarrow (10.26) \\
\xrightarrow{1} 4_C\ 2_L\ 2_R & \longrightarrow (10.27) \\
\xrightarrow{1} 4_C\ 2_L\ 1_R & \longrightarrow (10.27b) \\
\xrightarrow{1} 4_C\ 2_L\ 1_R\ 1_{V'} & \longrightarrow (10.26d) \\
\xrightarrow{1} 3_C\ 2_L\ 2_R\ 1_{B-L}\ 1_{V'} & \longrightarrow (10.26c) \\
\xrightarrow{1} 3_C\ 2_L\ 1_R\ 1_{B-L}\ 1_{V'} & \longrightarrow (10.26b) \\
\xrightarrow{1} 3_C\ 2_L\ 1_R\ 1_{B-L} & \xrightarrow{2} G_{SM}\ Z_2 \\
\xrightarrow{1} G_{SM}\ 1_V & \xrightarrow{2} G_{SM}\ Z_2 \\
\xrightarrow{1,2} G_{SM}\ 1_{V'}\ Z_2 & \xrightarrow{2} G_{SM}\ Z_2
\end{array}
\right.
\tag{10.29}
$$

The SSB schemes of $SU(6)$ and $SU(7)$ down to the G_{SM} which could accommodate an inflationary era with no defect (of any kind) at later times are inconsistent with proton lifetime measurements and minimal $SU(6)$ and $SU(7)$ do not predict neutrino masses [7], implying that these models are incompatible with high-energy physics phenomenology. Higher rank groups, namely $SO(14)$, $SU(8)$ and $SU(9)$, should in general lead to cosmic string formation at the end of hybrid inflation. In all these schemes, cosmic string formation is sometimes accompanied by the formation of embedded strings. The strings which form at the end of hybrid inflation have a mass which is proportional to the inflationary scale.

10.4 Braneworld Cosmology

One of our dreams in theoretical physics is to be able to unify all fundamental interactions into a unique theory. String theory offers one such attempt to unify gravity with the other interactions, in a self-consistent quantum theory. String theory is based on the proposal that one-dimensional extended objects (strings) are the fundamental constituents of matter. In the mid-1990s it was realised that higher dimensional extended membranes (*p-branes*, with $p > 1$) should also play a crucial role in string theory. In particular, branes offer the possibility of relating apparently different string theories. Of particular importance among p-branes are the Dp-branes on which open strings can end; they can describe matter fields living on the brane. Closed strings (e.g. graviton) live on the higher dimensional bulk; their excitations describe perturbations on the bulk geometry. Classically, matter and radiation fields are localised on the brane, with gravity propagating in the bulk.

Some of the extra dimensions could be far larger than what had been previously thought. If the extra dimensions were testable only via gravity then they might be relatively large, leading to a possible explanation for the weakness of gravity as compared to the other fundamental interactions. It has been proposed that the gravitational field of an object could leak out into the large but hidden extra dimensions, leading to a weaker gravity as perceived from an observer living in a four-dimensional universe. More precisely, the effective value of Newton's constant in a four-dimensional universe, $G_{(4)}$, can be written as $G_{(4)} \equiv G_{(D)}/R^{D-4}$, where D denotes the total dimensionality of spacetime and R stands for the radius of compactification (assumed, without loss of generality, to be the same in all extra dimensions). The absence of any observed deviation from the familiar Newton's law (in a four-dimensional spacetime) imposes an upper limit on the compactification radius. More precisely, the present experimental constraints yield $R \lesssim 0.2\,\mathrm{mm}$.

10.4.1 Inflation Within Braneworld Cosmologies

In the context of braneworld cosmology, brane inflation occurs in a similar way as hybrid inflation within supergravity, leading to string-like objects. In string theories, D-brane $\bar{\mathrm{D}}$-anti-brane annihilation leads generically to the production of lower dimensional D-branes, with D3- and D1-branes (D-strings) being predominant [29].

To sketch brane inflation (for example see [30]), consider a $\mathrm{D}p$–$\bar{\mathrm{D}}p$ system in the context of IIB string theory. Six of the spatial dimensions are compactified on a torus; all branes move relatively to each other in some directions. A simple and well-motivated inflationary model is brane inflation where the inflaton is simply the position of a $\mathrm{D}p$-brane moving in the bulk. As two branes approach, the open string modes between the branes develop a tachyon, indicating an instability. The relative $\mathrm{D}p$–$\bar{\mathrm{D}}p$-brane position is the inflation field and the inflaton potential comes from their tensions and interactions. Brane inflation ends by a phase transition mediated by open string tachyons. The annihilation of the branes releases the brane tension energy that heats up the universe so that the hot big bang epoch can take place. Since the tachyonic vacuum has a non-trivial π_1 homotopy group, there exist stable tachyonic string solutions with $(p-2)$ co-dimensions. These daughter branes have all dimensions compact; a four-dimensional observer perceives them as one-dimensional objects, the D-strings. Zero-dimensional defects (monopoles) and two-dimensional ones (domain walls), which are cosmologically undesirable, are not produced during brane intersections.

10.4.2 Cosmic Superstrings

The first to consider cosmic superstrings as playing the role of cosmic strings was Witten [31]. However, since for fundamental strings the linear mass density is proportional to (string energy scale)^2, it was realised that for a string energy scale of the order of the Planck mass, $G\mu$ becomes of the order of 1, and therefore this proposal was ruled out since observational data require $G\mu \lesssim 10^{-7}$. More recently, in the framework of braneworld scenarios the large compact dimensions and the large warp factors allow the string energy scale to be much smaller than the Planck scale. Thus, in models with large extra dimensions, cosmic superstring tensions could have values in the range between $10^{-13} < G\mu < 10^{-6}$, depending on the model. These cosmic superstrings are stable, or at least their lifetime is comparable to the age of the universe, so they can survive to form a cosmic superstring network.

Type IIB string theory, after compactification to 3+1 dimensions, has a spectrum of one-dimensional objects, the fundamental (F) strings, carrying charge under the Neveu Schwartz–Neveu Schwartz two-form potential, and the Dirichlet (D) strings carrying charge under the Ramond–Ramond two-form potential. Both these strings are individually $\frac{1}{2}$-BPS (Bogomol'nyi–Prasad–Sommerfield) objects, with, however, each type breaking a different half of the supersymmetry. F- and D-strings that survive the cosmological evolution become cosmic superstrings with interesting cosmological implications [32]. Thus, string theory offers two distinct candidates for playing the role of cosmic strings.

IIB string theory allows the existence of bound (p, q) states of p F-strings and q D-strings, where p and q are coprime. A (p, q) state is still a $\frac{1}{2}$-BPS object with tension

$$\mu_{(p,q)} = \mu_{\mathrm{F}} \sqrt{p^2 + q^2/g_{\mathrm{s}}^2} \ , \tag{10.30}$$

where μ_{F} denotes the effective F-string tension after compactification and g_{s} stands for the string coupling.

Cosmic superstrings share a number of properties with cosmic strings, but there are also differences which may lead to distinctive observational signatures. In general, string intersections lead to intercommutation and loop production. For cosmic strings the probability of intercommutation $\mathcal{P}$ is equal to 1, whereas this is not the case for F- and D-strings. Clearly, D-strings can miss each other in the compact dimension, leading to a smaller $\mathcal{P}$, while for F-strings the scattering has to be calculated quantum mechanically since these are quantum mechanical objects. The collisions between all possible pairs of superstrings have been studied in string perturbation theory [33]. For F-strings, the reconnection probability is of the order of g_{s}^2, where g_{s} stands for the string coupling. For F–F string collisions, it was found [33] that the reconnection probability $\mathcal{P}$ is $10^{-3} \lesssim \mathcal{P} \lesssim 1$. For D–D string collisions, one has $10^{-1} \lesssim \mathcal{P} \lesssim 1$. Finally, for F–D string collisions, the reconnection probability can take any value between 0 and 1. These results have been confirmed [34] by a quantum calculation of the reconnection probability for colliding D-strings. Similarly, the string self-intersection probability is reduced.

In contrast to the networks formed from Abelian strings, which consist of loops and long strings, (p, q) networks can also contain links which start and end at a three-point vertex. More precisely, when F- and D-strings meet they can form a three-string junction, with a composite FD-string. Such links could potentially lead to a frozen network, which could dominate the matter content of the universe.

Modelling the evolution of a (p,q) network is a challenging task, in particular due to the existence of junctions. Nevertheless, various attempts have been undertaken and they all conclude [35] that the network will reach a *scaling* regime, in which the length scales increase in proportion to time.

Cosmic superstrings interact with the standard model particles via gravity, implying that their detection involves gravitational interactions. Since the particular brane inflationary scenario remains unknown, the tensions of superstrings are only loosely constrained.

10.5 Observational Consequences

10.5.1 CMB Temperature Anisotropies

The CMB temperature anisotropies offer a powerful test for theoretical models aiming at describing the early universe. The characteristics of the CMB multipole moments can be used to discriminate among theoretical models and to constrain the parameters space.

The spherical harmonic expansion of the CMB temperature anisotropies, as a function of angular position, is given by

$$\frac{\delta T}{T}(\mathbf{n}) = \sum_{\ell m} a_{\ell m} \mathcal{W}_\ell Y_{\ell m}(\mathbf{n}) \quad \text{with} \quad a_{\ell m} = \int d\Omega_\mathbf{n} \frac{\delta T}{T}(\mathbf{n}) Y_{\ell m}^*(\mathbf{n}) \; ; \quad (10.31)$$

$\mathcal{W}_\ell$ stands for the ℓ-dependent window function of the particular experiment. The angular power spectrum of CMB temperature anisotropies is expressed in terms of the dimensionless coefficients C_ℓ, which appear in the expansion of the angular correlation function in terms of the Legendre polynomials P_ℓ:

$$\left\langle 0 \left| \frac{\delta T}{T}(\mathbf{n}) \frac{\delta T}{T}(\mathbf{n}') \right| 0 \right\rangle \bigg|_{(\mathbf{n}\cdot\mathbf{n}'=\cos\vartheta)} = \frac{1}{4\pi} \sum_\ell (2\ell + 1) C_\ell P_\ell(\cos\vartheta) \mathcal{W}_\ell^2 \; . \quad (10.32)$$

It compares points in the sky separated by an angle ϑ. In (10.31) the brackets denote spatial average, or expectation values if perturbations are quantised. Equation (10.32) holds only if the initial state for cosmological perturbations of quantum mechanical origin is the vacuum [36]. The value of C_ℓ is determined by fluctuations on angular scales of the order of π/ℓ. The angular power spectrum of anisotropies observed today is usually given by the power per logarithmic interval in ℓ, plotting $\ell(\ell+1)C_\ell$ versus ℓ.

On large angular scales, the main contribution to the CMB temperature anisotropies is given by the Sachs–Wolfe effect. Thus,

$$\frac{\delta T}{T}(\mathbf{n}) \simeq \frac{1}{3} \Phi[\eta_{\text{lss}}, \mathbf{n}(\eta_0 - \eta_{\text{lss}})] \; ; \quad (10.33)$$

$\Phi(\eta, \mathbf{x})$ denotes the Bardeen potential, η_0 and η_{lss} stand for the conformal time at present and at the last scattering surface, respectively.

Studies of the characteristics of the CMB spectrum (amplitude and position of acoustic peaks), in the framework of topological defect models, have been performed even before receiving any data. Let me discuss briefly the differences such models have, as compared to the adiabatic perturbations induced from the amplification of the quantum fluctuations of the inflation field at the end of inflation, and the difficulties one faces to extract the predictions.

For models with topological defects, perturbations are generated by *seeds* (sources), defined as any non-uniformly distributed form of energy, which contributes only a small fraction to the total energy density of the universe and which interacts with the cosmic fluid only gravitationally. Such models lead to isocurvature density perturbations, in the sense that the total density perturbation vanishes, but those of the individual particle species do not. Moreover, in models with topological defects, fluctuations are generated continuously and evolve according to inhomogeneous linear perturbation equations.

The energy-momentum tensor of defects is determined by their evolution which, in general, is a non-linear process. These perturbations are called *active* and *incoherent*, active since new fluid perturbations are induced continuously due to the presence of the defects and incoherent since the randomness of the non-linear seed evolution which sources the perturbations can destroy the coherence of fluctuations in the cosmic fluid. The highly non-linear structure of the topological defect dynamics makes the study of the evolution of these causal (there are no correlations on super-horizon scales) and incoherent initial perturbations much more complicated.

Within linear cosmological perturbation theory, structure formation induced by seeds is determined by the solution of the inhomogeneous equation

$$\mathcal{D}X(\mathbf{k}, t) = \mathcal{S}(\mathbf{k}, t) \ , \qquad (10.34)$$

where X is a vector containing all the background perturbation variables for a given mode specified by the wave-vector $\mathbf{k}$, like the a_{lm}s of the CMB anisotropies, the dark matter density fluctuation, the peculiar velocity potential, etc., $\mathcal{D}$ is a linear time-dependent ordinary differential operator, and the source term $\mathcal{S}$ is given by linear combinations of the energy-momentum tensor of the seed (the type of topological defects we are considering). The generic solution of this equation is given in terms of Green's function and has the following form [37]

$$X_i(\mathbf{k}, t_0) = \int_{t_{\mathrm{in}}}^{t_0} \mathcal{G}_{il}(\mathbf{k}, t_0, t)\mathcal{S}_l(\mathbf{k}, t)\mathrm{d}t \ . \qquad (10.35)$$

At the end, we need to determine expectation values, which are given by

$$\langle X_i(\mathbf{k}, t_0)X_j(\mathbf{k}, t_0)^* \rangle = \int_{t_{\mathrm{in}}}^{t_0} \int_{\eta_{\mathrm{in}}}^{\eta_0} \mathcal{G}_{il}(t_0, t)\mathcal{G}_{jm}^*(t_0, t')\langle \mathcal{S}_l(t)\mathcal{S}_m^*(t') \rangle \mathrm{d}t \, \mathrm{d}t'.$$
$$(10.36)$$

Thus, the only information we need from topological defects simulations in order to determine cosmic microwave background and large-scale structure power spectra is the *unequal time two-point correlators* [38], $\langle \mathcal{S}_l(t)\mathcal{S}_m^*(t')\rangle$, of the seed energy-momentum tensor. This problem can, in general, be solved by an eigenvector expansion method [39].

On large angular scales ($\ell \leq 50$), defect models lead to the same prediction as inflation, namely they both predict an approximately scale-invariant (Harrison–Zel'dovich) spectrum of perturbations. Their only difference concerns the statistics of the induced fluctuations. Inflation predicts generically Gaussian fluctuations, whereas in the case of topological defect models, even if initially the defect energy-momentum tensor would be Gaussian, non-Gaussianities will be induced from the non-linear defect evolution. Thus, in defect scenarios, the induced fluctuations are non-Gaussian, at least at sufficiently high angular resolution. This is an interesting fingerprint, even though difficult to test through the data.

On intermediate and small angular scales, however, the predictions of models with seeds are quite different than those of inflation, due to the different nature of the induced perturbations. In topological defect models, defect fluctuations are constantly generated by the seed evolution. The non-linear defect evolution and the fact that the random initial conditions of the source term in the perturbation equations of a given scale leak into other scales destroy perfect coherence. The incoherent aspect of active perturbations does not influence the position of the acoustic peaks, but it does affect the structure of secondary oscillations, namely secondary oscillations may get washed out. Thus, in topological defect models, incoherent fluctuations lead to a single bump at smaller angular scales (larger ℓ) than those predicted within any inflationary scenario. This incoherent feature is shared in common by local and global defects.

Let me briefly summarise the results: global $\mathcal{O}(4)$ textures lead to a position of the first acoustic peak at $\ell \simeq 350$ with an amplitude $\sim$1.5 times higher than the Sachs–Wolfe plateau [40]. Global $\mathcal{O}(N)$ textures in the large N limit lead to a quite flat spectrum, with a slow decay after $\ell \sim 100$ [41]. Similar are the predictions of other global $\mathcal{O}(N)$ defects [42]. (For a general study of the CMB anisotropies from scaling seed perturbations the reader is referred to [43]). Local cosmic strings lead to a power spectrum with a roughly constant slope at low multipoles, rising up to a single peak, with subsequent decay at small scales [44].

At this point, I would like to bring to the attention of the reader that the B-mode of the polarisation spectrum may be a smoking gun for the cosmic strings [44], since inflation gives just a weak contribution. The reason being that scalar modes may contribute to the B-mode only through the gravitational lensing of the E-mode. Thus, the large vector contribution from cosmic strings may lead in the future to the detection of strings.

The position and amplitude of the acoustic peaks, as found by the CMB measurements (see, e.g. [45]), are clearly in disagreement with the predictions of topological defect models. Thus, CMB measurements rule out pure topological defect models as the unique origin of initial density perturbations leading to the observed structure formation. However, since strings and string-like defects are generically formed, then one should consider them as a sub-dominant partner of inflation. Thus, one should study the compatibility between *mixed* perturbation models [46] and observational data.

Consider therefore a model in which a network of cosmic strings evolved independently of any pre-existing fluctuation background, generated by a standard cold dark matter with a non-zero cosmological constant (ΛCDM) inflationary phase. Restrict your attention to the angular spectrum, so that you are in the linear regime. Thus,

$$C_\ell = \alpha C_\ell^{\mathrm{I}} + (1 - \alpha)C_\ell^{\mathrm{s}} , \tag{10.37}$$

where C_ℓ^{I} and C_ℓ^{s} denote the (COBE normalised) Legendre coefficients due to adiabatic inflaton fluctuations and those stemming from the string network, respectively. The coefficient α in (10.37) is a free parameter giving the relative amplitude for the two contributions. Then one has to compare the C_ℓ, given by (10.37), with data obtained from CMB anisotropy measurements. The inflaton and string-induced uncorrelated spectra as a function of ℓ, both normalised on the COBE data, together with the weighted sum are shown in Fig. 10.5 (see [46]).

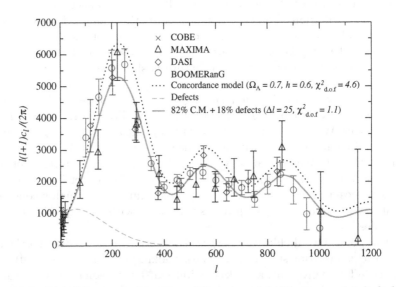

Fig. 10.5. $\ell(\ell + 1)C_\ell$ versus ℓ for three different models. The *upper dot-dashed line* represents the prediction of a ΛCDM model. The *lower dashed line* is a typical string spectrum. Combining both curves with the extra parameter α produces the *solid curve*, with a χ^2 per degree of freedom slightly above unity. The string contribution turns out to be some 18% of the total [46]

The quadrupole anisotropy due to *freezing in* of quantum fluctuations of a scalar field during inflation reads

$$\left(\frac{\delta T}{T}\right)_{\text{Q-infl}} = \left[\left(\frac{\delta T}{T}\right)_{\text{Q-scal}}^2 + \left(\frac{\delta T}{T}\right)_{\text{Q-tens}}^2\right]^{1/2} , \tag{10.38}$$

with the scalar and tensor contributions given by

$$\left(\frac{\delta T}{T}\right)_{\text{Q-scal}} = \frac{1}{4\sqrt{45}\pi} \frac{V^{3/2}(\varphi_Q)}{M_{\text{Pl}}^3 V'(\varphi_Q)} , \tag{10.39}$$

and

$$\left(\frac{\delta T}{T}\right)_{\text{Q-tens}} \sim \frac{0.77}{8\pi} \frac{V^{1/2}(\varphi_Q)}{M_{\text{Pl}}^2} , \tag{10.40}$$

respectively. Here V is the potential of the inflation field φ, with $V' \equiv dV(\varphi)/d\varphi$, M_{Pl} denotes the reduced Planck mass, $M_{\text{Pl}} = (8\pi G)^{-1/2} \simeq 2.43 \times 10^{18}$ GeV, and φ_Q is the value of the inflation field when the comoving scale corresponding to the quadrupole anisotropy became bigger than the Hubble radius.

Simulations of Goto–Nambu local strings in a Friedmann–Lemaître–Roberston–Walker spacetime lead to [47]

$$\left(\frac{\delta T}{T}\right)_{\text{cs}} \sim (9 - 10)G\mu \quad \text{with} \quad \mu = 2\pi\langle\chi\rangle^2 , \tag{10.41}$$

where $\langle\chi\rangle$ is the vacuum expectation value of the Higgs field responsible for the formation of cosmic strings.

Before discussing F- and D-term inflations, I would like to describe briefly the *curvaton* mechanism [48], according to which the primordial fluctuations could also be generated from the quantum fluctuations of a late-decaying scalar field, the *curvaton* field ψ, which does not play the role of the inflation field. During inflation the curvaton potential is very flat and the curvaton acquires quantum fluctuations, which are expressed in terms of the expansion rate during inflation, $H_{\text{infl}} = \sqrt{8\pi G/3V(\varphi)}$, through

$$\delta\psi_{\text{init}} = \frac{H_{\text{inf}}}{2\pi} . \tag{10.42}$$

They lead to entropy fluctuations at the end of inflation.

During the radiation-dominated era the curvaton decays and reheats the universe. The primordial fluctuations of the curvaton field are converted to purely adiabatic density fluctuations, thus the curvaton contribution in terms of the metric perturbation reads

$$\left(\frac{\delta T}{T}\right)_{\text{curv}} = -\frac{4}{27}\frac{\delta\psi_{\text{init}}}{\psi_{\text{init}}} . \tag{10.43}$$

If one assumes the additional contribution to the temperature anisotropies originated from the curvaton field, then

$$\left[\left(\frac{\delta T}{T}\right)_{\text{tot}}\right]^2 = \left[\left(\frac{\delta T}{T}\right)_{\text{infl}}\right]^2 + \left[\left(\frac{\delta T}{T}\right)_{\text{cs}}\right]^2 + \left[\left(\frac{\delta T}{T}\right)_{\text{curv}}\right]^2 . \tag{10.44}$$

The total quadrupole anisotropy, the l.h.s. of (10.44), is the one to be normalised to the cosmic background explore (COBE) data [49], namely $(\delta T/T)_Q^{\text{COBE}} \sim 6.3 \times 10^{-6}$.

F-Term Inflation

Considering only large angular scales one can calculate the contributions to the CMB temperature anisotropies analytically. The quadrupole anisotropy has one contribution coming from the inflation field, calculated using (10.5), and one contribution coming from the cosmic string network. Fixing the number of e-foldings to 60, the inflaton and cosmic string contributions to the CMB depend on the superpotential coupling κ, or, equivalently, on the symmetry breaking scale M associated with the inflaton mass scale, which coincides with the string mass scale.

The total quadrupole anisotropy, to be normalised to the COBE data, is found to be [10]

$$\left(\frac{\delta T}{T}\right)_{\text{Q-tot}} \sim \left\{ y_Q^{-4}\left(\frac{\kappa^2 \mathcal{N} N_Q}{32\pi^2}\right)^2 \left[\frac{64 N_Q}{45 \mathcal{N}} x_Q^{-2} y_Q^{-2} f^{-2}(x_Q^2) \right. \right.$$
$$\left. \left. + \left(\frac{0.77\kappa}{\pi}\right)^2 + 324 \right] \right\}^{1/2} . \tag{10.45}$$

In (10.45),

$$x_Q = \frac{|S_Q|}{M} \quad ; \quad y_Q^2 = \int_1^{x_Q^2} \frac{dz}{z f(z)} \tag{10.46}$$

and

$$N_Q = \frac{4\pi^2}{\kappa^2 \mathcal{N}} \frac{M^2}{M_{\text{Pl}}^2} y_Q^2 , \tag{10.47}$$

with

$$f(z) = (z+1)\ln(1+z^{-1}) + (z-1)\ln(1-z^{-1}) . \tag{10.48}$$

As noted earlier, the index Q denotes the scale responsible for the quadrupole anisotropy in the CMB.

The cosmic string contribution is consistent with the CMB measurements provided [10]

$$M \lesssim 2 \times 10^{15}\,\text{GeV} \quad \Leftrightarrow \quad \kappa \lesssim 7 \times 10^{-7} . \tag{10.49}$$

Strictly speaking the above condition was found in the context of SO(10) gauge group, but the conditions imposed in the case of other gauge groups are of the same order of magnitude since M is a slowly varying function of the dimensionality $\mathcal{N}$ of the representations to which the scalar components of the chiral Higgs superfields belong [10].

The superpotential coupling κ is also subject to the gravitino constraint, which imposes an upper limit to the reheating temperature to avoid gravitino overproduction. Within the framework of SUSY GUTs and assuming the see-saw mechanism to give rise to massive neutrinos, the inflation field decays during reheating into pairs of right-handed neutrinos. This constraint on the reheating temperature can be converted into a constraint on the superpotential coupling κ. The gravitino constraint on κ reads [10] $\kappa \lesssim 8 \times 10^{-3}$, which is a weaker constraint than the one obtained from the CMB, (10.49).

The tuning of the free parameter κ can be softened if one allows for the curvaton mechanism. Clearly, within supersymmetric theories such scalar fields are expected to exist. In addition, embedded strings, if they accompany the formation of cosmic strings, may offer a natural curvaton candidate, provided the decay product of embedded strings gives rise to a scalar field before the onset of inflation. Considering the curvaton scenario, the coupling κ is only constrained by the gravitino limit. More precisely, assuming the existence of a curvaton field there is an additional contribution to the temperature anisotropies. Calculating the curvaton contribution to the temperature anisotropies, one obtains the additional contribution [10]

$$\left[\left(\frac{\delta T}{T}\right)_{\text{curv}}\right]^2 = y_Q^{-4} \left(\frac{\kappa^2 \mathcal{N} N_Q}{32\pi^2}\right)^2 \left[\left(\frac{16}{81\pi\sqrt{3}}\right) \kappa \left(\frac{M_{\text{Pl}}}{\psi_{\text{init}}}\right)\right]^2 . \tag{10.50}$$

Normalising the total $(\delta T/T)_Q$ (i.e. the inflaton, cosmic string and curvaton contributions) to the data one gets [10] the following limit on the initial value of the curvaton field

$$\psi_{\text{init}} \lesssim 5 \times 10^{13} \left(\frac{\kappa}{10^{-2}}\right) \text{GeV} \quad \text{for} \quad \kappa \in [10^{-6}, 1] . \tag{10.51}$$

Finally, I would like to point out that in the case of F-term inflation,[1] the linear mass density μ (see (10.41)) gets a correction due to deviations from the Bogomol'nyi limit, enlarging the parameter space for F-term inflation [50]. More precisely, this correction to μ turns out to be proportional to $\ln(2/\beta)^{-1}$, where β is proportional to the square of the ratio between the superpotential and the GUT couplings. Thus, under the assumption that strings contribute less than 10% to the power spectrum at $\ell = 4$, the bound on κ reduces to the one imposed by the gravitino limit.

D-Term Inflation

D-term inflation leads to cosmic string formation at the end of the inflationary era. The total quadrupole temperature anisotropy, to be normalised to the COBE data, reads [10]

---

[1] This does not hold for D-term inflation; the strings formed at the end of D-term inflation are BPS objects.

$$\left(\frac{\delta T}{T}\right)_Q^{\text{tot}} \sim \frac{\xi}{M_{\text{Pl}}^2}\left(\frac{\pi^2}{90g^2}x_Q^{-4}f^{-2}(x_Q^2)\frac{W[x_Q^2(g^2\xi)(\lambda^2 M_{\text{Pl}}^2)]}{\left\{1+W[x_Q^2(g^2\xi)(\lambda^2 M_{\text{Pl}}^2)]\right\}^2}\right.$$

$$\left.+\left(\frac{0.77g}{8\sqrt{2\pi}}\right)^2+\left(\frac{9}{4}\right)^2\right)^{1/2}, \tag{10.52}$$

where the only unknown is the Fayet–Iliopoulos term ξ, for given values of g and λ. Note that $W(x)$ is the "W-Lambert function," i.e. the inverse of the function $F(x) = xe^x$. Thus, one can get ξ numerically, and then obtain x_Q, as well as the inflaton and cosmic string contribution, as a function of the superpotential and gauge couplings g and λ. In the case of minimal SUGRA, consistency between CMB measurements and theoretical predictions impose [10, 12]

$$g \lesssim 2 \times 10^{-2} \quad \text{and} \quad \lambda \lesssim 3 \times 10^{-5} , \tag{10.53}$$

which can be expressed as a single constraint on the Fayet–Iliopoulos term ξ,

$$\sqrt{\xi} \lesssim 2 \times 10^{15} \text{ GeV} . \tag{10.54}$$

These results are shown in Fig. 10.6.

The fine-tuning on the couplings can be softened if one invokes the curvaton mechanism. Calculating the curvaton contribution to the temperature anisotropies, one obtains the additional contribution [12]

$$\left[\left(\frac{\delta T}{T}\right)_{\text{curv}}\right]^2 = \frac{1}{6}\left(\frac{2}{27\pi}\right)^2\left(\frac{g\xi}{M_{\text{Pl}}\psi_{\text{init}}}\right)^2 . \tag{10.55}$$

Thus, the gauge coupling can reach the upper bound imposed from the gravitino mechanism, provided the initial value of the curvaton field is [12]

$$\psi_{\text{init}} \lesssim 3 \times 10^{14}\left(\frac{g}{10^{-2}}\right) \text{ GeV} \quad \text{for} \quad \lambda \in [10^{-1}, 10^{-4}] ; \tag{10.56}$$

for smaller values of λ, the curvaton mechanism is not necessary. This result is explicitly shown in Fig. 10.7.

Concluding, within minimal supergravity the couplings and masses must be fine-tuned to achieve compatibility between measurements on the CMB temperature anisotropies and theoretical predictions. Note that for minimal D-term inflation, one can neglect the corrections introduced by the superconformal origin of supergravity.

The constraints on the couplings remain qualitatively valid in non-minimal supergravity theories: the superpotential W given in (10.7) and we consider a non-minimal Kähler potential. Let us first consider D-term inflation based on Kähler geometry with shift symmetry. If we identify the inflation field with the real part of S then we obtain the same constraint for the superpotential coupling as in the minimal supergravity case. However, if the inflation field

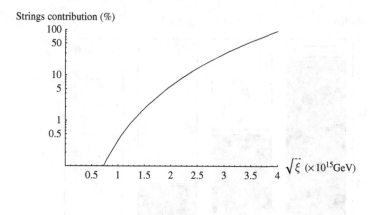

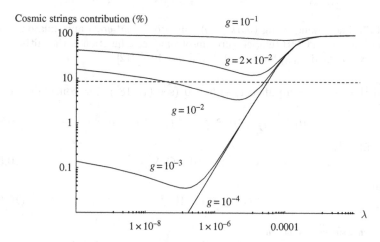

Fig. 10.6. At the *top*, the cosmic string contribution to the CMB, as a function of the mass scale $\sqrt{\xi}$ in units of 10^{15} GeV. At the *bottom*, cosmic string contribution to the CMB temperature anisotropies, as a function of the superpotential coupling λ, for different values of the gauge coupling g. The maximal contribution allowed by WMAP is represented by a *dotted line* [12]

is the imaginary part of S, then we get that the cosmic string contribution becomes dominant, in contradiction with the CMB measurements, unless the superpotential coupling is [14]

$$\lambda \lesssim 3 \times 10^{-5} \ . \tag{10.57}$$

We show this constraint in Fig. 10.8.

Considering D-term inflation based on a Kähler potential with non-renormalisable terms, the contribution of cosmic strings dominates if the superpotential coupling λ is close to unity. Setting $f_\pm(|S|^2/M_{\rm Pl}^2) = c_\pm(|S|^2/M_{\rm Pl}^2)$,

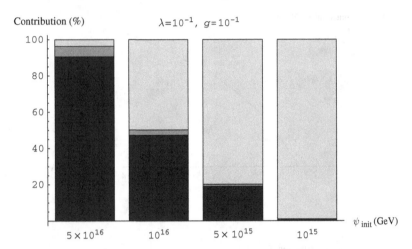

Fig. 10.7. The cosmic string (*dark grey*), curvaton (*light grey*) and inflaton (*grey*) contributions to the CMB temperature anisotropies as a function of the initial value of the curvaton field ψ_{init}, for $\lambda = 10^{-1}$ and $g = 10^{-1}$ [12]

we find that in the simplified case $b = 0$ (see (10.18)), the constraints on λ read [14]

$$(0.1 - 5) \times 10^{-8} \leq \lambda \leq (2 - 5) \times 10^{-5} \tag{10.58}$$

or equivalently

$$\sqrt{\xi} \leq 2 \times 10^{15} \text{ GeV} , \tag{10.59}$$

implying

$$G\mu \leq 8.4 \times 10^{-7} . \tag{10.60}$$

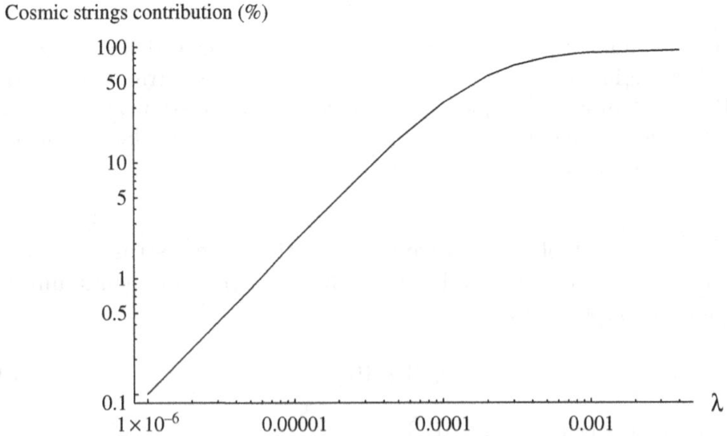

Fig. 10.8. Cosmic string contribution to the CMB temperature anisotropies as a function of λ, in the case of D-term inflation based on a Kähler geometry with shift symmetry. The inflation field is identified with the imaginary part [14]

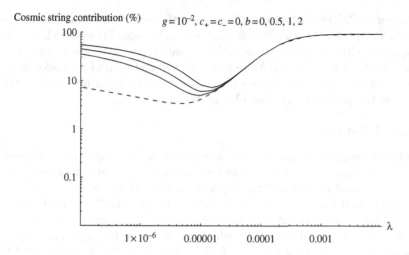

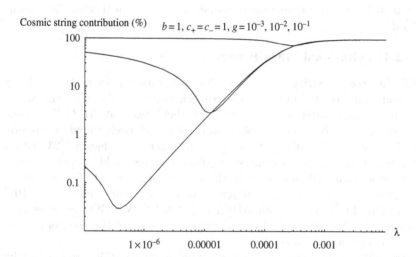

Fig. 10.9. Cosmic string contribution to the CMB temperature anisotropies as a function of λ, in the case of D-term inflation based on a Kähler potential with non-renormalisable terms [14]. In the *top panel* we set $g = 10^{-2}$ and $c_{\pm} = 0$; the simple case $b = 0$ is represented by the *dashed line*, while *plain lines* show the contributions for $b = 0.5, 1, 2$, going from the bottom to the top [14]. In the *bottom panel* we set $b = 1$ and $c_{\pm} = 1$; the *plain lines* show the contributions for $g = 10^{-3}, 10^{-2}, 10^{-1}$, going from bottom to the top [14]

In the general case, where $b \neq 0$, the constraints are shown in Fig. 10.9.

In conclusion, higher order Kähler potentials do *not* suppress cosmic string contribution, as it was incorrectly claimed in the literature. By allowing a small, but non-negligible, contribution of strings to the angular power spectrum of CMB anisotropies, we constrain the couplings of the inflationary models, or equivalently the dimensionless string tension. These models remain compatible with the most current CMB measurements, even when one calculates [51] the spectral index. More precisely, the inclusion of a sub-dominant string contribution to the large-scale power spectrum amplitude of the CMB increases the preferred value for the spectral index [51].

Brane Inflation

The CMB temperature anisotropies originate from the amplification of quantum fluctuations during inflation, as well as from the cosmic superstring network. If the scaling regime of the superstring network is the unique source of the density perturbations, the COBE data yield $G\mu \simeq 10^{-6}$. Using the latest WMAP data, the contribution from strings to the total CMB power spectrum on observed scales is at most 10%, which translates in the upper limit on the dimensionless string tension $G\mu \lesssim 1.8\,(2.7) \times 10^{-7}$ at 68 (95)% confidence [52]. Thus, the cosmic superstrings produced towards the end of inflation in the context of braneworld cosmological models is in agreement with the present CMB data.

10.5.2 Gravitational Wave Background

Oscillating cosmic string loops emit [53] gravitational waves (GW). Long strings are not straight but they have a superimposed wiggly small-scale structure due to string intercommutations, thus they also emit [54] GW. Cosmic superstrings can also generate [55] a stochastic GW background. Therefore, provided the emission of gravity waves is the efficient mechanism [23, 56] for the decay of string loops, cosmic strings/superstrings could provide a source for the stochastic GW spectrum in the low-frequency band. The stochastic GW spectrum has an almost flat region in the frequency range 10^{-8}–10^{10} Hz. Within this window, both ADVANCED LIGO/VIRGO (sensitive at a frequency $f \sim 10^2$ Hz) and LISA (sensitive at $f \sim 10^{-2}$ Hz) interferometers may have a chance of detectability.

Strongly focused beams of relatively high-frequency GW are emitted by cusps and kinks in oscillating strings/superstrings. The distinctive waveform of the emitted bursts of GW may be the most sensitive test of strings/superstrings. ADVANCED LIGO/VIRGO may detect bursts of GW for values of $G\mu$ as low as 10^{-13}, and LISA for values down to $G\mu \geq 10^{-15}$. At this point, I would like to remind the reader that there is still a number of theoretical uncertainties for the evolution of a string/superstring network [56].

Recently, limits have been imposed [57] on an isotropic gravitational wave background using pulsar timing observations, which offer a chance of studying low-frequency (in the range between 10^{-9} and 10^{-7} Hz) gravitational waves. The imposed limit on the energy density of the background per unit logarithmic frequency interval reads $\Omega_{\mathrm{GW}}^{\mathrm{cs}}(1/8\,\mathrm{yr})h^2 \leq 1.9 \times 10^{-8}$ (where h stands for the dimensionless amplitude in GW bursts).

If the source of the isotropic GW background is a cosmic string/superstring network, then it leads to an upper bound on the dimensionless tension of a cosmic string/superstring background. Under reasonable assumptions for the string network the upper bound on the string tension reads [57] $G\mu \leq 1.5 \times 10^{-8}$. This is a strongest limit than the one imposed from the CMB temperature anisotropies. Thus, F- and D-term inflations become even more fine-tuned, unless one invokes the curvaton mechanism.

This limit does not affect cosmic superstrings. However, it has been argued [57] that with the full Parkes Pulsar Timing Array (PPTA) project the upper bound will become $G\mu \leq 5 \times 10^{-12}$, which is directly relevant for cosmic superstrings. In conclusion, the full PPTA will either detect gravity waves from strings and string-like objects, or they will rule out a number of models.

10.6 Conclusions

Cosmic strings are generically formed at the end of hybrid inflation in a large number of models within supersymmetry and supergravity theories. String-like objects, which could play the role of cosmic strings, are also generically produced at the end of brane inflation, in many brane inflation models in the context of theories with large extra dimensions. These one-dimensional objects would contribute to the generation of fluctuations leading to the observed structure formation and the measured CMB temperature anisotropies. They would also source a stochastic gravity wave background.

Current measurements of the CMB spectrum, as well as of the gravitational wave background, impose severe constraints on the free parameters of the models. More precisely, the dimensionless parameter $G\mu$ must be small enough to avoid contradiction with the currently available data.

The role of strings can be suppressed by adding new terms in the superpotential [58], or by considering the curvaton mechanism [12, 59]. One can escape the *string problem* by complicating the models so that the produced strings (D-term strings formed at the end of D-term inflation, or D-strings formed at the end of brane collisions) become unstable (semilocal strings), along the lines of [11, 60]. To be more specific, it has been proposed [60] that by introducing additional matter multiplets one obtains a non-trivial global symmetry such as SU(2), leading to a simply connected vacuum manifold and the production of semilocal strings. Later on, it has been suggested [11] that if the waterfall Higgs fields are non-trivially charged under some other gauge symmetries H, such that the vacuum manifold, $[H \times \mathrm{U}(1)]/\mathrm{U}(1)$, is simply connected, then the strings are semilocal objects.

If the daily improved data require even more severe fine-tuning of the models, then I believe that one should develop and subsequently study models where the strings and string-like objects, formed at the end or after inflation, are indeed unstable.

References

1. A. H. Guth, Phys. Rev. D **23**, 347 (1981); A. D. Linde, Rep. Prog. Phys. **47**, 925 (1984).
2. T. Piran, Phys. Lett. B **181**, 238 (1986); D. S. Goldwirth, Phys. Rev. D **43**, 3204 (1991); E. Calzetta and M. Sakellariadou, Phys. Rev. D **45**, 2802 (1992); E. Calzetta and M. Sakellariadou, Phys. Rev. D **47**, 3184 (1993).
3. G. W. Gibbons and N. Turok, The measure problem in cosmology, hep-th/0609095; C. Germani, W. Nelson and M. Sakellariadou, On the onset of inflation in loop quantum cosmology, gr-qc/0701172.
4. A. D. Linde, Phys. Lett. B **129**, 177 (1983).
5. A. D. Linde, Phys. Lett. B **259**, 38 (1991).
6. L. A. Kofman and A. D. Linde, Nucl. Phys. B **282**, 555 (1987); A. D. Linde and A. Riotto, Phys. Rev. D **56**, 1841 (1997); D. H. Lyth and A. Riotto, Phys. Rep. **314**, 1 (1999).
7. R. Jeannerot, J. Rocher and M. Sakellariadou, Phys. Rev. D **68**, 103514 (2003).
8. E. Halyo, Phys. Lett. B **387**, 43 (1996); P. Binetruy and D. Dvali, Phys. Lett. B **388**, 241 (1996).
9. C. Coleman, and E. Weinberg, Phys. Rev. D **7**, 1888 (1973).
10. J. Rocher and M. Sakellariadou, JCAP **0503**, 004 (2005).
11. P. Binetruy, G. Dvali, R. Kallosh and A. Van Proeyen, Class. Quant. Grav. **21**, 3137 (2004).
12. J. Rocher and M. Sakellariadou, Phys. Rev. Lett. **94**, 011303 (2005).
13. J. P. Hsu and R. Kallosh, JHEP **0404**, 042 (2004).
14. J. Rocher and M. Sakellariadou, JCAP **0611**, 001 (2006).
15. A. Vilenkin and E. P. S. Shellard, *Cosmic Strings and Other Topological Defects* (Cambridge University Press, Cambridge, England, 2000); M. B. Hindmarsh and T. W. B. Kibble, Rep. Prog. Phys. **58**, 477 (1995).
16. T. W. B. Kibble, J. Phys. A **9**, 387 (1976).
17. N. Turok, Phys. Rev. Lett. **63**, 2625 (1989).
18. T. Vachaspati and M. Barriola, Phys. Rev. Lett. **69**, 1867 (1992).
19. M. Sakellariadou, Cosmic strings, in *Quantum Simulations via Analogues: From Phase Transitions to Black Holes* (Springer Lecture Notes in Physics) (to appear), arxiv:hep-th/0602276.
20. T. W. B. Kibble, Nucl. Phys. B **252**, 277 (1985).
21. A. Albrecht and N. Turok, Phys. Rev. Lett. **54**, 1868 (1985); A. Albrecht and N. Turok, Phys. Rev. D **40**, 973 (989).
22. M. Sakellariadou and A. Vilenkin, Phys. Rev. D **42**, 349 (1990); D. P. Bennett, in *Formation and Evolution of Cosmic Strings*, eds G. Gibbons, S. Hawking and T. Vachaspati (Cambridge University Press, Cambridge, England, 1990); F. R. Bouchet, ibid; E. P. S. Shellard and B. Allen, ibid.
23. C. Ringeval, M. Sakellariadou and F. R. Bouchet, JCAP **02**, 023 (2007).
24. V. Vanchurin, K. D. Olum and A. Vilenkin, Phys. Rev. D **74**, 063527 (2006); C. J. A. P. Martins and E. P. S. Shellard, Phys. Rev. D **73**, 043515 (2006); K. D. Olum and V. Vanchurin, Cosmic string loops in the expanding universe, arXiv:astro-ph/0610419.
25. Y. Fukuda et. al. [Super-Kamiokande Collaboration], Phys. Rev. Lett. **81**, 1562 (1998); Q. R. Ahmad et al. [SNO Collaboration], Phys. Rev. Lett. **87**, 071301 (2001); K. Eguchi et al. [KamLAND Collaboration], Phys. Rev. Lett. **90**, 021802 (2003).

26. P. Horava and E. Witten, Nucl. Phys. B **460**, 506 (1996).
27. R. Jeannerot and A.-C. Davis, Phys. Rev. D **52**, 7220 (1995).
28. M. Sakellariadou, Ann. Phys. **15**, 264 (2006).
29. R. Durrer, M. Kunz and M. Sakellariadou, Phys. Lett. B **614**, 12 (2005).
30. S.-H. Tye, Brane inflation: string theory viewed from the cosmos, hep-th/0610221.
31. E. Witten, Nucl. Phys. B **249**, 557 (1985).
32. J. Polchinski, Introduction to cosmic F- and D-strings, hep-th/0412244.
33. M. G. Jackson, N. T. Jones and J. Polchinski, JHEP **0510**, 013 (2005).
34. A. Hanany and K. Hashimoto, JHEP **0506**, 021 (2005).
35. M. Sakellariadou, JCAP **0504**, 003 (2005); E. Copeland and P. Saffin, JHEP **0511**, 023 (2005); S.-H. H. Tye, I. Wasserman and M. Wyman, Phys. Rev. D **71**, 103508 (2005) [Erratum-ibid. Phys. Rev. D **71**, 129906 (2005)]; M. Hindmarsh and P. M. Saffin, JHEP **0608**, 066 (2006).
36. J. Martin, A. Riazuelo and M. Sakellariadou, Phys. Rev. D **61**, 083518 (2002); A. Gangui, J. Martin and M. Sakellariadou, Phys. Rev. D **66**, 083502 (2002).
37. S. Veeraraghavan and A. Stebbins, Ap. J. **365**, 37 (1990).
38. G. Vincent, M. B. Hindmarsh and M. Sakellariadou, Phys. Rev. D **55**, 573 (1997).
39. U.-L. Pen, U. Seljak and N. Turok, Phys. Rev. Lett. **79**, 1611 (1997).
40. R. Durrer, A. Gangui and M. Sakellariadou, Phys. Rev. Lett. **76**, 579 (1996).
41. R. Durrer, M. Kunz, C. Lineweaver and M. Sakellariadou, Phys. Rev. Lett. **79**, 5198 (1997); R. Durrer, M. Kunz and A. Melchiorri, Phys. Rev. D **59**, 123005 (1999).
42. U.-L. Pen, U. Seljak and N. Turok, Phys. Rev. Lett. **79**, 1611 (1997); N. Turok, U.-L. Pen and U. Seljak, Phys. Rev. D **58**, 023506 (1998).
43. R. Durrer and M. Sakellariadou, Phys. Rev. D **56**, 4480 (1997).
44. N. Bevis, M. Hindmarsk, M. Kunz and J. Urrestilla, CMB power spectrum contribution from cosmic strings using field-evolution simulations of the Abelian Higgs model, arXiv:astro-ph/0605018.
45. D. N. Spergel et. al., Wilkinson microwave anisotropy probe (WMAP) three year results: implications for cosmology, arXiv:astro-ph/0603449.
46. F. R. Bouchet, P. Peter, A. Riazuelo and M. Sakellariadou, Phys. Rev. D **65**, 021301(R) (2001).
47. M. Landriau and E. P. S. Shellard, Phys. Rev. D **69**, 023003 (2004).
48. D. H. Lyth and D. Wands, Phys. Lett. B **524**, 5 (2002); T. Moroi and T. Takahashi, Phys. Lett. B **522**, 215 (2001) [Erratum-ibid. B **539**, 303 (2002)]; K. Enqvist, S. Kasuya and A. Mazumdar, Phys. Rev. Lett. **90**, 091302 (2003); K. Dimopoulos and D. Lyth, Phys. Rev. D **69**, 123509 (2004).
49. C. L. Bennett et. al. Astrophys. J. **464**, 1 (1996).
50. R. Jeannerot and M. Postma, JCAP **0607**, 012 (2006).
51. R. A. Battye, B. Garbrecht and A. Moss, JCAP **0609**, 007 (2006).
52. M. Wyman, L. Pogosian and I. Wasserman, Phys. Rev. D **72**, 023513 (2005) [Erratum-ibid. D**73**, 089904 (2006)].
53. A. Vilenkin, Phys. Lett. B **107**, 47 (1981).
54. M. Sakellariadou, Phys. Rev. D **42**, 354 (1990).
55. T. Damour and A. Vilenkin, Phys. Rev. D **71**, 063510 (2005); X. Siemens et. al., Phys. Rev. D **73**, 105001 (2006); X. Siemens, V. Mandic and J. Creighton, Gravitational wave stochastic background from cosmic (super)strings, astro-ph/0610920.

56. G. R. Vincent, M. Hindmarsh and M. Sakellariadou, Phys. Rev. D **56**, 637 (1997); G. R. Vincent, N. D. Antunes and M. Hindmarsh, Phys. Rev. Lett. **80**, 2277 (1998).
57. F. A. Jenet et al., Upper bounds on the low-frequency stochastic gravitational wave background from pulsar timing observations: current limits and future prospects, astro-ph/0609013.
58. C.-M. Lin and J. McDonald, Phys. Rev. D **74**, 063510 (2006).
59. M. Endo, M. Kawasaki and T. Moroi, Phys. Lett. B **569**, 73 (2003).
60. J. Urrestilla, A. Achúcarro and A. C. Davis, Phys. Rev. Lett. **92**, 251302 (2004).

Conceptual Problems of Inflationary Cosmology and a New Approach to Cosmological Structure Formation

Robert H. Brandenberger

Department of Physics, McGill University, Montreal, QC, H3A 2T8, Canada
rhb@hep.physics.mcgill.ca

Abstract. In spite of its great phenomenological success, current models of scalar field-driven inflation suffer from important unresolved conceptual issues. New fundamental physics will be required to address these questions. String theory is a candidate for a unified quantum theory of all four forces of nature. As will be shown, string theory may lead to a cosmological background quite different from an inflationary cosmology, and may admit a new stringy mechanism for the origin of a roughly scale-invariant spectrum of cosmological fluctuations.

11.1 Introduction

The inflationary universe scenario [1] (see also [2, 3, 4] for earlier ideas) has been extremely successful phenomenologically. In addition to providing answers to some key open questions of Standard big bang cosmology such as the horizon, flatness and entropy problems, inflation gives rise to a causal mechanism for structure formation [5] (see also [6] for more qualitative arguments and [7] for an early computation of the spectrum of gravitational waves in an inflationary background). Quantum vacuum fluctuations during the period of exponential expansion lead to a roughly scale-invariant spectrum of (in the simplest models) adiabatic fluctuations. These fluctuations are squeezed while their wavelength is larger than the Hubble radius, and thus re-enter the Hubble radius at late times as standing waves. As realized a long time ago in [8, 9], these features predict "acoustical" oscillations in the angular power spectrum of cosmic microwave background anisotropies. Both the approximate scale-invariance and the acoustical oscillations of the spectrum have recently, many years after these features were predicted, been confirmed by CMB anisotropy experiments [10, 11, 12].

On the theoretical front, the situation is much less satisfactory. In spite of over 20 years of research, no convincing theory of inflation has emerged. There are many models of inflation, but all of them involve new scalar fields. String theory and most other theories beyond the standard model do predict scalar

R. H. Brandenberger: *Conceptual Problems of Inflationary Cosmology and a New Approach to Cosmological Structure Formation*, Lect. Notes Phys. **738**, 393–424 (2008)
DOI 10.1007/978-3-540-74353-8_11   © Springer-Verlag Berlin Heidelberg 2008

fields, and thus may well eventually give rise to a good theory of inflation (see e.g. [13, 14, 15] for reviews on avenues to obtain inflation in string theory), but at the moment the question is not resolved. Furthermore, some of the conceptual issues which will be raised below (Sect. 11.2) are generic to any implementation of inflation by scalar fields in the context of Einstein gravity.

Thus, it is important to keep an open mind to the possibility that an early universe scenario which does not involve a period of cosmological inflation will emerge. As will be shown below, the new degrees of freedom and new symmetries of string theory give rise to the possibility of a cosmological background very different from that of inflationary cosmology (Sect. 11.3). Within this background cosmology, a stringy mechanism which can generate a scale-invariant spectrum of cosmological perturbations has recently been proposed [16] (see also [17, 18] for reviews). This mechanism, which yields a distinctive signature, namely a slight blue tilt in the spectrum of gravitational waves [19], will be discussed in Sect. 11.4. Sect. 11.4.5 reviews some results which were completed after the Colloque in Paris and appeared in [20].

11.2 Problems of Scalar Field-Driven Inflation

11.2.1 Review of the Inflationary Universe Scenario

Before discussing some key conceptual problems of conventional scalar field-driven inflationary cosmology, let us recall some of the main features of cosmological inflation. To set our notation, we use the following metric for the homogeneous and isotropic background space–time:

$$ds^2 = dt^2 - a(t)^2 dx^2 , \qquad (11.1)$$

where t is physical time, x denotes comoving coordinates on the spatial sections which we for simplicity assume to be $\mathcal{R}^3$, and $a(t)$ is the scale factor.

Figure 11.1 is a sketch of the space–time structure of an inflationary universe. The vertical axis is time, the horizontal axis is physical length. The time period between t_i and t_R is the period of inflation (here for simplicity taken to be exponential). During the period of inflation, the Hubble radius

$$\ell_H(t) \equiv H^{-1}(t) \text{ where } H(t) \equiv \frac{\dot{a}(t)}{a(t)} \qquad (11.2)$$

is constant. After inflation, the Hubble radius increases linearly in time. In contrast, the physical length corresponding to a fixed co-moving scale increases exponentially during the period of inflation, and then grows either as $t^{1/2}$ (radiation-dominated phase) or $t^{2/3}$ (matter-dominated phase), i.e. less fast than the Hubble radius.

The key feature of inflationary cosmology which can be seen from Fig. 11.1 is the fact that fixed comoving scales are red-shifted exponentially relative to the Hubble radius during the period of inflation. Provided that the

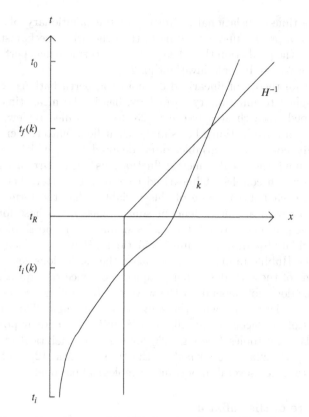

Fig. 11.1. Space–time diagram (sketch) showing the evolution of scales in inflationary cosmology. The vertical axis is time, and the period of inflation lasts between t_i and t_R, and is followed by the radiation-dominated phase of Standard big bang cosmology. During exponential inflation, the Hubble radius H^{-1} is constant in physical spatial coordinates (the horizontal axis), whereas it increases linearly in time after t_R. The physical length corresponding to a fixed comoving length scale labeled by its wavenumber k increases exponentially during inflation but increases less fast than the Hubble radius (namely as $t^{1/2}$), after inflation

period of inflation lasted more than about 50 Hubble expansion times (this number is obtained assuming that the energy scale of inflation is of the order of 10^{16}GeV), then modes with a wavelength today comparable to the current Hubble radius started out at the beginning of the period of inflation with a wavelength smaller than the Hubble radius at that time. Thus, it is possible to imagine a microscopic mechanism for creating the density fluctuations in the early universe which evolve into the cosmological structures we observe today.

Since during the period of inflation any pre-existing ordinary matter fluctuations are red-shifted, it is reasonable to assume that quantum vacuum fluctuations are the source of the currently observed structures [5] (see also [6]). The time-translational symmetry of the inflationary phase leads, independent of a precise understanding of the generation mechanism for the fluctuations, to the prediction that the spectrum of cosmological perturbations should be approximately scale-invariant [6, 2].

The quantum theory of linearized cosmological perturbations [21, 22], in particular applied to inflationary cosmology, has in the mean time become a well-developed research area (see e.g. [23] for a detailed review, and [24] for a pedagogical introduction). For simple scalar field matter, there is a single canonically normalized variable, often denoted by v, which carries the information about the "scalar metric fluctuations", the part of the metric perturbations which couples at linearized level to the matter. The equation of motion for each Fourier mode of this variable v has the form of a harmonic oscillator with a time-dependent square mass m^2, whose form is set by the cosmological background. On scales smaller than the Hubble radius, the modes oscillate (quantum vacuum oscillations). However, on length scales larger than the Hubble radius, m^2 is negative, the oscillations cease, and the wave functions of these modes undergo squeezing. Since the squeezing angle in phase space does not depend on the wave number, all modes re-enter the Hubble radius at late times with the same squeezing angle. This then leads to the prediction of "acoustic" oscillations [8, 9] in the angular power spectrum of CMB anisotropies (see e.g. [25] for a recent analytical treatment), a prediction spectacularly confirmed by the WMAP data [12], and allowing cosmologists to fit for several important cosmological parameters.

11.2.2 Nature of the Inflaton

In the context of General Relativity as the theory of space–time, matter with an equation of state

$$p \simeq -\rho \tag{11.3}$$

(with ρ denoting the energy density and p the pressure) is required in order to obtain almost exponential expansion of space. If we describe matter in terms of fields with canonical kinetic terms, a scalar field is required since in the context of usual field theories it is only for scalar fields that a potential energy function in the Lagrangian is allowed, and of all energy terms only the potential energy can yield the equation of state (11.3).

In order for scalar fields to generate a period of cosmological inflation, the potential energy needs to dominate over the kinetic and spatial gradient energies. It is generally assumed that spatial gradient terms can be neglected. This is, however, not true in general. Assuming a homogeneous field configuration, we must ensure that the potential energy dominates over the kinetic energy. This leads to the first "slow-roll" condition. Requiring the period of inflation to last sufficiently long leads to a second slow-roll condition, namely that the $\ddot{\varphi}$ term in the Klein–Gordon equation for the inflaton φ be negligible. Scalar fields charged with respect to the Standard Model symmetry groups do not satisfy the slow-roll conditions.

Assuming that both slow-roll conditions hold, one obtains a "slow-roll trajectory" in the phase space of homogeneous φ configurations. In large-field inflation models such as "chaotic inflation" [26] and "hybrid inflation" [27], the slow-roll trajectory is a local attractor in initial condition space [28] (even when linearized metric perturbations are taken into account [29]), whereas this is not the case [30] in small-field models such as "new inflation" [31]. As shown in [32], this leads to problems for some models of inflation which have recently been proposed in the context of string theory. To address this problem, it has been proposed that inflation may be future-eternal [33] and that it is hence sufficient that there be configurations in initial condition space which give rise to inflation within one Hubble patch, inflation being then self-sustaining into the future. However, one must still ensure that slow-roll inflation can locally be satisfied.

Many models of particle physics beyond the Standard Model contain a plethora of new scalar fields. One of the most conservative extensions of the Standard Model is the MSSM, the "Minimal Supersymmetric Standard Model". According to a recent study, among the many scalar fields in this model, only a hand-full can be candidates for a slow-roll inflaton, and even then very special initial conditions are required [34]. The situation in supergravity and superstring-inspired field theories may be more optimistic, but the issues are not settled.

11.2.3 Hierarchy Problem

Assuming for the sake of argument that a successful model of slow-roll inflation has been found, one must still build in a hierarchy into the field theory model in order to obtain an acceptable amplitude of the density fluctuations (this is sometimes also called the "amplitude problem"). Unless this hierarchy is observed, the density fluctuations will be too large and the model is observationally ruled out.

In a wide class of inflationary models, obtaining the correct amplitude requires the introduction of a hierarchy in scales, namely [35]

$$\frac{V(\varphi)}{\Delta\varphi^4} \leq 10^{-12}\,, \tag{11.4}$$

where $\Delta\varphi$ is the change in the inflation field during one Hubble expansion time (during inflation), and $V(\varphi)$ is the potential energy during inflation.

This problem should be contrasted with the success of topological defect models (see e.g. [36, 37, 38] for reviews) in predicting the right order of magnitude of density fluctuations without introducing a new scale of physics. The GUT scale as the scale of the symmetry breaking phase transition (which produces the defects) yields the correct magnitude of the spectrum of density fluctuations [39]. Topological defects, however, cannot be the prime mechanism for the origin of fluctuations since they do not give rise to acoustic oscillations in the angular power spectrum of the CMB anisotropies [40].

At first sight, it also does not appear to be necessary to introduce a new scale of physics into the string gas cosmology structure formation scenario which will be described in Sect. 11.4.

11.2.4 Trans-Planckian Problem

A more serious problem is the "trans-planckian problem" [41]. Returning to the space–time diagram of Fig. 11.1, we can immediately deduce thatprovided the period of inflation lasted sufficiently long (for GUT scale inflation the number is about 70 e-foldings), all scales inside of the Hubble radius today started out with a physical wavelength smaller than the Planck scale at the beginning of inflation. Now, the theory of cosmological perturbations is based on Einstein's theory of General Relativity coupled to a simple semi-classical description of matter. It is clear that these building blocks of the theory are inapplicable on scales comparable and smaller than the Planck scale. Thus, the key successful prediction of inflation (the theory of the origin of fluctuations) is based on suspect calculations since new physics *must* enter into a correct computation of the spectrum of cosmological perturbations. The key question is as to whether the predictions obtained using the current theory are sensitive to the specifics of the unknown theory which takes over on small scales.

One approach to study the sensitivity of the usual predictions of inflationary cosmology to the unknown physics on trans-planckian scales is to study toy models of ultraviolet physics which allow explicit calculations. The first approach which was used [42, 43] is to replace the usual linear dispersion relation for the Fourier modes of the fluctuations by a modified dispersion relation, a dispersion relation which is linear for physical wavenumbers smaller than the scale of new physics, but deviates on larger scales. Such dispersion relations were used previously to test the sensitivity of black hole radiation on the unknown physics of the UV [44, 45]. It was found [42] that if the evolution of modes on the trans-planckian scales is non-adiabatic, then substantial deviations of the spectrum of fluctuations from the usual results are possible. Non-adiabatic evolution turns an initial state minimizing the energy density into a state which is excited once the wavelength becomes larger than the cutoff scale. Back-reaction effects of these excitations may limit the magnitude of the trans-planckian effects, but – based on our recent study [46] – not to the extent initially assumed [47, 48]. Other approaches to study the trans-planckian problem have been pursued, e.g. based on implementing the space–space [49] or space–time [50] uncertainty relations, on a minimal length hypothesis [51], on "minimal trans-planckian" assumptions (taking as initial conditions some vacuum state at the mode-dependent time when the wavelength of the mode is equal to the Planck scale [52], or on effective field theory [53], all showing the possibility of trans-planckian corrections (see also [54] for a review of some of the previous work on the trans-planckian problem).

From the point of view of fundamental physics, the *trans-planckian problem* is not a problem. Rather, it yields a window of opportunity to probe new fundamental physics in current and future observations, even if the scale of the new fundamental physics is close to the Planck scale. The point is that if the universe in fact underwent a period of inflation, then trans-planckian physics leaves an imprint on the spectrum of fluctuations. The exponential expansion of space amplifies the wavelength of the perturbations to observable scales. At the present time, it is our ignorance about quantum gravity which prevents us from making any specific predictions. For example, we do not understand string theory in time-dependent backgrounds sufficiently well to be able to at this time make any predictions for observations.

11.2.5 Singularity Problem

The next problem is the "singularity problem". This problem, one of the key problems of Standard Cosmology, has not been resolved in models of scalar field-driven inflation.

As follows from the Penrose–Hawking singularity theorems of General Relativity (see e.g. [55] for a textbook discussion), an initial cosmological singularity is unavoidable if space–time is described in terms of General Relativity, and if the matter sources obey the weak energy conditions . Recently, the singularity theorems have been generalized to apply to Einstein gravity coupled to scalar field matter, i.e. to scalar field-driven inflationary cosmology [56]. It is shown that in this context, a past singularity at some point in space is unavoidable.

In the same way that the appearance of an initial singularity in Standard Cosmology told us that Standard Cosmology cannot be the correct description of the very early universe, the appearance of an initial singularity in current models of inflation tell us that inflationary cosmology cannot yield the correct description of the very, very early universe. At sufficiently high densities, a new description will take over. In the same way that inflationary cosmology contains late-time standard cosmology, it is possible that the new cosmology will contain, at later times, inflationary cosmology. However, one should keep an open mind to the possibility that the new cosmology will connect to present observations via a route which does not contain inflation.

11.2.6 Breakdown of Validity of Einstein Gravity

The Achilles heel of scalar field-driven inflationary cosmology is, however, the use of intuition from Einstein gravity at energy scales not far removed from the Planck and string-scales, scales where correction terms to the Einstein–Hilbert term in the gravitational action dominate and where intuition based on applying the Einstein equations break down (see also [57] for arguments along these lines).

All approaches to quantum gravity predict correction terms in the action which dominate at energies close to the Planck scale – in some cases in fact even much lower. Semiclassical gravity leads to higher curvature terms, and may (see e.g. [58, 59]) lead to bouncing cosmologies without a singularity). Loop quantum cosmology leads to similar modifications of early universe cosmology (see e.g. [60] for a recent review). String theory, the theory we will focus on in the following sections, has a maximal temperature for a string gas in thermal equilibrium [61], which may lead to an almost static phase in the early universe – the Hagedorn phase [62].

Common to all of these approaches to quantum gravity corrections to early universe cosmology is the fact that a transition from a contracting (or quasi-static) early universe phase to the rapidly expanding radiation phase of standard cosmology can occur *without* violating the usual energy conditions for matter. In particular, it is possible (as is predicted by the string gas cosmology model discussed below) that the universe in an early high-temperature phase is almost static. This may be a common feature to a large class of models which resolve the cosmological singularity.

Closely related to the above is the "cosmological constant problem" for inflationary cosmology. We know from observations that the large quantum vacuum energy of field theories does not gravitate today. However, to obtain a period of inflation one is using precisely the part of the energy–momentum tensor of the inflation field which looks like the vacuum energy. In the absence of a convincing solution of the cosmological constant problem it is unclear whether scalar field-driven inflation is robust, i.e. whether the mechanism which renders the quantum vacuum energy gravitationally inert today will not also prevent the vacuum energy from gravitating during the period of slow-rolling of the inflaton field.[1]

11.3 String Gas Cosmology

11.3.1 Preliminaries

Since string theory is the best candidate we have for a unified theory of all forces at the highest energies, we will in the following explore the possible implications of string theory for early universe cosmology.

An immediate problem which arises when trying to connect string theory with cosmology is the *dimensionality problem*. Superstring theory is perturbatively consistent only in 10 space–time dimensions, but we only see three large spatial dimensions. The original approach to addressing this problem was to assume that the six extra dimensions are compactified on a very small space which cannot be probed with our available energies. However, from the point of view of cosmology, it is quite unsatisfactory not to be able to understand why it is precisely three dimensions which are not compactified and why the

---

[1] Note that the approach to addressing the cosmological constant problem making use of the gravitational back-reaction of long range fluctuations (see [63] for a summary of this approach) does not prevent a long period of inflation in the early universe.

compact dimensions are stable. Brane world cosmology [64] provides another approach to this problem: it assumes that we live on a three-dimensional brane embedded in a large nine-dimensional space. Once again, a cosmologically satisfactory theory should explain why it is likely that we will end up exactly on a three-dimensional brane (for some interesting work addressing this issue see [65, 66, 67]).

Finding a natural solution to the dimensionality problem is thus one of the key challenges for superstring cosmology. This challenge has various aspects. First, there must be a mechanism which singles out three dimensions as the number of spatial dimensions we live in. Second, the moduli fields which describe the volume and the shape of the unobserved dimensions must be stabilized (any strong time-dependence of these fields would lead to serious phenomenological constraints). This is the *moduli problem* for superstring cosmology. As mentioned above, solving the *singularity problem* is another of the main challenges. These are the three problems which *string gas cosmology* [62, 68, 69] explicitly addresses at the present level of development.

In order to make successful connection with late time cosmology, any approach to string cosmology must also solve the *flatness problem*, namely make sure that the three large spatial dimensions obtain a sufficiently high entropy (size) to explain the current universe. Finally, it must provide a mechanism to produce a nearly scale-invariant spectrum of nearly adiabatic cosmological perturbations. If string theory leads to a successful model of inflation, then these two issues are automatically addressed. In Sect. 11.5, we will discuss a cosmological scenario which does not involve inflation but nevertheless leads to a viable structure formation scenario [16].

11.3.2 Heuristics of String Gas Cosmology

In the absence of a non-perturbative formulation of string theory, the approach to string cosmology which we have suggested, *string gas cosmology* [62, 68, 69] (see also [70] for early work, and [71, 72, 73] for reviews), is to focus on symmetries and degrees of freedom which are new to string theory (compared to point particle theories) and which will be part of a non-perturbative string theory, and to use them to develop a new cosmology. The symmetry we make use of is *T-duality*, and the new degrees of freedom are *string winding modes* and *string oscillatory modes*.

We take all spatial directions to be toroidal, with R denoting the radius of the torus. Strings have three types of states: *momentum modes* which represent the center of mass motion of the string, *oscillatory modes* which represent the fluctuations of the strings, and *winding modes* counting the number of times a string wraps the torus. Both oscillatory and winding states are special to strings. Point particle theories do not contain these modes.

The energy of an oscillatory mode is independent of R, momentum mode energies are quantized in units of $1/R$, i.e.

$$E_n = n\frac{1}{R},$$

(11.5)

whereas the winding mode energies are quantized in units of R, i.e.

$$E_m = mR,$$

(11.6)

where both n and m are integers. The energy of oscillatory modes does not depend on R.

The T-duality symmetry is the invariance of the spectrum of string states under the change

$$R \to 1/R$$

(11.7)

in the radius of the torus (in units of the string length ℓ_S). Under such a change, the energy spectrum of string states is not modified if winding and momentum quantum numbers are interchanged

$$(n, m) \to (m, n).$$

(11.8)

The string vertex operators are consistent with this symmetry, and thus T-duality is a symmetry of perturbative string theory. Postulating that T-duality extends to non-perturbative string theory leads [74] to the need of adding D-branes to the list of fundamental objects in string theory. With this addition, T-duality is expected to be a symmetry of non-perturbative string theory. Specifically, T-duality will take a spectrum of stable Type IIA branes and map it into a corresponding spectrum of stable Type IIB branes with identical masses [75].

Since the number of string oscillatory modes increases exponentially as the string mode energy increases, there is a maximal temperature of a gas of strings in thermal equilibrium, the *Hagedorn temperature* T_H [61]. If we imagine taking a box of strings and compressing it, the temperature will never exceed T_H. In fact, as the radius R decreases below the string radius, the temperature will start to decrease, obeying the duality relation [62]

$$T(R) = T(1/R).$$

(11.9)

This argument shows that string theory has the potential of taming singularities in physical observables. Similarly, the length L measured by a physical observer will be consistent with the symmetry (11.7), hence realizing the idea of a minimal physical length [62]. Figure 11.2 provides a sketch of how the temperature T changes as a function of R.

If we imagine that there is a dynamical principle that tells us how R evolves in time, then Fig. 11.2 can be interpreted as depicting how the temperature changes as a function of time. If R is a monotonic function of time, then two interesting possibilities for cosmology emerge. If $\ln R$ decreases to zero at some fixed time (which without loss of generality we can call $t = 0$), and continues to decrease, we obtain a temperature profile which is symmetric with respect to $t = 0$ and which (since small R is physically equivalent to large R)

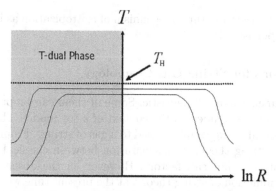

Fig. 11.2. Sketch (based on the analysis of [62] of the evolution of temperature T as a function of the radius R of space of a gas of strings in thermal equilibrium. The top curve is characterized by an entropy higher than the bottom curve, and leads to a longer region of Hagedorn behavior

represents a bouncing cosmology (see [76] for a concrete recent realization of this scenario). If, on the other hand, it takes an infinite amount of time to reach $R = 0$, an *emergent universe* scenario [77] is realized.

It is important to realize that in both of the cosmological scenarios which, as argued above, seem to follow from string theory symmetry considerations alone, a large energy density does *not* lead to rapid expansion in the Hagedorn phase, in spite of the fact that the matter sources we are considering (namely a gas of strings) obeys all of the usual energy conditions discussed e.g. in [55]). These considerations are telling us that intuition drawn from Einstein gravity may give us a completely incorrect picture of the early universe. This provides a response to the main criticism raised in [78].

Any physical theory requires both a specification of the equations of motion and of the initial conditions. We assume that the universe starts out small and hot. For simplicity, we take space to be toroidal, with radii in all spatial directions given by the string-scale. We assume that the initial energy density is very high, with an effective temperature which is close to the Hagedorn temperature, the maximal temperature of perturbative string theory.

In this context, it was argued [62] that in order for spatial sections to become large, the winding modes need to decay. This decay, at least on a background with stable one cycles such as a torus, is only possible if two winding modes meet and annihilate. Since string world sheets have measured zero probability for intersecting in more than four space–time dimensions, winding modes can annihilate only in three spatial dimensions (see, however, the recent caveats to this conclusion based on the work of [79, 80]). Thus, only three spatial dimensions can become large, hence explaining the observed dimensionality of space–time. As was shown later [69], adding branes to the system does not change these conclusions since at later times the strings dominate the cosmological dynamics. Note that in the three dimensions, which are becoming large, there is a natural mechanism of isotropization as long as some winding modes persist [81].

11.3.3 Equations for String Gas Cosmology

The above arguments were all heuristic. Some of them can be put on a more firm mathematical basis, albeit in the context of a toy model. The toy model consists of a classical background coupled to a gas of strings. From the point of view of rigorous string theory, this separation between classical background and stringy matter is not satisfactory. However, in the absence of a non-perturbative formulation of string theory, at the present time we are forced to make this separation. Note that this separation between classical background geometry and string matter is common to all current approaches to string cosmology.

As our background we choose dilaton gravity. It is crucial to include the dilaton in the Lagrangian, firstly since the dilaton arises in string perturbation theory at the same level as the graviton (when calculating to leading order in the string coupling and in α'), and secondly because it is only the action of dilaton gravity (rather than the action of Einstein gravity) which is consistent with the T-duality symmetry. We will see, however, that the background dynamics inevitably drives the system into a parameter region where the dilaton is strongly coupled and hence beyond the region of validity of the approximations made.

For the moment, however, we consider the dilaton gravity action coupled to a matter action S_m

$$S = \frac{1}{2\kappa^2} \int d^{10}x \sqrt{-\hat{g}} e^{-2\phi} \left[\hat{R} + 4\partial^\mu \phi \partial_\mu \phi \right] + S_m , \tag{11.10}$$

where $\hat{g}$ is the determinant of the metric, $\hat{R}$ is the Ricci scalar, ϕ is the dilaton, and κ is the reduced gravitational constant in ten dimensions. The metric appearing in the above action is the metric in the string frame.

In the case of a homogeneous and isotropic background given by

$$ds^2 = dt^2 - a(t)^2 d\mathbf{x}^2 , \tag{11.11}$$

the three resulting equations (the generalization of the two Friedmann equations plus the equation for the dilaton) in the string frame are [68] (see also [82])

$$-d\dot{\lambda}^2 + \dot{\varphi}^2 = e^\varphi E \tag{11.12}$$

$$\ddot{\lambda} - \dot{\varphi}\dot{\lambda} = \frac{1}{2} e^\varphi P \tag{11.13}$$

$$\ddot{\varphi} - d\dot{\lambda}^2 = \frac{1}{2} e^\varphi E , \tag{11.14}$$

where E and P denote the total energy and pressure, respectively, d is the number of spatial dimensions, and we have introduced the logarithm of the scale factor

$$\lambda(t) = \log[a(t)] \tag{11.15}$$

and the rescaled dilaton

$$\varphi = 2\phi - d\lambda . \tag{11.16}$$

From the second of these equations it follows immediately that a gas of strings containing both stable winding and momentum modes will lead to the stabilization of the radius of the torus: windings prevent expansion, momenta prevent the contraction. The right hand side of the equation can be interpreted as resulting from a confining potential for the scale factor.

Note that this behavior is a consequence of having used dilaton gravity rather than Einstein gravity as the background. The dilaton is evolving at the time when the radius of the torus is at the minimum of its potential. In fact, for the branch of solutions we are considering, the dilaton is increasing as we go into the past. At some point, therefore, it becomes greater than zero. At this point, we enter the region of strong coupling. As already discussed in [83], a different dynamical framework is required to analyze this phase. In particular, the fundamental strings are no longer the lightest degrees of freedom. We will call this phase the "strongly coupled Hagedorn phase" for which we lack an analytical description. Since the energy density in this phase is of the string-scale, the background equations should also be very different from the dilaton gravity equations used above. In the following section, we will make the assumption that the dilaton is in fact frozen in the strongly coupled Hagedorn phase. This could be a consequence of S-duality (see e.g. [84] for a recent work).

11.3.4 String Gas Cosmology and Moduli Stabilization

One of the outstanding issues when dealing with theories with extra dimensions is the question of how the size and shape moduli of the extra-dimensional spaces are stabilized. String gas cosmology provides a simple and string-specific mechanism to stabilize most of these moduli (see [73] for a recent review). The outstanding issue is how to stabilize the dilaton.

It is easiest to first understand radion stabilization in the string frame [85]. The mechanism can be immediately seen from the basic equations (11.12), (11.13), (11.14) of dilaton gravity. From (11.13) it follows that if the string gas contains both winding and momentum modes, then there is a preferred value for the scale factor. Winding modes prevent the radion from increasing, momentum modes prevent it from decreasing. If the number of winding and momentum modes is the same, then the fixed point of the dynamics corresponds to string-scale radion. As can be seen from (11.14), the dilaton is evolving when the radion is at the fixed point.

In order to make contact with late-time cosmology, we need to assume that the dilaton is fixed by some non-perturbative mechanism. The issue of radion stabilization is then no longer this simple. A string matter state which minimizes the effective potential in the Einstein frame is only consistent with stabilization if the energy density vanishes at the minimum – otherwise, the state will in fact lead to inflationary expansion. It thus turns out that [86, 87] (see also [85, 88]) states which are massless and have an enhanced symmetry at a particular radius play a crucial role. If such states exist, then the radion can be fixed at this particular radius (in our case it will be the self-dual radius). Such states exist in the heterotic string theory, but not in Type II string theories where they are projected out by the GSO projection.

To understand stabilization in the Einstein frame, let us consider the equations of motion which arise from coupling the Einstein action (as opposed to the dilaton gravity action) to a string gas. In the anisotropic setting when the metric is taken to be

$$ds^2 = dt^2 - a(t)^2 dx^2 - \sum_{\alpha=1}^{6} b_\alpha(t)^2 dy_\alpha^2, \tag{11.17}$$

where the y_α are the internal coordinates, the equations of motion for b_α becomes

$$\ddot{b}_\alpha + \left(3H + \sum_{\beta=1,\beta\neq\alpha}^{6} \frac{\dot{b}_\beta}{b_\beta}\right)\dot{b}_\alpha = \sum_{n,m} 8\pi G \frac{\mu_{m,n}}{\sqrt{g}\epsilon_{m,n}} \mathcal{S} \tag{11.18}$$

where $\mu_{m,n}$ is the number density of (m, n) strings, $\epsilon_{m,n}$ is the energy of an individual (m, n) string, and g is the determinant of the metric. The source term $\mathcal{S}$ depends on the quantum numbers of the string gas, and the sum runs over all momentum numbers and winding number vectors m and n, respectively (note that n and m are six-vectors, one component for each internal dimension). If the number of right-moving oscillator modes is given by N, then the source term for fixed m and n is

$$\mathcal{S} = \sum_\alpha \left(\frac{m_\alpha}{b_\alpha}\right)^2 - \sum_\alpha n_\alpha^2 b_\alpha^2 + \frac{2}{D-1}\left[(n,n) + (n,m) + 2(N-1)\right]. \tag{11.19}$$

To obtain this equation, we have made use of the mass spectrum of string states and of the level matching conditions. In the case of the bosonic superstring, the mass spectrum for fixed m, n, N and $\tilde{N}$, where $\tilde{N}$ is the number of left-moving oscillator states, on a six-dimensional torus whose radii are given by b_α is

$$m^2 = \left(\frac{m_\alpha}{b_\alpha}\right)^2 - \sum_\alpha n_\alpha^2 b_\alpha^2 + 2(N + \tilde{N} - 2), \tag{11.20}$$

and the level matching condition reads

$$\tilde{N} = (n, m) + N , \qquad\qquad (11.21)$$

where (n, m) indicates the scalar product of n and m in the trivial internal metric.

There are modes which are massless at the self-dual radius $b_\alpha = 1$. One such mode is the graviton with $n = m = 0$ and $N = 1$. The modes of interest to us are modes which contain winding and momentum, namely

- $N = 1$, $(m, m) = 1$, $(m, n) = -1$ and $(n, n) = 1$;
- $N = 0$, $(m, m) = 1$, $(m, n) = 1$ and $(n, n) = 1$;
- $N = 0$ $(m, m) = 2$, $(m, n) = 0$ and $(n, n) = 2$.

Note that some of these modes survive in the Heterotic string theory, but they do not survive the GSO [74] truncation in Type II string theories.

In string theories which admit massless states (i.e. states which are massless at the self-dual radius), these states will dominate the initial partition function. The background dynamics will then also be dominated by these states. To understand the effect of these strings, consider the equation of motion (11.18) with the source term (11.19). The first two terms in the source term correspond to an effective potential with a stable minimum at the self-dual radius. However, if the third term in the source does not vanish at the self-dual radius, it will lead to a positive potential which causes the radion to increase. Thus, a condition for the stabilization of b_α at the self-dual radius is that the third term in (11.19) vanish at the self-dual radius. This is the case if and only if the string state is a massless mode.

The massless modes have other nice features which are explored in detail in [87]. They act as radiation from the point of view of our three large dimensions and hence do not lead to an over-abundance problem. As our three spatial dimensions grow, the potential which confines the radion becomes shallower. However, rather surprisingly, it turns out that the potential remains steep enough to avoid fifth force constraints.

Key to the success in simultaneously avoiding the moduli overclosure problem and evading fifth force constraints is the fact that the stabilization mechanism is an intrinsically stringy one, as opposed to an effective field theory mechanism. In the case of effective field theory, both the confining force and the overdensity in the moduli field scale as $V(\varphi)$, where $V(\varphi)$ is the potential energy density of the field φ. In contrast, in the case of stabilization by means of massless string modes, the energy density in the string modes (from the point of view of our three large dimensions) scales as p_3, whereas the confining force scales as p_3^{-1}, where p_3 is the momentum in the three large dimensions. Thus, for small values of p_3, one simultaneously gets large confining force (thus satisfying the fifth force constraints) and small energy density [87].

In the presence of massless string states, the shape moduli also can be stabilized, at least in the simple toroidal backgrounds considered so far [89] (see also [90]). The stabilization mechanism is once again dynamical, and makes use of the massless states with enhanced symmetry (see also [91, 92] for more on the use of massless states with enhanced symmetries in cosmology).

There has been other recent work on string gas cosmology [93]. In particular, the difficulties of dilaton stabilization have been addressed in [94]. For recent attempts to stabilize the dilaton see [95] and [84].

11.4 String Gas Cosmology and Structure Formation

11.4.1 Overview

Let us recall the key aspects of the dynamics of string gas cosmology. First of all, note that in thermal equilibrium at the string scale ($R \simeq l_s$), the self-dual radius, the number of winding and momentum modes will be equal. Since winding and momentum modes give an opposite contribution to the pressure, the pressure of the string gas in thermal equilibrium at the self-dual radius will vanish. From the dilaton gravity equations of motion (11.12), (11.13), (11.14) it then follows that a static phase $\lambda = 0$ will be a fixed point of the dynamical system. This phase is the Hagedorn phase.

On the other hand, for large values of R in thermal equilibrium the energy will be exclusively in momentum modes. These act as usual radiation. Inserting the radiative equation of state into the above equations (11.12), (11.13), (11.14) it follows that the source in the dilaton equation of motion vanishes and the dilaton approaches a constant as a consequence of the Hubble damping term in its equation of motion. Consequently, the scale factor expands as in the usual radiation-dominated universe.

The transition between the Hagedorn phase and the radiation-dominated phase with fixed dilaton is achieved via the annihilation of winding modes, as studied in detail in [96]. The main point is that, starting in a Hagedorn phase, there will be a smooth transition to the radiation-dominated phase of standard cosmology with fixed dilaton.

Our new cosmological background is obtained by following our currently observed universe into the past according to the string gas cosmology equations. The radiation phase of standard cosmology is unchanged. In particular, the dilaton is fixed in this phase.[2] However, as the temperature of the radiation bath approaches the Hagedorn temperature, the equation of state of string gas matter changes. The equation of state parameter $w = P/E$ decreases towards a pressureless state and the string frame metric becomes static. Note that, in order for the present size of the universe to be larger than our current Hubble radius, the size of the spatial sections in the Hagedorn phase must be at least 1 mm.[3] We will denote the time when the transition from the Hagedorn phase

---

[2] The dilaton comes to rest, but it is not pinned to a particular value by a potential. Thus, in order to obtain consistency with late time cosmology, an additional mechanism operative at late times which fixes the dilaton is required.

[3] How to obtain this initial size starting from string-scale initial conditions constitutes the *entropy problem* of our scenario. A possible solution making use of an initial phase of bulk dynamics is given in [97].

to the radiation phase of standard cosmology occurs by t_R, to evoque the analogy with the time of reheating in inflationary cosmology. As we go back in time in the Hagedorn phase, the dilaton increases. At the time t_c when the dilaton equals zero, a second transition occurs, the transition to a "strongly coupled Hagedorn phase" (using the terminology introduced in [20]). We take the dilaton to be fixed in this phase. In this case, the strongly coupled Hagedorn phase may have a duration which is very long compared to the Hubble time immediately following t_R. It is in this cosmological background that we will study the generation of fluctuations.

It is instructive to compare the background evolution of string gas cosmology with the background of inflationary cosmology. Figure 11.3 is a sketch of the space–time evolution in string gas cosmology. For times $t < t_R$, we are in the quasi-static Hagedorn phase, for $t > t_R$ we have the radiation-dominated

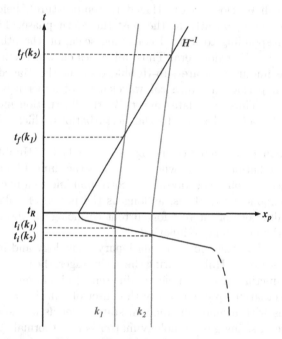

Fig. 11.3. Space–time diagram (sketch) showing the evolution of fixed comoving scales in string gas cosmology. The vertical axis is time, the horizontal axis is physical distance. The Hagedorn phase ends at the time t_R and is followed by the radiation-dominated phase of standard cosmology. The solid curve represents the Hubble radius H^{-1} which is cosmological during the quasi-static Hagedorn phase, shrinks abruptly to a microphysical scale at t_R and then increases linearly in time for $t > t_R$. Fixed comoving scales (the *dotted lines* labeled by k_1 and k_2) which are currently probed in cosmological observations have wavelengths which are smaller than the Hubble radius during the Hagedorn phase. They exit the Hubble radius at times $t_i(k)$ just prior to t_R, and propagate with a wavelength larger than the Hubble radius until they reenter the Hubble radius at times $t_f(k)$

period of standard cosmology. To understand why string gas cosmology can lead to a causal mechanism of structure formation, we must compare the evolution of the physical wavelength corresponding to a fixed comoving scale (fluctuations in early universe cosmology correspond to waves with a fixed wavelength in comoving coordinates) with that of the Hubble radius $H^{-1}(t)$, where $H(t)$ is the expansion rate. The Hubble radius separates scales on which fluctuations oscillate (wavelengths smaller than the Hubble radius) from wavelengths where the fluctuations are frozen in and cannot be effected by microphysics. Causal microphysical processes can generate fluctuations only on sub-Hubble scales (see e.g. [24] for a concise overview of the theory of cosmological perturbations and [23] for a comprehensive review).

In string gas cosmology, the Hubble radius is infinitely large in the Hagedorn phase. As the universe starts expanding near t_R, the Hubble radius rapidly decreases to a microscopic value (set by the temperature at $t = t_R$ which will be close to the Hagedorn temperature), before turning around and increasing linearly in the post-Hagedorn phase. The physical wavelength corresponding to a fixed comoving scale, on the other hand, is constant during the Hagedorn era. Thus, all scales on which current experiments measure fluctuations are sub-Hubble deep in the Hagedorn phase. In the radiation period, the physical wavelength of a perturbation mode grows only as $\sqrt{t}$. Thus, at a late time $t_f(k)$ the fluctuation mode will re-enter the Hubble radius, leading to the perturbations which are observed today.

In contrast, in inflationary cosmology (Fig. 11.1) the Hubble radius is constant during inflation ($t < t_R$, where here t_R is the time of inflationary reheating), whereas the physical wavelength corresponding to a fixed comoving scale expands exponentially. Thus, as long as the period of inflation is sufficiently long, all scales of interest for current cosmological observations are sub-Hubble at the beginning of inflation.

In spite of the fact that both in inflationary cosmology and in string gas cosmology, scales are sub-Hubble during the early stages, the actual generation mechanism for fluctuations is completely different. In inflationary cosmology, any thermal fluctuations present before the onset of inflation are red-shifted away, leaving us with a quantum vacuum state, whereas in the quasi-static Hagedorn phase of string gas cosmology matter is in a thermal state. Hence, whereas in inflationary cosmology the fluctuations originate as quantum vacuum perturbations, in string gas cosmology the inhomogeneities are created by the thermal fluctuations of the string gas.

As we have shown in [16, 17, 18], string thermodynamical fluctuations in the Hagedorn phase of string gas cosmology yield an almost scale-invariant spectrum of both scalar and tensor modes. This result stems from the holographic scaling of the specific heat $C_V(R)$ (evaluated for fixed volume) as a function of the radius R of the box

$$C_V(R) \sim R^2 \,. \tag{11.22}$$

As derived in [98], this result holds true for a gas of closed strings in a space–time in which the three large spatial dimensions are compact. The scaling (11.22) is an intrinsically stringy result: thermal fluctuations of a gas of particles would lead to a very different scaling.

Since the primordial perturbations in our scenario are of thermal origin (and there are no non-vanishing chemical potentials), they will be adiabatic. The spectrum of scalar metric fluctuations has a slight red tilt. As a distinctive feature [19], our scenario predicts a slight blue tilt for the spectrum of gravitational waves. The red tilt for the scalar modes is due to the fact that the temperature when short wavelength modes exit the Hubble radius is slightly lower than the temperature when longer wavelength modes exit. The gravitational wave amplitude, in contrast, is determined by the pressure. Since the pressure is closer to zero the deeper in the Hagedorn phase we are, a slight blue tilt for the tensor fluctuations results.

These results are explained in more detail in the following subsections.

11.4.2 Extracting Metric Fluctuations from Matter Perturbations

In this subsection, we show how the scalar and tensor metric fluctuations can be extracted from knowledge of the energy-momentum tensor of the string gas.

Working in conformal time η [defined via $\mathrm{d}t = a(t)\mathrm{d}\eta$], the metric of a homogeneous and isotropic background space–time perturbed by linear scalar metric fluctuations and gravitational waves can be written in the form

$$\mathrm{d}s^2 = a^2(\eta)\left\{(1+2\Phi)\,\mathrm{d}\eta^2 - \left[(1-2\Phi)\,\delta_{ij} + h_{ij}\right]\mathrm{d}x^i dx^j\right\}. \tag{11.23}$$

Here, Φ (which is a function of space and time) describes the scalar metric fluctuations. The tensor h_{ij} is transverse and traceless and contains the two polarization states of the gravitational waves. In the above, we have fixed the coordinate freedom by working in the so-called "longitudinal" gauge in which the scalar metric fluctuation is diagonal. We have also assumed that there is no anisotropic stress. Note that to linear order in the amplitude of the fluctuations, scalar and tensor modes decouple, and the tensor modes are gauge-invariant.

Our approximation scheme for computing the cosmological perturbations and gravitational wave spectra from string gas cosmology is as follows (the analysis is similar to how the calculations were performed in [99, 100] in the case of inflationary cosmology). For a fixed comoving scale k we follow the matter fluctuations until the time $t_i(k)$ shortly before the end of the Hagedorn phase when the scale exits the Hubble radius[4] At the time of Hubble radius crossing, we use the Einstein constraint equations (discussed below) to

---

[4] Recall that on sub-Hubble scales, the dynamics of matter is the dominant factor in the evolution of the system, whereas on super-Hubble scales, matter fluctuations freeze out and gravity dominates. Thus, it is precisely at the time of

compute the values of $\Phi(k)$ and $h(k)$ (h is the amplitude of the gravitational wave tensor), and then we propagate the metric fluctuations according to the standard gravitational perturbation equations until scales re-enter the Hubble radius at late times. Note that since the dilaton is fixed in the radiation phase, we are justified in using the perturbed Einstein equations after the time t_R.

Since the dilaton comes to rest at the end of the Hagedorn phase, we will use the Einstein equations to relate the matter fluctuations to the metric perturbations at the time that the scales k exit the Hubble radius at times $t_i(k)$. Since $t_i(k) < t_R$, there are potentially important terms coming from the dilaton velocity which we are neglecting [20, 78]. We will return to this issue later.

Inserting the metric (11.23) into the Einstein equations, subtracting the background terms and truncating the perturbative expansion at linear order leads to the following system of equations

$$-3\mathcal{H}\left(\mathcal{H}\Phi + \Phi'\right) + \nabla^2\Phi = 4\pi G a^2 \delta T^0{}_0$$
$$\left(\mathcal{H}\Phi + \Phi'\right)_{,i} = 4\pi G a^2 \delta T^0{}_i$$
$$\left(2\mathcal{H}' + \mathcal{H}^2\right)\Phi + 3\mathcal{H}\Phi' + \Phi'' = -4\pi G a^2 \delta T^i{}_i ,$$
$$-\frac{1}{2}\left(\mathcal{H}' + \frac{1}{2}\mathcal{H}^2\right)h_{ij} + \frac{1}{4}\mathcal{H}h'_{ij} + \left(\frac{\partial^2}{\partial\eta^2} - \nabla^2\right)h_{ij} = -4\pi G a^2 \delta T^i{}_j ,$$

$$\text{for } i \neq j \,,(11.24)$$

where $\mathcal{H} = a'/a$, a prime denotes the derivative with respect to conformal time η, and G is Newton's gravitational constant.

In the Hagedorn phase, these equations simplify substantially and allow us to extract the scalar and tensor metric fluctuations individually. Replacing comoving by physical coordinates, we obtain from the 00 equation

$$\nabla^2\Phi = 4\pi G \delta T^0{}_0 \qquad (11.25)$$

and from the $i \neq j$ equation

$$\nabla^2 h_{ij} = -4\pi G \delta T^i{}_j . \qquad (11.26)$$

The above equations (11.25) and (11.26) allow us to compute the power spectra of scalar and tensor metric fluctuations in terms of correlation functions of the string energy-momentum tensor. Since the metric perturbations are small in amplitude we can consistently work in Fourier space. Specifically,

$$\langle|\Phi(k)|^2\rangle = 16\pi^2 G^2 k^{-4} \langle \delta T^0{}_0(k)\delta T^0{}_0(k)\rangle , \qquad (11.27)$$

where the pointed brackets indicate expectation values, and

Hubble radius crossing that we must extract the metric fluctuations from the matter perturbations. Since the concept of an energy density fluctuation is gauge-dependent on super-Hubble scales, one cannot extrapolate the matter spectra to larger scales as was suggested in Sect. 11.3 of [78].

$$\langle |h(k)|^2 \rangle = 16\pi^2 G^2 k^{-4} \langle \delta T^i_{\ j}(k)\delta T^i_{\ j}(k)\rangle , \qquad (11.28)$$

where on the right hand side of (11.28) we mean the average over the correlation functions with $i \neq j$.

11.4.3 String Thermodynamics Fluctuations

Since the Hagedorn phase is quasi-static and dominated by a gas of strings, fluctuations in our scenario are the thermal fluctuations of a gas of strings. We will consider a gas of closed strings in a compact space, i.e. our three-dimensional space is considered to be large but compact. Specifically, it is important to have winding modes in the spectrum of string states.

The correlation functions of the energy-momentum tensor can be obtained from the partition function Z, which determines the free energy F via

$$F = \frac{-1}{\beta} \ln Z , \qquad (11.29)$$

where β is the inverse temperature.

The expectation value of the energy-momentum tensor T^ν_μ is then given in terms of the free energy by

$$\langle T^\mu_\nu \rangle = \frac{-2g_{\nu\lambda}}{\sqrt{-g}} \frac{\delta F}{\delta g_{\lambda\mu}} . \qquad (11.30)$$

Taking one additional variational derivative of (11.30) we obtain the following expression for the fluctuations of T^ν_μ (see [17, 18] for more details):

$$\langle T^\mu_\nu T^\kappa_\beta \rangle - \langle T^\mu_\nu \rangle \langle T^\kappa_\beta \rangle = \frac{1}{\beta} \frac{2g_{\beta\tau}}{\sqrt{-g}} \frac{\delta}{\delta g_{\tau\kappa}} \left(\frac{-2g_{\nu\lambda}}{\sqrt{-g}} \frac{\delta F}{\delta g_{\lambda\mu}} \right) . \qquad (11.31)$$

In particular, the energy density fluctuation is

$$\langle \delta\rho^2 \rangle = -\frac{1}{R^6} \frac{\partial}{\partial\beta} \left(F + \beta\frac{\partial F}{\partial\beta} \right) = \frac{T^2}{R^6} C_V , \qquad (11.32)$$

where C_V is the specific heat. The off-diagonal pressure fluctuations, in turn, are given by

$$\langle \delta T^i_{\ j}{}^2 \rangle = \langle T^i_{\ j}{}^2 \rangle - \langle T^i_{\ j} \rangle^2 \qquad (11.33)$$

$$= \frac{1}{\beta R^3} \frac{\partial}{\partial \ln R} \left(-\frac{1}{R^3} \frac{\partial F}{\partial \ln R} \right) = \frac{1}{\beta R^2} \frac{\partial p}{\partial R} ,$$

where the string pressure is given by

$$p \equiv -V^{-1} \left(\frac{\partial F}{\partial \ln R} \right) = T \left(\frac{\partial S}{\partial V} \right)_E . \qquad (11.34)$$

So far, the analysis has been general thermodynamics. Let us now specialize to the thermodynamics of strings. In [98], the thermodynamical properties of a gas of closed strings in a toroidal space of radius R were computed. To compute the fluctuations in a region of radius R which forms part of our three-dimensional compact space, we will apply the results of [98] for a box of strings in a volume $V = R^3$.

The starting point of the computation is the formula for the density of states $\Omega(E, R)$ which determines the entropy S via

$$S(E, R) = \ln \Omega(E, R) . \tag{11.35}$$

The entropy, in turn, determines the free energy F from which the correlation functions can be derived. In the Hagedorn phase, the density of states has the following form

$$\Omega(E, R) \simeq \beta_{\mathrm{H}} e^{\beta_{\mathrm{H}} E + n_{\mathrm{H}} V} [1 + \delta \Omega_{(1)}(E, R)] , \tag{11.36}$$

where $\beta_{\mathrm{H}} = T_{\mathrm{H}}^{-1}$, n_{H} is a (constant) number density of order ℓ_{S}^{-3} (ℓ_{S} being the string length), ρ_{H} is the "Hagedorn Energy density" of the order ℓ_{S}^{-4}, and

$$\ell_{\mathrm{S}}^3 \delta \Omega_{(1)} \simeq -\frac{R^2}{T_{\mathrm{H}}} \left(1 - \frac{T}{T_{\mathrm{H}}} \right) . \tag{11.37}$$

In addition, we find

$$\langle E \rangle \simeq \ell_{\mathrm{S}}^{-3} R^2 \ln \left[\frac{\ell_{\mathrm{S}}^3 T}{R^2 (1 - T/T_{\mathrm{H}})} \right] . \tag{11.38}$$

Note that to ensure that $|\delta \Omega_{(1)}| \ll 1$ and $\langle E \rangle \gg \rho_{\mathrm{H}} R^3$, one should demand

$$(1 - T/T_{\mathrm{H}}) R^2 \ell_{\mathrm{S}}^{-2} \ll 1 . \tag{11.39}$$

The results (11.36) and (11.37) now allow us to compute the correlation functions (11.32) and (11.33). We first compute the energy correlation function (11.32). Making use of (11.38), it follows from (11.22) that

$$C_V \approx 2 \frac{R^2 / \ell_{\mathrm{S}}^3}{T (1 - T/T_{\mathrm{H}})} . \tag{11.40}$$

The "holographic" scaling $C_V(R) \sim R^2$ is responsible for the overall scale-invariance of the spectrum of cosmological perturbations. The factor $(1 - T/T_{\mathrm{H}})$ in the denominator is responsible for giving the spectrum a slight red tilt. It comes from the differentiation with respect to T.

For the pressure, we obtain

$$p(E, R) \approx n_{\mathrm{H}} T_{\mathrm{H}} - \frac{2}{3} \frac{(1 - T/T_{\mathrm{H}})}{\ell_{\mathrm{S}}^3 R} \ln \left[\frac{\ell_{\mathrm{S}}^3 T}{R^2 (1 - T/T_{\mathrm{H}})} \right] , \tag{11.41}$$

which immediately yields

$$\langle \delta T^i{}_j{}^2 \rangle \simeq \frac{T(1 - T/T_H)}{\ell_S^3 R^4} \ln^2 \left[\frac{R^2}{\ell_S^2}(1 - T/T_H) \right] . \tag{11.42}$$

Note that since no temperature derivative is taken, the factor $(1 - T/T_H)$ remains in the numerator. This leads to the slight blue tilt of the spectrum of gravitational waves. As mentioned earlier, the physical reason for this blue tilt is that larger wavelength modes exit the Hubble radius deeper in the Hagedorn phase where the pressure is smaller and thus the strength of the tensor modes is less.

11.4.4 Power Spectra

The power spectrum of scalar metric fluctuations is given by

$$\begin{aligned}
P_\Phi(k) &\equiv k^3 |\Phi(k)|^2 \tag{11.43} \\
&= 16\pi^2 G^2 k^{-1} \langle |\delta\rho(k)|^2 \rangle . \\
&= 16\pi^2 G^2 k^2 \langle (\delta M)^2 \rangle_R \\
&= 16\pi^2 G^2 k^{-4} \langle (\delta\rho)^2 \rangle_R ,
\end{aligned}$$

where in the first step we have used (11.27) to replace the expectation value of $|\Phi(k)|^2$ in terms of the correlation function of the energy density, and in the second step we have made the transition to position space (note that $k = R^{-1}$).

According to (11.32), the density correlation function is given by the specific heat via $T^2 R^{-6} C_V$. Inserting the expression from (11.40) for the specific heat of a string gas on a scale R yields to the final result

$$P_\Phi(k) = 16\pi^2 G^2 \frac{T}{\ell_S^3} \frac{1}{1 - T/T_H} \tag{11.44}$$

for the power spectrum of cosmological fluctuations. In the above equation, the temperature T is to be evaluated at the time $t_i(k)$ when the mode k exits the Hubble radius. Since modes with larger values of k exit the Hubble radius slightly later when the temperature is slightly lower, a small red tilt of the spectrum is induced. The amplitude $\mathcal{A}_S$ of the power spectrum is given by

$$\mathcal{A}_S \sim \left(\frac{\ell_{Pl}}{\ell_S} \right)^4 \frac{1}{1 - T/T_H} . \tag{11.45}$$

Taking the last factor to be of order unity, we find that a string length three orders of magnitude larger than the Planck length, a string length which was assumed in early studies of string theory, gives the correct amplitude of the spectrum. Thus, it appears that the string gas cosmology structure formation mechanism does not have a serious amplitude problem.

Similarly, we can compute the power spectrum of the gravitational waves and obtain

$$P_h(k) \sim 16\pi^2 G^2 \frac{T}{\ell_S^3}(1 - T/T_H)\ln^2\left[\frac{1}{\ell_S^2 k^2}\left(1 - \frac{T}{T_H}\right)^{-1}\right]. \qquad (11.46)$$

This shows that the spectrum of tensor modes is – to a first approximation, namely neglecting the logarithmic factor and neglecting the k-dependence of $T(t_i(k))$ – scale-invariant.[5] The k-dependence of the temperature at Hubble radius crossing induces a small blue tilt for the spectrum of gravitational waves.

Comparing (11.44) and (11.46) we see that the tensor to scalar ratio is suppressed by the factor $(1 - T/T_H)^2$. Given a good understanding of the exit from the Hagedorn phase we would be able to compute this ratio as well as the magnitude of the spectral tilts for both scalar and tensor modes.

11.4.5 The Strongly Coupled Hagedorn Phase

In the previous discussion we have assumed that during the Hagedorn phase, the kinetic energy of the dilaton has negligible effects. However, this is not the case [20, 78] if the background is described in terms of dilaton gravity. However, as stressed in Sect. 11.2.6, it is unrealisitc to expect that pure dilaton gravity is a good approximation to the dyanamics of the Hagedorn phase. Both string loop and α' corrections will be important. Another aspect of this issue is that – according to the dilaton gravity equations – the string coupling constant quickly becomes greater than unity as we go back in time from t_R. At that point, the separation between classical dilaton background and stringy matter becomes untenable.

Therefore, in order to put our scenario on a firm basis, we need a consistent description of the Hagedorn phase. It is crucial that there be a phase before or immediately leading up to t_R, the time when the winding string modes decay to string loops, when the dilaton is fixed. In this case, the calculations we have done above are well justified.

Let us assume, for example, that the dilaton gets fixed once it reaches its self-dual value, at a time which we denote by t_c. Freezing the dilaton allows the Hagedorn phase to be of sufficiently long duration for thermal string equilibrium to be established on scales up to 1 mm. It will also put out structure formation scenario on a more solid footing. In this case, the space–time diagram is given by Fig. 11.4, where we are now plotting scales consistently in the Einstein frame. As is apparent, the overall structure of the diagram is the same as that of Fig. 11.3, except for the fact that scales exit the

---

[5] We believe that the calculation of Sect. 11.4 in [78] which yields a result with different slope and much smaller amplitude is based on a temporal Green function calculation which misses the initial condition term which dominates the spectrum.

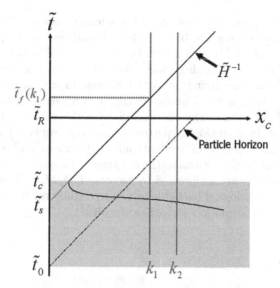

Fig. 11.4. Space–time diagram (sketch) showing the evolution of fixed comoving scales in string gas cosmology. The vertical axis is Einstein frame time, the horizontal axis is comoving distance. The *solid curve* represents the Einstein frame Hubble radius $\tilde{H}^{-1}$ which is linearly increasing after $\tilde{t}_c$. Fixed comoving scales (the *dotted lines* labeled by k_1 and k_2) which are currently probed in cosmological observations have wavelengths which are larger than the Einstein frame Hubble radius during the part of the Hagedorn phase in which the dilaton is rolling. However, due to the presence of the initial strong coupling Hagedorn phase, the horizon becomes much larger than the Hubble radius. The shaded region corresponds to the strong coupling Hagedorn phase

Einstein frame Hubble radius at a time $t_i(k)$ immediately before t_c instead of immediately before t_R. A valid concern is that we might not be allowed to neglect the higher derivative terms in the gravitational action at such early times during the Hagedorn phase.

A background in which our string gas structure formation scenario can be implemented [76] is the ghost-free and asymptotically free higher derivative gravity model proposed in [59] given by the gravitational action

$$S = \int \mathrm{d}^4 x \sqrt{-g} F(R) \tag{11.47}$$

with

$$F(R) = R + \sum_{n=0}^{\infty} \frac{c_n}{M_s^{2n}} R \left(\frac{\partial^2}{\partial^2 t} - \nabla^2 \right)^n R , \tag{11.48}$$

where M_s is the string mass scale (more generally, it is the scale where non-perturbative effects start to dominate), and the c_n are coefficients of order unity.

As shown in [59] and [76], this action has bouncing cosmological solutions. If the temperature during the bounce phase is sufficiently high, then a gas of strings will be excited in this phase. In the absence of initial cosmological perturbations in the contracting phase, our string gas structure formation scenario is realized. The string network will contain winding modes in the same way that a string network formed during a cosmological phase transition will contain infinite strings. The dilaton is fixed in this scenario, thus putting the calculation of the cosmological perturbations on a firm basis. There are no additional dynamical degrees of freedom compared to those in Einstein gravity. The higher derivative corrections to the equations of motion (in particular to the Poisson equation) are suppressed by factors of $(k/M_s)^2$. Thus, all of the conditions on a cosmological background to successfully realize the string gas cosmology structure formation scenario are realized.

11.5 Discussion

In spite of the great phenomenological successes at solving key problems of Standard big bang cosmology such as the horizon, entropy and flatness problems, and at providing a simple and predictive structure formation scenario, current realizations of inflation are beset by important conceptual problems. Most importantly, the applicability of effective field theory techniques and Einstein gravity intuition at energy scales close to the string and Planck scales are questionable.

New input from fundamental physics is required to address the problems of our current models of inflation. It is likely that superstring theory will lead to a new paradigm of early universe cosmology. Such a new paradigm may yield a convincing realization of inflation, but it may also give rise to a quite different scenario for early universe cosmology.

We have presented a new cosmological background motivated by key symmetries and making use of new degrees of freedom of string theory. This background is characterized by an early quasi-static Hagedorn phase during which the matter content of the universe is a thermal gas of strings. The decay of string winding modes leads to a smooth transition to the radiation phase of standard cosmology, without an intervening period of inflation. However, thermal fluctuations of strings during the Hagedorn phase yields a new structure formation scenario. The holographic scaling of the specific heat of a gas of closed strings on a compact three-dimensional space then leads to an almost scale-invariant spectrum of cosmological perturbations. A key prediction of this scenario is a slight blue tilt of the spectrum of gravitational waves. Note that the spectrum of the scalar modes has a slight red tilt.

There are many outstanding issues. We have presented a potential alternative solution for the origin of structure in the universe. Inflation, however, has other key successes. Most importantly, it generates a large, homogeneous, high entropy and spatially flat universe starting from initial conditions where space is small and has only a low entropy. Are there alternatives to solving these problems which arise from string gas cosmology? If the universe starts out cold and large (natural initial conditions in the context of a bouncing universe scenario), the above problems do not arise. Since string gas cosmology may well lead to a bouncing cosmology, as in the setup of [59], this possibility should be kept in mind. Alternatively, starting from initial conditions where all dimensions of space start out small, there may be an initial phase of bulk dynamics which telescopes an initial string scale into the required scale of 1mm during the Hagedorn phase (see [97] for an example).

Another outstanding issue is to obtain a better understanding of the dynamics of the Hagedorn phase. Dilaton gravity does not provide the adequate background equations, in particular at early stages during the Hagedorn phase. An improved understanding of the background dynamics is also required in order to be able to calculate the temperature $T[t_i(k)]$ at the time when scales k exit the Hubble radius. The value of the temperature and its k-dependence are crucial in order to be able to calculate the magnitude of the tilt of the power spectra of fluctuations as well as the tensor to scalar ratio.

Acknowledgements

I would like to thank the organizers of the Colloque 2006 of the IAP for inviting me to speak and for their generous hospitality during this stimulating conference. I wish to thank my collaborators Ali Nayeri, Subodh Patil and Cumrun Vafa for a most enjoyable collaboration. I also thank Lev Kofman, Andrei Linde and Slava Mukhanov for vigorous discussions which helped clarify some of the points presented in Sect. 11.4.5. Some of the arguments in that subsection were finalized after the Colloque, and appeared in [20]. I am grateful to Sugumi Kanno, Jiro Soda, Damien Easson, Justin Khoury, Patrick Martineau, Ali Nayeri and Subodh Patil for their collaboration on this paper, and I in particular thank Sugumi Kanno for allowing me to use three figures of [20] in this writeup. I also with to thank Tirthabir Biswas for extensive discussions. My research is supported by an NSERC Discovery Grant and by the Canada Research Chairs program.

References

1. A. H. Guth, Phys. Rev. D **23**, 347 (1981).
2. K. Sato, Mon. Not. Roy. Astron. Soc. **195**, 467 (1981).
3. A. A. Starobinsky, Phys. Lett. B **91**, 99 (1980).
4. R. Brout, F. Englert and E. Gunzig, Annals Phys. **115**, 78 (1978).
5. V. F. Mukhanov and G. V. Chibisov, JETP Lett. **33**, 532 (1981) [Pisma Zh. Eksp. Teor. Fiz. **33**, 549 (1981)].
6. W. Press, Phys. Scr. **21**, 702 (1980).

7. A. A. Starobinsky, JETP Lett. **30**, 682 (1979) [Pisma Zh. Eksp. Teor. Fiz. **30**, 719 (1979)].
8. R. A. Sunyaev and Y. B. Zeldovich, Astrophys. Space Sci. **7**, 3 (1970).
9. P. J. E. Peebles and J. T. Yu, Astrophys. J. **162**, 815 (1970).
10. G. F. Smoot et al., Astrophys. J. **396**, L1 (1992).
11. P. de Bernardis et al. [Boomerang Collaboration], Nature **404**, 955 (2000), arXiv:astro-ph/0004404.
12. C. L. Bennett et al., Astrophys. J. Suppl. **148**, 1 (2003), arXiv:astro-ph/0302207.
13. C. P. Burgess, Pramana **63**, 1269 (2004), arXiv:hep-th/0408037.
14. J. M. Cline, arXiv:hep-th/0501179.
15. A. Linde, eConf **C040802**, L024 (2004), arXiv:hep-th/0503195.
16. A. Nayeri, R. H. Brandenberger and C. Vafa, Phys. Rev. Lett. **97**, 021302 (2006), arXiv:hep-th/0511140.
17. A. Nayeri, arXiv:hep-th/0607073.
18. R. H. Brandenberger, A. Nayeri, S. P. Patil and C. Vafa, arXiv:hep-th/0608121.
19. R. H. Brandenberger, A. Nayeri, S. P. Patil and C. Vafa, arXiv:hep-th/0604126.
20. R. H. Brandenberger et al., JCAP **0611**, 009 (2006), arXiv:hep-th/0608186.
21. M. Sasaki, Prog. Theor. Phys. **76**, 1036 (1986).
22. V. F. Mukhanov, Sov. Phys. JETP **67**, 1297 (1988) [Zh. Eksp. Teor. Fiz. **94N7**, 1 (1988)].
23. V. F. Mukhanov, H. A. Feldman and R. H. Brandenberger, Phys. Rept. **215**, 203 (1992).
24. R. H. Brandenberger, Lect. Notes Phys. **646**, 127 (2004), arXiv:hep-th/0306071.
25. V. Mukhanov, arXiv:astro-ph/0303072.
26. A. D. Linde, Phys. Lett. B **129**, 177 (1983).
27. A. D. Linde, Phys. Rev. D **49**, 748 (1994), arXiv:astro-ph/9307002.
28. R. H. Brandenberger and J. H. Kung, Phys. Rev. D **42**, 1008 (1990).
29. H. A. Feldman and R. H. Brandenberger, Phys. Lett. B **227**, 359 (1989).
30. D. S. Goldwirth and T. Piran, Phys. Rept. **214**, 223 (1992).
31. A. D. Linde, Phys. Lett. B **108**, 389 (1982);
    A. Albrecht and P. J. Steinhardt, Phys. Rev. Lett. **48**, 1220 (1982).
32. R. Brandenberger, G. Geshnizjani and S. Watson, Phys. Rev. D **67**, 123510 (2003), arXiv:hep-th/0302222.
33. A. D. Linde, Mod. Phys. Lett. A **1**, 81 (1986);
    A. D. Linde and D. A. Linde, Phys. Rev. D **50**, 2456 (1994), arXiv:hep-th/9402115.
34. R. Allahverdi, K. Enqvist, J. Garcia-Bellido and A. Mazumdar, arXiv:hep-ph/0605035.
35. F. C. Adams, K. Freese and A. H. Guth, Phys. Rev. D **43**, 965 (1991).
36. A. Vilenkin and E.P.S. Shellard, *Cosmic Strings and Other Topological Defects* (Cambridge University Press, Cambridge, 1994).
37. M. B. Hindmarsh and T. W. B. Kibble, Rept. Prog. Phys. **58**, 477 (1995), arXiv:hep-ph/9411342.
38. R. H. Brandenberger, Int. J. Mod. Phys. A **9**, 2117 (1994), arXiv:astro-ph/9310041.
39. N. Turok and R. H. Brandenberger, Phys. Rev. D **33**, 2175 (1986);
    H. Sato, Prog. Theor. Phys. **75**, 1342 (1986);
    A. Stebbins, Ap. J. (Lett.) **303**, L21 (1986).

40. A. Albrecht, D. Coulson, P. Ferreira and J. Magueijo, Phys. Rev. Lett. **76**, 1413 (1996), arXiv:astro-ph/9505030;
    J. Magueijo, A. Albrecht, D. Coulson and P. Ferreira, Phys. Rev. Lett. **76**, 2617 (1996), arXiv:astro-ph/9511042;
    U. L. Pen, U. Seljak and N. Turok, Phys. Rev. Lett. **79**, 1611 (1997), arXiv:astro-ph/9704165.

41. R. H. Brandenberger, in proc. of IPM School On Cosmology 1999: Large Scale Structure Formation, arXiv:hep-ph/9910410.

42. R. H. Brandenberger and J. Martin, Mod. Phys. Lett. A **16**, 999 (2001), arXiv:astro-ph/0005432;
    J. Martin and R. H. Brandenberger, Phys. Rev. D **63**, 123501 (2001), arXiv:hep-th/0005209.

43. J. C. Niemeyer, Phys. Rev. D **63**, 123502 (2001), arXiv:astro-ph/0005533;
    S. Shankaranarayanan, Class. Quant. Grav. **20**, 75 (2003), arXiv:gr-qc/0203060;
    J. C. Niemeyer and R. Parentani, Phys. Rev. D **64**, 101301 (2001), arXiv:astro-ph/0101451.

44. W. G. Unruh, Phys. Rev. D **51**, 2827 (1995).

45. S. Corley and T. Jacobson, Phys. Rev. D **54**, 1568 (1996), arXiv:hep-th/9601073.

46. R. H. Brandenberger and J. Martin, Phys. Rev. D **71**, 023504 (2005), arXiv:hep-th/0410223.

47. T. Tanaka, arXiv:astro-ph/0012431.

48. A. A. Starobinsky, Pisma Zh. Eksp. Teor. Fiz. **73**, 415 (2001), [JETP Lett. **73**, 371 (2001)], arXiv:astro-ph/0104043.

49. C. S. Chu, B. R. Greene and G. Shiu, Mod. Phys. Lett. A **16**, 2231 (2001), arXiv:hep-th/0011241;
    R. Easther, B. R. Greene, W. H. Kinney and G. Shiu, Phys. Rev. D **64**, 103502 (2001), arXiv:hep-th/0104102;
    R. Easther, B. R. Greene, W. H. Kinney and G. Shiu, Phys. Rev. D **67**, 063508 (2003), arXiv:hep-th/0110226;
    F. Lizzi, G. Mangano, G. Miele and M. Peloso, JHEP **0206**, 049 (2002), arXiv:hep-th/0203099;
    S. F. Hassan and M. S. Sloth, Nucl. Phys. B **674**, 434 (2003), arXiv:hep-th/0204110.

50. R. Brandenberger and P. M. Ho, Phys. Rev. D **66**, 023517 (2002), [AAPPS Bull. **12N1**, 10 (2002)], arXiv:hep-th/0203119.

51. A. Kempf and J. C. Niemeyer, Phys. Rev. D **64**, 103501 (2001), arXiv:astro-ph/0103225.

52. U. H. Danielsson, Phys. Rev. D **66**, 023511 (2002), arXiv:hep-th/0203198;
    V. Bozza, M. Giovannini and G. Veneziano, JCAP **0305**, 001 (2003), arXiv:hep-th/0302184;
    J. C. Niemeyer, R. Parentani and D. Campo, Phys. Rev. D **66**, 083510 (2002), arXiv:hep-th/0206149.

53. C. P. Burgess, J. M. Cline, F. Lemieux and R. Holman, JHEP **0302**, 048 (2003), arXiv:hep-th/0210233;
    K. Schalm, G. Shiu and J. P. van der Schaar, AIP Conf. Proc. **743**, 362 (2005), arXiv:hep-th/0412288.

54. J. Martin and R. Brandenberger, Phys. Rev. D **68**, 063513 (2003), arXiv:hep-th/0305161.

55. S. Hawking and G. Ellis, *The Large-Scale Structure of Space-Time* (Cambridge University Press, Cambridge, 1973).
56. A. Borde and A. Vilenkin, Phys. Rev. Lett. **72**, 3305 (1994), arXiv:gr-qc/9312022.
57. N. Arkani-Hamed, L. Motl, A. Nicolis and C. Vafa, arXiv:hep-th/0601001.
58. R. H. Brandenberger, V. F. Mukhanov and A. Sornborger, Phys. Rev. D **48**, 1629 (1993), arXiv:gr-qc/9303001;
    V. F. Mukhanov and R. H. Brandenberger, Phys. Rev. Lett. **68**, 1969 (1992).
59. T. Biswas, A. Mazumdar and W. Siegel, JCAP **0603**, 009 (2006), arXiv:hep-th/0508194.
60. M. Bojowald, Living Rev. Rel. **8**, 11 (2005), arXiv:gr-qc/0601085.
61. R. Hagedorn, Nuovo Cim. Suppl. **3**, 147 (1965).
62. R. H. Brandenberger and C. Vafa, Nucl. Phys. B **316**, 391 (1989).
63. R. H. Brandenberger, in Proc. of the 18th IAP Colloquium on the Nature of Dark Energy: Observational and Theoretical Results on the Accelerating Universe, arXiv:hep-th/0210165.
64. P. Brax, C. van de Bruck and A. C. Davis, Rept. Prog. Phys. **67**, 2183 (2004), arXiv:hep-th/0404011.
65. M. Majumdar and A.-C. Davis, JHEP **0203**, 056 (2002), arXiv:hep-th/0202148.
66. R. Durrer, M. Kunz and M. Sakellariadou, Phys. Lett. B **614**, 125 (2005), arXiv:hep-th/0501163.
67. A. Karch and L. Randall, Phys. Rev. Lett. **95**, 161601 (2005), arXiv:hep-th/0506053.
68. A. A. Tseytlin and C. Vafa, Nucl. Phys. B **372**, 443 (1992), arXiv:hep-th/9109048.
69. S. Alexander, R. H. Brandenberger and D. Easson, Phys. Rev. D **62**, 103509 (2000), arXiv:hep-th/0005212.
70. J. Kripfganz and H. Perlt, Class. Quant. Grav. **5**, 453 (1988).
71. T. Battefeld and S. Watson, Rev. Mod. Phys. **78**, 435 (2006), arXiv:hep-th/0510022.
72. R. H. Brandenberger, in Proc. of the 59th Yamada Conference on Inflating Horizon of Particle Astrophysics and Cosmology, arXiv:hep-th/0509099.
73. R. H. Brandenberger, Prog. Theor. Phys. Suppl. **163**, 358 (2006), arXiv:hep-th/0509159.
74. J. Polchinski, *String Theory, Vols. 1 and 2* (Cambridge University Press, Cambridge, 1998).
75. T. Boehm and R. Brandenberger, JCAP **0306**, 008 (2003), arXiv:hep-th/0208188.
76. T. Biswas, R. Brandenberger, A. Mazumdar and W. Siegel, arXiv:hep-th/0610274.
77. G. F. R. Ellis and R. Maartens, Class. Quant. Grav. **21**, 223 (2004), arXiv:gr-qc/0211082.
78. N. Kaloper, L. Kofman, A. Linde and V. Mukhanov, JCAP **0610**, 006 (2006), arXiv:hep-th/0608200.
79. R. Easther, B. R. Greene, M. G. Jackson and D. Kabat, JCAP **0502**, 009 (2005), arXiv:hep-th/0409121.
80. R. Danos, A. R. Frey and A. Mazumdar, Phys. Rev. D **70**, 106010 (2004), arXiv:hep-th/0409162.
81. S. Watson and R. H. Brandenberger, Phys. Rev. D **67**, 043510 (2003), arXiv:hep-th/0207168.

82. G. Veneziano, Phys. Lett. B **265**, 287 (1991).
83. M. Maggiore and A. Riotto, Nucl. Phys. B **548**, 427 (1999), arXiv:hep-th/9811089.
84. S. Arapoglu, A. Karakci and A. Kaya, arXiv:hep-th/0611193.
85. S. Watson and R. Brandenberger, JCAP **0311**, 008 (2003), arXiv:hep-th/0307044.
86. S. P. Patil and R. Brandenberger, Phys. Rev. D **71**, 103522 (2005), arXiv:hep-th/0401037.
87. S. P. Patil and R. H. Brandenberger, JCAP **0601**, 005 (2006), arXiv:hep-th/0502069.
88. T. Battefeld and S. Watson, JCAP **0406**, 001 (2004), arXiv:hep-th/0403075.
89. R. Brandenberger, Y. K. Cheung and S. Watson, JHEP **0605**, 025 (2006), arXiv:hep-th/0501032.
90. S. Kanno and J. Soda, Phys. Rev. D **72**, 104023 (2005), arXiv:hep-th/0509074.
91. S. Watson, Phys. Rev. D **70**, 066005 (2004), arXiv:hep-th/0404177.
92. L. Kofman, A. Linde, X. Liu, A. Maloney, L. McAllister and E. Silverstein, JHEP **0405**, 030 (2004), arXiv:hep-th/0403001.
93. G. B. Cleaver and P. J. Rosenthal, Nucl. Phys. B **457**, 621 (1995), arXiv:hep-th/9402088;

    M. Sakellariadou, Nucl. Phys. B **468**, 319 (1996), arXiv:hep-th/9511075;

    R. Easther, B. R. Greene and M. G. Jackson, Phys. Rev. D **66**, 023502 (2002), arXiv:hep-th/0204099;

    R. Easther, B. R. Greene, M. G. Jackson and D. Kabat, Phys. Rev. D **67**, 123501 (2003), arXiv:hep-th/0211124;

    R. Easther, B. R. Greene, M. G. Jackson and D. Kabat, JCAP **0401**, 006 (2004), arXiv:hep-th/0307233;

    S. H. S. Alexander, JHEP **0310**, 013 (2003), arXiv:hep-th/0212151;

    B. A. Bassett, M. Borunda, M. Serone and S. Tsujikawa, Phys. Rev. D **67**, 123506 (2003), arXiv:hep-th/0301180;

    D. A. Easson, Int. J. Mod. Phys. A **18**, 4295 (2003), arXiv:hep-th/0110225;

    A. Campos, Phys. Rev. D **68**, 104017 (2003), arXiv:hep-th/0304216;

    T. Biswas, JHEP **0402**, 039 (2004), arXiv:hep-th/0311076;

    A. Kaya and T. Rador, Phys. Lett. B **565**, 19 (2003), arXiv:hep-th/0301031;

    A. Kaya, Class. Quant. Grav. **20**, 4533 (2003), arXiv:hep-th/0302118;

    A. Campos, Phys. Lett. B **586**, 133 (2004), arXiv:hep-th/0311144;

    S. Watson and R. Brandenberger, JHEP **0403**, 045 (2004), arXiv:hep-th/0312097;

    S. Watson, Phys. Rev. D **70**, 023516 (2004), arXiv:hep-th/0402015;

    R. Brandenberger, D. A. Easson and A. Mazumdar, Phys. Rev. D **71**, 083514 (2005), arXiv:hep-th/0307043;

    A. Kaya, JCAP **0408**, 014 (2004), arXiv:hep-th/0405099;

    S. Arapoglu and A. Kaya, Phys. Lett. B **603**, 107 (2004), arXiv:hep-th/0409094;

    T. Rador, Phys. Lett. B **621**, 176 (2005), arXiv:hep-th/0501249;

    T. Rador, JHEP **0506**, 001 (2005), arXiv:hep-th/0502039;

    T. Rador, arXiv:hep-th/0504047;

    A. Kaya, Phys. Rev. D **72**, 066006 (2005), arXiv:hep-th/0504208;

    F. Ferrer and S. Rasanen, JHEP **0602**, 016 (2006), arXiv:hep-th/0509225;

    M. Borunda and L. Boubekeur, JCAP **0610**, 002 (2006), arXiv:hep-th/0604085;

    A. Chatrabhuti, arXiv:hep-th/0602031;

    J. Y. Kim, arXiv:hep-th/0608131.

94. A. J. Berndsen and J. M. Cline, Int. J. Mod. Phys. A **19**, 5311 (2004), arXiv:hep-th/0408185;
    A. Berndsen, T. Biswas and J. M. Cline, JCAP **0508**, 012 (2005), arXiv:hep-th/0505151;
    D. A. Easson and M. Trodden, Phys. Rev. D **72**, 026002 (2005), arXiv:hep-th/0505098;
    S. Cremonini and S. Watson, Phys. Rev. D **73**, 086007 (2006), arXiv:hep-th/0601082.
95. S. P. Patil, arXiv:hep-th/0504145.
96. R. Brandenberger, D. A. Easson and D. Kimberly, Nucl. Phys. B **623**, 421 (2002), arXiv:hep-th/0109165.
97. R. Brandenberger and N. Shuhmaher, JHEP **0601**, 074 (2006), arXiv:hep-th/0511299.
98. N. Deo, S. Jain, O. Narayan and C. I. Tan, Phys. Rev. D **45**, 3641 (1992).
99. J. M. Bardeen, P. J. Steinhardt and M. S. Turner, Phys. Rev. D **28**, 679 (1983).
100. R. H. Brandenberger and R. Kahn, Phys. Rev. D **29**, 2172 (1984);
    R. H. Brandenberger, Nucl. Phys. B **245**, 328 (1984).

Index